Climate Impacts and Challenges in Agriculture, Forests and Food Systems

Vincent Blanfort • Marie Hrabanski
Julien Demenois
Editors

Climate Impacts and Challenges in Agriculture, Forests and Food Systems

Perspectives on the Global South

Second Edition

Editors
Vincent Blanfort
Cirad joint research unit Selmet "Tropical and Mediterranean Animal Production Systems", Montpellier Univ, Inrae Institut Agro
Montpellier, France

Marie Hrabanski
Cirad UMR Actors Resources Territories & Development (ART-Dev)
Montpellier, France

Julien Demenois
CIRAD UPR Agroecology and sustainable intensification of annual crops (Aïda)
Montpellier, France

ISBN 978-3-032-04330-6   ISBN 978-3-032-04331-3   (eBook)
https://doi.org/10.1007/978-3-032-04331-3

Translation from the French language edition: "L'agriculture et les systèmes alimentaires du monde face au changement climatique" by Vincent Blanfort et al., © Éditions Quæ 2025. Published by Éditions Quæ. All Rights Reserved.

Jointly published with Éditions Quæ, Versailles, France

The original submitted manuscript has been translated into English. The translation was done using artificial intelligence. A subsequent revision was performed by the author(s) to further refine the work and to ensure that the translation is appropriate concerning content and scientific correctness. It may, however, read stylistically different from a conventional translation.

This book is an open access publication.

© Editions Quae 2016, 2026

Open Access   This book is licensed under the terms of the Creative Commons Attribution-NonCommercial-NoDerivatives 4.0 International License (http://creativecommons.org/licenses/by-nc-nd/4.0/), which permits any noncommercial use, sharing, distribution and reproduction in any medium or format, as long as you give appropriate credit to the original author(s) and the source, provide a link to the Creative Commons license and indicate if you modified the licensed material. You do not have permission under this license to share adapted material derived from this book or parts of it.

The images or other third party material in this book are included in the book's Creative Commons license, unless indicated otherwise in a credit line to the material. If material is not included in the book's Creative Commons license and your intended use is not permitted by statutory regulation or exceeds the permitted use, you will need to obtain permission directly from the copyright holder.

The use of general descriptive names, registered names, trademarks, service marks, etc. in this publication does not imply, even in the absence of a specific statement, that such names are exempt from the relevant protective laws and regulations and therefore free for general use.

The publisher, the authors and the editors are safe to assume that the advice and information in this book are believed to be true and accurate at the date of publication. Neither the publisher nor the authors or the editors give a warranty, expressed or implied, with respect to the material contained herein or for any errors or omissions that may have been made. The publisher remains neutral with regard to jurisdictional claims in published maps and institutional affiliations.

This Springer imprint is published by the registered company Springer Nature Switzerland AG
The registered company address is: Gewerbestrasse 11, 6330 Cham, Switzerland

If disposing of this product, please recycle the paper.

Foreword by Laurence Tubiana

President and CEO of the European Climate Foundation

Founder of the Institute for Sustainable Development and International Relations (Iddri)

In a complex and uncertain geopolitical context, COP30, to be held in Brazil in November 2025, represents a significant opportunity to revitalise multilateral commitment to the fight against climate change. Under the presidency of Lula, Brazil is indeed seeking to position itself as a climate leader and to demonstrate that multilateralism remains essential to address global challenges.

Brazil embodies both the consequences of climate disruption and the economic, social and financial opportunities inherent in climate action. The country is already suffering heavy losses related to natural disasters; floods, fires and droughts are affecting its GDP and financial stability. Moreover, hosting COP30 in Belém symbolises Brazil's desire to put the Amazon rainforest and its inhabitants at the centre of the debates. Deforestation in the Amazon is a major concern, and the tropical forest is close to a tipping point that could have catastrophic repercussions for the entire planet. President Lula wants the world to consider the Amazon not only as a crucial carbon sink but also as a region inhabited by millions of people whose survival depends on the protection of the ecosystem. Politically, Brazil also illustrates the impact of governmental leadership on climate action. After several years of increasing greenhouse gas emissions, President Lula's return marked a commitment to reducing these emissions and preserving the Amazon. Finally, Brazil, as a member of the G20 and BRICS, plays a key diplomatic role and could help advance the climate agenda despite geopolitical tensions between the United States and China.

The challenge of COP30 will be to put climate back at the centre of international concerns. The history of the COPs shows that the commitment of citizens and nongovernmental actors is essential to advance political decisions. However, recent COPs, notably those in Baku and Dubai, have marginalised civil society. Brazil wants to reverse this trend by integrating citizens into the decision-making process, while placing social justice at the heart of the debates. Global public opinion remains overwhelmingly in favour of strong climate action, and indigenous and scientific knowledge must contribute to enlightening the debates. More than ever, scientific knowledge, which is currently under unprecedented attack internationally, must continue to guide policy directions.

At the interface between climate, energy, food security, development and biodiversity issues, agricultural, food and forestry issues have now become essential topics within the COPs. Because it brings together scientific knowledge from researchers specialising in these issues and produced in the diverse territories of the Global South particularly affected by climate change (Africa, Latin and Central America, Asia, etc.), this work is an essential reference for understanding some of the challenges of COP30 and future climate negotiations.

Laurence Tubiana

Foreword by Élisabeth Claverie de Saint Martin

President and CEO of Cirad

This work comes 10 years after the publication of the collective book co-edited by Cirad and titled *Climate Change and World Agriculture*,[1] in a similar context characterised by the publication of an IPCC report.[2] The renewal of this work is motivated by the conclusions of the 6th IPCC report, which reaffirms the urgency to combat global warming and the necessity to adapt quickly. The sectors of agriculture, food systems and forests are particularly affected by ongoing and future climate disruptions, especially in the Southern countries, which are more vulnerable than industrialised countries. It is important for Cirad and its partners to share recent knowledge produced on these issues in the tropical and Mediterranean countries, where uncertainties are much greater than in industrialised countries. This work arrives at a unique moment in the history of Cirad, which celebrated its fortieth anniversary in 2024 with conferences around the theme: "40 years of agricultural research in the South to feed the planet in 2050". The primary and universal purpose of agriculture is indeed to enable humanity to feed itself and be in good health, and therefore to ensure food security. To do this, it is crucial to intersect social, economic and political issues and integrate the dimension of climate disruption.

Globally, agricultural, food and forest systems are major contributors to climate change by producing more than a third of greenhouse gas (GHG) emissions, when deforestation and upstream and downstream supply chains are included. They are also victims of the effects of these changes, especially in the Southern countries. However, the land sector distinguishes itself from other sectors by multiple adaptation capacities, combined with a significant mitigation potential with the sequestration of carbon in soils and biomass. The land sector therefore plays a crucial role in ensuring the development of societies and their transition in a context dominated by climate change, even though the latter regularly threatens agricultural production,

---

[1] Torquebiau E. (coord.), 2015. *Climate change and world agriculture*, Versailles, Quæ editions, 328 p.
[2] IPCC: Intergovernmental Panel on Climate Change.

thus weakening global food systems, particularly in the Southern countries. Nevertheless, agriculture was absent from climate negotiations in international negotiation arenas until 2011. Since then, the consideration of agricultural issues in climate negotiations has been increasing, but these issues are not considered or treated in the same way in the Northern and Southern countries.

In the Southern countries, the priority is mainly given to adaptation to climate change, due to the issues related to food security. In the Northern countries, which are largely responsible for historical GHG emissions and which have a priori better technical adaptation capacities, the focus is on mitigation within the framework of global climate governance. Beyond these differences, the global history of agriculture can appear as a constant adaptation to a "natural" intra- and interannual climate variability. But the ongoing climate change, linked to the increase in atmospheric carbon concentration, is of a completely different magnitude; it introduces increasingly intense and frequent hazards: irregular seasonality, precipitations shifted in time or distributed differently, extreme events, temperatures modifying harvest dates, new or more active pests. These various impacts have strong consequences on food security, constituting globally a new constraint for farmers and livestock keepers worldwide, even if in some regions the impacts can currently prove positive (increase in yields under certain latitudes in particular). In the Southern countries, the effects of climate change on the agricultural sector are particularly severe. For example, the vast majority of agricultural areas in Africa are not irrigated and depend entirely on precipitation. The use of mineral fertilisers is not widespread, which implies relying only on the natural fertility of the soils or on that which trees can provide, or often, on organic amendments when livestock activities are present. Beyond these agronomic and ecological vulnerabilities, it is the entire agricultural and food systems that must adapt, which also requires innovations in economic, social, political and financial matters at the territorial level. Concurrently, agricultural and food systems contribute to climate change mitigation efforts by influencing the reduction of GHG emissions and/or increasing carbon sequestration (capture and storage).

Faced with major challenges and global changes that are differently declined at regional, national and territorial scales, new challenges call on research to contribute to the transition of agricultural and food systems. It is about analysing, evaluating and designing sustainable and integrative solutions to mitigate climate change and adapt to it.

In this context, Cirad made the choice in 2018, for the first time in its history, to inscribe climate issues as a strategic scientific priority. This was reflected in the formulation of concrete commitments in its strategic vision document 2018–2028. The strategic thematic field (STF) "Supporting all Southern agricultures in climate change" is considered a priority for the establishment and is the subject of interdisciplinary scientific animation, in connection with Cirad's partners.

The current diversity of Cirad's actions with its partners is structured notably through five science fronts covering several fields (agronomy, genetic improvement, ecology, socio-economics considering territorial actors, etc.): (1) implement sustainable public agricultural policies for adaptation to climate change and its mitigation; (2) diversify agricultural production systems for greater sustainability and resilience in the face of climate change; (3) understand the resilience of tropical forests and the people who live from them in the face of climate change; (4) reposition livestock farming in the challenges of climate change; (5) develop integrated water management approaches for greater resilience in the face of climate change.

Given the climate emergency in the countries of the South, relying solely on models from the North is insufficient and even inappropriate. With its presence in about 50 countries in Africa, Latin America, Asia and the Pacific, Cirad contributes to the production of knowledge on agricultural systems in the South, in various contexts, to meet specific development challenges identified with its partners. In view of the alarming climate projections, the expertise developed by Cirad and its partners for over 20 years for the agricultures of the South is also expected to benefit the development of sustainable solutions in the North. This implies positioning the research questions addressed in a perspective of balanced partnership between the North and the South, where the scientific expertise of Cirad and its Southern partners will play an enlightening role.

This collective work was conceived in a spirit of sharing the knowledge obtained by Cirad and its partners, within the framework of the scientific animation of the CTS "Supporting all the agricultures of the South in the face of climate change". This thematic field brings together an interdisciplinary community of 150 scientists from the North and the South, carrying knowledge and know-how built with the partners from the South, in response to their climate issues, in a wide variety of pedoclimatic situations and socio-ecosystems.

Beyond the scientific community, this work is aimed at policymakers and civil society. It contributes to strengthening the role of Cirad as a reference interlocutor in research on issues linking climate and agriculture.

"Avoid the unmanageable, manage the inevitable."
François Gemenne, co-author of the 6th
IPCC report.

Élisabeth Claverie de Saint Martin

Foreword by Vincent Blanfort, Marie Hrabanski and Julien Demenois

Another book on climate change?

Because at a time when science—climate science in particular—is too often ignored, questioned or dismissed as mere opinion, it is more urgent than ever to affirm these scientific truths, to explain them clearly, and to make them accessible to all. With a measure of hope, we trust that these facts will help shape public policies, guide the actions of key stakeholders, contribute to education, and, quite simply, raise awareness about processes that affect every part of society—not just farmers and agricultural workers.

Yes, also, because this book highlights the specific situations of countries in the Global South, which remain under-represented in much scientific research, despite being among the most vulnerable to climate change. At a time when international solidarity is under strain and multilateralism is being questioned, the scientific community must stand firm. By focusing on the South, and by making this book freely available in digital form, we are contributing—however modestly—to this collective resistance.

Lastly, this collaborative work brings together the scientific expertise of women and men who have carried out fieldwork in the South, addressing key issues related to agriculture, livestock, forests, and food systems.

In the face of an unprecedented climate challenge, these sectors occupy a critical position. Unlike many others, they are directly tied to a fundamental and daily need: feeding humanity. This vital responsibility, in the context of the climate emergency, calls for solutions that are sustainable and scientifically grounded. Meeting this challenge requires more integrated approaches that also take into account the growing diversity and complexity of food systems, and their deep connections with both the environment and society.

In this setting, science must not stand aside. It must engage, collaborate, and help drive meaningful change alongside those most affected.

Montpellier, France

Vincent Blanfort
Marie Hrabanski
Julien Demenois

Acknowledgements

The global dimension that we wanted to give to this work, as well as the diversity of contexts and agricultural, food and forestry systems, required a multidisciplinary and cross-cutting perspective. We therefore naturally solicited authors from diverse and complementary horizons and disciplines. It is to them that we first express our thanks. They are nearly 150 (see the list at the end of the book)! We therefore extend a big and warm collective thank you to them.

The project for this work was built in close collaboration with the Directorate for Research and Strategy (DGD-RS) of Cirad which financed it. We particularly want to sincerely thank Philippe Petithuguenin, who gave us unfailing support as Deputy Director General for Research and Strategy.

Writing a book is one thing, organising its edition is another. We were able to measure this during the different stages of setting up, designing, producing and correcting. First and foremost, a big thank you to the editorial team at Quæ who accompanied us with great efficiency throughout this project, to Sophie De Decker for her careful and enlightened proofreading of the form of the manuscript, and to Hélène Bonnet for the layout work.

We finally thank the scientific reviewers, whose contributions were essential to improving the content of the book.

Introduction: Agricultural and Food Systems, Contributors, Victims of Climate Change and Bearers of Solutions

Agriculture, food systems, and forests play a significant role in both ongoing and future climate disruptions. The "land sector"—encompassing agriculture, forestry, and other land uses (AFOLU, for its acronym in English)—is among the most severely affected by climate change (CC). Agricultural, livestock, and forestry production systems are already facing the consequences of CC, notably through the increasing frequency and intensity of extreme events such as heatwaves, droughts, wildfires, and floods. These acute impacts are compounded by long-term trends, including rising temperatures and shifts in both intra- and interannual precipitation patterns. These multifaceted effects jeopardize key resources (such as water, soil, and biodiversity) and significantly destabilize global food systems, particularly in countries of the Global South, which are more vulnerable than countries of the Global North.[1]

At the same time, when considering the entire spectrum of agri-food systems[2]—which includes pre-production and post-production activities, on-farm production (farm gate), and land-use change—agriculture and food systems account for nearly one-third of all global anthropogenic greenhouse gas (GHG) emissions (Crippa et al. 2021; Rosenzweig et al. 2020; Tubiello et al. 2022). Between 1990 and 2019, these emissions increased by 17% globally, primarily due to a doubling of emissions from pre- and post-production processes. Emissions at the farm level rose by 9%, while land-use change emissions decreased by 25% (Tubiello et al. 2022). However, substantial disparities exist between industrialized and non-industrialized countries.

---

[1] The "Global North", or "Northern countries", once referred to all so-called "developed" countries. The term "Global South", or "Southern countries", refers to countries formerly known as "third world". The concept now encompasses a heterogeneous group of non-aligned Southern states. https://www.lemonde.fr/idees/article/2022/10/26/le-sud-global-cet-ensemble-heterogene-de-pays-non-alignes_6147333_3232.html

[2] We refer to the FAOSTAT methodology (https://files-faostat.fao.org/production/GT/GT_en.pdf) to characterise the greenhouse gas (GHG) emissions of different sectors of agricultural and food systems or *agri-food systems*.

Unlike other sectors such as transportation and energy, agricultural and food systems, along with forests, provide essential services to humanity, including food provision, ecosystem services, and climate regulation. These systems thus face a major challenge: to feed a growing global population—expected to reach around 10 billion by 2100—while operating under the constraints imposed by climate change. Furthermore, these systems demonstrate undeniable adaptive capacities to CC impacts. They also hold considerable potential for GHG emissions reduction, further enhanced by the land sector's ability to sequester carbon in soils and biomass, positioning AFOLU as a key component of the solution to climate change.

Although adaptation and mitigation strategies are often portrayed as conflicting, in agriculture they can offer substantial synergies. In the Global South, adaptation has been prioritized, particularly in light of food security concerns. In contrast, the Global North—primarily responsible for historical GHG emissions and generally possessing higher technical adaptive capacities—has focused more on mitigation within the global climate governance framework. Nevertheless, a growing consensus now recognizes the necessity of addressing adaptation and mitigation in an integrated and synergistic manner.

Analyzing, proposing, and identifying sustainable solutions to mitigate climate change and to adapt agricultural and food systems to its impacts represents a major challenge for the scientific community. This volume seeks to contribute to these objectives by drawing on inter- and transdisciplinary scientific perspectives at the interface between environmental sciences (agronomy, plant breeding, ecology, etc.) and social sciences (economics, geography, sociology, political science, etc.). It also integrates the complexity arising from the multilevel dynamics of climate disruption and the diversity of agricultural and food systems and their contextual specificities around the world.

Building upon the latest IPCC Assessment Report (AR6), the contributions presented by CIRAD researchers and their partners go a step further by offering a comprehensive volume devoted to the biophysical, genetic, agronomic, social, institutional, and political dimensions of agricultural, food, and forest-related challenges in the context of climate change. These chapters collectively aim to support the development of inclusive and sustainable agricultural and food systems, across various spatial and temporal scales, with particular attention to the most vulnerable regions of the Global South.

This introductory chapter is organized into three sections, reflecting the overall structure of the book. First, we provide definitions of the key concepts necessary to understand the challenges of adaptation and mitigation in agricultural, food, and forest systems. We then outline the structure of the volume. The first part examines the growing prominence of agricultural and food issues in both scientific and policy agendas at the international level. It includes an "Atlas" section, featuring climate change maps designed to illustrate the challenges of climate change at multiple levels: (i) at the continental scale (Europe, the Americas, Africa, Oceania, Asia); (ii) through the lens of GHG emissions (by sector, by country, etc.); and (iii) by highlighting the effects of climate change and the related mitigation challenges. The second part of the volume explores the dual role of agricultural and food systems (including forests) as both victims and contributors to climate change. The third and final part addresses the critical question of *solutions*, from various angles—technical, place-based, and policy-oriented—with the overarching aim of adapting agricultural and food systems to climate change while also contributing to greenhouse gas mitigation.

The Land Sector and Food Systems Facing Climate Change: Semantic and Epistemological Challenges

Several terms and acronyms attempt to account for the place of food systems, agriculture, or even forests in GHG emissions. To clarify these terms and their issues, it seems necessary to present them and possibly identify their complementarity and overlaps.

Agri-food systems refer to the full set of processes required to feed a population. They encompass what the literature commonly defines as "land-use change," "pre-harvest stages" (e.g., fertilizer production, irrigation), "production" (at the farm gate), and "post-harvest stages" (including transportation, processing, distribution, consumption, and waste management). The Food and Agriculture Organization (FAO 2024) also employs the term *agri-food system* to capture the entirety of these interconnected stages—this is the terminology adopted throughout the present volume. Alongside this term, other concepts and acronyms are also in use, each referring to distinct and/or complementary issues.

Since the publication of the IPCC's Fifth Assessment Report in 2014, the land sector has been referred to as the AFOLU sector. By incorporating agriculture into the previously defined LULUCF sector (Land Use, Land-Use Change, and Forestry), the acronym AFOLU (Agriculture, Forestry and Other Land Use) has become widely adopted. Nevertheless, it is important to understand what the term LULUCF encompasses (Table 1).

Anthropogenic CO_2 emissions originate primarily from two sources: the combustion of fossil fuels and land-use practices. Greenhouse gas (GHG) inventories group CO_2 emissions and removals related to land use under the LULUCF category. The net balance of these inventories is calculated by subtracting CO_2 removals from CO_2 emissions across various terrestrial carbon pools (such as biomass and soils), relative to the managed land area within a given territory.

A positive balance indicates that the territory is a net *source* of GHGs—typically due to deforestation or land degradation. Conversely, a negative balance signifies that the territory functions as a *carbon sink*—as in the case of a growing forest—resulting in what are referred to as *negative emissions*. The LULUCF sector remains the only sector currently capable of achieving such negative emissions through *natural carbon sinks*, namely, biomass (forests, hedgerows, agroforestry systems) and soils (including agricultural soils).

Table 1 Details of the categories used by IPCC, the FAO and their aggregation used in the FAOSTAT database

IPCC for NGHGI		FAO categories and gases		FAO aggregates	
LULUCF	Forest land	Forestland	CO_2	Land use change	Agri-food systems
	Burning biomass	Fires, other forest	CH_4; N_2O		
		Fires, organic soils	CO_2, CH_4		
		Fires, humid tropical forest	CH_4; N_2O		
	Forest land converted to other land uses (CL, GL, Settlement, Wetlands, etc.)	Net forest conversion	CO_2		
Agriculture (AFOLU)	Drained organic soils	Drained organic soils	CO_2	Farm gate (Agricultural land)	
	Cultivation of histosols		N_2O		
	Inorganic N fertilizers	Synthetic fertilizers	N_2O		
	Crop residues	Crop residues	N_2O		
	Manure deposited on pasture, range and paddock	Manure left on pasture	N_2O		
	Manure applied to soils	Manure applied to soils	N_2O		
	Manure management	Manure management	CH_4; N_2O		
	Enteric fermentation	Enteric fermentation	CH_4		
	Prescribed burning of savanna	Savanna fires	CH_4; N_2O		
	Burning crop residues	Burning- crop residues	CH_4; N_2O		
	Rice cultivation	Rice cultivation	CH_4		
Energy		On-farm energy use	CO_2; CH_4; N_2O		
		Fertilizers manufacturing	CO_2; N_2O		
		Pesticides manufacturing	CO_2; CH_4; N_2O		
		Food household consumption	CO_2; CH_4; N_2O		
		Food packaging	CO_2; CH_4; N_2O		
		Food processing	CO_2; CH_4; N_2O		
		Food transport	CO_2; CH_4; N_2O		
		Food retail	CO_2; CH_4; N_2O		
IPPU			F-gases		
Waste		Food waste disposal	CO_2; CH_4; N_2O		

The acronym *AFOLU*, introduced in 2014, thus extends the scope of LULUCF by including agriculture, and is now used to refer more broadly to the *land sector*. However, it is important to note that the land sector does *not* fully encompass *agri-food systems*, as it excludes GHG emissions from the post-harvest stages of food systems (transport, processing, distribution, consumption, and waste management).

Accordingly, in this volume we use the term *agricultural and food system* when referring to both the AFOLU sector and the post-harvest stages. In contrast, the term *agricultural system* refers exclusively to the AFOLU sector (Table 1).

Finally, it is important to recall that agricultural and food systems are responsible for the emission of three main types of greenhouse gases (GHGs). Nitrous oxide (N_2O) is primarily generated during the production and application of synthetic fertilizers. Methane (CH_4) originates mainly from the natural digestive processes of ruminants (enteric fermentation), manure management, and flooded rice cultivation.

Carbon dioxide (CO_2), meanwhile, has a variety of sources, including the combustion of fossil fuels, deforestation, the microbial decomposition of soil organic matter, and the respiration of plants and animals. Within food systems—especially during the post-harvest stages (transportation, processing, distribution, consumption, and waste management)—CO_2 is the predominant emission.

Beyond these technical clarifications and the semantic considerations surrounding the AFOLU sector, it is necessary to address the contentious use of the term *"climate change,"* which is employed throughout most chapters of this volume. Several scholars have argued that the term risks obscuring the political responsibilities of actors and countries that are historically accountable for the crisis (Bonneuil and Fressoz 2013). In other words, the expression *climate change* may serve to euphemize the structural and political drivers behind what could more accurately be described as *climate disruption*. Language shapes the way problems are defined and risks are perceived; it is part of the broader political framing that either assigns or erases responsibility. Framing everyone—or no one—as responsible for global warming fosters the notion that it is inevitable and beyond contestation.

In reality, the contributions to climate change are highly unequal. Different actors—such as industrialized countries, large corporations, and the wealthiest individuals—contribute disproportionately to rising global temperatures. This disparity becomes even more pronounced when one considers not only current emissions but the *cumulative CO_2 emissions* associated with human activity since the onset of the industrial era. This results in a form of *double injustice*: those who have contributed the least to the problem—primarily populations in the Global South—are the ones most exposed to its impacts (Althor et al. 2016; IPCC 2019). Conversely, those most responsible—mainly in the Global North—are often the least vulnerable (Guivarch and Taconet 2020).

Far from depoliticizing climate change, this volume contributes to its *repoliticization* by analyzing the challenges climate change poses for agricultural and food systems across diverse political contexts, with particular attention to the most vulnerable regions. It explicitly highlights the vulnerability of countries in the Global South and emphasizes the historical responsibility of past emitters, as well as the accountability of the major contemporary contributors to GHG emissions.

An Acceleration of the Climate Focus on Agricultural and Food Issues in Scientific and Political Arenas

Following these semantic and definitional clarifications, we now turn to the presentation of the first part of this volume. This section examines how agricultural issues have been addressed within climate science forums since the creation of the Intergovernmental Panel on Climate Change (IPCC) in 1988, and within political arenas since the adoption of the United Nations Framework Convention on Climate Change (UNFCCC) at the Rio Earth Summit in 1992. Since then, the climate issue has progressively become "globalized," in the sense that it has become intertwined with a broad array of other concerns—particularly development and energy—thereby drawing them into climate debates (a centripetal dynamic). At the same time, an expanding range of actors, human communities, natural environments, and economic sectors (e.g., oceans, fisheries, forests, agriculture, food security) have increasingly claimed a stake in climate negotiations, seeking to express their specific concerns and articulate their interests in climate terms (a centrifugal dynamic) (Dahan 2016, 2018; Aykut et al. 2017). This dual movement of "climatization" also applies to agricultural and food-related issues. The gradual integration of these concerns into international scientific and political agendas (Hrabanski and Le Coq 2022) is the central focus of Part I of this volume.

Until the late 1990s, climate negotiations were primarily focused on mitigation, and consequently, only forests—recognized as carbon sinks—were explicitly included. From the 2000s onward, the growing salience of adaptation, particularly advocated by countries from the G77, brought increased attention to the impacts of climate change on land sectors and food systems—both in the Global North and, even more so, in the Global South (Hrabanski 2020). This dynamic found expression both within the IPCC—through the publication of a second special report on land in 2019 (see Chap. 1)—and within the international negotiations under the UNFCCC. Dedicated working groups on agriculture first emerged within the framework of the *Koronivia Joint Work on Agriculture*, which ultimately led to the *Sharm el-Sheikh Four-Year Joint Work on Implementation of Climate Action on Agriculture and Food Security*, adopted at COP27 in 2023 (see Chap. 2). The final text recognized the importance of both adaptation and mitigation in agriculture for all signatory countries, though without recommending specific pathways or strategies.

Several competing concepts—such as *climate-smart agriculture*, *agroecology*, and *nature-based solutions* (see Chap. 3)—have emerged as potential responses to the dual challenge of climate change and food production. These approaches offer alternatives to both input-intensive industrial agriculture and so-called traditional forms of agriculture, which are often less productive and/or economically viable (Duru et al. 2015; Lipper et al. 2014). However, no single strategic direction has yet been endorsed to guide the evolution of the land sector or, more broadly, food systems.

Finally, given the diversity and complexity of the contemporary world, we considered an *Atlas* essential to addressing the issue of climate change in agricultural and food systems (Chap 4). To conclude the first part of this volume, we therefore present *17 thematic maps* designed to capture the challenges associated with GHG emissions from agri-food systems, the impacts of climate change on the land sector, and carbon storage potential. In addition, *six chapters* summarizing projected climate trends across the world's major regions (Africa, Southeast Asia, Central and South America, Oceania, Europe, and North America) have been included to complement and contextualize the Atlas.

Agricultural and Food Systems, Both Contributors and Victims of Climate Change

The second part of the book begins with the observation that agricultural and food systems are both contributors to climate change and severely affected by it. Faced with intense anthropogenic pressure and extreme climatic events, the land sector has never encountered such a wide range of risks since the Neolithic era and the advent of agriculture. The exceedances of the 2 °C limit by 2100 (the target set by the Paris Agreement relative to the pre-industrial era) envisaged by the IPCC scenarios (ranging from +1.1 to +6.4 °C) are increasingly plausible, with major repercussions on agricultural productivity and global food security. Beyond this very concerning global observation, climate change affects the world's agricultural systems in a highly diverse and unequal manner in terms of intensity. These risks will be higher in Southern countries (despite being the lowest emitters), where an average temperature increase of 2 °C has a much greater impact than in temperate zones, particularly on crop cycles, due to a greater dependence on natural resources and higher vulnerability (Blanfort and Demenois 2019).

Simultaneously, according to the latest IPCC report on land (IPCC 2019), agricultural and food systems directly affect more than 70% of the ice-free land surface, with 50% dedicated to food production. They therefore hold a unique and now predominant place in ongoing and future climatic changes and are among the highest contributing sectors in terms of greenhouse gas (GHG) emissions. The AFOLU sector is responsible for 23% (12 ± 2.9 Gt CO2eq/year) of the net total of anthropogenic GHG emissions (IPCC 2019). Including deforestation as well as the increasingly emissive upstream and downstream supply chains of food systems, between 20% and 37% of the net total of anthropogenic GHG emissions are attributable to them. These emissions, due to the increase in the global population as well as changes in dietary patterns, are expected to further increase to reach 30–65% by 2030 (IPCC 2022) of total GHG emissions, thus becoming the largest emitting sector. In its roadmap for the adaptation of global food systems to climate change published during COP28 in 2023, the FAO warned that without measures, these emissions could explode by 2050. Substantial efforts are therefore essential in terms of adaptation and mitigation to achieve increased productivity while reducing resource consumption to reach "net zero carbon"[3] and even make food systems a net carbon sink of −1.5 Gt CO2eq per year.

Beyond these global figures on the impact of climate change on agri-food systems and their contribution to climate change, the book aims to decipher "what lies beneath." Emissions come from multiple sources and vary depending on

---

[3] While the term "carbon neutral" refers to balancing the total amount of carbon emissions (the balance between carbon emission and absorption of carbon emissions by carbon sinks), "net zero carbon" means that no carbon has been emitted and therefore it is not necessary to capture or offset carbon.

biophysical and biogeographical factors, production systems, agricultural practices, types of food produced and consumed, and waste management systems. For example, there are significant disparities between industrialized countries and others: China, Indonesia, the United States, Brazil, the European Union, and India are the countries (or groups of countries) with the highest emissions in volume (see Atlas maps and graphs 2, 3, 4). Similarly, different agricultural sectors do not contribute in the same way or to the same extent to GHG emissions.

Given this observation, the second part of the book is devoted to the major issues of agriculture, food systems, and forests in the face of climate change: family farming (Chap. 5), migration, employment, and land tenure (Chap. 6). Then, four major challenges are analyzed: water and agriculture (Chap. 7), food systems (Chap. 8), forests (Chap. 9), animal production (Chap. 10), health (Chap. 11), and finally a chapter on what pastoralism tells us about climate change, particularly in sub-Saharan Africa (Chap. 12). An analysis by sector is also proposed and is based in particular on the research work of CIRAD and its partners. It contextualizes the often very significant transformations of agricultural sectors over the last few decades, these changes being driven in particular by developments in food demand. The entire chain concerning a product, from the producer to the consumer, has been addressed to varying degrees for (i) major crops (the cases of rice, sorghum, sugarcane, and cotton sectors) (Chap. 13), (ii) the oil palm (Chap. 14), (iii) horticultural production (Chap. 15), and (iv) different livestock systems (Chap. 16).

For each of these issues and for each of these sectors, the aim is to identify the challenges both in terms of contribution to GHG emissions, introducing explanatory variables such as geographical, climatic, political, institutional, and territorial variables, and also to understand how these questions and/or sectors are affected by climate change.

Agricultural and Food Systems: Solutions to Address Climate Change

The final part of the book focuses on various possible solutions for the adaptation and mitigation of agricultural and food systems. Three types of solutions are identified.

The first type concerns solutions related to resource management, whether they involve carbon sequestration in agricultural and forest soils (Chap. 17), better management of water resources (Chap. 18), or the development of renewable energy production through the land sector (Chap. 19).

The second type of solution pertains to innovations in agricultural practices for the adaptation and mitigation of agricultural and food systems (Chap. 20), as well as innovations in terms of species and varieties, particularly highlighting the major role played by cultivated and natural diversity in adaptation and mitigation (Chap.

21). However, these innovative solutions must be adapted to the diversity of territories (Chap. 22).

Finally, a third type of solution, which is more political, must be supported and developed in both the Global North and South. To achieve this, the focus is first placed on food systems (Chap. 23). The commitments made under the "Global Methane Pledge," aimed at reducing methane emissions, particularly those from livestock and especially in the industrial livestock systems of Northern countries, are then analyzed (Chap. 24). More broadly, it is the entire set of public policies that must be better coordinated to promote adaptive governance and thus respect the commitments made by states as part of their nationally determined contributions (NDCs) (Chap. 25). The issues of financing these policies constitute a central point (Chap. 26). Finally, the spaces for dialogue between science and decision-makers, and all stakeholders, must be further strengthened within the framework of more inclusive and participatory science-policy interfaces (Chap. 27).

Marie Hrabanski
Cirad, UMR Actors Resources Territories &
Development (ART-Dev)
Montpellier, France
marie.hrabanski@cirad.fr

Vincent Blanfort
Cirad, UMR SELMET, Univ. Montpellier,
INRAE, Institut Agro
Montpellier, France

Julien Demenois
Cirad, UPR Agroecology and Sustainable
Intensification of Annual Crops (Aïda),
Montpellier, France

Emmanuel Torquebiau
Cirad
Montpellier, France

Box 1 Thinking About Adaptation and Mitigation in Synergy

Although the issues of mitigation and adaptation stem from distinct biophysical and political challenges, they must now be addressed jointly. The Sixth Assessment Report of the IPCC (IPCC 2023) (IPCC: Intergovernmental Panel on Climate Change) confirms that the simultaneous implementation of mitigation and adaptation measures—as well as the trade-offs between them—can foster co-benefits and synergies for human health and well-being, while contributing to sustainable development. The IPCC Working Group II report on impacts, adaptation, and vulnerability (IPCC 2022) identifies numerous adaptation options that exhibit varying degrees of synergy with mitigation, particularly in the land sector (e.g., forest-based adaptation, agroforestry, biodiversity management, and improved crop and livestock management).

Mitigation can serve as a function of adaptation—for instance, through technical pathways that enhance crop or livestock productivity while simultaneously increasing carbon stocks in soils and biomass. Conversely, the design of mitigation policies may also facilitate adaptation, such as through the creation of financial incentives for agroecological practices. There thus exist a range of biotechnical, institutional, and policy solutions capable of addressing the interconnected challenges of food security, mitigation, and adaptation simultaneously.

However, achieving this balance is complex. Given the time and resource constraints imposed by climate change, it is crucial to ensure that adaptation measures adopted by decision-makers do not inadvertently worsen GHG emissions. When this occurs, it is referred to as *maladaptation*—a term used to describe outcomes that are detrimental or contrary to the intended benefits of adaptation (Boutroue et al. 2022; Magnan et al. 2016). For example, the establishment of fast-growing clonal tree plantations may appear to be an effective adaptation strategy, but can ultimately degrade soil quality and increase local vulnerability to pests and disease outbreaks.

To mitigate the risk of maladaptation, it is essential to prioritize *"no-regret" strategies*, which provide benefits regardless of how future climate conditions unfold. The protection of ecosystems rich in biodiversity and carbon, as well as the restoration of degraded landscapes, are considered no-regret options that effectively combine adaptation and mitigation objectives.

References

Althor G, Watson JE, Fuller RA (2016) Global mismatch between greenhouse gas emissions and the burden of climate change. Sci Rep 6(1):20281

Aykut S, Foyer J, Morena E (eds) (2017) Globalising the climate: COP21 and the climatisation of global debates. Taylor & Francis

Blanfort V, Demenois J (2019) Changements climatiques et agriculture: quels enjeux, quels impacts aujourd'hui et demain? Diplomatie. Les Grands Dossiers 49:85–89

Bonneuil C, Fressoz JB (2013) The Anthropocene event: the earth, history and us, Media Diffusion

Boutroue B, Bourblanc M, Mayaux PL, Ghiotti S, Hrabanski M (2022) The politics of defining maladaptation: enduring contestations over three (mal)adaptive water projects in France, Spain and South Africa. Int J Agric Sustain 20(5):892–910

Crippa M, Solazzo E, Guizzardi D, Monforti-Ferrario F, Tubiello FN, Leip AJNF (2021) Food systems are responsible for a third of global anthropogenic GHG emissions. Nat Food 2(3):198–209

Dahan A (2016) Climate governance: between world air conditioning and reality schism. Man Soc 1:79–90

Dahan A (2018) The air conditioning of the world. Spirit 1:75–86

Duru M, Therond O, Fares M (2015) Designing agroecological transitions; a review. Agron Sustain Dev 35:1237–1257. https://doi.org/10.1007/s13593-015-0318-x

FAO (2024) FAO knowledge portal. https://www.fao.org/faostat/fr/#data

Guivarch C, Taconet N (2020) Global inequalities and climate change. Rev OFCE 165(1):35–70

Hrabanski M, Le Coq J-F (2022) Climatisation of agricultural issues in the international agenda through three competing epistemic communities: climate-smart agriculture, agroecology, and nature-based solutions. Environ Sci Pol 127:311–320. https://doi.org/10.1016/j.envsci.2021.10.022

Hrabanski M (2020) A climatisation of agricultural issues by the science? Controversies surrounding climate-smart agriculture. Int Critique 1:189–208

Hrabanski M, Le Coq J-F (2025) Agriculture at COP27: antagonistic political framing and fragmentation of agricultural issues within climate negotiations and beyond. Global Environ Politic:1–14. https://doi.org/10.1162/glep_a_00778

IPCC (2019) Summary for policymakers. In: Climate change and land. https://www.ipcc.ch/srccl/

IPCC (2022) Summary for policymakers. In: AR6 climate change 2022: impacts, adaptation and vulnerability. https://www.ipcc.ch/report/sixth-assessment-report-working-group-ii/

IPCC (2023) AR6 synthesis report: climate change 2023. https://www.ipcc.ch/report/sixth-assessment-report-cycle/

Lipper L, Thornton P, Campbell BM, Baedeker T, Braimoh A, Bwalya M et al (2014) Climate-smart agriculture for food security. Nat Clim Chang 4(12):1068–1072. https://doi.org/10.1038/nclimate2437

Magnan AK, Schipper ELF, Burkett M, Bharwani S, Burton I, Eriksen S et al (2016) Addressing the risk of maladaptation to climate change. Wiley Interdiscip Rev Clim Chang 7(5):646–665

Rosenzweig C, Mbow C, Barioni LG, Benton TG, Herrero M, Krishnapillai M et al (2020) Climate change responses benefit from a global food system approach. Nat Food 1(2):94–97

Tubiello FN, Karl K, Flammini A, Gütschow J, Obli-Laryea G, Conchedda G et al (2022) Pre- and post-production processes increasingly dominate greenhouse gas emissions from agri-food systems. Earth Syst Sci Data 14(4):1795–1809

Contents

Part I Agricultural and Food Systems Facing Climate Change: Global Overview

1. **A Global Overview of Agricultural and Food Systems in the Context of Climate Change** 3
   Vincent Blanfort, Julien Demenois, and Adèle Gaveau

2. **Agricultural, Food and Forestry Issues in International Climate Negotiations: Setting the Agenda and Challenges** 17
   Marie Hrabanski, Valérie Dermaux, Alexandre K. Magnan, Adèle Tanguy, Anaïs Valance, and Roxane Moraglia

3. **Tensions and Synergies Between the Concepts of Climate-Smart Agriculture, Agroecology, and Nature-Based Solutions** 31
   Nadine Andrieu, Audrey Naulleau, Jean-François Le Coq, and Marie Hrabanski

4. **Atlas of World Agriculture and Food Systems in the Face of Climate Change** 39
   Vincent Blanfort, Julien Demenois, Marie Hrabanski, Nicolas Viovy, Jacques André Ndione, Moussa Waongo, Maguette Kaire, Lilian Blanc, Sylvain Schmitt, Séverine Bouard, Catherine Sabinot, Pierre-François Duyck, Philippe Birnbaum, Audrey Leopold, Julien Drouin, Fabian Carriconde, Laurent L'Huillier, and Christophe Menkès

Part II Agricultural and Food Systems and the Land Sector: Contributors and Victims of Climate Change

5. **Family Farming in the Face of Climate Change: Potential for Adaptation Through Agroecology** 101
   Jean-Michel Sourisseau and Jean-François Le Coq

6	**Climate Change, (Im)Mobility and Land Tenure: Challenges for Family Farming in the Global South**. . . . . Sara Mercandalli, Hadrien Di Roberto, and Pierre Girard	113
7	**Water, Agriculture and Climate Change: Global Perspectives**. . . . . . Magalie Bourblanc, Caroline Lejars, and Pierre-Louis Mayaux	127
8	**Food Systems: Both Responsible for and Victims of Climate Change** . . . . Hélène David-Benz, Arlène Alpha, Victoria Bancal, Carine Barbier, Damien Beillouin, Yannick Biard, Daniel Fonceka, Franck Galtier, Sandra Payen, Ninon Sirdey, and Mathieu Weil	141
9	**Forests and Climate Change** . . . . Jacques Tassin, Alexandre Caron, Vincent Freycon, Bruno Hérault, Bruno Locatelli, Marie Ange Ngo Bieng, Régis Peltier, and Camille Piponiot	155
10	**Agriculture and Climate Change Debates: The Case of Livestock Production** . . . . Christian Corniaux, Vincent Blanfort, Mathieu Vigne, Jonathan Vayssières, and Guillaume Duteurtre	167
11	**Agriculture, Health and Climate Change: Towards a "One Health" Vision** . . . . Marisa Peyre, Didier Lesueur, Flavie L. Goutard, Alexandre Hobeika, Alexis Delabouglise, Maxime Tesch, Daan Vink, and François Roger	177
12	**What Pastoralism Tells Us About Climate Change** . . . . Saverio Krätli, Véronique Ancey, and Der Dabire	189
13	**Major Crops and Climate Change: The Cases of Rice, Sorghum, Sugarcane and Cotton**. . . . . Alexia Prades, Patricio Mendez del Villar, Didier Tharreau, Edward Gérardeaux, Raphaëlle Ducrot, Aude Ripoche, David Pot, Mohamed Lamine Tékété, Cyril Diatta, Laurent Laplaze, Boris Parent, Isabelle Basile-Doelsch, Christine Granier, Myriam Adam, Julie Dusserre, Michel Vaksmann, Mathias Christina, Christophe Poser, Bruno Bachelier, and Romain Loison	205
14	**Oil Palm: Building Climate Resilience** . . . . Alain Rival and Cécile Chéron-Bessou	229
15	**Horticultural Production in the Face of Climate Change** . . . . Éric Malézieux, Damien Beillouin, Raphaël Belmin, Isabelle Grechi, Rémi Kahane, Thibaud Martin, and Fabrice Le Bellec	241

Contents

16 Livestock Systems Facing the Challenges of Climate Change 253
Vincent Blanfort, Christian Corniaux, Véronique Alary,
and Guillaume Duteurtre

Part III Mitigating and Adapting Agricultural and Food Systems: What Solutions, What Synergies?

17 Soil Carbon Sequestration: A Solution to Mitigate and Adapt to Climate Change? 269
Julien Demenois, Damien Beillouin, David Berre, Vincent Blanfort,
Rémi Cardinael, Abigail Fallot, Frédéric Feder, Christophe Jourdan,
Dominique Masse, and Tom Wassenaar

18 What Solutions for Agricultural Water Management in the Face of Climate Change? 283
Caroline Lejars, Koladé Akakpo, Magalie Bourblanc,
Emeline Hassenforder, and Pierre-Louis Mayaux

19 Energy Production in Agriculture to Tackle Climate Change 295
François Pinta, Antoine Ducastel, Patrice Dumas, Marie Hrabanski,
and Grâce Chidikofan

20 Adapting to Climate Change: What Innovative Practices in Tropical Production Systems? 305
Éric Justes, Benoit Bertrand, Hervé Etienne, Frédéric Gay,
Bruno Rapidel, Philippe Thaler, and François-Xavier Côte

21 Adapting and Innovating in Terms of Cultivated Species and Varieties: A Key Role for Cultivated and Natural Diversity? AFS 319
Sophie Léran, Myriam Adam, Mathieu Gonin, Cécile Grenier,
Pierre Marraccini, Fabienne Micheli, Maria Camila Rebolledo,
Clément Rigal, Mohamed Lamine Tékété, Michel Vaksmann,
Hervé Étienne, and Delphine Luquet

22 Territorializing Climate Action 333
Camille Jahel, Amandine Adamczewski, Jérémy Bourgoin,
Guillaume Lestrelin, Ronan Mugelé, René Poccard-Chapuis,
Fatma Rostom, Tiago Teixeira Da Silva Siqueira, and Elodie Valette

23 Food Systems and Climate Change: Mitigation and Adaptation in Agri-Food Chains and Consumption 345
Marie Walser, Carine Barbier, Nicolas Bricas, and Patrice Dumas

24 Agricultural Methane: A Lever for Reducing Greenhouse Gas Emissions to Comply with the Paris Agreement 357
Rémi Prudhomme, Myriam Adam, and Mohamed Habibou Assouma

25　**The Heterogeneity of Institutionalization Pathways for Climate Policies and Instruments in Agriculture: A Comparative Analysis of Senegal, Colombia, Brazil, and France**............... 371
Marie Hrabanski, Jean François Le Coq, Gilles Massardier, Carolina Milhorance, Yves Montouroy, and Eric Sabourin

26　**Finance, Agriculture and Climate**............................ 383
Antoine Ducastel

27　**Science–Policy Interfaces in the Face of Climate Change**.......... 393
Carolina Milhorance, Antoine Perrier, Julien Demenois, Vincent Freycon, Camille Piponiot, Paul Luu, Adèle Gaveau, Marie Hrabanski, and Sélim Louafi

Conclusion: Strengthening Independent National Scientific Institutions in a World Under Geopolitical and Financial Tension...... 407
Michel Eddi, and Sébastien Treyer

Afterword by Stéphane Le Foll.................................. 411

Contributors

Myriam Adam Cirad, UMR AGAP Institute, Montpellier, France

UMR AGAP Institute, Univ Montpellier, Cirad, INRAE, Institut Agro, Montpellier, France

Faculty of Agriculture and Food Processing, National University of Battambang, Battambang, Cambodia

Amandine Adamczewski Cirad, UMR G-EAU, Montpellier, France

Koladé Akakpo Cirad, UMR G-EAU, Meknes, Morocco

Véronique Alary Cirad joint research unit Selmet "Tropical and Mediterranean Animal Production Systems", Montpellier Univ, Inrae, Institut Agro, Montpellier, France

Arlène Alpha MoISA, Univ Montpellier, CIHEAM-IAMM, Cirad, INRAE, Institut Agro, IRD, Montpellier, France

Cirad, UMR MoISA, Montpellier, France

Véronique Ancey Cirad, UMR Actors Resources Territories & Development (ART-Dev), Montpellier, France

Nadine Andrieu Cirad, UMR Innovation, Capesterre-Belle-Eau, Guadeloupe, France

Mohamed Habibou Assouma UMR SELMET, University of Montpellier, Cirad, INRAE, Montpellier Institut Agro, Montpellier, France

Cirad, UMR SELMET, dP ASAP (Agro-sylvo-pastoral systems in West Africa), Bobo Dioulasso, Burkina Faso

International Research-Development Centre on Livestock in Subhumid Zone (CIRDES), Bobo-Dioulasso, Burkina Faso

Bruno Bachelier Cirad, UPR Agroecology and Sustainable Intensification of Annual Crops (Aïda), Montpellier, France

Aïda, Univ. Montpellier, Cirad, Montpellier, France

Victoria Bancal Cirad, UMR Qualisud, Abidjan, Ivory Coast

UFR Sciences and Technologies of Food, University Nangui Abrogoua, Abidjan, Ivory Coast

Qualisud, Univ Montpellier, Avignon University, Cirad, Institut Agro, University of La Réunion, Montpellier, France

Carine Barbier UMR International Centre for Research on Environment and Development, CNRS, ENPC, Cirad, AgroParisTech, EHESS, Montpellier, France

Isabelle Basile-Doelsch Aix-Marseille University, CNRS, IRD, INRAE, CEREGE, Aix-en-Provence, France

Damien Beillouin Cirad, UPR HortSys, Caribbean Agro-Environmental Campus (CAEC), Martinique, France

HortSys, Cirad, University of Montpellier, Montpellier, France

Fabrice Le Bellec Cirad, UPR HortSys, University of Montpellier, Montpellier, France

Raphaël Belmin Cirad, UPR HortSys, Senegalese Institute of Agricultural Research, Dakar, Senegal

HortSys, Cirad, University of Montpellier, Montpellier, France

David Berre Cirad, UPR Agroecology and Sustainable Intensification of Annual Crops (Aïda), Montpellier, France

Aïda, Univ. Montpellier, Cirad, Montpellier, France

Benoît Bertrand Cirad, UMR DIADE, Montpellier, France

DIADE, University of Montpellier, IRD, Cirad, Montpellier, France

Yannick Biard Cirad, UPR HortSys, Montpellier, France

HortSys, Cirad, University of Montpellier, Montpellier, France

ELSA (Environmental Life cycle and Sustainability Assessment), Montpellier, France

Marie Ange Ngo Bieng UPR Forests and Societies, Cirad, Montpellier, France

Unit Forests and Biodiversity in Productive Landscapes, CATIE, Cdad. de Guatemala, Guatemala

Philippe Birnbaum Cirad, UMR AMAP, Univ Montpellier, CNRS, INRAE, IRD, Montpellier, France

Lilian Blanc Cirad, UPR Forests and Societies, Montpellier, France

Forests and Societies, Univ. Montpellier, Cirad, Montpellier, France

Vincent Blanfort Cirad joint research unit Selmet "Tropical and Mediterranean Animal Production Systems", Montpellier Univ, Inrae, Institut Agro, Montpellier, France

Contributors

Séverine Bouard Department of Environmental Management, Lincoln University, Christchurch, New Zealand

Magalie Bourblanc Cirad, Joint Research Unit Water Management, Actors, Uses (G-EAU), Montpellier, France

Jérémy Bourgoin Cirad, UMR TETIS, International Land coalition, Rome, Italy

Nicolas Bricas Cirad, UMR MoISA, Montpellier, France

MoISA, University of Montpellier, CIHEAM, Cirad, INRAE, Institut Agro Montpellier, IRD, Montpellier, France

Rémi Cardinael Cirad, UPR Aïda, Univ Montpellier, Montpellier, France

Department of Plant Production Sciences and Technologies, University of Zimbabwe, Harare, Zimbabwe

Alexandre Caron UPR Forests and Societies, UMR ASTRE, Cirad, Montpellier, France

Fabian Carriconde IAC (New Caledonian Agronomic Institute), New Caledonia, France

René Poccard-Chapuis Cirad, UMR SELMET, EMBRAPA Eastern Amazon, Belém, Brazil

Cécile Chéron-Bessou Cirad, UMR ABSys, Montpellier, France

UMR ABSys, University of Montpellier, Cirad, INRAE, Institut Agro, Montpellier, France

Grâce Chidikofan National University of Sciences, Technologies, Engineering and Mathematics of Abomey (ENSGEP/UNSTIM), Abomey, Benin

Mathias Christina Cirad, UPR Agroecology and Sustainable Intensification of Annual Crops (Aïda), Montpellier, France

Aïda, Univ. Montpellier, Cirad, Montpellier, France

Jean-François Le Coq Cirad, UMR Actors Resources Territories & Development (ART-Dev), Montpellier, France

Christian Corniaux Cirad joint research unit Selmet "Tropical and Mediterranean Animal Production Systems", Montpellier Univ, Inrae, Institut Agro, Montpellier, France

François-Xavier Côte Cirad, DGD-RS, Montpellier, France

Der Dabire International Research-Development Centre on Livestock in the Subhumid Zone (CIRDES), Bobo-Dioulasso, Burkina Faso

Hélène David-Benz Cirad, UMR MoISA, Saint-Pierre, La Réunion, France

MoISA, Univ Montpellier, CIHEAM-IAMM, Cirad, INRAE, Institut Agro, IRD, Montpellier, France

Alexis Delabouglise Cirad, UMR ASTRE, Montpellier, France

Julien Demenois Cirad, UPR Agroecology and Sustainable Intensification of Annual Crops (Aïda), Montpellier, France

Aïda, Univ. Montpellier, Cirad, Montpellier, France

Valérie Dermaux DGPE, Ministry of Agriculture and Food Sovereignty, Paris, France

Cyril Diatta Regional Study Centre for Drought Adaptation Improvement, Senegalese Institute of Agricultural Research, Thiès, Senegal

Julien Drouin IAC (New Caledonian Agronomic Institute), New Caledonia, France

Antoine Ducastel Cirad, UMR Actors Resources Territories & Development (ART-Dev), FAO, Rome, Italy

Raphaëlle Ducrot Cirad, UMR G-EAU, Royal University of Agriculture, Phnom Penh, Cambodia

Patrice Dumas Cirad, UMR Cired, Montpellier, France

Julie Dusserre Cirad, UPR Agroecology and Sustainable Intensification of Annual Crops (Aïda), Montpellier, France

Aïda, Univ. Montpellier, Cirad, Montpellier, France

Guillaume Duteurtre Cirad joint research unit Selmet "Tropical and Mediterranean Animal Production Systems", Montpellier Univ, Inrae, Institut Agro, Montpellier, France

Pierre-François Duyck Cirad, BIOS, UMR PVBMT, New Caledonia, France

Michel Eddi Ministry of Higher Education and Research, France. Research, Iddri, Paris, France

Hervé Etienne Cirad, UMR DIADE, Montpellier, France

DIADE, University of Montpellier, IRD, Cirad, Montpellier, France

Abigail Fallot Cirad, UMR SENS, Montpellier, France

SENS, Univ Montpellier, Cirad, Montpellier, France

Frédéric Feder Cirad, UPR Recycling and Risk, Univ. Montpellier, Montpellier, France

Daniel Fonceka Cirad, UMR AGAP, Montpellier, France

Cirad, INRAE, AGAP, University of Montpellier, Institut Agro, Montpellier, France

Vincent Freycon UPR Forests and Societies, Cirad, Montpellier, France

Franck Galtier MoISA, Univ Montpellier, CIHEAM-IAMM, Cirad, INRAE, Institut Agro, IRD, Montpellier, France

Cirad, UMR MoISA, Montpellier, France

Adèle Gaveau University of Lausanne, Lausanne, Switzerland

Frédéric Gay Cirad, UMR ABSys, Montpellier, France

UMR ABSys, University of Montpellier, Cirad, INRAE, Institut Agro, Montpellier, France

Edward Gérardeaux Cirad, UPR Agroecology and Sustainable Intensification of Annual Crops (Aïda), Montpellier, France

Aïda, Univ. Montpellier, Cirad, Montpellier, France

Pierre Girard Cirad, UMR ART-Dev, Institute of Statistical Social and Economic Research (ISSER), Accra, Ghana

Mathieu Gonin Cirad, UMR DIADE, Montpellier, France

DIADE, University of Montpellier, IRD, Cirad, Montpellier, France

National Coffee Research Institute, National Agricultural Research Organization, Mukono, Uganda

Flavie L. Goutard Cirad, UMR ASTRE, National Institute of Animal Science (NIAS), Hanoi, Vietnam

Christine Granier INRAE, UMR AGAP Institute, Montpellier, France

Isabelle Grechi Cirad, UPR HortSys, University of Montpellier, Montpellier, France

Cécile Grenier Cirad, UMR AGAP Institute, Montpellier, France

UMR AGAP Institute, Univ Montpellier, Cirad, INRAE, Institut Agro, Montpellier, France

Emeline Hassenforder Cirad, UMR G-EAU, National Agronomic Institute of Tunisia (INAT), Tunis, Tunisia

Bruno Hérault UPR Forests and Societies, Cirad, Montpellier, France

Alexandre Hobeika Cirad, UMR MoISA, Montpellier, France

MoISA, Univ Montpellier, Cirad, CIHEAM-IAMM, INRAE, Institut Agro, IRD, Montpellier, France

Marie Hrabanski Cirad, UMR Actors Resources Territories & Development (ART-Dev), Montpellier, France

Camille Jahel Cirad, UMR TETIS, Montpellier, France

Christophe Jourdan Cirad, UMR Eco&Sols, Montpellier, France

Eco&Sols, Univ Montpellier, Cirad, INRAE, IRD, InstitutAgro, Montpellier, France

Éric Justes Cirad, Département PERSYST, Montpellier, France

Rémi Kahane Cirad, UPR HortSys, University of Montpellier, Montpellier, France

Maguette Kaire AGRHYMET, Regional Climate Centre for West Africa and the Sahel (CCR-AOS), Niamey, Niger

Saverio Krätli International Union of Anthropological and Ethnological Sciences (IUAES), Commission on Nomadic Peoples (CNP), Lewes, UK

Laurent Laplaze UMR DIADE, University of Montpellier, IRD (Research Institute for Development), Montpellier, France

Caroline Lejars Cirad, joint research unit Water Management, Actors, Uses (G-EAU), Université de Montpellier, Montpellier, France

Audrey Leopold IAC (New Caledonian Agronomic Institute), New Caledonia, France

Sophie Léran Cirad, UMR DIADE, Montpellier, France

DIADE, University of Montpellier, IRD, Cirad, Montpellier, France

Nicafrance Foundation, Finca La Cumplida, Matagalpa, Nicaragua

Guillaume Lestrelin Cirad, UMR TETIS, National Agronomic Institute of Tunisia (INAT), Tunis, Tunisia

Didier Lesueur Cirad, UMR Eco&Sols, Montpellier, France

Alliance of Biodiversity International and International Center for Tropical Agriculture (Ciat), Asia hub, Common Microbial Biotechnology Platform (CMBP), Hanoi, Vietnam

Laurent L'Huillier IAC (New Caledonian Agronomic Institute), New Caledonia, France

Bruno Locatelli UPR Forests and Societies, Cirad, Montpellier, France

Romain Loison Cirad, UPR Agroecology and Sustainable Intensification of Annual Crops (Aïda), Montpellier, France

Aïda, Univ. Montpellier, Cirad, Montpellier, France

Sélim Louafi Cirad, DGD-RS, Montpellier, France

Delphine Luquet Cirad, Bios department, Montpellier, France

Paul Luu International Initiative "4 per 1000", Montpellier, France

Alexandre K. Magnan Iddri, Paris, France

Éric Malézieux Cirad, UPR HortSys, Capesterre-Belle-Eau, Guadeloupe, France

HortSys, Cirad, University of Montpellier, Montpellier, France

Pierre Marraccini Cirad, UMR DIADE, Montpellier, France

DIADE, University of Montpellier, IRD, Cirad, Montpellier, France

Thibaud Martin Cirad, UPR HortSys, University of Montpellier, Montpellier, France

Gilles Massardier Cirad, UMR Actors Resources Territories & Development (ART-Dev), Montpellier, France

Contributors

Dominique Masse IRD, UMR Eco&Sols, Montpellier, France

Eco&Sols, Univ Montpellier, Cirad, INRAE, IRD, InstitutAgro, Montpellier, France

Research Pole Ecological Intensification of Cultivated Soils in West Africa (IESOL), IRD, ISRA, Dakar, Senegal

Pierre-Louis Mayaux Cirad, joint research unit Water Management, Actors, Uses (G-EAU), Université de Montpellier, Montpellier, France

Christophe Menkès Research Institute for Development (IRD), UMR ENTROPIE, Ifremer, IRD, Univ. La Réunion, Univ. New Caledonia, CNRS, New Caledonia, France

Sara Mercandalli Cirad, UMR ART-Dev, Montpellier, France

Fabienne Micheli Cirad, UMR AGAP Institute, Montpellier, France

UMR AGAP Institute, Univ Montpellier, Cirad, INRAE, Institut Agro, Montpellier, France

Carolina Milhorance Cirad, UMR Actors Resources Territories & Development (ART-Dev), Montpellier, France

Yves Montouroy LC2S, University of the Antilles, Guadeloupe, France

Roxane Moraglia DGPE, Ministry of Agriculture and Food Sovereignty, Paris, France

Ronan Mugelé UMR Prodig, Aubervilliers, France

Audrey Naulleau Innovation, Université de Montpellier, Cirad, INRAE, Institut Agro, Montpellier, France

Cirad, UMR Innovation, Tunis, Tunisia

Jacques André Ndione Regional Agency for Agriculture and Food (ARAA), ECOWAS, Lomé, Togo

Boris Parent UMR LEPSE, University of Montpellier, INRAE, Institut Agro, Montpellier, France

Sandra Payen Cirad, UMR ABSys, Montpellier, France

UMR ABSys, University of Montpellier, Cirad, INRAE, Institut Agro, Montpellier, France

Régis Peltier Cirad, UPR Forests and Societies, Montpellier, France

Antoine Perrier Institut Agro Angers, Angers, France

Marisa Peyre Cirad, UMR ASTRE, Montpellier, France

François Pinta Cirad, UPR BioWooEB (Biomass, Wood, Energy, Bioproducts) University of Montpellier, Montpellier, France

Camille Piponiot Cirad, UPR Forests and Societies, Montpellier, France

Christophe Poser Cirad, UPR Agroecology and sustainable intensification of annual crops (Aïda), Montpellier, France

Aïda, Univ. Montpellier, Cirad, Montpellier, France

David Pot Cirad, UMR AGAP Institute, Montpellier, France

UMR AGAP Institute, Univ Montpellier, Cirad, INRAE, Institut Agro, Montpellier, France

Alexia Prades Cirad, DGD-RS, Montpellier, France

Rémi Prudhomme Cirad, UMR CIRED, Nogent-sur-Marne, France

Bruno Rapidel Cirad, UMR ABSys, Montpellier, France

UMR ABSys, University of Montpellier, Cirad, INRAE, Institut Agro, Montpellier, France

Maria Camila Rebolledo Cirad, UMR AGAP Institute, Montpellier, France

UMR AGAP Institute, Univ Montpellier, Cirad, INRAE, Institut Agro, Montpellier, France

Alliance Bioversity Ciat, Palmira, Colombia

Clément Rigal Cirad, UMR ABSys, Montpellier, France

UMR ABSys, University of Montpellier, Cirad, INRAE, Institut Agro, Montpellier, France

WorldAgroforestry, Vietnam Office, Hanoi, Vietnam

Aude Ripoche Cirad, UR GECO, Station de La Bretagne, Saint-Denis, La Réunion, France

Alain Rival Cirad, UMR ABSys, Montpellier, France

UMR ABSys, University of Montpellier, Cirad, INRAE, Agro Institute of Montpellier, Montpellier, France

Cirad DRASEI, Jakarta, Indonesia

Hadrien Di Roberto Cirad, UMR ART-Dev, Houphouet Bouany University, Abidjan, Ivory Coast

François Roger Cirad, DGD-RS, Van Phuc Diplomatic Compound, Hanoi, Vietnam

Fatma Rostom Cirad, UMR ART-Dev, Montpellier, France

Catherine Sabinot Research Institute for Development (IRD), Espacedev IRD, Univ Montpellier, Univ Guyane, Univ La Réunion, Univ Antilles, Univ New Caledonia, Univ Perpignan, Perpignan, France

Eric Sabourin Cirad, UMR Actors Resources Territories & Development (ART-Dev), Montpellier, France

Contributors

Sylvain Schmitt Cirad, UPR Forests and Societies, Montpellier, France
Forests and Societies, Univ. Montpellier, Cirad, Montpellier, France

Tiago Teixeira Da Silva Siqueira Cirad, UMR SELMET, Saint-Pierre, La Réunion, France

Ninon Sirdey MoISA, Univ Montpellier, CIHEAM-IAMM, Cirad, INRAE, Institut Agro, IRD, Montpellier, France
Cirad, UMR MoISA, Montpellier, France

Jean-Michel Sourisseau Cirad, UMR Actors Resources Territories & Development (ART-Dev), Montpellier, France

Adèle Tanguy Iddri, Paris, France

Jacques Tassin Cirad, UPR Forests and Societies, Montpellier, France

Mohamed Lamine Tékété Rural Economy Institute, Bamako, Mali

Maxime Tesch Cirad, UMR ASTRE, Montpellier, France

Philippe Thaler Cirad, UMR Eco&Sols, Montpellier, France
Eco&Sols, Univ Montpellier, Cirad, INRAE, IRD, Agro Institute, Montpellier, France

Didier Tharreau Cirad, UMR PHIM, Montpellier, France

Emmanuel Torquebiau Cirad, Montpellier, France

Sébastien Treyer Iddri, Paris, France

Michel Vaksmann Cirad, UMR AGAP Institute, Montpellier, France
UMR AGAP Institute, Univ Montpellier, Cirad, INRAE, Agro Institute, Montpellier, France

Anaïs Valance High Council for Climate, Paris, France

Elodie Valette Cirad, UMR ART-Dev, Montpellier, France

Jonathan Vayssières Cirad, UMR SELMET, Univ. Montpellier, INRAE, Agro Institute, Montpellier, France

Mathieu Vigne Cirad, UMR SELMET, Univ. Montpellier, INRAE, Agro Institute, Montpellier, France

Patricio Mendez del Villar Cirad, UMR TETIS, Univ Montpellier, AgroParisTech, Cirad, CNRS, INRAE, Montpellier, France

Daan Vink Cirad, UMR ASTRE, Montpellier, France

Nicolas Viovy Laboratory of Climate and Environmental Sciences (LSCE/IPSL), CEA-CNRS-UVSQ, University Paris-Saclay, Gif-sur-Yvette, France

Marie Walser Unesco Chair in World Food Systems, Cirad, Institut Agro Montpellier, Montpellier, France

Moussa Waongo AGRHYMET, Regional Climate Centre for West Africa and the Sahel (CCR-AOS), Niamey, Niger

Tom Wassenaar Cirad, UPR Recycling and Risk, Univ. Montpellier, Montpellier, France

Mathieu Weil Cirad, UMR Qualisud, Montpellier, France

Qualisud, University of Montpellier, Cirad, Agro Institute, Avignon University, University of La Réunion, Montpellier, France

Abbreviations

ACC	Adaptation to climate change
AEMC	Agri-environmental and climate measures
AFOLU	Agriculture, Forestry and Other Land Use
AFS	Agroforestry System
ANR	Assisted natural regeneration
AR	Assessment Report
BRC	Biological resource centre
C	Carbon
CC	Climate change
CDM	Clean Development Mechanism
CGIAR	Consultative Group on International Agricultural Research
Cive	Intermediate crop for energy purposes
COP	Conference of the Parties (e.g. COP27)
Corsia	Carbon Offsetting and Reduction Scheme for International Aviation
CSA	Climate-smart agriculture
ECOWAS	Economic Community of West African States
ENSO	El Niño-Southern Oscillation
ERC	"Avoid, reduce, compensate" sequence
EU	European Union
FA	Family farming
FA	Farm operation
FAO	Food and Agriculture Organization of the United Nations
FAOSTAT	Food and Agriculture Organization Corporate Statistical Database
GHG	Greenhouse gases
GMP	Global Methane Pledge
GWP	Global Warming Potential
HLPE	High Level Panel of Experts on Food Security and Nutrition
IEA	International Energy Agency
Ifad	International Fund for Agricultural Development
IOM	International Organisation for Migration

IPBES	Intergovernmental Science-Policy Platform on Biodiversity and Ecosystem Services
IPCC	Intergovernmental Panel on Climate Change
Ippu	Industrial Processes and Product Use
ITMO	Internationally Transferred Mitigation Outcome or carbon credits
IUCN	International Union for Conservation of Nature
IWRM	Integrated Water Resources Management
JI	Joint Implementation Mechanism
LCA	Life cycle analysis
LULUCF	Land Use, Land Use Change and Forestry
MENA	Middle East and North Africa
NAP	National Adaptation Plan
NBS	Nature-based solutions
NDC	Nationally determined contribution
NGO	Non-governmental organisation
OECD	Organisation for Economic Co-operation and Development
PES	Payments for Environmental Services
RCP	Representative Concentration Pathways
RE	Renewable energies
REDD+	Reducing Emissions from Deforestation and Forest Degradation
RSPO	Roundtable on Sustainable Palm Oil
SBSTA	Subsidiary Body for Scientific and Technological Advice
SDG	Sustainable Development Goal
SELMET	Mediterranean and Tropical Livestock Systems
SOC	Soil organic carbon
SPI	Science-Policy Interface
SRCCL	Special Report on Climate Change and Land
SSP	Shared Socio-economic Pathways
UNCCD	United Nations Convention to Combat Desertification
UNEP	United Nations Environment Programme
UNFCCC	United Nations Framework Convention on Climate Change
USDA	United States Department of Agriculture
WHO	World Health Organization
WMO	World Meteorological Organization
WTO	World Trade Organisation

Part I
Agricultural and Food Systems Facing Climate Change: Global Overview

Chapter 1
A Global Overview of Agricultural and Food Systems in the Context of Climate Change

Vincent Blanfort, Julien Demenois, and Adèle Gaveau

Abstract Since its creation in 1988, the IPCC has become the key scientific authority informing international climate negotiations under the UNFCCC. Through its assessment reports, the IPCC shapes political decisions by providing credible, expert-backed climate science. This chapter traces how agricultural and land issues have evolved in IPCC reports, highlighting the physical and agronomic impacts of climate change on agriculture globally and regionally. It also examines how agricultural and food systems contribute to climate change.

Over the decades, climate science has evolved and strengthened, particularly through the publications of the Intergovernmental Panel on Climate Change (IPCC). Established in 1988 under the auspices of the World Meteorological Organization (WMO) and the United Nations Environment Programme (UNEP), the IPCC has since the beginning of climate negotiations been the scientific and technical basis on which diplomats integrated into the United Nations Framework Convention on Climate Change (UNFCCC) rely. Through its scientific and technical framing and its legitimisation of certain solutions evaluated within the reports, the expert group indirectly influences the political sphere (De Pryck 2022). Thanks to its unique status and hybrid operation involving the expertise of volunteer scientists from around the world, the IPCC has over the years established itself as a credible and legitimate intergovernmental organisation for all its 195 member states. To date, six

---

V. Blanfort (✉)
Cirad joint research unit Selmet "Tropical and Mediterranean Animal Production Systems", Montpellier Univ, Inrae, Institut Agro, Montpellier, France
e-mail: vincent.blanfort@cirad.fr

J. Demenois
Cirad, UPR Agroecology and Sustainable Intensification of Annual Crops (Aïda), Montpellier, France

Aïda, Univ. Montpellier, Cirad, Montpellier, France

A. Gaveau
University of Lausanne, Lausanne, Switzerland

© The Author(s) 2026
V. Blanfort et al. (eds.), *Climate Impacts and Challenges in Agriculture, Forests and Food Systems*, https://doi.org/10.1007/978-3-032-04331-3_1

assessment reports have been produced, referred to in this chapter by the acronym AR (Assessment Report) from the first report (AR1) to the sixth (AR6), published respectively in 1990, 1995, 2001, 2007, 2014 and 2023.[1]

This chapter presents the evolution of the land sector and agricultural issues within the IPCC reports. Without detailing what constitutes the subject of many chapters of the book, it then develops at a very global level the physical and agronomic impacts of climate change on the agricultural sector at the global and regional scale. Finally, it describes the impacts of agricultural and food systems on climate change.

1.1 The Evolution of the Land Sector and Agricultural Issues within the IPCC Reports

Since 1990, the IPCC has published two special reports on the land sector. A first special report on land use, land-use change and forestry was commissioned by the Subsidiary Body for Scientific and Technological Advice (SBSTA) of the UNFCCC and was published in 2000 (IPCC 2000a, b) after the adoption of provisions relating to the sequestration of biotic carbon within the Kyoto Protocol. Following its adoption, many terms and provisions relating to these new carbon sinks remained to be defined within the protocol, with various interpretations on their use by member states, notably within the framework of the Clean Development Mechanism (CDM) (Fogel 2005). With a different mandate from the first, a second special report was subsequently published by the IPCC in August 2019, focusing more broadly on climate change, desertification, land degradation, sustainable land management, food security, and greenhouse gas (GHG) fluxes in terrestrial ecosystems (SRCCL, or Special Report on Climate Change and Land). This special report marks a turning point in the treatment of agricultural issues within the reports. It indeed provides a more in-depth and systemic focus on options for reducing GHGs from agricultural systems, including various aspects such as soil degradation and desertification, the link with biodiversity, agroecology, *climate-smart agriculture* (see Chap. 3) and different dietary choices (*dietary options*), including meat consumption (see Chap. 23). This land report also fits more broadly into a burgeoning context of publications at the same time on the issue of land use.[2]

Unlike previous reports, the SRCCL was jointly developed by the three working groups of the IPCC and for the first time in the history of the IPCC, by more authors

---

[1] All reports can be consulted here: https://www.ipcc.ch/reports/.

[2] In parallel with the SRCCL, the first edition of the Global Land Outlook report from the United Nations Convention to Combat Desertification (UNCCD), the thematic assessment of the Intergovernmental Science-Policy Platform on Biodiversity and Ecosystem Services (IPBES) on land degradation and restoration (Assessment Report on Land Degradation and Restoration) in March 2018, and the IPBES report on the global assessment of biodiversity and ecosystem services (Global Assessment Report on Biodiversity and Ecosystem Services) in May 2019 were also published.

from the Global South (53%) than the Global North.³ While the summary for policymakers of the first special report on land, published in 2000, established a direct link between the scientific content of the report and certain articles of the Kyoto Protocol, the link with the articles of the Paris Agreement is not explicitly mentioned in that of the 2019 report. However, after being presented at COP25 in Madrid in 2019, the SRCCL served as the basis for organising a dialogue between member states in November and December 2020 focused on the relationship between lands and adaptation to climate change within the framework of the SBSTA.⁴

After the publication of the 2000 special report, the term LULUCF (Land Use, Land Use Change and Forestry) was introduced into the IPCC vocabulary and also appears in the 3rd report (AR3) in 2001 and the working group dedicated to mitigation.⁵ Later, the establishment of the IPCC *guidelines* for GHG inventories in 2006 proposed the term AFOLU (Agriculture, Forestry and Other Land Use) integrating the agricultural variable: this concept then gradually took over from the LULUCF category.⁶ AFOLU is then integrated into the third working group on mitigation in the 5th IPCC report (AR5) as well as in the following one (AR6).⁷ The concept of LULUCF does not, however, completely disappear, and now mainly refers to a type of GHG: the "CO_2 –LULUCF".⁸

Historically, all IPCC reports (from AR1 to AR6) have highlighted the importance of the role played by the land sector in mitigating and adapting to climate change (Fig. 1.1). Unlike other sectors responsible for large-scale GHG emissions, this sector has been identified as the only one for which mitigation could be possible in the short term, for example through afforestation and reforestation or the management of soil organic carbon. The methods of evaluating mitigation have, however, significantly evolved over the course of the reports and modelling scenarios have demonstrated how considerable the potential role of the sector is in reducing GHGs. The 3rd assessment report (AR3 in 2001) innovated methodologically for the analysis of the land sector with the introduction of integrated *top-down* assessment models, which have the advantage of taking into account all mitigation options from all sectors. These approaches select the least costly mitigation options, but from a limited number of agricultural options, focusing on other GHGs than CO_2, such as methane.

Regarding the more precise treatment of the agricultural issue by the IPCC, from the publication of the first report in 1990, the analysis of climate impacts and adaptation of the agricultural and food sectors has greatly evolved. Over the course of the

---

[3] https://www.ipcc.ch/srccl/authors/
[4] https://unfccc.int/event/dialogue-on-the-relationship-between-land-and-climate-change-adaptation-related-matters
[5] https://www.ipcc.ch/site/assets/uploads/2018/03/WGIII_TAR_full_report.pdf
[6] https://www.ipcc-nggip.iges.or.jp/public/2006gl/pdf/4_Volume4/V4_01_Ch1_Introduction.pdf
[7] https://www.ipcc.ch/site/assets/uploads/2018/02/ipcc_wg3_ar5_full.pdf
[8] https://www.ipcc.ch/report/ar6/wg3/downloads/report/IPCC_AR6_WGIII_FullReport.pdf

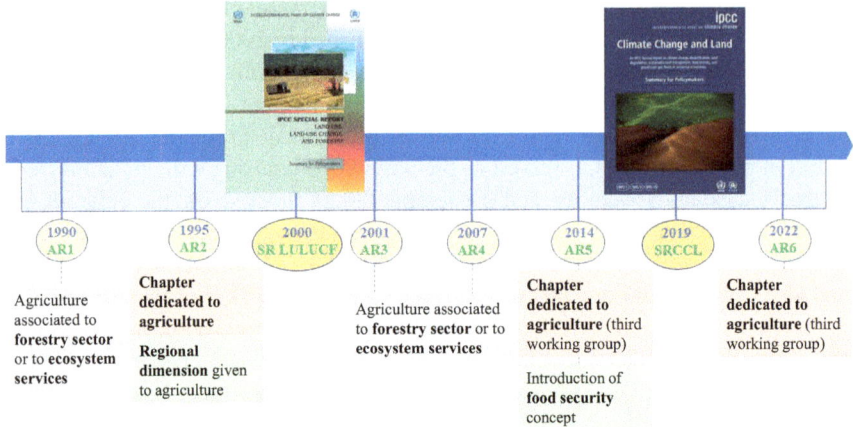

Fig. 1.1 Evolution of the treatment of the agriculture sector in the IPCC reports

reports, the number of studies dealing with the climate consequences on the production of main crops such as wheat, rice, maize, and soy has greatly increased. However, the focus has been more on crops rather than on livestock, and predominantly in the temperate zones of developed countries. Similarly, while quantitative studies focused on the adaptation of the agricultural sector have been increasing over the course of the reports, few quantitative data on the impacts and on the adaptation modalities of the livestock sector have been produced. The first three reports, for example, contain no quantitative data on this subject, and only eighteen studies are counted in the AR4 and AR5 (Porter et al. 2017, 2019; Rivera-Ferre et al. 2016). This can be explained by the fact that simulation models were little or not used for the livestock sector at this time, with more partial data, especially in the Southern countries. Regarding adaptation, in terms of the "capacity [of agricultural systems] to withdraw from an unsuitable climate regime".[9] AR4 frames the adaptation of the sector under autonomous and planned modalities, and AR5, in an "incremental, even transformative" manner (Porter et al. 2019).

Prior to the publication of the 5th report (AR5), the link remained very partial between, on the one hand, emission mitigation and, on the other hand, food production and security. However, from AR5, concrete suggestions regarding policy options were proposed for the AFOLU sector, thus moving beyond the analysis paradigm from "agricultural production" to an analysis of "food systems". In 2019, the SRCCL addressed for the first time within the same report a wide range of issues related to the land sector, highlighting action on integrated response options to address climate change, taking into account synergies and *trade-offs*. In the 6th report (AR6), the Chap. 11 of the third working group also emphasises the inherent intersectorality of the land sector, and indicates that this sector maintains significant links with others such as human consumption and societal well-being, bioenergy,

---

[9] In the text: "The ability of community constituents to migrate away from an unsuitable climate regime." (AR1, p. 321.)

building, transportation or industry, the mitigation of these sectors can thus be heavily dependent on the contribution of the AFOLU sector.

1.2 The Effects of Climate Change on the Land Sector

The land sector — already heavily impacted by climate change — also faces significant pressures due to an unprecedented expansion of human activities that directly affect more than 70% of the ice-free surface (IPCC 2019). Climate change can exacerbate land degradation processes (such as deforestation, unsustainable agricultural practices, overexploitation of natural resources, urbanisation) by increasing the intensity of precipitation, floods, the frequency and severity of droughts, thermal stress, periods of drought, wind, sea-level rise, and permafrost thawing. Faced with extreme climatic hazards and an anthropogenic pressure that has never been so strong, the land sector has thus never been confronted with such a range of risks since the Neolithic revolution with the invention of agriculture. The IPCC estimates that about a quarter of the ice-free land surface is undergoing degradation induced by human activities. The functioning of the land sector is, of course, by nature governed by the climate and its variations, and agriculture is one of the human activities most dependent on these processes, particularly under tropical latitudes characterised by significant intra- and interannual climate variability. However, due to the magnitude of the changes, the current situation is of a completely different nature. It is becoming increasingly difficult to limit the rise in temperatures to 2 °C by 2100; the scenarios considered by the IPCC beyond this limit will lead to major changes in the land sector, particularly negative impacts on agricultural productivity and global food security (Figs. 1.2a and 1.2b). According to these scenarios, the IPCC predicts that 8% to 30% of current agricultural land will become climatically unsuitable by 2100. The FAO, for its part, warns that without effective adaptation, yields of major crops could lose an average of 2% per decade, while production will need to increase by 14% every 10 years to meet global demand.

Beyond this very concerning global observation, climate change affects the world's agriculture in a very diverse and uneven manner in terms of intensity. In general, climate change induces negative impacts on the land sector; however, in some regions at middle and high latitudes, certain beneficial effects of the lengthening of the growing season and CO_2 fertilisation are observed. They lead to an increase in the productivity of ecosystems in these geographical areas (greening of vegetation in certain regions of Asia, Europe, South America, central North America and southwest Australia) (IPCC 2019). The yields of certain crops specific to these latitudes (for example maize, wheat and sugar beet) have increased in recent decades; however, their stagnation, or even decrease, is observed with recurring summer droughts. In these regions, the increase in frequency and intensity of extreme events (droughts or heavy rains, high summer temperatures) induces increasingly critical negative impacts for the agricultural sector and for water resources. Satellite observations show a browning of vegetation over the last three

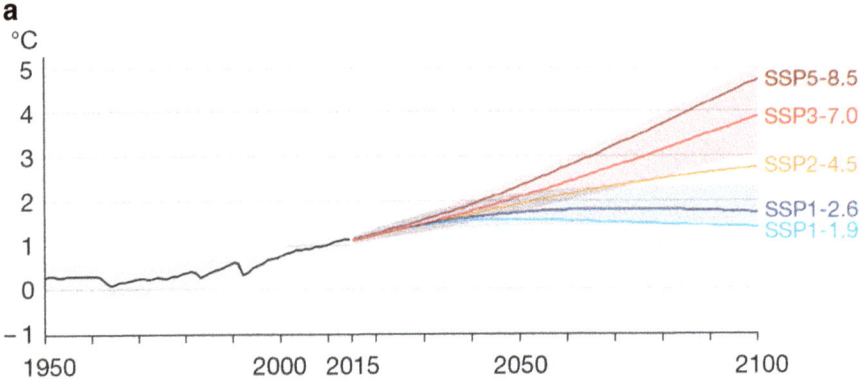

Fig. 1.2a The five main scenarios of the IPCC (AR6). (Source: IPCC 2021)

In the 6th IPCC report, the so-called SSP* (*shared socio-economic pathways) scenarios* replace the so-called RCP (*representative concentration pathways) scenarios*. The new SSP scenarios illustrate different socio-economic developments in connection with the different trajectories of GHG concentrations in the atmosphere. These new scenarios can be used in addition to the previously defined RCPs in the 5 th IPCC report. The IPCC emphasises that the impacts on our societies will not be the same depending on the levels of global warming.

* The different so-called SSP scenarios. SSP1: the sustainable and "green" path, common goods preserved, human well-being rather than economic growth, reduced income inequalities, minimisation of the use of material resources and energy; SSP2: the "median" path, incomes in different countries diverge, maintenance of cooperation between states, moderate and stabilised global population growth after 2025, moderate degradation of environmental systems; SSP3: regional rivalries, resurgence of nationalism and regional conflicts, policies focus on security issues, investments in education and technological development decrease, inequalities increase, considerable environmental damage in certain regions; SSP4: inequality, the gap widens between developed societies and those with low incomes and a low level of education, environmental problems solved in some regions, but not in others; SSP5: development from fossil fuels, integrated global markets, with innovations and technological progress, energy-intensive social and economic development worldwide, local environmental problems such as air pollution are successfully addressed.

decades in northern Eurasia and in certain parts of North America and Central Asia (IPCC 2019). But the generally favourable economic context of the states in these regions, their production structures and relatively active public policies often allow for easier adjustments.

Due to a greater dependence on natural resources, a higher vulnerability (larger populations in a context socio-economic poverty, and often political instability), the major expected effects of climate change are generally more significant in countries within the intertropical and Mediterranean zones. Small family farms (see Chap. 5) typical of these contexts often have limited access to technologies, services, and innovative inputs, which restricts their adaptive capacity. In a report on the impact of disasters and crises on agriculture between 2006 and 2016, the FAO emphasises that agriculture in the Global South (crops, livestock, fishing, aquaculture, and forestry) is affected by 25% of all damage and losses caused by climate-related

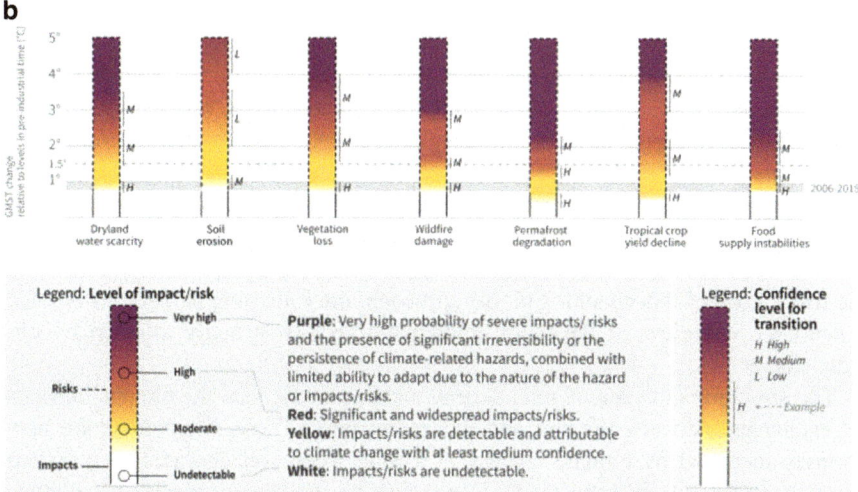

Fig. 1.2b Risks posed by the effects of climate change on terrestrial processes for human populations and ecosystems. (Source: IPCC 2019)
The rise in average global surface temperature compared to pre-industrial levels affects processes related to desertification (water scarcity), land degradation (soil erosion, vegetation loss, uncontrolled fires, permafrost thawing) and food security (crop yield variation, instability of food supply). This results in risks for food systems, livelihoods, infrastructure, land value, and the health of human populations and ecosystems. The alteration of a process (fires or water scarcity, for example) can lead to combined risks. Risks are location-specific and vary by region

disasters (floods, droughts, tropical storms), this figure rising to 83% for drought. Crops are the most affected sub-sector, constituting 50% of all these impacts (crop losses due to sudden events or slow changes that reduce yields). Livestock farming in the Global South suffers 36% of these impacts, mainly due to the weakening of the animals' physical condition and the reduction of their productivity, particularly in pastoral systems in Africa (see Chaps. 10 and 16). There are also significant consequences on animal health, due to changes in habitats, the spread of diseases, heat stress, and the alteration of feeding cycles. These risks also affect wildlife.

However, within this tropical and Mediterranean geographical ensemble, significant variations are observed due to the extent of these regions, the mode of land occupation, and their great socio-economic and bioclimatic diversity. Nevertheless, climate change significantly affects them, and the IPCC's evolution scenarios predict an intensification of these impacts for global temperature increases of more than 1.5 °C. (IPCC 2022). In many regions of the lower latitudes, yields of certain crops (such as maize and wheat) have already decreased. Regions heavily dependent on rain-fed agriculture are particularly exposed to climatic hazards. In sub-Saharan Africa, they represent nearly 93% of cultivated land. For some regions with particularly pessimistic scenarios, such as the Sudan-Sahel zone, annual temperatures are expected to increase by about 3 °C by 2050 compared to the 1980–2009 period, with a critical impact on crop yields.

Finally, sea-level rise, while not often discussed in this context, is nonetheless a significant factor in relation to the land sector, causing saltwater intrusion into coastal areas, which increases soil salinity and reduces their agricultural value.

This observation of the high diversity of climate change impacts in the land sector has fundamental implications for the relevant scale at which decisions should be taken and public policies formulated. The atlas of this book (see Chap. 4) describes in detail the impacts by major region. Beyond this biogeographical vision, within the framework of this chapter's global approach, the following paragraphs provide a synthetic overview of the main emblematic sectors strongly affected by climate change.

The structure and state of soils depend on the climate, and the climate depends on exchanges between the soil and the atmosphere. These interactions are now strongly modified by climate change, at a time when soil degradation threatens more than 40% of emerged lands. Depending on the region considered, climate change could accelerate this soil degradation process and compromise their ability to provide ecosystem services. Extreme weather events such as storms and torrential rains can, for example, lead to accelerated soil erosion, causing a direct loss of soil mass or essential nutrients for soil fertility, a decrease in water quality, and the destruction of natural habitats. Excessive rainfall also causes floods and landslides, affecting a land sector with often already fragile infrastructures in the Global South. Soil health, including its biodiversity, also suffers from changes in environmental conditions such as temperature, humidity, and precipitation. These changes can affect the distribution of soil organisms, which can lead to a reduction in soil biological diversity. They disrupt the biogeochemical cycles[10] of soils such as the nitrogen, carbon, and phosphorus cycles, which are essential for ecosystem functioning (see Chap. 17) and for the fertility of crop, herbivore rearing, and forest areas. This soil degradation can lead to desertification. According to the IPCC, a third of emerged lands could be affected by these processes by 2050, even with warming contained to 2 °C.

During the period 1961–2013, the average proportion of arid and semi-arid areas affected by drought increased by just over 1% per year (IPCC 2021). This dramatic increase (30% in total!) increasingly affects populations, with nearly 500 million people affected by these consequences in 2015, particularly in South and East Asia, and in the peripheral areas of the Sahara. Generally, these territories with limited

---

[10] Biogeochemical cycle: a process of transport and cyclical transformation (recycling) of an element or chemical compound between the major reservoirs, which are the geosphere, atmosphere, and hydrosphere, within which the biosphere is found. The prefix *bio* refers to organisms and biological mechanisms of the cycle, while the term *geo* denotes the environment, i.e., the atmosphere, lithosphere, and hydrosphere. The most important cycles are the nitrogen, carbon, water, and hydrogen cycles.

natural resources (water, arable land) are the site of transhumant livestock activities that alone can valorise them. They indeed demonstrate remarkable adaptation and flexibility that can buffer the effects of climate variability (see Chaps. 12 and 16). This climate resilience is weakened by the sedentarisation policies and by the reduction and fragmentation of traditional pastoral areas at the interfaces with more agricultural zones, whose expansion is itself correlated with the reduction in agricultural yields due to climate change. In Africa, these regions are sometimes dramatically affected by conflicts over the control of these resources, conducive to the rise of terrorism and extremism.

Among the alterations of the major global biogeochemical cycles, those of the water cycle are certainly the most noticeable due to their direct effects on water resources (recharge of groundwater, rivers, drought, evapotranspiration). These effects can be exacerbated by human activities such as deforestation, which can affect runoff and water infiltration into the soil, altering the availability of water in terrestrial and aquatic ecosystems. It is already observed in many regions that overexploitation of water resources linked to climate change increases the vulnerability of these territories to drought. This often leads to the implementation of unsustainable agricultural practices such as inefficient or excessive irrigation. This maladaptation then contributes to a decrease in water availability in arid and semi-arid areas. These significant changes in water resources can also contribute to soil erosion, salinisation and desertification (see Chaps. 7 and 17).

As for ecosystems particularly dependent on the water cycle, forests are among those most disturbed to varying degrees by climate change. Even though the rate of deforestation tends to decrease globally, climate change poses a new threat to their functioning and sustainability (increase in temperatures and alteration of precipitation regimes (see Chap. 9). Combined with the increased risk of fires and deforestation, these disturbances have significant consequences on the ability of forests to provide ecosystem services such as climate regulation, regional rain regimes and soil protection. In 2019, nearly one million hectares of Amazonian forest burned (see Chap. 4). Tropical forests represent more than 50% of the forest area and the global forest carbon sink. Among these, the three major tropical forest basins - Amazon, Congo and Borneo-Mekong-Southeast Asia - play a major role in climate regulation and also host an exceptional fauna and flora.

Climate change can influence the spread and impacts of pests, disease vectors for plants and animals, and invasive species, by altering environmental conditions, promoting the expansion of their range, altering natural habitats, and intensifying the effects of invasions (IPCC 2019).

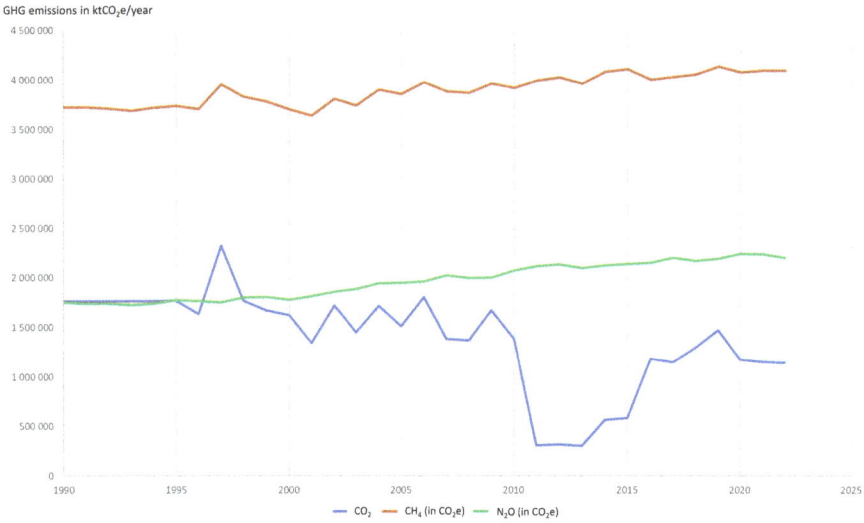

Fig. 1.3 Evolution of CO_2, CH_4 and N_2O emissions from the AFOLU sector between 1990 and 2022, expressed in kt eqCO_2/year. (Source: FAOSTAT data)

1.3 The Impact of Agricultural and Food Systems on Climate Change

Food systems are one of the main drivers of human impacts on the environment, especially on GHG emissions and climate change. They have contributed to the crossing of several of the "planetary boundaries" that define a safe operating space for humanity on a stable Earth system, particularly those concerning climate change (Springmann et al. 2018). Food systems are thus responsible for about a third of global GHG emissions, including deforestation[11] as well as increasingly emitting upstream and downstream supply chains (IPCC 2022). According to the latest IPCC report on land (IPCC 2019), the AFOLU sector (agriculture, forestry and other land uses) is responsible for, respectively, 13%, 44% and 81% of anthropogenic emissions of CO_2, CH_4 and N_2O, or 23% (12 ± 2.9 Gt eqCO_2/year) of the total net anthropogenic GHG emissions. These are mainly divided between direct GHG emissions attributable to agricultural production (12%) and deforestation (9%). Methane (CH_4) (see Chap. 24) and nitrous oxide (N_2O) emissions have been steadily increasing since 1990 (Fig. 1.3).

If emissions associated with pre- and post-production activities (2.4–4.8 Gt eqCO_2/year) in the global food system are included, these estimates range from 11 to 19 Gt eqCO_2/year, or 20–37% of the total net anthropogenic GHG emissions.

---

[11] According to the latest assessment of global forest resources by the FAO, agriculture is responsible for nearly 90% of deforestation worldwide.

This distribution needs to be nuanced according to the level of development of countries. Agricultural production is by far the main source of anthropogenic emissions in LI (*low income*) and LMI (*low to moderate income*) countries, while post-production stages emit as much GHG as production stages in HI (*high income*) countries (Vermeulen et al. 2012).

Although there is little information available, it is estimated that a third of the food produced worldwide for human consumption is lost or wasted between the stages of production and consumption (Gustavsson et al. 2011; HLPE 2014). Regarding food loss and waste (FLW, including harvest, storage, transport, and processing), the global carbon footprint for all countries combined is estimated at 3.3 Gt eqCO$_2$, which represents a huge portion of global GHG emissions (8–10%). It is difficult to differentiate between countries beyond a trend. Although it is not possible to be very precise, losses appear to be more significant in LMI countries. In these countries, losses and waste would primarily be due to infrastructural, financial, and technical constraints (Gustavsson et al. 2011). *Conversely*, in HI countries, waste would be linked to overproduction and overconsumption.

Currently, more than 50% of emerged lands are dedicated to food production (IPCC 2019). Changes in land use and rapid intensification of these uses have significantly increased the production of food, fodder, and fibres (+240% for cereals between 1961 and 2017, for example). Here too, these developments vary according to sectors, geographical areas, or levels of development. If we consider the issue of deforestation, which has been the subject of political and media debates since the 20th century, it mainly concerns LI countries and LMI countries, where the link between food systems and deforestation is crucial. Studies (Feintrenie et al. 2019) show that deforestation in the tropics increased between 2000 and 2012, mainly due to the increase in urban populations and the expansion of export agriculture. Commercial agriculture is therefore the main driver of deforestation (68% in Latin America, 35% in Africa and Asia), followed by subsistence agriculture (27% and 40% of deforestation on each continent (Hosonuma et al. 2012).

Another sector is the subject of controversy: livestock farming, which has undergone a real revolution over the past four decades, due to the high demand for animal products linked to demographic changes, economic growth, and urbanisation (see Chaps. 4 and 16). These major changes have had a significant impact on the increase in GHG emissions from this sector, which, according to the latest FAO estimates (FAO 2023), generates 12% of global anthropogenic GHG emissions (6.2 Gt eqCO$_2$/year). More than 70% of these emissions are due to ruminants (because of methane emissions) (see Chap. 24), but grassland farming systems would be responsible for "only" 20% of total livestock emissions (Gerber et al. 2013), while presenting a significant potential for mitigation through soil carbon sequestration.

While technical and behavioural changes supported by public policies can help limit GHG emissions from agricultural and food systems, more systemic changes are needed to adapt to climate change without compromising its mitigation. Dominant agricultural models must evolve (reduction of chemical inputs, reassociation of animal and plant productions, changes in agricultural practices towards agroecology, etc.) and our food systems need to be adapted (change in diets and eating habits, reduction of losses, etc.) and reterritorialised. It should also be noted that 31% of anthropogenic GHG emissions are absorbed by emerged lands in vegetation and soils, and thus constitute a larger carbon sink than the oceans.[12] Indeed, the AFOLU sector has a significant potential for carbon sequestration, whether through soils (see Chap. 17) or forests (see Chap. 9). Moreover, any change in surface, whether resulting from land use or climate change, affects the regional climate through biophysical effects such as albedo or evapotranspiration. Thus, in boreal regions, where the forest limit will migrate northwards and/or the growing season will lengthen, winter warming will be increased due to the reduction in snow cover and albedo, while warming will be reduced during the growing season due to increased evapotranspiration (IPCC 2019). These interactions between the carbon cycle, surface change, and biophysical effects make the assessment of the impacts of the land sector on climate change particularly complex.

Beyond a simplified carbon accounting, it is essential to integrate into the assessment of the agricultural and food sector its contribution to poverty reduction and food security. Agriculture is indeed the world's largest employer: 40% of global jobs and 60% in Africa. Food insecurity affects 30% of farmers in the South according to the FAO and also contributes to population displacement and the multiplication of conflicts. By 2050, it is predicted that the majority of African countries will be under the influence of climates currently still unknown on more than half of the arable land.

References

De Pryck K (2022) IPCC: the voice of climate. Presses de Sciences Po, p 240

FAO (2023) Pathways towards lower emissions—a global assessment of the greenhouse gas emissions and mitigation options from livestock agrifood systems. FAO, Rome. https://doi.org/10.4060/cc9029en

Feintrenie L, Betbeder J, Piketty M-G, Gazull L (2019) Deforestation for food production. In: Dury S, Bendjebbar P, Hainzelin E, Giordano T, Bricas N (eds) Food systems at risk: new trends and challenges. FAO/Cirad/European Commission, pp 43–46. https://doi.org/10.19182/agritrop/00080

Fogel C (2005) Biotic carbon sequestration and the Kyoto protocol: the construction of global knowledge by the intergovernmental panel on climate change. Int Environ Agreem: Politics Law Econ 5(2):191–210. https://doi.org/10.1007/s10784-005-1749-7

Gerber PJ, Steinfeld H, Henderson B, Mottet A, Opio C, Dijkman J et al (2013) Tackling climate change through livestock—a global assessment of emissions and mitigation opportunities. FAO, p 139

Gustavsson J, Cederberg C, Sonesson U, van Otterdijk R, Meybeck A (2011) Global food losses and food waste: extent, causes and prevention. FAO, p 37

HLPE (2014) Food losses and waste in the context of sustainable food systems. A report by the High Level Panel of Experts on Food Security and Nutrition of the Committee on World Food Security, 9 p

---

[12] Global Carbon Budget 2023: https://globalcarbonbudget.org/carbonbudget2023/

Hosonuma N, Herold M, Sy VD, Fries RS, Brockhaus M, Verchot L et al (2012) An assessment of deforestation and forest degradation drivers in developing countries. Environ Res Lett 7:044009

IPCC (2000a) Land use, land use change and forestry. A special report. https://www.ipcc.ch/report/land-use-land-use-change-and-forestry/

IPCC (2000b) Summary for policymakers. In: Land use, land use change and forestry. A Special Report of IPCC. Cambridge University Press, 30 p

IPCC (2019) Summary for policymakers. In: Climate change and land. https://doi.org/10.1017/9781009157988.001

IPCC (2021) Summary for policymakers. In: Climate change 2021: the physical science basis. https://www.ipcc.ch/report/ar6/wg1/

IPCC (2022) Summary for policymakers. In: Climate change 2022: impacts, adaptation and vulnerability. https://www.ipcc.ch/report/ar6/wg2/chapter/summary-for-policymakers/

Porter JR, Howden M, Smith P (2017) Considering agriculture in IPCC assessments. Nat Clim Chang 7(10):680–683. https://doi.org/10.1038/nclimate3404

Porter JR, Challinor AJ, Henriksen CB, Howden SM, Martre P, Smith P (2019) Invited review: intergovernmental panel on climate change, agriculture, and food—a case of shifting cultivation and history. Glob Chang Biol 25(8):2518–2529. https://doi.org/10.1111/gcb.14700

Rivera-Ferre MG, López-i-Gelats F, Howden M, Smith P, Morton JF, Herrero M (2016) Re-framing the climate change debate in the livestock sector: mitigation and adaptation options. Mitigation and adaptation options in the livestock sector. Wiley Interdiscip Rev Clim Chang 7(6):869–892. https://doi.org/10.1002/wcc.421

Springmann M, Clark M, Mason-D'Croz D, Wiebe K, Bodirsky BL, Lassaletta L et al (2018) Options for keeping the food system within environmental limits. Nature 562:519–525. https://doi.org/10.1038/s41586-018-0594-0

Vermeulen SJ, Campbell BM, Ingram JSI (2012) Climate change and food systems. Annu Rev Environ Resourc 37:195–222

Open Access This chapter is licensed under the terms of the Creative Commons Attribution-NonCommercial-NoDerivatives 4.0 International License (http://creativecommons.org/licenses/by-nc-nd/4.0/), which permits any noncommercial use, sharing, distribution and reproduction in any medium or format, as long as you give appropriate credit to the original author(s) and the source, provide a link to the Creative Commons license and indicate if you modified the licensed material. You do not have permission under this license to share adapted material derived from this chapter or parts of it.

The images or other third party material in this chapter are included in the chapter's Creative Commons license, unless indicated otherwise in a credit line to the material. If material is not included in the chapter's Creative Commons license and your intended use is not permitted by statutory regulation or exceeds the permitted use, you will need to obtain permission directly from the copyright holder.

Chapter 2
Agricultural, Food and Forestry Issues in International Climate Negotiations: Setting the Agenda and Challenges

Marie Hrabanski, Valérie Dermaux, Alexandre K. Magnan, Adèle Tanguy, Anaïs Valance, and Roxane Moraglia

Abstract Since 1992, the UNFCCC Conferences of the Parties (COP) have guided global climate action, but agriculture was initially excluded as negotiations focused on reducing greenhouse gas emissions. From the 2010s, agriculture and food systems gained attention as both major emitters and potential solutions to climate change. States now include agriculture in their Nationally Determined Contributions (NDCs) for climate action. This chapter reviews the evolution of agriculture's role in climate negotiations, carbon markets for land use, and the challenges of assessing national commitments, highlighted by the 2023 Global Stocktake.

Since 1992 and the signing of the United Nations Framework Convention on Climate Change (UNFCCC), governments or parties gather annually at the Conferences of the Parties (COP) to provide guidance on and operationalise States' commitments to climate change action. Agriculture has long been absent from these negotiations, which until the late 1990s focused on mitigating greenhouse gas (GHG) emissions (Caron and Treyer 2016; Hrabanski 2020; Hrabanski and Le Coq 2022). However, agricultural and food systems are significant emitters of GHGs, and are both "victims" and "solutions" to climate change. From the 2010s, agricultural and then food issues gradually entered the international climate agenda (Chandra et al. 2016; Soto and Visseren-Hamakers 2018). States are tasked with implementing climate actions for agriculture and food, which are detailed in their

M. Hrabanski (✉)
Cirad, UMR Actors Resources Territories & Development (ART-Dev), Montpellier, France
e-mail: marie.hrabanski@cirad.fr

V. Dermaux · R. Moraglia
DGPE, Ministry of Agriculture and Food Sovereignty, Paris, France

A. K. Magnan · A. Tanguy
Iddri, Paris, France

A. Valance
High Council for Climate, Paris, France

© The Author(s) 2026
V. Blanfort et al. (eds.), *Climate Impacts and Challenges in Agriculture, Forests and Food Systems*, https://doi.org/10.1007/978-3-032-04331-3_2

national climate commitments[1] known as Nationally Determined Contributions (NDCs).[2]

This chapter synthesises the challenges of climate negotiations for agriculture. The first part presents the stages of setting the agricultural agenda in these negotiations up to the creation of the quadrennial Sharm el-Sheikh joint work on implementation of climate action on agriculture and food security at COP27 in 2022. The second part deals with the issues related to carbon markets for the land sector (agriculture and forests). Finally, we address the delicate question of evaluating the commitments made by States, a challenge recently debated in the context of the first Global Stocktake (concluded in 2023) reviewing the progress made towards the Paris Agreement goals, which will take place every 5 years.

2.1 From 1992 to 2022: The Difficult Task of Setting the Agricultural Agenda in Climate Negotiations

Articles 2 and 4 of the convention (UNFCCC) adopted in 1992 mention the link between climate change and agriculture. However, the focus is on mitigation, notably through negotiations on the REDD+ framework,[3] which concluded in 2013 in Warsaw after several years of very laborious and divisive discussions, particularly between developed and developing countries. The Kyoto Protocol, adopted in 1997, refers to agriculture and forests, highlighting that the land use, land-use change and forestry (LULUCF) sector can be a source of GHGs. This protocol sets ambitious emission reduction targets only for industrialised (so-called "Annex I") countries, in a top-down approach, unlike the Paris Agreement. It covered methane and nitrous oxide,[4] the main gases emitted by the agricultural sector, and established forest reference levels to be respected. However, this approach showed its limitations, with the United States not ratifying this protocol and Canada withdrawing from it. Under this protocol, two carbon offset project certification mechanisms were developed: the Joint Implementation (JI) mechanism and the Clean Development Mechanism (CDM), within which the agricultural and forestry sectors were not integrated until the mid-2000s (Vespa 2002).

It was not until COP17 in Durban, in 2011 (Fig. 2.1), that agriculture was addressed as a global issue, being framed both through mitigation and adaptation to climate change (Hrabanski 2020; Hrabanski and Le Coq 2025). Indeed, following

---

[1] https://www.lemonde.fr/planete/article/2023/12/01/cop28-134-countries-commit-to-include-agriculture-and-food-in-their-climate-plans_6203425_3244.html

[2] In 2020, over 90% of these nationally determined contributions included adaptation to climate change and made agriculture a priority sector, and about 80% of them identified climate change mitigation objectives in the agricultural sector.

[3] REDD+: Reducing emissions from deforestation and forest degradation.

[4] https://unfccc.int/process-and-meetings/the-kyoto-protocol/what-is-the-kyoto-protocol/kyoto-protocol-targets-for-the-first-commitment-period

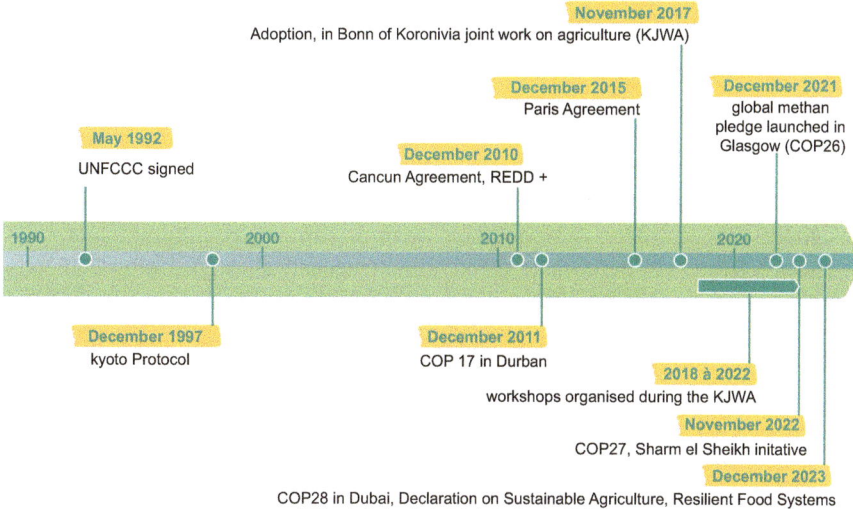

Fig. 2.1 Agricultural issues in climate negotiations between 1992 and 2023
The timeline illustrates the inclusion of agricultural issues in climate negotiations between 1992 and 2023. From the adoption of the Climate Convention, in 1992, until 2011, agricultural issues are rarely addressed within the negotiations. Starting from the Durban Cop, in 2011, these issues have, however, been gaining increasing interest as illustrated by—the launch of the Koronovia process in 2017- prolonged in 2022 through the Charm and Cheikh initiative on agriculture and food security and—the Declaration on food systems in 2023 at the COP28 in Dubai. KJWA: Koronovia Joint Work on Agriculture (KJWA)

the mobilisation of heterogeneous actors in favour of the concept of climate-smart agriculture[5] and within a renewed political context (Hrabanski 2020), agriculture is incorporated into the official agenda of the COP body responsible for scientific and technical issues (SBSTA, Subsidiary Body for Scientific and Technological Advice) (Fleurant 2021). Five workshops will take place between 2013 and 2016.[6] However, while there is indeed a day dedicated to agriculture during COP21[7] (in 2015) parallel to the negotiations, the Paris Agreement mainly addresses agriculture from the perspective of food security and the vulnerability of food production systems.[8]

---

[5] The FAO has been promoting *climate-smart agriculture* since the late 2000s. This concept aims to address three main objectives: the sustainable increase in agricultural productivity and income (food security); adaptation and strengthening of resilience to the impacts of climate change (adaptation); and the reduction and/or removal of greenhouse gas emissions (mitigation), where applicable.

[6] https://unfccc.int/topics/land-use/workstreams/agriculture/agriculture-workshops-and-documents

[7] http://sdg.iisd.org/events/farmers-day-at-cop-21/

[8] Considering: Recognising the fundamental priority of protecting food security and eradicating hunger, and the particular vulnerability of food production systems to the adverse effects of climate change; Article 2.1b: Enhancing the ability to adapt to the adverse effects of climate change and promoting resilience to these changes and low GHG emission development, in a way that does not threaten food production.

Agriculture and forest ecosystems are only covered by Article 5 of the Paris Agreement, which emphasises the importance of preserving and enhancing natural carbon sinks and highlights tools such as REDD+ results-based payments and the joint mechanism for forest mitigation and adaptation (JMA).[9] A significant step is taken in 2017, with the creation of the Koronivia Joint Work on Agriculture (KJWA).[10] From 2018 to 2021, seven workshops had been organised (on adaptation assessment methods, soil carbon health and fertility, etc.) to allow all states and stakeholders (stakeholders) to share their views on various agricultural issues.

The acceleration of the climate agenda simultaneously allowed, during COP26 in Glasgow, to address, in parallel with the negotiations, the issue of methane emissions, nearly 40% of which are agricultural according to the IEA[11] (International Energy Agency). A "Global Methane Pledge" was launched in 2021 by the European Union (EU) and the United States, with the aim of reducing global methane emissions by 30% by 2030 compared to 2020. It now includes 158 countries, although China, India and Russia are not among the signatories (see Chap. 24).

In 2022, the Koronivia Joint Work on Agriculture came to an end (Fig. 2.1). The analysis of submissions made by countries and observers, including research, highlighted the diversity of ways of thinking about how to address the link between agriculture and climate change under the UNFCCC, which resulted in significant tensions between Northern and Southern countries in the negotiations at COP27 in Sharm el-Sheikh, Egypt (2022).

Three main points of contention were identified between, globally, Northern and Southern countries. Other divides also emerged, thus relativising the existence of a global North and a global South, which would necessarily oppose each other. The first relates to the use of the term mitigation in the text of the COP decision (Hrabanski and Le Coq 2025). Indeed, while all parties agreed that the text should include the importance of adapting agriculture to climate change, India, supported by other emerging countries that remained more in the background, was particularly reluctant to also include the term mitigation. For this large agricultural country, the challenges of mitigation should not hinder the food security of developing and emerging countries. A few hours before the close of negotiations, India agreed that the term mitigation be included in the COP3/CP27 decision, creating "the "Sharm el-Sheikh Quadrennial Joint Initiative on the Implementation of Climate Action for Agriculture and Food Security". This episode shows how it is not a given to tackle in synergy, the challenges of 'adaptation and mitigation for many emerging and Southern countries. A second sticking point was the creation of a permanent structure dedicated to agricultural issues within the UNFCCC. This request, which

---

[9] JMA: Joint Mitigation and Adaptation Mechanism for the Integral and Sustainable Management of Forests.

[10] Workshops are now conducted in cooperation with bodies established under the convention, such as the Green Climate Fund. Observers, including NGOs and research institutions, also participate in the workshops.

[11] https://www.iea.org/reports/global-methane-tracker-2023/overview

remains a point of contention, is mainly supported by the G77 countries, even if notable differences exist between the proposals made. Finally, an issue can be identified related to the place of food systems in climate action. For many European and emerging countries, the reflection must be made at the scale of food systems: our eating practices are closely linked to the modes of production of agricultural products, and an approach taking into account the upstream with the production of inputs and possible deforestation, and downstream, with transport, cooling, processing, and therefore also the losses and waste and diets, is more likely to allow the emergence of win-win solutions at all levels (see Chap. 23). However, on the one hand, the Africa group preferred to focus on the agricultural sector, a question already complex to instruct. On the other hand, some Northern countries and transition economies refused to see the term food system appear, the most likely hypothesis being the fear of questioning overconsumption of meat, deforestation, or even trade, which they wish to avoid at all costs. The term food system(s) was therefore rejected in the text of the joint quadrennial initiative of Sharm el-Sheikh.

Despite these tensions, the joint quadrennial initiative of Sharm el-Sheikh on the implementation of a climate action for agriculture and food security was adopted and this COP decision (3/CP.27) therefore marks a decisive step in the negotiations. It should be noted, however, that this text does not promote either agroecology, which would have paved the way for a holistic overhaul of agricultural systems, or climate-smart agriculture, which is more oriented towards technological solutions. No specific target for reducing agricultural GHG emissions is discussed in the COPs; no practice has been encouraged or stigmatised (massive use of chemical inputs, etc.). The Emirati presidency of COP28 then put this issue at the top of the political agenda, in parallel with the negotiations, by proposing the Declaration on Sustainable Agriculture, Resilient Food Systems and Climate Action, signed by 160 countries.[12] It calls on the countries that join it to strengthen the place of agriculture and food systems in nationally determined contributions and in national adaptation plans and in biodiversity plans. Following COP28, the FAO proposed a roadmap that establishes 120 measures (including so-called agroecological measures) and key steps in ten areas for adaptation and mitigation for agriculture and food systems. It aims to reduce agricultural and food-origin emissions by 25%, to achieve carbon neutrality by 2035, and to transform by 2050 these systems into carbon sinks capturing 1.5 Gt of GHGs per year. Ultimately, the Sharm el-Sheikh initiative focuses on agriculture and not on food systems, but will lead to a workshop, in June 2025, on systemic and holistic approaches in agriculture and food systems, and in September 2025 the forum of the Standing Committee on Finance will focus on sustainable agriculture and food systems.[13] The subject is therefore making its way into the UNFCCC arena.

---

[12] https://www.tappcoalition.eu/images/COP28-UAE-Declaration-on-Sustainable-Agriculture-Resilient-Food-Systems-and-Climate-Action-1701436580.pdf

[13] https://unfccc.int/event/2025-forum-of-the-standing-committee-on-finance

2.2 Carbon Markets for Agriculture and Forests: What Are the Issues, What Progress Has Been Made?

The "carbon markets" have been booming for several years. For some observers (scientists and NGOs), subject to integrity and robust rules, they are considered a tool to accelerate action and achieve climate neutrality as soon as possible (Schilling et al. 2023). Others, more critical, highlight the risk that these mechanisms divert certain actors from immediate and drastic reductions in their GHG emissions (Paul et al. 2023; Kreibich and Hermwille 2021), or denounce the very functioning of these mechanisms (Aykut 2017). In fact, they allow actors (state or non-state) to exchange mitigation results through bilateral agreements between countries or by financing offset projects in a host country. The buyer then benefits from the credits provided for in Article 6, called ITMO[14] to offset emissions that would exceed their target (country climate targets in their NDCs, Corsia target[15] for international airlines, etc.). The seller, on the other hand, benefits from financing, but will have to subtract the sold mitigation result from the host country's inventory (as long as it is used for international compliance purposes). This operation, called "corresponding adjustment", prevents double counting of units exchanged under Article 6.

The first devices regulated by the United Nations such as CDM and Moc (see Chap. 1),[16] the project certification instruments from the Kyoto Protocol, apply respectively to non-industrialised and industrialised countries (so-called "Annex I"). Alongside the compliance market (or regulated), numerous standards are established on the voluntary market, certifying projects with their own requirements and quality criteria. There is not one, but several carbon markets, heterogeneous, several standards are indeed facing recent criticism in the press,[17] which notably question the additionality of credits. However, additionality is a basic criterion of carbon credits: this giving rise to a "right" to emit by the buyer, it must generate a flow of emission reductions or carbon absorptions, in order to have a "neutral" balance for the atmosphere.

The remuneration of existing carbon stocks, such as the conservation of existing forests or the conservation of fossil stocks (coal, oil, etc.), is the subject of requests by countries with high forest cover. But it is not eligible for carbon markets, as it is not additional. Article 6 of the Paris Agreement, concluded in 2024 at COP29, sets the new UN rules for regulated carbon markets (replacing those of the Kyoto Protocol). Many hope that these new rules will lead to the establishment of a standard or benchmark positively influencing the voluntary market.

---

[14] ITMO: *internationally transferred mitigation outcome* or internationally transferred mitigation results, in French.

[15] Corsia: Carbon Offsetting and Reduction Scheme for International Aviation.

[16] https://agriculture.gouv.fr/protocole-de-kyoto-et-marche-carbone-europeen-comment-les-emissions-des-secteurs-de-lagrofourniture

[17] https://www.lemonde.fr/planete/article/2023/08/25/la-credibilite-de-plusieurs-programmes-de-compensation-carbone-mise-en-doute-par-des-chercheurs_6186554_3244.html; https://theconversation.com/histoire-des-credits-carbone-vie-et-mort-dune-fausse-bonne-idee-212903

The latest synthesis report on NDCs (NDC Synthesis Report, 2023)[18] indicates that 76% of States, mainly developing countries wishing to sell offset projects, plan to use the Article 6 market to receive financing. The buyer will use the credit to achieve the target set in their NDC at a lower cost if it is a country, or their Corsia target for airlines subject to international regulation, to achieve their "net zero" target, or to communicate on their "contribution" to climate neutrality for businesses and other actors. Some, like the EU, want to achieve their NDC target on their own, without resorting to this flexibility of carbon markets regulated by Article 6. Thus, the EU is equipping itself with a voluntary domestic framework: carbon absorption certificates, which notably define the rules for mitigation projects on EU territory, which will contribute to achieving climate neutrality and the EU's NDC target.

The general rules of these new carbon markets of Article 6 were adopted by the parties at COP26 in 2021: the exchanged mitigation results (or ITMOs) can concern all sectors (energy, transport, agriculture, forest, etc.), be GHG emission reductions or carbon absorptions. Regarding the Specifically, in the AFOLU sector, (1) emission reductions can be linked, for example, to a decrease in deforestation or methane emissions from cattle, improvements in the fertilisation of agricultural soils (reductions in N_2O emissions), etc., and (2) carbon sequestration corresponds to carbon farming with practices that increase carbon storage in soils and biomass, promoted by the international initiative "4 per 1000", such as agroecological practices, agroforestry, sustainable forest management, restoration of degraded lands, etc.

Since 2021, UNFCCC negotiations have focused on operationalising the rules adopted at COP26: the construction of registries to ensure the traceability of credits, the content of authorisation letters from host countries, the definition of methodologies for offset projects, the treatment of carbon sequestration, etc.

At COP29 in Baku, the operationalisation rules of Article 6 were finally adopted, allowing the market to officially launch, even though additional work is planned for 2025 (particularly on how to construct reference scenarios for projects or take into account the risk of non-permanence).

For the AFOLU sector, the standard on the treatment of carbon sequestration has crystallised tensions and blockages between countries, some promoting technological solutions (CCU/S)[19] and others natural sequestration in ecosystems (or nature-based solutions, NBS). This standard was adopted at COP29, but the additional work to be carried out in 2025 will be crucial for the AFOLU sector.

The AFOLU sector has levers on emission reductions and sequestration (sources and sinks of GHGs), and can therefore play on these two aspects in the carbon markets. Its sequestration is based on a natural process, photosynthesis (no anthropogenic energy consumption to store carbon unlike CCU/S). It is a sector with multiple assets and synergies to value (biodiversity, water, air, soils, etc.), but also limitations, primarily the high risk of non-permanence (linked for example to fires, storms or other disturbances releasing stored carbon into the atmosphere) which must be appropriately taken into account (Box 2.1).

---

[18] https://unfccc.int/ndc-synthesis-report-2023

[19] CCU/S : *carbon capture, utilisation and storage.*

Box 2.1 The Three Tools of Article 6 Defining a Framework for Cooperation Between Parties under the Paris Agreement

Two market mechanisms (Articles 6.2 and 6.4)

Article 6.2 establishes accounting rules for carbon credit exchanges between countries and defines the conditions for participation. ITMOs (internationally transferred mitigation outcome) represent a reduction in GHG emissions (or sequestration) achieved in one country that can be transferred to another country to contribute to its own objectives. The transferred results are removed from the host country's inventory, which cannot use them to achieve its own NDC, this is the "corresponding adjustment". However, some countries wish to automatically recognise the Warsaw Framework for REDD+ under Article 6.2, delivering payments based on results. This is a red line for many other countries, as it contradicts the rules adopted at COP26, particularly in terms of additionality and reference scenarios which must be "much better" than historical ones, aligned with the host country's NDC and with the long-term objective of the Paris Agreement. The rules on registries, authorisations, transparency of information, etc., which were the subject of the latest negotiations, were finalised at COP29. The UNFCCC Secretariat is tasked with creating an international registry that will ensure the tracking and traceability of exchanged ITMOs.

Article 6.4 aims to establish methodologies to certify carbon offset projects; it could serve as an international standard, also for the voluntary market. Much of the operationalisation work is entrusted to a UN expert group (Supervisory Body)[20] tasked with providing guidelines for the methodologies to be applied to offset projects and for carbon sequestration. After two failures, during the negotiations of COP27 and then COP28, the 6.4 standards on methodologies, carbon sequestration and the clean development mechanism (a tool ensuring socio-environmental safeguards and respect for sustainable development objectives on offset projects) were validated on the first day of COP29. These standards set more ambitious rules than the texts proposed at previous COPs (particularly on reference scenarios, additionality, the risk of non-permanence, etc.). However, there are still missing pieces that will be the subject of additional work by the Supervisory Body in 2025 (particularly on the construction of reference scenarios and the risk of non-permanence).

A non-market mechanism (Article 6.8)

Outside of carbon markets, Article 6.8 establishes a UN showcase allowing the highlighting of non-market cooperation approaches with a systemic vision linking mitigation, adaptation and biodiversity. This mechanism, strongly supported by Bolivia as alternatives to market-based regulation, could

(continued)

---

[20] https://unfccc.int/process-and-meetings/bodies/constituted-bodies/article-64-supervisory-body

Box 2.1 (continued)

highlight solutions for the conservation of stocks (not eligible for the carbon market) or result-based payments (of the REDD+ type) that would not meet the requirements of the markets. Carbon (particularly in terms of additionality), but which would value other benefits than mitigation. A decision was also validated at COP29 on this Article 6.8, with a view to improving the UNFCCC web platform created to disseminate non-market approach projects.

The land sector under the cooperative approaches of Article 6 is therefore at the heart of UNFCCC negotiations, the stakes having grown with the operationalisation of markets and expectations being high, particularly from forest countries. However, controversies persist between the actors: countries, NGOs, private sector, etc. Some also point out the risk of greenwashing diverting buyers from adhering to the "avoid, reduce, compensate" (ARC) sequence. Challenges remain to be met in future negotiations: how to ensure the permanence of offset projects? How to ensure that the project does not have negative impacts on the environment, food security, local populations? How to ensure sufficient income for farmers and foresters, "producers" of carbon credits? Many other questions remain to be explored.

2.3 Assessing Adaptation International Efforts for the Agricultural Sector

While international climate negotiations have gradually integrated agricultural issues, the question remains of evaluating the mitigation and adaptation efforts of countries in this sector. Under the Paris Agreement, Parties have committed to submitting their NDCs, and to updating them every 5 years; and to increase their ambition in terms of mitigation and adaptation. In 2023, the first Global Stocktake was completed at COP28 in Dubai to assess progress towards the 3 main objectives of the Paris Agreement (mitigation, adaptation and means of implementation). This last part focuses on the adaptation dimension to show that it requires new methodological approaches.

Beyond the agricultural sector, efforts to strengthen policies, implementation and financing of adaptation are clearly underway globally, through various financing channels (for example, multi and bilateral financing organisations and the private sector) (see Chap. 26). However, it is estimated that in view of the increasing climate risks the change of scale that is required has not yet taken place (Berrang-Ford et al. 2021; IPCC 2022; UNEP 2023). Overall, adaptation strategies remain confined to a relatively short-term view (a horizon of one to three decades at most) and have a limited scope in that they do not address the root causes of exposure and vulnerability to climate change. Their geographical coverage remains incomplete, and their rate of progress too slow. The IPCC concludes that the observed impacts, projected risks, trends in vulnerability, limits to adaptation and associated losses and damages demonstrate that " transformation for sustainable and climate resilient development is more urgent than previously assessed (very high confidence)" (IPCC 2023; p. 89).

Within the UNFCCC, "adequacy" and "effectiveness" are the key dimensions through which the question of adaptation efforts is raised. Adequacy refers to the consistency between instruments (such as financing) and identified adaptation needs. Effectiveness relates to the results produced by these instruments in terms of disbursement and/or reduction of climate risk. For example, payments in support of specific commodities to reduce producers' short-term vulnerability may discourage adjustments or shifts in production, which would be beneficial in the medium or long term (OECD 2023), creating maladaptation (Boutroue et al. 2022). On this basis, various adaptation evaluation initiatives have been developed for several years within the UNFCCC (synthesis reports from the Secretariat or reports from the Adaptation Committee) or in support of it (such as the UNEP's Adaptation Gap Report). All rely on NDCs, adaptation communications, national adaptation plans (NAPs)[21] and the financing flows reported by bi and multilateral partners (UNEP 2023). However, the data seem insufficient to understand in detail the adequacy and effectiveness of adaptation strategies and interventions globally. Other necessary information relate to governance processes at subnational levels, the effective reduction of natural and anthropogenic climate risk factors in different contexts and sectors, and even elements on cross-border climate risks (Anisimov and Magnan 2023). However, neither national communications based on averaged national statistics nor databases on projects funded under international cooperation offer such granularity (Magnan et al. 2023). It is therefore necessary to rely, in addition, on other methods of adaptation evaluation such as expert judgement methods (see for example the following works: Hallegatte et al. 2020; Browne et al. 2021; World Bank 2021; Magnan et al. 2025), to integrate qualitative and quantitative information. These can prove to be highly useful in informing adaptation efforts within the frame of the Global goal on adaptation and the Global Stocktake. These methods allow for minimising the problem of defining the "right" quantitative adaptation indicators, a complex discussion that refers to divergent views among parties to the UNFCCC (such as universally applicable indicators vs those that reflect the contextual specifics of each country; there are various sources of information on these indicators).[22] They also allow for addressing the problem of access to data, and question the risk of a significant additional burden that would fall on both the statistical services of countries and their negotiation teams. Thus, an externalised,

---

[21] NAP: National Adaptation Plan.
[22] 2023 Report of the Food and Agriculture Organization: https://www.fao.org/documents/card/en/c/cc2038en; 2023 OECD Report: https://www.oecd-ilibrary.org/agriculture-and-food/agricultural-policy-monitoring-and-evaluation-2023_b14de474-en; Agriculture Committee of the International Platform on Adaptation Metrics: https://adaptationmetrics.org/committees

scientifically-based assessment process would perhaps have a role to play in the work that needs to be undertaken to inform reflections on adaptation progress globally.

2.4 Conclusion

Discussions on agriculture in climate negotiations have so far been laborious and often tense. However, the process is gradually being built and is starting to bear fruit. The decision in Sharm el-Sheikh (COP27) is a major step forward: for the first time, the outcome of the negotiations is operationalised in the form of recommendations to the financial entities of the UNFCCC and to countries. It extends agricultural negotiations until the end of 2026, and sends a strong political message. The agricultural sector is a major emitter of greenhouse gases, while according to the latest figures from the World Bank,[23] only 4% of climate finance is allocated to it. This decision is therefore a call to international and national, public and private funders, to target their actions more towards the agricultural sector, especially in developing and least developed countries, whose food security is particularly threatened by climate change. In this vein, the workshop on systemic and holistic approaches in agriculture and food systems, the one on access to means of implementation (such as finance), and the two annual reports of the UNFCCC secretariat on the work of bodies under the UNFCCC should improve understanding and highlight these issues. In parallel, the negotiations within Article 6 have confirmed the central place of the land sector in carbon markets, delivering emission reductions (urgent and drastic in the short term) and absorptions (necessary to offset residual emissions in the long term). With COP29 having adopted the rules for operationalising the carbon markets of Article 6, the first methodologies for certifying projects should be submitted in 2025, alongside additional UN work that will need to clarify the construction of reference scenarios, the consideration of non-permanence risk, etc. Finally, the first global stocktake completed at COP28 allowed for a review of countries' commitments in relation to targets, and demonstrated the need to strengthen climate action. It led to a COP decision, which underlines how essential evaluation methods of commitments are, in connection with financing issues.

---

[23] https://www.worldbank.org/en/topic/climate-smart-agriculture. If climate finance and development funding are included in the calculation, then 22% of these funds were allocated to agriculture in 2020; however, this figure is only decreasing, as it was 45% in 2000 (Galbiati et al. 2023). https://openknowledge.fao.org/server/api/core/bitstreams/cb1ced5c-f9a5-48b1-962c-82ac719d2722/content

References

Anisimov A, Magnan AK (eds) (2023) The global transboundary climate risk report 2023. The Institute for Sustainable Development and International relations & Adaptation Without Borders, 144 p. https://www.iddri.org/en/publications-and-events/report/global-transboundary-climate-risk-report-2023

Aykut, S. (2017). La "gouvernance incantatoire". L'accord de Paris et les nouvelles formes de gouvernance globale. La pensée écologique, 1(1). https://lapenseeecologique.com/la-gouvernance-incantatoire-laccord-de-paris-et-les-nouvelles-formes-degouvernance-globale/

Berrang-Ford L, Siders AR, Lesnikowski A et al (2021) A systematic global stocktake of evidence on human adaptation to climate change. Nat Clim Chang 11:989–1000. https://doi.org/10.1038/s41558-021-01170-y

Boutroue B, Bourblanc M, Mayaux PL, Ghiotti S, Hrabanski M (2022) The politics of defining maladaptation: enduring contestations over three (mal) adaptive water projects in France, Spain and South Africa. Int J Agric Sustain 20(5):892–910. https://doi.org/10.1080/14735903.2021.2015085

Browne N, Rozenberg J, De Vries Robbé S, Kappes M, Lee W, Prasad A (2021) 360° resilience: a guide to preparing the Caribbean for a new generation of shocks. World Bank, Washington, DC. http://hdl.handle.net/10986/36405. License: CC BY 3.0 IGO

Caron P, Treyer S (2016) Climate-smart agriculture and international climate change negotiation forums. In: Torquebiau E (ed) Climate change and agriculture worldwide. Springer, pp 325–336. https://doi.org/10.3917/crii.086.0189

Chandra KE, McNamara P, Dargusch et al (2016) Bridging the UNFCCC divide on climate-smart agriculture. Carbon Manag 7(5–6):295–299. https://doi.org/10.1080/17583004.2016.1235420

Fleurant MM (2021) Agriculture in the international legal climate regime. Law J 62(3):935–965. https://doi.org/10.7202/1080617ar

Galbiati GM, Yoshida M, Benni N, Bernoux M (2023) Climate-related development finance to agrifood systems – Global and regional trends between 2000 and 2021. FAO, Rome. https://doi.org/10.4060/cc9010en

Hallegatte S, Rentschler J, Rozenberg J (2020) Adaptation principles: a guide for designing strategies for climate change adaptation and resilience. World Bank, Washington, DC. https://openknowledge.worldbank.org/handle/10986/34780

Hrabanski, M. (2020). Une climatisation des enjeux agricoles par la science? Les controverses relatives à la climate-smart agriculture. Critique internationale, 86(1), 189–208

Hrabanski M, Le Coq J-F (2022) Climate conditioning of agricultural issues in the international agenda through three competing epistemic communities: climate-smart agriculture, agroecology, and nature-based solutions. Environ Sci Pol 127:311–320. https://doi.org/10.1016/j.envsci.2021.10.022

Hrabanski M, Le Coq J-F (2025) Agriculture at COP27: antagonistic political framing and fragmentation of agricultural issues within climate negotiations and beyond. Glob Environ Polit:1–14. https://doi.org/10.1162/glep_a_00778

IPCC (2022) Summary for policymakers. In: Pörtner H-O et al (eds) Climate change 2022: impacts, adaptation and vulnerability

IPCC (2023) In: Lee H et al (eds) Climate change 2023: synthesis report. Contribution of working groups I, II and III to the sixth assessment report of the intergovernmental panel on climate change. IPCC

Kreibich N, Hermwille L (2021) Caught in between: credibility and feasibility of the voluntary carbon market post-2020. Clim Pol 21(7):939–957. https://doi.org/10.1080/14693062.2021.1948384

Magnan AK, Anisimov A, Vallejo L (2023) The potential of expert judgment-based approaches to assessing adaptation under the GST: the case of the GAP-Track. In: Perspectives: adequacy and effectiveness of adaptation in the global Stocktake. UNEP Copenhagen Climate Centre, pp 48–64. https://unepccc.org/wp-content/uploads/2023/02/perspectives-adequacy-and-effectiveness-of-adaptation-in-the-global-stocktake-web.pdf

Magnan AK, Li J, Tanguy A, Hallegatte S, Buffet C (2025) The value of structured expert judgment to help assess climate adaptation. Clim Risk Manag 47:100692. https://doi.org/10.1016/j.crm.2025.100692

OECD (2023) Agricultural policy monitoring and evaluation 2023: adapting agriculture to climate change. OECD Publishing, Paris. https://doi.org/10.1787/b14de474-en

Paul C, Bartkowski B, Dönmez C, Don A, Mayer S, Steffens M et al (2023) Carbon farming: are soil carbon certificates a suitable tool for climate change mitigation? J Environ Manag 330:117142. https://doi.org/10.1016/j.jenvman.2022.117142

Schilling F, Baumüller H, Ecuru J, von Braun J (2023) Carbon farming in Africa: opportunities and challenges for engaging smallholder farmers. Center for Development Research (ZEF). https://doi.org/10.48565/bonndoc-122

Soto GC, Visseren-Hamakers IJ (2018) Framing and integration in the global forest, agriculture and climate change nexus. Environ Plann C: Polit Space 36(8):1415–1436. https://doi.org/10.1177/2399654418788566

UNEP (2023) Adaptation gap report 2023: underfinanced. Underprepared. Inadequate investment and planning on climate adaptation leaves world exposed, Nairobi. https://www.unep.org/resources/adaptation-gap-report-2023

Vespa M (2002) Climate change 2001: Kyoto at Bonn and Marrakech. Ecology LQ, 29, 395

World Bank (2021) Resilience rating system. The World Bank, Washington, DC. https://openknowledge.worldbank.org/han-dle/10986/35039

Open Access This chapter is licensed under the terms of the Creative Commons Attribution-NonCommercial-NoDerivatives 4.0 International License (http://creativecommons.org/licenses/by-nc-nd/4.0/), which permits any noncommercial use, sharing, distribution and reproduction in any medium or format, as long as you give appropriate credit to the original author(s) and the source, provide a link to the Creative Commons license and indicate if you modified the licensed material. You do not have permission under this license to share adapted material derived from this chapter or parts of it.

The images or other third party material in this chapter are included in the chapter's Creative Commons license, unless indicated otherwise in a credit line to the material. If material is not included in the chapter's Creative Commons license and your intended use is not permitted by statutory regulation or exceeds the permitted use, you will need to obtain permission directly from the copyright holder.

Chapter 3
Tensions and Synergies Between the Concepts of Climate-Smart Agriculture, Agroecology, and Nature-Based Solutions

Nadine Andrieu, Audrey Naulleau, Jean-François Le Coq, and Marie Hrabanski

Abstract Climate-smart agriculture, agroecology, and nature-based solutions offer alternative approaches to address climate change challenges, aiming to replace both industrial and traditional farming. While industrial agriculture increased production, it caused environmental harm, and some traditional practices have led to resource overexploitation. These concepts share similarities but also differ, especially regarding the inclusion of traditional knowledge and risks of greenwashing. Critics warn that nature-based solutions may exclude rural communities or benefit polluting industries. This chapter explores the tensions and synergies between these approaches at policy and production system levels.

Climate-smart agriculture, agroecology, and nature-based solutions are all concepts that propose new approaches to address the challenges of climate change. These approaches aim to replace both input-intensive industrial agriculture and traditional farming, which is often less productive and/or less profitable (Duru et al. 2015; Lipper et al. 2014). While industrial agriculture and livestock farming, promoted by the green revolution in various southern countries, have led to significant increases in production, they have also resulted in negative environmental externalities, particularly in terms of greenhouse gas (GHG) emissions generated by the intensive use of synthetic inputs, heavy mechanisation, and livestock growth. Concurrently, some forms of agriculture and livestock farming that did not benefit from the green

---

N. Andrieu (✉)
Cirad, UMR Innovation, Capesterre-Belle-Eau, Guadeloupe, France
e-mail: nadine.andrieu@cirad.fr

A. Naulleau
Innovation, Université de Montpellier, Cirad, INRAE, Institut Agro, Montpellier, France

Cirad, UMR Innovation, Tunis, Tunisia

J.-F. Le Coq · M. Hrabanski
Cirad, UMR Actors Resources Territories & Development (ART-Dev), Montpellier, France

© The Author(s) 2026
V. Blanfort et al. (eds.), *Climate Impacts and Challenges in Agriculture, Forests and Food Systems*, https://doi.org/10.1007/978-3-032-04331-3_3

revolution remain low users of synthetic inputs, but in certain contexts, they have led to overexploitation of natural resources, contributing to deforestation and soil erosion (Tittonell 2020). The concepts of climate-smart agriculture, agroecology, or nature-based solutions, championed by different epistemic communities, have points of convergence, but also divergences in how to respond to climate challenges for these different agricultural models.

Several authors have sought to find synergies between these concepts. They have highlighted possible synergies between the concepts of climate-smart agriculture (CSA) and agroecology. Saj et al. (2017) and Torquebiau et al. (2018) suggest formally promoting the principles of agroecology in CSA approaches. Other authors have emphasised differences, such as the insufficient consideration in the CSA concept of traditional farmers' knowledge and practices that have enabled the resilience of agricultural systems to multiple shocks over time, based on the principles of agroecology (Altieri et al. 2015). Another strong criticism, voiced by NGOs, is the risk of co-optation of this concept by some major global industries, contributors to GHG emissions, and ultimately greenwashing (Pimbert 2015). The same type of criticism is made about the concept of nature-based solutions (NBS), which promotes ecosystem restoration approaches that can exclude rural communities or be interpreted as pollution rights for large industries.

This chapter defines and describes the tensions and synergies between the three concepts at two scales of climate action, namely (1) the scale of climate change adaptation policies and (2) the scale of designing production systems adapted to climate change.

3.1 Origins and Definitions

Agroecology is the oldest of the three concepts, having emerged in 1920. It has evolved over time, transitioning from a discipline at the interface between agronomy and ecology to a transdisciplinary science explicitly considering the sociotechnical changes to be made within production systems and food systems (Wezel et al. 2020). The concept thus advocates a profound transformation of production and food systems, rather than a simple substitution of conventional practices with new ones (Gliessman 2016; Duru et al. 2015). Agroecology is not described by a set of predefined agricultural practices. It is based on biotechnical principles such as the valorisation of ecological processes, the promotion of biodiversity within agricultural systems, the recycling of biomass and nutrients, and the synergies between the components of these systems. Agroecology also relies on sociotechnical principles such as the co-creation of knowledge between producers and scientists, and the promotion of mechanisms favouring shared governance of productive resources and contributing to the resilience of agricultural and food systems to multiple shocks. Climate change is not a key principle; however, a growing number of studies, particularly since the early 2000s, have shown that agriculture based on the principles of agroecology offers benefits in terms of adaptation and mitigation (Saj et al. 2017; Altieri et al. 2015; Quintero et al. 2024).

To better integrate agriculture into international climate negotiations, the FAO (2013) proposed the concept of CSA, which aims (1) to sustainably increase the productivity of agricultural systems to ensure food security and increase income, (2) to strengthen their resilience to climate, and (3) to reduce GHG emissions, while contributing to the preservation of resources (FAO 2013). The proposal of this concept is based on the principle that agriculture worldwide is both a source of GHG emissions and also part of the solution to limit these emissions. The so-called "land" sector (agriculture, forestry, pastoralism and other Land use) indeed stands out from other sectors due to its ability to adapt to climate change combined with a significant potential for mitigating greenhouse gas emissions through the reduction of greenhouse gas emissions, but also through the sequestration of carbon in soils and biomass. The concept originating from the political sphere (Hrabanski 2020) was then adopted by the scientific community and in particular by the CGIAR's cross-cutting Climate Change and Food Security programme (*global research partnership for a food-secure future dedicated to transforming food, land, and water systems in a climate crisis*) lending credibility to the concept. The three objectives or pillars of CSA can be addressed at different levels, from local to global. Despite various attempts to list CSA practices, the concept aims more at finding synergies between food security and climate change issues (Torquebiau et al. 2018).

The concept of NBS also emerged in the early 2000s in internal documents of the International Union for Conservation of Nature (IUCN). It is defined as actions aimed at protecting, sustainably managing, and restoring natural or modified ecosystems, which respond to societal challenges (such as climate change, food and water security or natural disasters) in an effective and adaptive manner, while simultaneously providing benefits in terms of human well-being and biodiversity (Cohen-Shacham et al. 2016). The concept aims to meet the demand of international organisations to find "alternatives" to conventional engineering solutions to adapt and mitigate the effects of climate change, while improving livelihoods and protecting natural ecosystems and biodiversity (Cohen-Shacham et al. 2016). While nature conservation is the primary principle, the founders recognise the need to integrate other societal issues, the cultural specificity of each context and the production of societal benefits equitably, while promoting transparency and broad participation.

3.2 Incorporating Climate Change into the Political Agenda with the Three Concepts

Although promoted by three different epistemic communities, these concepts have played a decisive role in the emergence of agricultural issues in the global climate agenda (Hrabanski and Le Coq 2022). Until the Conference of the Parties (COP) of the United Nations Framework Convention on Climate Change in Durban in 2011, agriculture was almost absent from international climate negotiations (see Chap. 2). However, towards the end of the 2000s, a window of opportunity opened in favour of incorporating agricultural issues into climate governance. It was in this context that the concept of CSA and then NBS emerged. Following this, other concepts were re-mobilised such as agroecology to bring together climate issues and agricultural questions. Since the early 2010s, agriculture has gradually become one of the major issues of the COPs, as debates on how to approach the sensitive issue of agricultural models multiply.

However, the work of the IPCC (Intergovernmental Panel on Climate Change) remained for a long time particularly cautious about the use of these concepts. The concept of agroecology only appears in the bibliographic references of the 1995 and 2007 reports of Group II[1]. As for CSA, it is mentioned in the 2014 report of Group II. This relative absence of concepts probably underlines the authors' desire to keep at a distance the political issues related to their use. However, since the publication of the IPCC's special report on land in 2019, the principles of agroecology, CSA and NBS have been more openly discussed and highlighted (IPCC 2019). The latest IPCC report of 2023 confirms this trend as it encourages both the technologies derived from CSA, agroecology and NBS.

The use of these concepts illustrates the importance of scientific communities in the policy-making process at the international and transnational levels. Despite the shared interest in integrating agriculture into the climate agenda, these three concepts are in competition for the promotion of distinct agricultural models. Thus these conceptual debates should not obscure the associated political issues (Hrabanski and Le Coq 2022).

Hrabanski and Le Coq (2022) show that political tensions limit the promotion of these terms at the international level. Since the Paris COP21 in 2015, there has been an exponential increase in the number of events associated with them at the COPs. Similarly, these three terms can be used by political actors in their speeches and statements, as was the case with Kofi Annan's opening speech at COP17 in favour of the CSA concept. However, the controversies and issues are such that these concepts do not appear in official decisions. Thus, the final text of the joint 4-year initiative of Charm el-Cheikh on the implementation of climate action for agriculture and food security, which nevertheless enshrines the place of agricultural issues in climate negotiations, does not promote either agroecology, which would have paved the way for a holistic overhaul of agricultural systems, nor CSA, which is more oriented towards the promotion of technological solutions. The concept of NBS made its first appearance in a COP decision, but only in Sharm el-Sheikh Implementation Plan, of which Article 14 concerns forests, and which is in fact far removed from agricultural issues. On the contrary, states themselves, widely mobilise these concepts in their nationally determined contributions (NDCs). In 2019, out of the 197 states that drafted an NDC, thirteen addressed the concept of agroecology and 29, of CSA. However, the concept of CSA was mainly associated with adaptation and the need for technology transfer (Hrabanski and Le Coq 2022). Furthermore,

---

[1] Group II studies the impacts, vulnerability and adaptation to climate change.

three countries simultaneously addressed the concepts of CSA and agroecology in their NDCs. Although SFNs were not directly mentioned, twelve countries stated "ecosystem management", twenty stipulated "ecosystem-based adaptation", and two countries mentioned "ecosystem-based mitigation".

3.3 Co-designing Innovative Practices to Address Climate Challenges

In the co-design processes of innovative systems to address climate change carried out with field actors (producers, technicians, researchers), the main challenge is to be able to articulate these issues with other problems faced by producers. These can be other environmental issues (Acosta-Alba et al. 2019) or the improvement of productive capacities (Jules et al. 2023). In these processes, producers do not always explicitly claim from one epistemic community or another. However, these processes are carried by researchers or technicians who can claim from one or another community depending on their personal or academic trajectory. In several works, the carriers have sought to pragmatically or opportunistically promote synergies between the concepts of agroecology and CSA, promoting technical approaches based on the principles of agroecology and measuring their performance in terms of food security, adaptation and mitigation (for example Osorio-Garcia et al. 2019; Selbonne et al. 2022). These approaches are based on the principles of diversification of agricultural and livestock systems (such as agroforestry, associated crops, multi-species livestock systems), recycling (such as the integration of agriculture-livestock to reduce the production, transport and use of synthetic inputs), more efficient use of water and nutrient resources through landscape or plot-scale developments. They are locally implemented in participatory processes that rely on innovation platforms as a place for knowledge exchange or co-construction of solutions with the actors. But tensions can exist locally between these two concepts, as the CSA constraint framework is less restrictive and can lead to the promotion of practices that may not align with the principles of agroecology. For example, the new more resilient varieties are often presented as CSA, without specifying the level of intensification (in synthetic inputs or not) required for these varieties to express their potential (Akinyi et al. 2022). Also tensions emerge when the design process is centred on the technical dimension underestimating the importance of integrating the knowledge of field actors and accompanying the transitions of agricultural and livestock systems at the scale of the farm or territory (Scherr et al. 2012; Mockshell and Kamanda 2018).

SFNs applied to agricultural systems include the development of agroforestry systems, which can play a role in carbon storage and in maintaining biodiversity (Getnet et al. 2023), and the resilience of systems (Tham-Agyekum et al. 2023). Conservation agriculture is also presented as an SFN. The promoted practices can therefore appear similar to those promoted in the different scientific communities. But again the evaluations focus mainly on the technical and environmental dimensions and little attention is paid to the accompaniment of transitions (Soterroni et al. 2023).

3.4 An Impossible Integration of Concepts for Climate Action?

Table 3.1 summarises the tensions and synergies identified between the concepts in the political agenda setting of climate change and in the co-design processes of innovative agricultural systems.

Coordinated climate action would aim to limit divisions between epistemic communities by leveraging identified synergies at these two scales. Politically, tensions prevail and are explained by different trajectories and visions of the actors carrying these policies. These tensions will have repercussions on the nature of the political instruments envisaged to support local transitions. However, at the local level, synergies are possible when the approach adopted by field actors is participatory and allows for the co-construction of solutions. This also implies adopting multi-criteria methods to assess the effects that co-designed solutions can have on the range of criteria considered in these three communities (food security, ecosystem services, conservation of adjacent natural ecosystems).

Table 3.1 Tensions and synergies between the concepts of climate-smart agriculture, nature-based solutions and agroecology

	Synergies	Tensions
For the political agenda	The three concepts offer an opportunity to consider agriculture as a solution to address climate change.	The proponents of the concepts have very different agricultural development models. Conservationists who advocate for Nature-Based Solutions (NBS) often view agricultural systems as destructive to nature, making it difficult to align with proponents of agroecology or Climate-Smart Agriculture (CSA). Academic proponents of CSA (for example, CGIAR) are now integrating agroecology more extensively. Advocates and social movements within the field of agroecology are generally suspicious of aligning with proponents of CSA.
For the design of innovative systems	Evaluations of proposed technical solutions are multi-criteria, with different prevalence depending on the concept: food security for CSA, nature conservation for NBS, and ecosystem services for agroecology. When the design approach is participatory, it promotes the co-creation of knowledge and synergies between the three concepts.	The agroecological movement, due to its origins, is the most advanced in terms of supporting transitions. CSA and NBS have a more technical entry point. Agricultural systems face multiple risks other than climate, which are not always considered by the three concepts.

References

Acosta-Alba I, Chia E, Andrieu N (2019) The LCA4CSA framework: Using life cycle assessment to strengthen environmental sustainability analysis of climate smart agriculture options at farm and crop system levels. Agric Syst 171:155–170. https://doi.org/10.1016/j.agsy.2019.02.001

Akinyi DP, Karanja Ng'ang'a S, Ngigi M, Mathenge M, Girvetz E (2022) Cost-benefit analysis of prioritized climate-smart agricultural practices among smallholder farmers: evidence from selected value chains across sub-Saharan Africa. Heliyon 8. https://doi.org/10.1016/j.heliyon.2022.e09228

Altieri M, Nicholls C, Henao A, Lana MA (2015) Agroecology and the design of climate change-resilient farming systems. Agron Sustain Dev 35:869–890. https://doi.org/10.1007/s13593-015-0285-2

Cohen-Shacham E, Walters G, Janzen C, Maginnis S (eds) (2016) Nature-based solutions to address global societal challenges. IUCN, Gland, Switzerland. 97p

Duru M, Therond O, Fares M (2015) Designing agroecological transitions; a review. Agron Sustain Dev 35:1237–1257. https://doi.org/10.1007/s13593-015-0318-x

FAO (2013) Climate-smart Agriculture Sourcebook. Rome, Italy. 570 p. https://openknowledge.fao.org/server/api/core/bitstreams/b21f2087-f398-4718-8461-b92afc82e617/content

Getnet D, Mekonnen Z, Anjulo A (2023) The potential of traditional agroforestry practices as nature-based carbon sinks in Ethiopia. Nature-Based Solutions 4. https://doi.org/10.1016/j.nbsj.2023.100079

IPCC (2019) Climate Change and Land: an IPCC special report on climate change, desertification, land degradation, sustainable land management, food security, and greenhouse gas fluxes in terrestrial ecosystems. https://www.ipcc.ch/site/assets/uploads/2019/08/4.-SPM_Approved_Microsite_FINAL.pdf

Gliessman S (2016) Transforming food systems with agroecology. Agroecol Sustain Food 40(3):187–189. https://doi.org/10.1080/21683565.2015.1130765

Hrabanski M (2020) A climate focus on agricultural issues by science? The controversies surrounding climate-smart agriculture. Critique Int 1:189–208

Hrabanski M, Le Coq J-F (2022) Climatisation of agricultural issues in the international agenda through three competing epistemic communities: climate-smart agriculture, agroecology, and nature-based solutions. Environ Sci Pol 127:311–320. https://doi.org/10.1016/j.envsci.2021.10.022

Jules J, Paul B, Adam M, Andrieu N (2023) Co-designing with producers climate change adaptation strategies: the case agricultural operations in Haiti. Cah Agric 3227. https://doi.org/10.1051/cagri/2023020

Lipper L, Thornton P, Campbell BM, Baedeker T, Braimoh A, Bwalya M et al (2014) Climate-smart agriculture for food security. Nat Clim Chang 4(12):1068–1072. https://doi.org/10.1038/nclimate2437

Mockshell J, Kamanda J (2018) Beyond the agroecological and sustainable agricultural intensification debate: is blended sustainability the way forward? Int J Agric Sustain 16(2):127–159

Osorio-García M, Paz L, Howland F, Ortega LA, Acosta-Alba I, Arenas L et al (2019) Can an innovation platform support a local process of climate-smart agriculture implementation? A case study in Cauca, Colombia. Agroecol Sustain Food Syst 44(3):378–411. https://doi.org/10.1080/21683565.2019.1629373

Pimbert M (2015) Agroecology as an alternative vision to conventional development and climate-smart agriculture. Development 58:286–298. https://doi.org/10.1057/s41301-016-0013-5

Quintero C, Arce A, Andrieu N (2024) Evidence of agroecology's contribution to mitigation, adaptation, and resilience under climate variability and change in Latin America. Agroecol Sustain Food Syst 48(2):228–252. https://doi.org/10.1080/21683565.2023.2273835

Saj S, Torquebiau E, Hainzelin E, Pages J, Maraux F (2017) The way forward: an agroecological perspective for climate-smart agriculture. Agric Ecosyst Environ 250:20–24. https://doi.org/10.1016/j.agee.2017.09.003

Scherr SJ, Shames S, Friedman R (2012) From climate-smart agriculture to climate-smart landscapes. Agric Food Secur 1:12. https://doi.org/10.1186/2048-7010-1-12

Selbonne S, Guindé L, Belmadani A, Bonine C, Causeret FL, Duval M et al (2022) Designing scenarios for upscaling climate-smart agriculture on a small tropical island. Agric Syst 199. https://doi.org/10.1016/j.agsy.2022.103408

Soterroni AC, Império M, Scarabello MC, Seddon N, Obersteiner M, Rochedo PRR et al (2023) Nature-based solutions are critical for putting Brazil on track towards net-zero emissions by 2050. Glob Chang Biol:1–17. https://onlinelibrary.wiley.com/doi/abs/10.1111/gcb.16984

Tham-Agyekum EK, Ntem S, Sarbah E, Anno-Baah K, Asiedu P et al (2023) Resilience against climate variability: the application of nature based solutions by cocoa farmers in Ghana. Environ Sustain Indic 20:100310. https://doi.org/10.1016/j.indic.2023.100310

Tittonell P (2020) Assessing resilience and adaptability in agroecological transitions. Agric Syst 184:102862. https://doi.org/10.1016/j.agsy.2020.102862

Torquebiau E, Rosenzweig C, Chatrchyan AM, Andrieu N, Khosla R (2018) Identifying Climate-smart agriculture research needs. Cah Agric 27(2):e26001. https://doi.org/10.1051/cagri/2018010

Wezel A, Herren BG, Kerr RB, Barrios E, Gonçalves ALR, Sinclair F (2020) Agroecological principles and elements and their implications for transitioning to sustainable food systems. A review. Agron Sustain Dev 40(6). https://doi.org/10.1007/s13593-020-00646-z

Open Access This chapter is licensed under the terms of the Creative Commons Attribution-NonCommercial-NoDerivatives 4.0 International License (http://creativecommons.org/licenses/by-nc-nd/4.0/), which permits any noncommercial use, sharing, distribution and reproduction in any medium or format, as long as you give appropriate credit to the original author(s) and the source, provide a link to the Creative Commons license and indicate if you modified the licensed material. You do not have permission under this license to share adapted material derived from this chapter or parts of it.

The images or other third party material in this chapter are included in the chapter's Creative Commons license, unless indicated otherwise in a credit line to the material. If material is not included in the chapter's Creative Commons license and your intended use is not permitted by statutory regulation or exceeds the permitted use, you will need to obtain permission directly from the copyright holder.

Chapter 4
Atlas of World Agriculture and Food Systems in the Face of Climate Change

Vincent Blanfort, Julien Demenois, Marie Hrabanski, Nicolas Viovy, Jacques André Ndione, Moussa Waongo, Maguette Kaire, Lilian Blanc, Sylvain Schmitt, Séverine Bouard, Catherine Sabinot, Pierre-François Duyck, Philippe Birnbaum, Audrey Leopold, Julien Drouin, Fabian Carriconde, Laurent L'Huillier, and Christophe Menkès

Abstract The atlas presented in this work analyses the challenges of climate change for agri-food systems, first through the lens of (1) greenhouse gas emissions, then (2) the impacts of climate change on the land sector, followed by (3) the question of the effects of climate change and mitigation challenges, and finally (4) by adopting an analysis by major regions.

V. Blanfort (✉)
Cirad joint research unit Selmet "Tropical and Mediterranean Animal Production Systems", Montpellier Univ, Inrae, Institut Agro, Montpellier, France
e-mail: vincent.blanfort@cirad.fr

J. Demenois
Cirad, UPR Agroecology and Sustainable Intensification of Annual Crops (Aïda), Montpellier, France; Aïda, Univ. Montpellier, Cirad, Montpellier, France

Aïda, Univ. Montpellier, CIRAD, Montpellier, France

M. Hrabanski
Cirad, UMR Actors Resources Territories & Development (ART-Dev), Montpellier, France

N. Viovy
Laboratory of Climate and Environmental Sciences (LSCE/IPSL), CEA-CNRS-UVSQ, University Paris-Saclay, Gif-sur-Yvette, France

J. A. Ndione
Regional Agency for Agriculture and Food (ARAA), ECOWAS, Lomé, Togo

M. Waongo · M. Kaire
AGRHYMET, Regional Climate Centre for West Africa and the Sahel (CCR-AOS), Niamey, Niger

L. Blanc · S. Schmitt
Cirad, UPR Forests and Societies, Montpellier, France; Forests and Societies, Univ. Montpellier, Cirad, Montpellier, France

S. Bouard
Department of Environmental Management, Lincoln University, Christchurch, New Zealand

© The Author(s) 2026
V. Blanfort et al. (eds.), *Climate Impacts and Challenges in Agriculture, Forests and Food Systems*, https://doi.org/10.1007/978-3-032-04331-3_4

4.1 Introduction and Methods

Vincent Blanfort, Julien Demenois, and Marie Hrabanski

The atlas presented in this work analyses the challenges of climate change for agri-food systems, first through the lens of (1) greenhouse gas emissions, then (2) the impacts of climate change on the land sector, followed by (3) the question of the effects of climate change and mitigation challenges, and finally (4) by adopting an analysis by major regions.

4.1.1 Greenhouse Gas Emissions from Agri-Food Systems

We first wanted to emphasise that although agri-food systems are responsible for nearly a third of greenhouse gas (GHG) emissions (Crippa et al. 2021; Rosenzweig et al. 2020; Tubiello et al. 2022), this global figure conceals strong and sometimes counter-intuitive disparities, as illustrated in Fig. 4.1. This figure indeed reminds us that the main continent emitting agricultural greenhouse gases is Asia, with nearly 39% of all agricultural greenhouse gases. Rice cultivation there generates significant amounts of methane, as does the intensive use of nitrogen fertilisers with nitrous oxide. The export of agricultural products results in additional emissions due to transport. The Americas are the second largest continent emitting greenhouse gases (30%).

Within this region, intensive agricultural systems in the United States and Canada—particularly maize and soybean production for animal feed—are associated with a large carbon footprint. This is further increased by the long-distance

---

C. Sabinot
Research Institute for Development (IRD), Espacedev IRD, Univ Montpellier, Univ Guyane, Univ La Réunion, Univ Antilles, Univ New Caledonia, Univ Perpignan, Perpignan, France

P.-F. Duyck
Cirad, BIOS, UMR PVBMT, New Caledonia, France

P. Birnbaum
Cirad, UMR AMAP, Univ Montpellier, CNRS, INRAE, IRD, Montpellier, France

A. Leopold · J. Drouin · F. Carriconde · L. L'Huillier
IAC (New Caledonian Agronomic Institute), New Caledonia, France

C. Menkès
Research Institute for Development (IRD), UMR ENTROPIE, Ifremer, IRD, Univ. La Réunion, Univ. New Caledonia, CNRS, New Caledonia, France

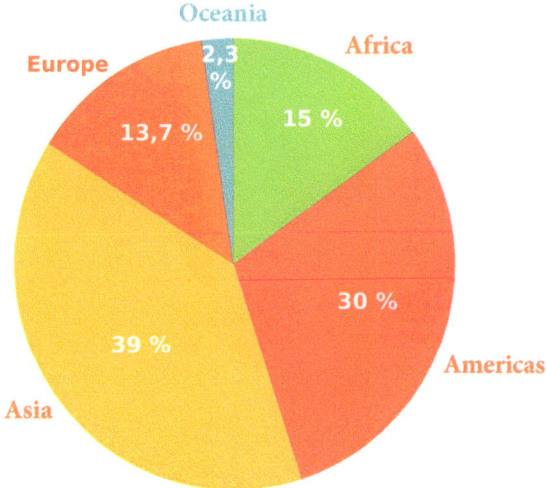

Fig. 4.1 Share of average eq CO_2 emissions (AR5) from agricultural and food systems by continent between 1990 and 2021. (Source: FAO (2024))

transport of these products and by high levels of consumption of processed foods. In Latin America, between 50% and 60% of total emissions come from the agricultural sector, especially in Brazil and Argentina, both major exporters of beef and soybeans.

In Africa, the carbon footprint of agricultural and food systems, in relation to the population and food production, is surprisingly high (15% of global sector emissions), and is mainly explained by land use change (deforestation in Central Africa) and low agricultural productivity. These global figures, however, mask a great geographical diversity. Central Africa is responsible for 35% of the greenhouse gas emissions linked to the continent's food systems, ahead of West Africa, responsible for 28% of these. North Africa accounts for only 12% of the total. Despite more efficient agricultural practices in Europe (13.7% of agricultural emissions) and effective EU climate policies, livestock remains a major source of greenhouse gas emissions also linked to food processing and its transport chains. Finally, in Oceania, the use of land for extensive grazing is a source of emissions, although the relatively small population limits its overall impact.

Having outlined these global figures, we then sought to deepen the analysis in the following section in order to better understand the various challenges linked to emissions from agri-food systems. The first ten maps proposed (Maps 4.1, 4.2, 4.3, 4.4, 4.5, 4.6, 4.7, 4.8, 4.9 and 4.10) are from the FAOSTAT database (FAO 2024) and represent the average emissions for the period 1990–2021 by country. They aim

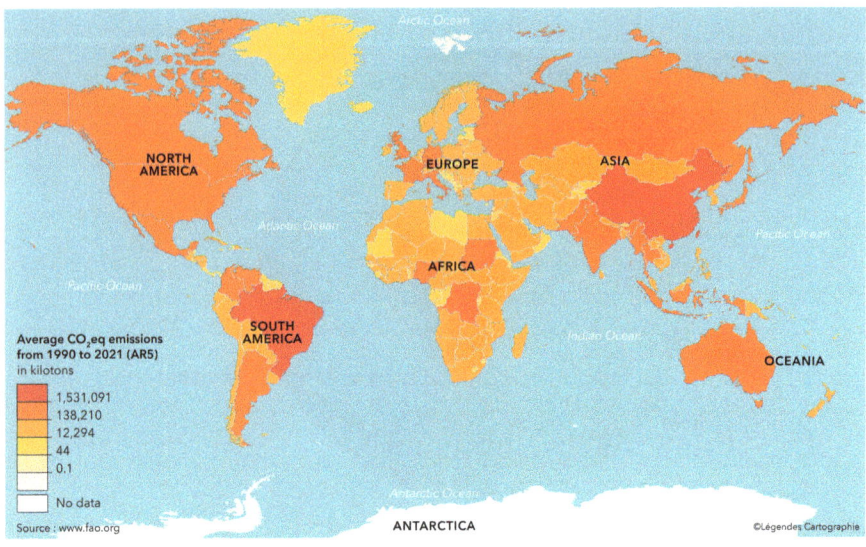

Map 4.1 Average CO_2 eq emissions from 1990 to 2021 (AR5)

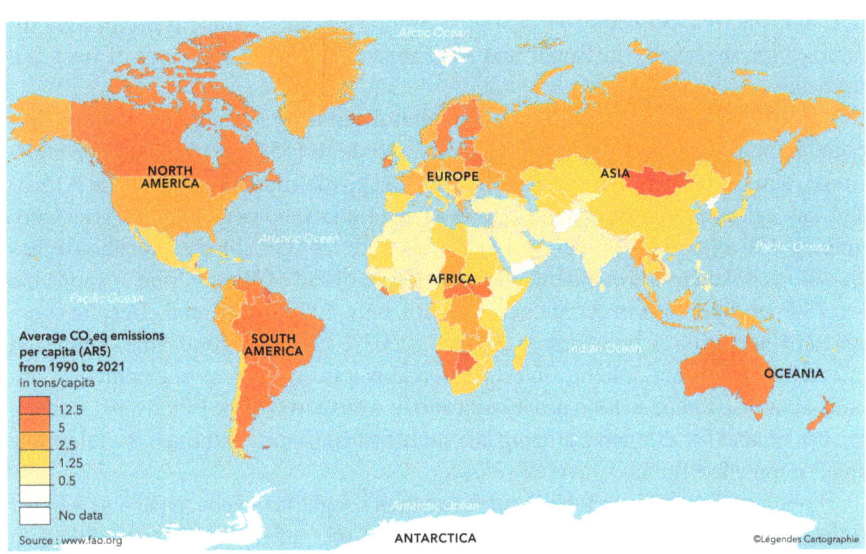

Map 4.2 Average CO_2 eq emissions per capital from 1990 to 2021 in tons/capita

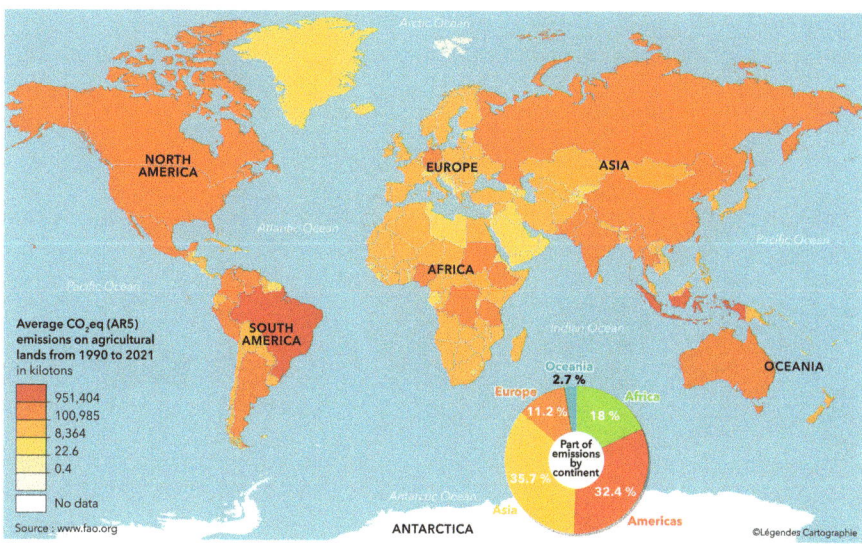

Map 4.3 Average CO_2 eq emissions on agricultural lands from 1990 to 2021 in kilotons (AR5)

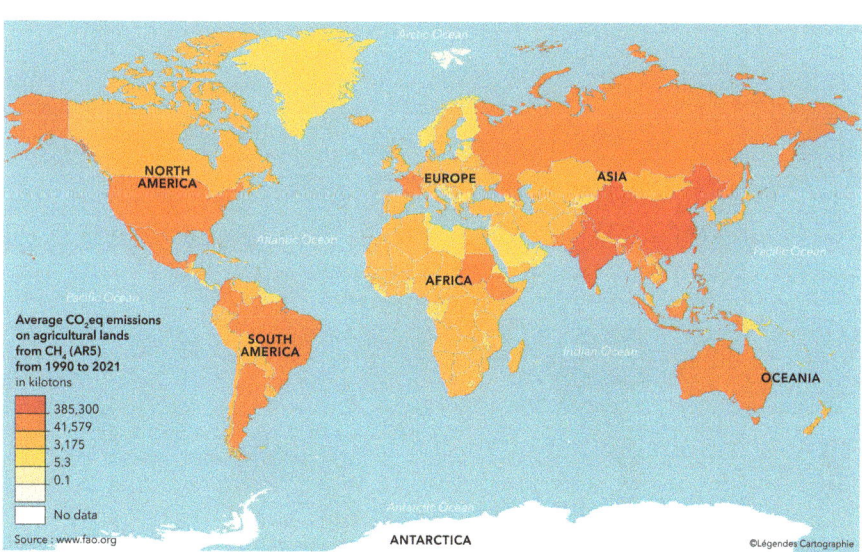

Map 4.4 average CO_2 eq emissions on agricultural lands from CH_4 (methane) from 1990 to 2021 in kilotons (AR5)

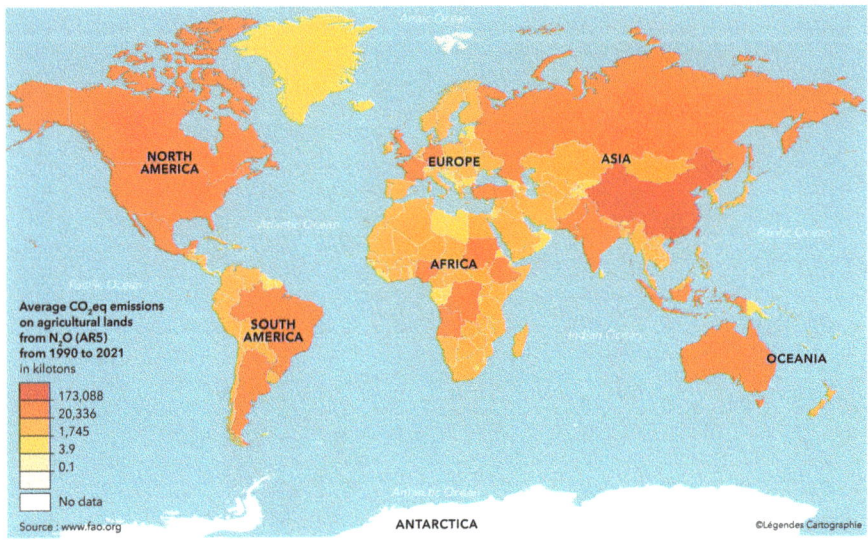

Map 4.5 Average CO_2 eq emissions on agricultural lands from N_2O (Nitrous oxide) from 1990 to 2021 in kilotons (AR5)
Carbon sequestration potential from natural forest regrowth

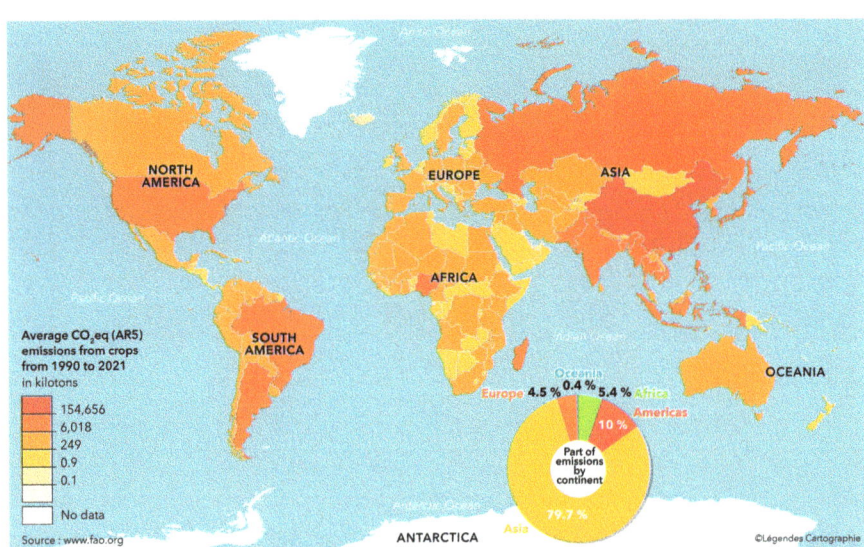

Map 4.6 Average CO_2 eq emissions from crops 1990 to 2021 (AR5) in kilotons

4 Atlas of World Agriculture and Food Systems in the Face of Climate Change   45

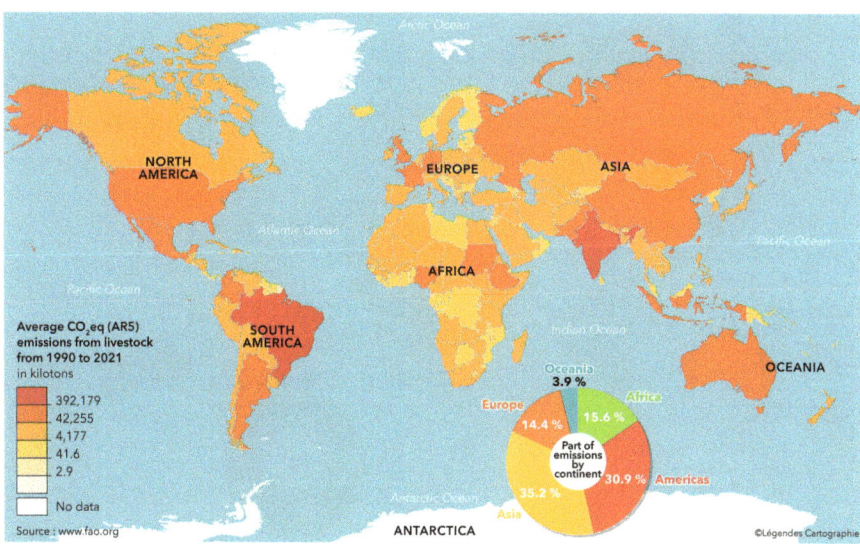

Map 4.7 Average CO_2 eq emissions from livestock 1990 to 2021 (AR5) in kilotons

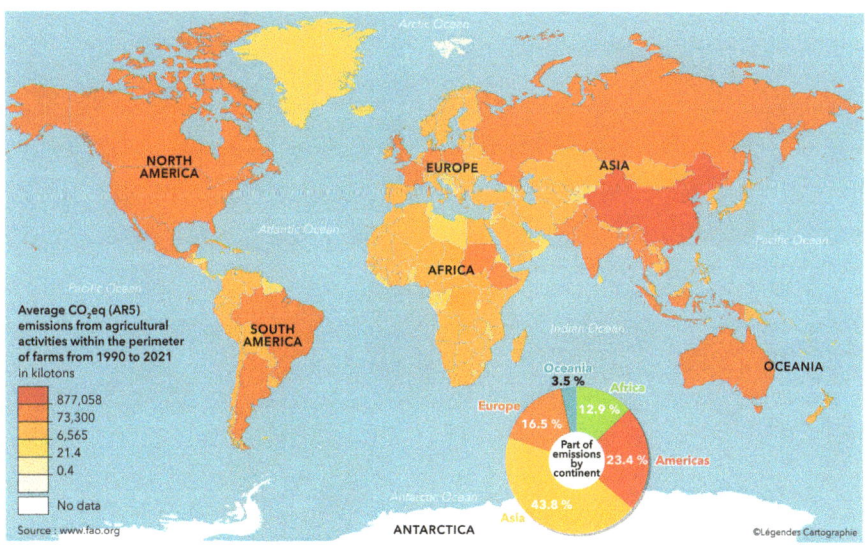

Map 4.8 Average CO_2 eq emissions from agricultural activities within the perimeter of farms from 1990 to 2021 in kilotons

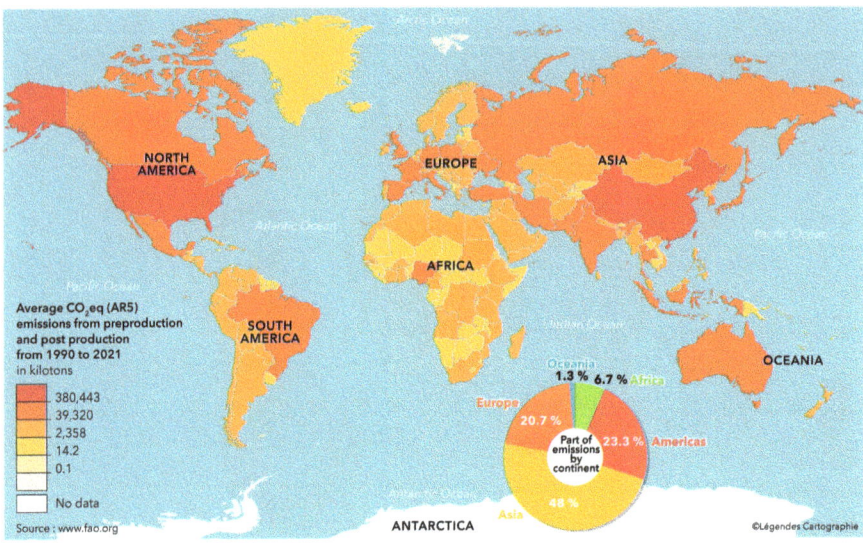

Map 4.9 Average CO_2 eq emissions from preproduction and post production from 1990 to 2021 in kilotons

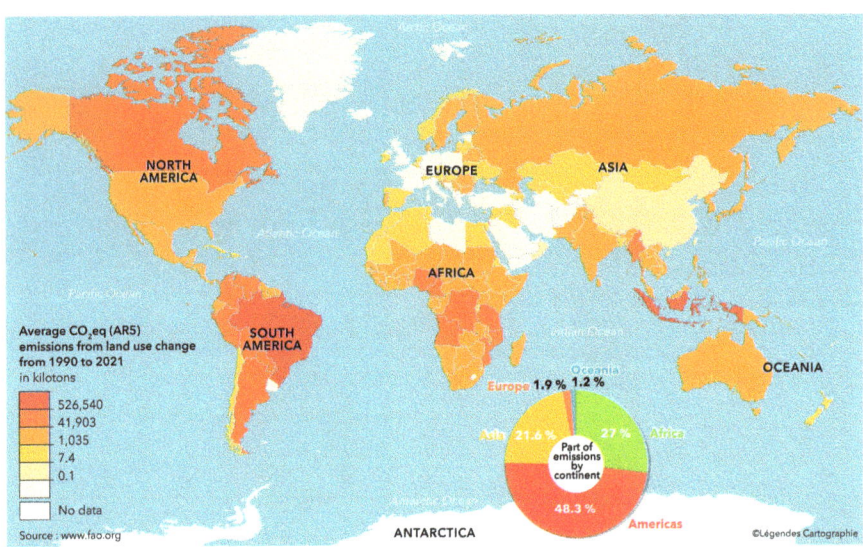

Map 4.10 Average CO_2 eq emissions from land use change from 1990 to 2021 in kilotons

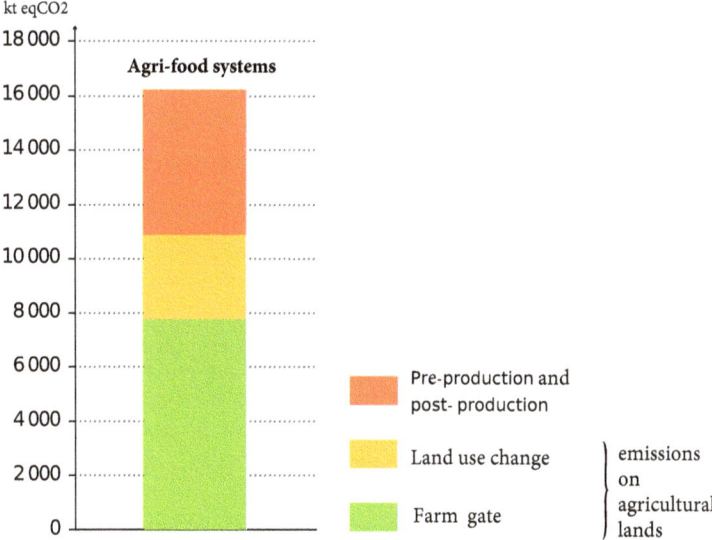

Fig. 4.2 Average CO_2 emissions by sector (in kt eq CO_2), between 1990 and 2021. (Source: FAO (2024))

to clarify the issues related to emissions from the agri-food sector (from pre-production stages, to production and post-production stages) using different variables (emissions by country, per capita, at each stage of the agri-food system, etc.).

Emissions indeed come from multiple sources (Box 4.1) and vary according to biophysical factors, biogeographical factors, production systems, agricultural practices, types of food produced and consumed, and waste management systems. As the FAO chart below (Fig. 4.2) reminds us, half of the emissions are attributable to production activities within the farm (*farm gate*). Added to the GHG emissions related to land use change, emissions on agricultural lands produce nearly two-thirds of the emissions from agri-food systems. However, pre-production and post-production activities constitute an increasing share of these system emissions.

The categories used to distinguish between emissions from land-use change, on-farm emissions, and emissions related to the pre-production and post-production stages are based on the classifications employed by the FAO (see Table 1 in Introduction).

4.1.2 The Impacts of Climate Change on the Land Sector

The following section, focusing on the impacts of climate change on the land sector, proposes through Map 4.11 (based on the IPCC's 2023 report) a more in depth understanding of the impacts of climate change on agricultural productivity and climate vulnerabilities.

4.1.3 Carbon Storage

The following section addresses the issue of carbon—its storage, fluxes, and sequestration by the land sector—through six distinct maps (Maps 4.12, 4.13, 4.14, 4.15, 4.16 and 4.17). These maps are based on data from the Global Soil Partnership, the Impact4Soil platform, and Global Forest Watch platforms.

> **Box 4.1 Types of GHGs and global warming potentials**
>
> The tonne of carbon dioxide equivalent (CO_2e) is a unit introduced in the IPCC First Assessment Report (IPCC 1990), allowing emissions from different greenhouse gases (GHGs) to be aggregated after conversion.
>
> This unit is based on the Global Warming Potential (GWP), an index that compares the warming effect of a greenhouse gas relative to carbon dioxide (CO_2) over a chosen time period. In other words, the GWP of a gas is the ratio of the warming caused by the release of a given mass of that gas at the beginning of the period, compared to the warming caused by the same mass of carbon dioxide. By definition, the GWP of CO_2 is always equal to 1. The warming effects are calculated over a set time frame, beyond which any residual effects are deliberately excluded. The most commonly used time horizon is 100 years, referred to as GWP100.
>
> Other gases, such as methane (CH_4) and nitrous oxide (N_2O), have much higher GWPs than CO_2 because they are more efficient at trapping heat in the atmosphere, even though they are present in smaller quantities:
>
> - Methane (CH_4): GWP ≈ 28 times higher than that of CO_2 over 100 years (according to the IPCC Sixth Assessment Report). This means that the release of one tonne of CH_4 is equivalent to the release of 28 tonnes of CO_2. Among the major GHGs, CH_4 has a relatively short atmospheric lifetime. The IPCC (2021) estimates this at 11.8 years;
> - Nitrous oxide (N_2O): GWP ≈ 273 times higher than that of CO_2 over 100 years (according to the IPCC Sixth Assessment Report). In other words, emitting one tonne of N_2O is equivalent to releasing 273 tonnes of CO_2. N_2O has a long atmospheric lifetime, estimated by the IPCC at 109 years.
>
> Furthermore, GHG emissions can be expressed in kilotonnes (kt), in megatonnes (Mt or million tonnes), in gigatonnes (Gt or billion tonnes), or even in petagrams (Pg) equivalent to Gt.

4 Atlas of World Agriculture and Food Systems in the Face of Climate Change   49

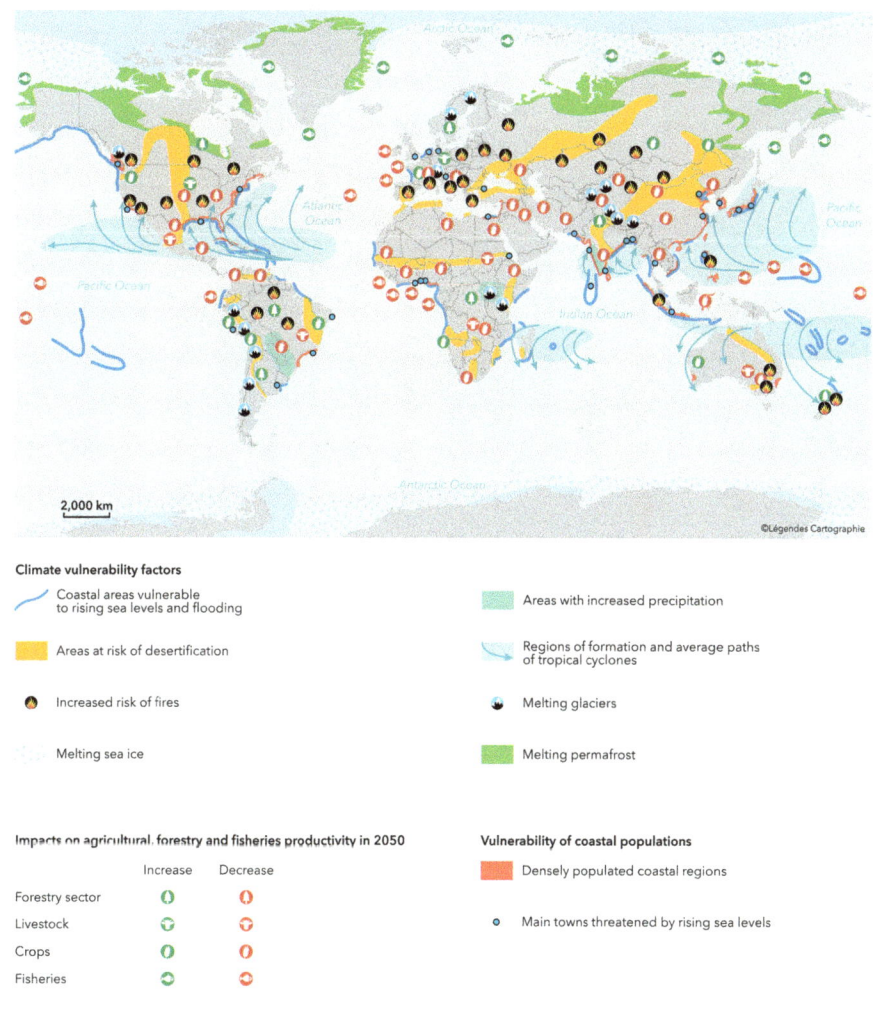

Map 4.11 Climte vulnerability factors, Impacts on agricultural, forestry and fisheries productivity in 2050 and vulnerability of coastal populations

4.1.4 Evolution and Impacts of Climate Change on the Land Sector by Continent

Finally, in the last section of this chapter, six analyses at the scale of continents are proposed. For Africa, Southeast Asia, Central and South America, small island states and territories of Oceania, Europe and North America, the atlas analyses the

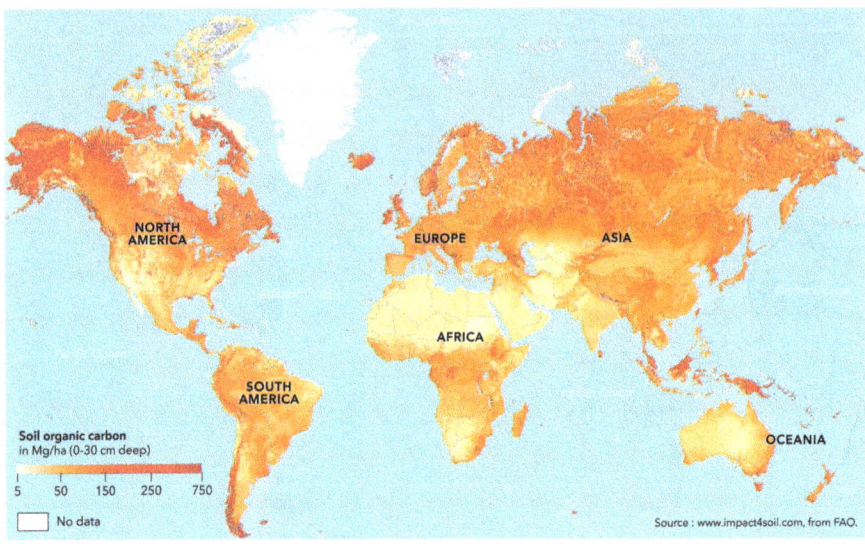

Map 4.12 Soil organic carbon stock in Mg/ha (0-30cm depth)

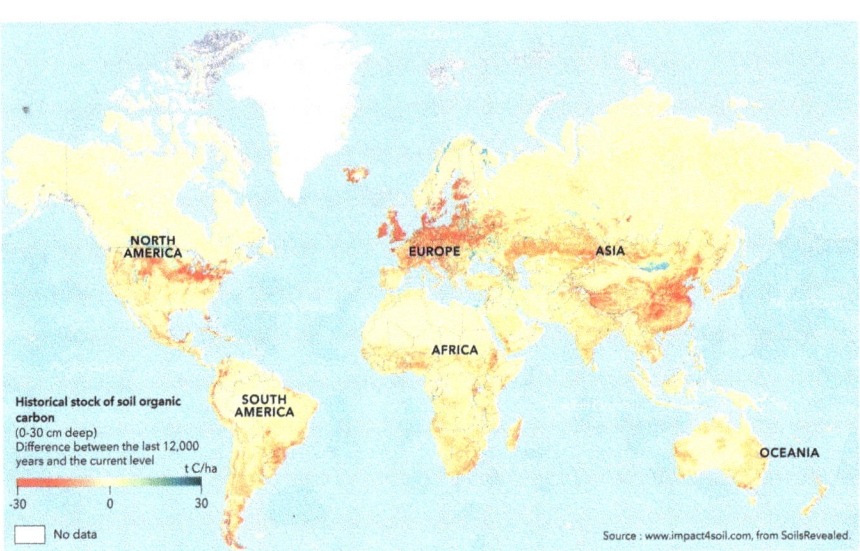

Map 4.13 Historic soil organic carbon stocks in tC/ha (0-30 cm depth)

4 Atlas of World Agriculture and Food Systems in the Face of Climate Change

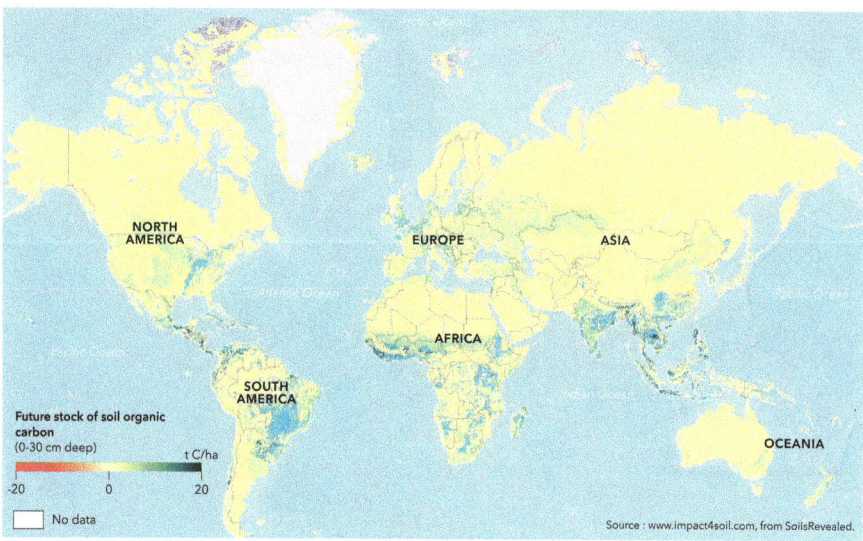

Map 4.14 Future stock of soil organic carbon in tC/ha (0-30 cm depth)

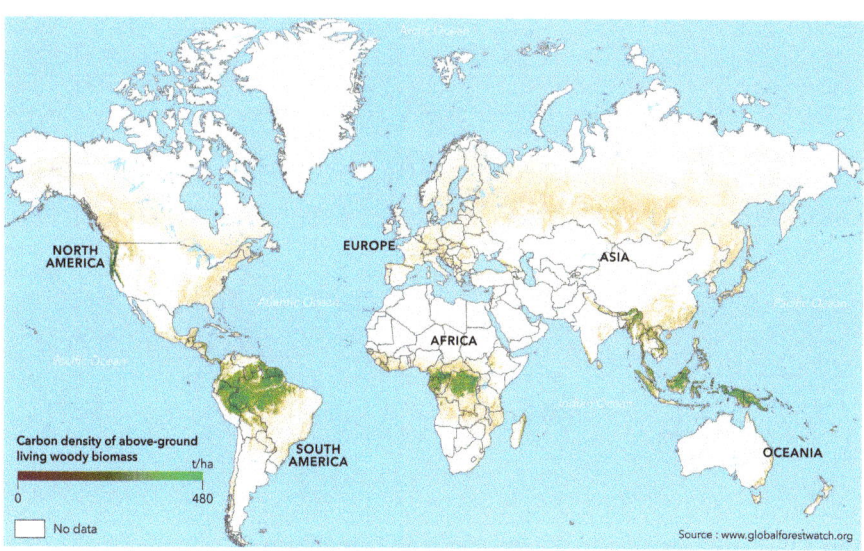

Map 4.15 Carbon density of above-ground living woody biomass

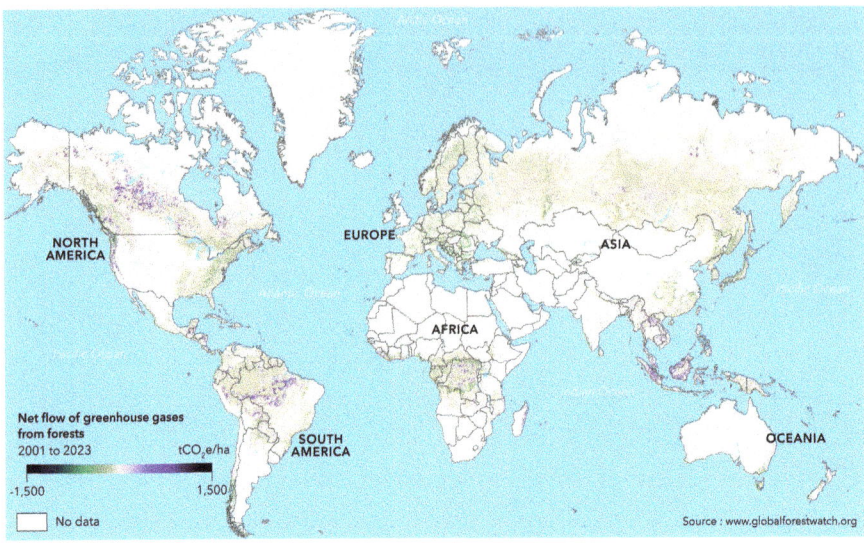

Map 4.16 Net flux of greenhouse gases from forests between 2001 and 2023

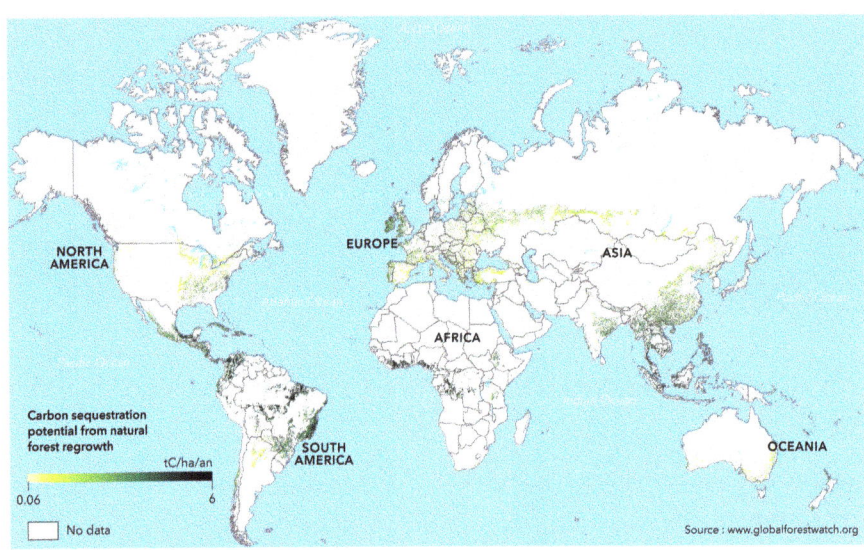

Map 4.17 Carbon sequestration potential from natural forest regrowth

past, current and future changes in climatic parameters (temperature, precipitation, occurrence of fires, sea levels). It also briefly discusses the consequences on agriculture, forests, livestock or even on fishery resources. It is therefore clear that, while all continents are affected by the overall rise in temperatures, some regions within them will be more severely impacted than others, along with the agricultural production systems found there.

4.2 Atlas of Greenhouse Gas Emissions from Agri-Food Systems

Marie Hrabanski, Vincent Blanfort, and Julien Demenois

4.2.1 Map of Agri-Food System Emissions by Country Between 2019 and 2021

Emissions from the agri-food sector refer to three different posts: emissions emitted at the plot scale (*farm gate*), emissions related to land use change and emissions related to pre-production and post-production activities. This representation of GHG emissions from the agri-food sector combines the total of all GHGs by country expressed by averaging the period 1990–2021 (Map 4.1). There are significant disparities between industrialised countries and others. The Chinese and Brazilian agri-food systems are the largest GHG emitters. Emissions produced by China are particularly high due to its intensive agricultural production (the world's largest rice producer) and its livestock industry (40% of agricultural emissions). In Brazil, it is also the intensive agri-food production (about 70% of Brazil's total emissions) and the conversion of vast areas of forests into agricultural land that explain these high emissions.

In second place are Indonesia, the United States, the European Union (EU), India, Russia and Australia. The agri-food systems of these countries (or group of countries) are also particularly emitting, due to the volume of their agricultural productions and associated sectors.

In Africa, the situation varies, but most countries are at medium to low emission levels. However, Africa was Originally, in 2019, 17% (15% for Sub-Saharan Africa and 2% for North Africa) of global GHG emissions were linked to agri-food systems, more than the United States (9%) and the EU (7%). However, there are significant disparities from one region or even one country to another. In Sub-Saharan Africa, these systems emit slightly less GHG per capita (2.3 t CO_2 eq) than in the EU (2.6 t CO_2 eq). But food production per capita is also much lower there (which partly explains the food insecurity of the sub-region). In proportion to the value of food production, the quantities of GHGs emitted south of the Sahara are therefore more than twice as high as in the EU. They are even five times higher if we take into account land use change (deforestation in Nigeria and the Congo basin and cultivation of grazing areas).

This initial global analysis is complemented by a comparison of global emissions by major sector (Fig. 4.3). Livestock activities (Chaps. 10, 16 and 24), and net forest conversion (Chap. 9) are the most impactful, but other sectors can also significantly increase these emissions if we add up the sub-sectors they involve (example of crops with fertiliser manufacturing, energy consumption, etc.).

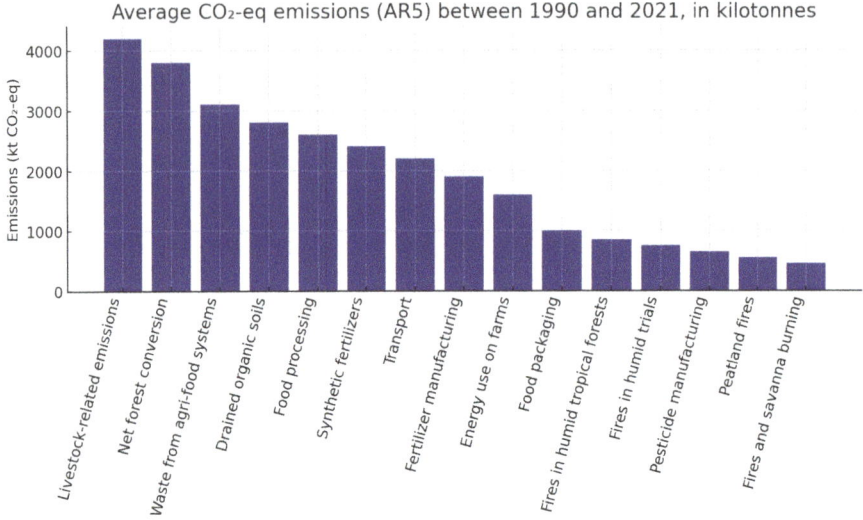

Fig. 4.3 Comparison of global emissions by major sector. (Source: www.fao.org)

4.2.2 Map of Agri-Food Sector Emissions Per Capita

It is essential to put national emission figures into perspective through the lens of each country's population. The map of agri-food sector emissions per capita (Map 4.2) thus presents a completely different vision than the previous one: the countries with the highest emissions in volume are no longer at the top of the ranking. South America, Canada, Northern Europe still have a high ratio due to the impact of agricultural expansion and forestry exploitation at the expense of forests. Western Europe, Asia and Africa have lower per capita scores than their gross emission volume due to their high population density.

Some countries with average low national emissions are propelled to the top of the ranking due to a very low population. Notably, Mongolia, due to a very high ratio of livestock per capita. For the same reasons, Botswana and Namibia are characterised by an agricultural sector (especially livestock) that represents 10–15% of their total GHG emissions.

4.2.3 Map and Graph of Agricultural Land Emissions

In reference to Fig. 4.2 (see the introduction to this chapter), the synthetic definition of agricultural land includes: emissions emitted at the farm gate (from entry to exit, excluding prior and subsequent stages) thus excluding pre-harvest and post-harvest processes; and land use change.

The emissions from this sector therefore reflect the impacts of strict farm production activities combined with land use changes aimed at expanding agricultural areas.

According to the FAO, agricultural lands include areas used for cultivation, for pastures, and for related agricultural activities. These include arable lands (temporary crops, fallow lands), areas allocated to perennial crops (fruit trees, plantations), and permanent meadows used for livestock farming. These lands can vary according to agricultural systems and local conditions, also encompassing natural resources essential for agricultural productivity, such as soils, vegetation, and water resources.

This definition reflects an integrated approach, taking into account not only soils, but also environmental and socio-economic factors related to sustainable land use. Agricultural lands play a key role in global food supply, ecosystem conservation, and rural development.

The Asia region alone contributes to nearly 36% of agricultural land emissions, mainly due to methane emissions, due to rice cultivation and intensive livestock farming (Maps 4.3 and 4.4). There are very high emissions in insular Southeast Asia linked to deforestation for palm oil production.

It is closely followed by the American continent with a strong impact of deforestation for crops and pastures in the Amazon. The United States and Canada have highly mechanised agricultural systems, which results in CO_2 emissions related to the use of fossil fuels. The production of beef, pork, and poultry is intensive, which increases methane emissions associated with the intensive use of synthetic fertilisers for major crops such as corn, soybeans, and wheat.

Africa, with 18% of emissions, is characterised by regional trends. In West and East Africa, livestock activities are a source of methane and extensive agricultural practices such as slash-and-burn agriculture, a major source of N_2O and CO_2. In East Africa, livestock farming is dominant, with high methane emissions due to enteric fermentation. Deforestation in Central Africa is a significant source of emissions. Monocultures and increased use of fertilisers in Southern Africa contribute to nitrous oxide emissions (Map 4.5).

The analysis of greenhouse gas (GHG) emissions from agricultural lands in Europe (11.2%) reflects dynamics specific to this industrialised and technologically advanced region, particularly in Northern Europe. Intensive land use and high mechanisation are controlled by advanced sustainability policies (CAP, Common Agricultural Policy) and practices such as organic farming and conservation agriculture.

Eastern Europe is still in a post-Soviet transition characterised by a gradual increase in agricultural intensity. Intensive practices in horticultural and wine crops can be associated with high input emissions in Southern Europe.

4.2.4 Map and Graph of Emissions Attributable to Crops Worldwide, Between 1990 and 2021

Map 4.6 analyses the GHG emissions produced between 1990 and 2021 by agricultural crops, and in particular by the following productions: wheat, barley, oats, soybeans, potatoes, rice, corn, sugar cane, millet, sorghum, dry beans, rye.

Nearly 80% of GHG emissions from crops are attributable to Asia. Taking into account all types of GHGs, methane emissions from rice cultivation partly explain this particularly high figure. The Americas are the second emitting continent, with 10% of emissions (soybeans, corn, wheat). Europe and Africa are each responsible for about 5% of GHG emissions from crops; these emissions per capita on each of these two continents indicate that GHG emissions from crops in Africa are particularly low. Finally, only 0.4% of GHG emissions from crops are attributable to Oceania.

4.2.5 Map and Graph of Emissions Attributable to Livestock Farming

According to an FAO (2023) assessment, livestock farming accounts for 40% of total emissions from agri-food systems, or about 12% of total anthropogenic GHG emissions globally (see Chaps. 10 and 16). It is methane (CH_4), mainly from the enteric fermentation of ruminants (cows, goats, sheep) and the management of animal waste, which is the main cause with more than half of the emissions linked to animal production. It is followed by carbon dioxide (CO_2, 31%), attributable to land use changes (deforestation for pastures or fodder crops) as well as energy consumption for production and transport. Finally, 15% of these emissions are attributable to nitrous oxide (N_2O) produced by the use of fertilisers and soil management.

Beyond these global figures, GHG emissions from the livestock sector vary greatly depending on the regions of the world. (Map 4.7).

An analysis by major world region shows that Asia and the Americas contribute to two-thirds of global emissions. Asia emits a third of the emissions due to the rapid growth in demand for animal products linked to population growth and urbanisation. China is thus the world's leading producer of pork, and India is the world's leading producer of milk. The American continent also contributes to more than 30% of global emissions, particularly due to the significant numbers of ruminants (Brazil in particular, with the conversion of primary forests into pasture and crops for animal feed).

In Africa, animal production plays a crucial role in food security, livelihoods, and the economy, particularly in pastoral areas: the continent consequently contributes to 15.6% of global GHG emissions. These emissions represent about 10% of the continent's total GHG emissions. Europe is almost on par, despite a more regulated livestock sector in terms of regulations and agricultural practices. The sector represents a dominant share of agricultural emissions in Europe estimated at over 70%, or 10% of the total GHG emissions of the European Union.

The main sources of emissions related to livestock are:

- Enteric fermentation: ruminants (cows, sheep, goats) produce methane (CH_4) during their digestion;
- Manure management: animal waste emits methane (CH_4) and nitrous oxide (N_2O) when it decomposes;
- Animal feed production: the cultivation of cereals and forage used to feed animals generates GHG emissions (CO_2 linked to mechanisation and the use of synthetic fertilisers, N_2O released by soils treated with fertilisers);
- Land-use change: deforestation to create pastures or cultivated land to produce animal feed releases CO_2;
- Transport and processing: the transport of food, live animals, and processed products contributes to CO_2 emissions.

4.2.6 Map of Emissions from Farm Gate Productions

The emissions from the agri-food sector refer to three different categories: emissions emitted at the plot level (farm gate), emissions related to land-use change, and emissions related to pre-production and post-production activities. The map presented here (Map 4.8) focuses only on the emissions produced at the pre-production and post-production stages between 1990 and 2021. The pre-harvest stages refer notably to the production of fertilisers and irrigation, and the post-harvest stages to the following five categories: transport, processing, distribution, consumption, and finally waste management. As a reminder, Tubiello's (2022) study shows that since 2019, pre-production and post-production processes have surpassed farm processes to become the largest component of GHG emissions from agri-food systems, as confirmed by Map 4.8. The leading country in terms of agricultural GHG emissions at the plot level is China. A second group of countries includes major agricultural producers such as North America (United States and Canada), major European agricultural countries (France, Germany, Poland), Brazil and Argentina, and part of Asia (India, Indonesia), Russia and, more surprisingly, Sudan and Ethiopia (mainly due to agricultural methane emissions related to livestock).

4.2.7 Emissions from the Agri-Food Sector

The proposed map (Map 4.9) is based on the same sectors as Map 4.8 and focuses only on the emissions produced at the pre-production and post-production stages emitted between 1990 and 2021. As a reminder, since 2019, emissions during the pre-production and post-production stages have now surpassed emissions at the farm level (Tubiello et al. 2022), as confirmed by the map below. This shows that, as with emissions produced at the plot level, China and the United States are the leading producers of GHGs at the pre-production and post-production stages. A second group of countries includes most European countries (more countries than when only considering emissions at the plot level), which is mainly due to the transport of agricultural products. Brazil and Australia also belong to this second group, as well as a large part of Asia and Russia. In Africa, only Nigeria is part of this second group.

4.2.8 GHG Emissions Related to Land Use Change

By land use change, we refer here to the burning of forest biomass and the conversion of forest lands (into agricultural lands in particular). Between 1990 and 2021, we observe here (Map 4.10) that nearly half of the emissions related to land use change originate from the Americas, and substantially from Latin America, due notably to deforestation in the Amazon basin. Emissions induced by land use change are also high in Asia (21.6%) and in Africa (27%), also due to deforestation in Central Africa and Southeast Asia, particularly in Indonesia. The period studied (1990–2021) explains why Europe is an exception with emissions related to land use change, as the emissions related to land use change date mainly from the nineteenth and twentieth centuries.

4.3 Map of the Impacts of Climate Change on the Land Sector

Vincent Blanfort, Marie Hrabanski, and Julien Demenois

This map (Map 4.11) was established based on the IPCC synthesis report published in 2023 (AR6). It covers:

- the expected impacts of climate change in 2050 on agricultural, forestry and fishery productivity (forestry sector, livestock, crops, fishing);
- climate vulnerability factors (coastal areas vulnerable to sea level rise and flooding, areas exposed to desertification and increased fire risks, etc.);
- and the vulnerability of coastal populations (regions with high population density and cities threatened by rising waters).

The map unsurprisingly highlights the extent of the impacts of climate change on agricultural productivity and the increasing vulnerabilities to climate change in many areas of the planet.

More specifically, what stands out first when reading this map is the projected decline in productivity in both livestock and crop sectors across many areas and on every continent. However, some regions are also expected to benefit, with increases in crop productivity anticipated in the north-western United States, north-western Brazil, Namibia, Peru, Bolivia, and eastern Russia. As for livestock, productivity gains are expected in the central United States and Central Europe.

The situation for fisheries is much more geographically contrasted. A rise in fishery productivity is projected across the fishing grounds north of the Arctic Circle, while decreases are expected in temperate and tropical zones, particularly in Europe, along the west coast of Africa, in the Oceanian basin, and in the temperate Pacific region. Forest productivity, meanwhile, is expected to increase in Central Africa, the Amazon, Canada, and the Scandinavian countries, while a decline is projected in the eastern United States.

Beyond these initial observations, climate-related vulnerabilities are expected to multiply. Notably, there is likely to be an increase in the regions where tropical cyclones form, which will affect densely populated coastal areas. This concerns both the eastern and western coasts of the United States, as well as the waters off China, Japan, southern Africa, and the north-east of Australia. Wildfires are also expected to become more frequent across all continents, with the exception of Africa. Finally, drought- and desertification-prone areas are projected to expand.

4.4 Atlas of Land Carbon and Its Evolutions

Julien Demenois, Vincent Blanfort, and Marie Hrabanski

4.4.1 Map of Soil Organic Carbon Stocks

The soil is the largest terrestrial carbon reservoir: 2400 Gt of carbon between 0 and 2 m depth on a global scale (see Chap. 17). This stock is mainly found in the soil of forests (30%) and grasslands (30–35%), and to a lesser extent in the soil of croplands (15%). This is in addition to the carbon of permafrost (over 1600 Gt between 0 and 3 m depth).

The map of soil organic carbon stocks (SOC) between 0 and 30 cm depth is the result of the work of the Global Soil Partnership (GSOCmap, *global soil organic carbon map*). This map (Map 4.12) is the first global map of soil organic carbon ever produced in a consultative and participatory process involving member countries, making it absolutely new and unique. Maps 4.12, 4.13 and 4.14 can be viewed in digital and interactive version on the Impact4Soil platform (www.impact4soil.com) developed by Cirad and Vizzuality within the framework of the European ORCaSa project.

The highest SOC (Soil Organic Carbon) stocks are found in high or low latitude areas (above 40°) and in the intertropical zone. Ten countries hold more than 60% of these stocks, with Russia leading the way. Brazil, Indonesia, and the Democratic Republic of Congo are also on this list. *Conversely*, the dry regions of North Africa, the Middle East, or central Australia have lower SOC stocks.

4.4.2 Map of Past Changes in Soil Organic Carbon Stocks

Land use changes are a major source of net GHG emissions. Thus, the conversion of natural ecosystems into agricultural land has resulted in a cumulative loss of 116 Gt of carbon in the top two metres of soil since the Neolithic era (see Chap. 17).

The map of past changes in SOC stocks between 0 and 30 cm depth (Map 4.13) is based on the analysis by Sanderman et al. (2017). It shows the differences in SOC stocks between historical levels (−12,000 years) and current levels. This map is a statistical representation of the amount of soil organic carbon that may have been lost due to agriculture. It is a global product derived at a resolution of about 10 km.

The losses of SOC stocks (shown in red on the map) have been highest in Europe, the United States, and China. *Conversely*, some regions have seen their SOC stocks increase (shown in green on the map), notably in southern Mongolia or in the Nile delta.

4.4.3 Map of Future Changes in Soil Organic Carbon Stocks

Regarding climate regulation, carbon sequestration in soils, through an increase in SOC stocks, would represent 47% of the mitigation potential of agriculture and grasslands (see Chap. 17).

Using the IPCC's Level 1 accounting approach[1], the map of future changes in SOC stocks (Map 4.14) allows visualising the differences in SOC stocks between 2018 and 2038 between 0 and 30 cm depth, if land restoration practices for cultivated lands and grasslands were implemented. The IPCC's Level 1 accounting approach uses average regional response functions to estimate changes in soil organic carbon due to changes in land use. As such, this approach does not take into account smaller scale variations in the evolution of soil organic carbon, which will likely be observed depending on local soil types, local climate, and the local implementation of specific management strategies. The map does not take into account

---

[1] IPCC Glossary: "A level represents a degree of methodological complexity. Generally, three levels are proposed. Level 1 is the basic method, level 2 is intermediate, and level 3 is the most demanding in terms of complexity and data requirements." https://www.ipcc.ch/site/assets/uploads/2019/12/19R_V0_02_Glossary.pdf.

local feasibility or current adoption rates. This analysis is based on the assumption of instant adoption on all lands, even though in reality, the change will occur gradually over time.

This map highlights the significant potential for increasing SOC stocks (shown in green on the map), particularly in South America, Africa, and Southeast Asia, thanks to the restoration of cultivated lands and grasslands.

4.4.4 Current Carbon Stock of Above-Ground Biomass

The carbon stocks of terrestrial lands are mainly found in soils and in shrub and tree vegetation. Forests cover the planet over 3900 Mha (million hectares) and represent around 860 Gt of carbon (Gt: billion tonnes), of which more than 40% is in living biomass (see Chap. 9). The tropical forest represents more than half of the stocks. Forests, therefore, play a crucial role in climate change.

The map of above-ground biomass stocks (Map 4.15) represents the carbon stocks in above-ground biomass, which is on average composed of 50% carbon. This map is based on the methodology presented in Baccini et al. (2012) to generate a global map of the density of living woody above-ground biomass at a resolution of about 30 m for the year 2000. The map can be viewed in digital and interactive form on the Global Forest Watch platform (www.globalforestwatch.org).

The three major tropical forest masses, namely the Amazon, Central Africa, and the Borneo-Mekong basin, hold the majority of these above-ground biomass stocks.

4.4.5 Current Net GHG Fluxes at the Forest Level

Forests are the main carbon sink in terrestrial environments, the second sink after the oceans. They, therefore, play a crucial role in climate change. They would capture, according to the estimates, between 70% and 100% of the carbon annually captured by the emerged lands (1.8 Gt of carbon per year) (see Chap. 9).

The map of current net GHG fluxes in forests (Map 4.16) shows the difference between the carbon emitted by forests and the carbon absorbed by forests between 2001 and 2023. Negative values (in green on the map) correspond to net carbon sinks and positive values (in purple on the map) to net carbon sources. Net fluxes are calculated based on the IPCC guidelines for national greenhouse gas inventories. The map can be viewed in digital and interactive format on the Global Forest Watch platform (www.globalforestwatch.org).

This map clearly highlights GHG emissions due to deforestation and forest degradation in North America (particularly due to mega-fires), in the Amazon, and in Indonesia. Conversely, GHG absorptions by forests in Europe and China are noticeable.

4.4.6 The Potential for GHG Sequestration in Forests

Forests could store 226 Gt of carbon in addition to the current stocks, if regenerated. Such restoration could be implemented in areas often on the fringes of large tropical forest masses, which are no longer used for agriculture and remain free from urbanisation (see Chap. 9).

The map of the potential for GHG sequestration in forests (Map 4.17) displays the rate at which forests could capture carbon from the atmosphere and store it in above-ground living biomass over the first 30 years of natural forest regrowth. These rates are presented for reforestable areas and exclude areas of grasslands and indigenous crops in order to preserve food and fibre production as well as biodiversity habitat. The map can be viewed in digital and interactive format on the Global Forest Watch platform (www.globalforestwatch.org).

This map highlights the importance of this potential in Central America and the Caribbean, as well as in South America, Southeast Asia, and Europe. On the other hand, the potential appears more limited in Africa, with the exception of West Africa.

4.5 Changes and Impacts of Climate Change by Major Region on the Land Sector

4.5.1 The Anticipated Climate Changes and Impacts in Africa

Nicolas Viovy, Jacques André Ndione, Moussa Waongo, and Maguette Kaire

4.5.1.1 The Anticipated Climate Changes in Africa

Temperature

As with all continents, the temperature in Africa will increase, but less strongly, on average, than on the Eurasian and North American continents. Moreover, climate change will affect the different African regions unevenly: in the high latitudes, the temperature increase will be greater than in the tropical zones, with the exception of the Sahel (Fig. 4.4). Consequently, this increase (of about 0.8 °C) will be more significant in the north and south of the continent than in the equatorial zone. However, as the temperature is already high, particularly in the savannah and semi-desert areas, many regions will reach temperature peaks above 50 °C, a critical threshold for humans and all species.

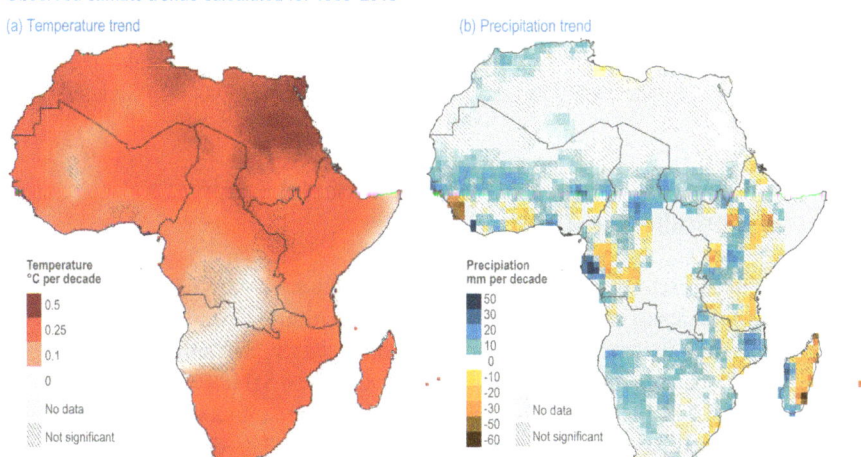

Fig. 4.4 Trends in temperature and precipitation changes over the last few decades (1980–2015) in Africa (**a**) Températures (**b**) Précipitations. (Source: IPCC AR6 (2022))

Precipitation

The uncertainty about the evolution of precipitation in Africa is significant. Projections indicate strong contrasts with a drying of the entire Mediterranean rim as well as part of the Atlantic coast and throughout the south of the continent (Fig. 4.4). On the other hand, the rest of the continent should see its precipitation increase. Unlike other regions, such as India, the duration of the monsoon period should not vary significantly. Interannual variability should increase leading to both an increase in episodes of intense precipitation, but also periods of drought during the rainy season. These effects could unfortunately offset the positive effect of the average increase in precipitation over part of the continent.

Winds

As in the entire tropical zone, if there is no clear trend on the change in the frequency of cyclones (which could even decrease slightly), their intensity has, however, increased and should continue to grow in the future. This phenomenon is linked to the increase in the temperature of the oceans and the atmosphere, which is becoming more humid.

Sea Level

As with the rest of the globe, an increase in sea levels is expected, which will affect coastal areas. The rise in water levels will continue, even if we manage to stabilise atmospheric CO_2, the rise will continue on a long time after the CO_2 decrease. The sea level has risen by about 23 cm since 1880 and is expected to rise another 30 cm by 2050 (IPCC 2014). This effect, combined with the increase in the intensity of cyclones, raises concerns about risks increased submersion of certain heavily populated coastal regions (such as the Nile Delta, the city of Lagos in Nigeria or Dakar in Senegal).

4.5.1.2 The Impacts of Climate Change on Different Sectors of Agriculture in Africa

Climate change poses complex challenges for agriculture in Africa, requiring robust adaptation strategies to ensure the resilience of agricultural systems and food security. The increase in precipitation variability is a major problem, particularly in areas heavily dependent on the rainfall regime such as the Sahel (Box 4.2), Southern Africa and the Horn of Africa. On the one hand, the unpredictability of rainfall in some places complicates the planning of crop cycles, and late or irregular rains can lead to a decrease in yields or the total loss of some crops. On the other hand, the risk of prolonged droughts will increase and will have impacts on agricultural production. It should be noted that the increase in temperatures increases evapotranspiration and therefore the risk of agricultural drought. This increase also poses a direct risk to crops by reinforcing the risk of heat stress and therefore of yield decreases. Climate change also risks inducing an extension of diseases and pests affecting both crops and livestock. At the same time, extreme weather events such as torrential rains and droughts accelerate soil erosion and land degradation, reducing soil fertility and long-term agricultural production capacity. The increase in the irregularity of rainfall as well as the increase in agricultural commodity needs linked to that of the population (and a possible change in diet) increase the need for irrigation, which can exacerbate conflicts between farmers, livestock farmers and other water users, complicating the sustainable management of water resources. Finally, climate constraints sometimes force farmers to abandon traditional diversified crops in favour of more resistant crops or monocultures, which can reduce agricultural biodiversity and increase vulnerability to other threats, such as pests. These various risks are exacerbated by the fact that the majority of farmers in Africa are smallholders, often poor: their capacity (financial and technological) to adapt to the impacts of climate change is limited, which makes them particularly vulnerable.

> **Box 4.2 West Africa will be heavily affected by climate change**
> Currently representing only 1.8% of global greenhouse gas (GHG) emissions, the countries of the ECOWAS region (Economic Community of West African States) contribute very little to climate change. According to the most pessimistic scenarios, West Africa will experience, by 2060, a temperature increase of +2.3 °C, i.e. a warming of +0.6 °C per decade. Indeed, in West Africa, the number of potentially lethal heat days could reach 50 to 150 days per year for

(continued)

Box 4.2 (continued)

a climate warming of 1.6 °C and 100 to 250 days per year for a climate warming of 2.5 °C, with the strongest increases in coastal areas. Children born in West Africa in 2020 will, in the event of climate warming of 1.5 °C, be exposed to 4 to 6 times more heatwaves during their lifetime than those born in 1960. For the tropical zone of West Africa, the risk of heat-related mortality is 6 to 9 times higher than the average for the years 1950–2005 at 2 °C of climate warming. With increasing urbanisation, cities like Lagos, Niamey, Kano and Dakar are particularly exposed (Trisos et al. 2022). As for rainfall, it is expected to be more erratic and should lead to an increase in the frequency and intensity of extreme climate hazards already known in the region: floods, increased rainfall variability (translation of isohyets), pockets of extremely long drought, among other corollaries, with dramatic human (ecosystem degradation) and economic consequences on all economic sectors and on the most vulnerable populations, including women, young people and the elderly. Regarding the emblematic region of the Sahel, recent analyses on the current evolution of the climate show an even more acute intensification of extremes. The Droughts are becoming increasingly intense, as are floods, to the point that researchers are discussing the advent of a new climatic era (Panthou et al. 2018).

The consequences are already dramatic. In October 2022, heavy rains and floods killed hundreds of people, displaced thousands more, and destroyed over a million hectares of cultivated land in Niger, Mali, and Burkina Faso (UN News[2]). Compared to the period 1986–2005, a global warming of 3 °C is expected to reduce work capacity in agriculture by 30–50% in Sub Saharan Africa, due to rising temperatures. Climate change has increased extreme poverty in the West Africa region by nearly 3% in 2021. Gender inequalities in the Sahel and West Africa are among the highest in the world. Multiple interconnected crises (climate, security, health, and food) exacerbate these inequalities and increase the vulnerability of women.

Regarding agricultural sectors, the livestock sector is particularly affected due to the decrease in water availability, a decline in biomass potential, degradation of pastures (Savadogo et al. 2011), and an increase in cases of thermal diseases. With a global warming of 1.5 °C, sorghum yields are expected to decrease in West Africa, and a 9% decrease in maize yields is expected (Fig. 4.5). For cocoa trees, which grow at optimal temperatures around 25 °C in West Africa, temperature increases will cause drastic drops in cocoa production by 2030 in Ghana and Ivory Coast.

---

[2] UN News Global perspective Human stories. https://news.un.org/en/story/2022/10/1129997.
[3] https://cordex.org/.

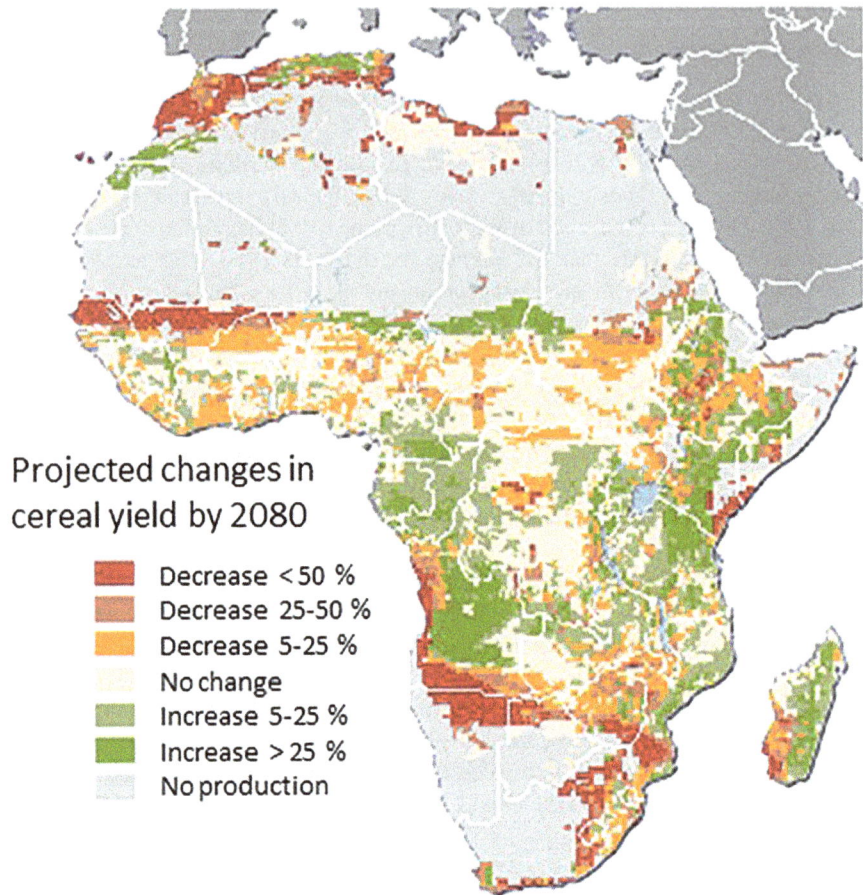

Fig. 4.5 Change in cereal yields by 2080 in Africa compared to the year 2000 in a high emission scenario (A2). (Source: Fisher et al. (2005))

4.5.1.3 Contribution of the Agricultural Sector to the Greenhouse Gas Balance

While Africa is one of the regions most vulnerable to climate change, the direct impact of its agriculture on GHG emissions is smaller than in developed regions (Fig. 4.6). Agriculture is indeed much less mechanised and uses less fertiliser, and the diet is less meat-based than in developed countries. However, if we take into

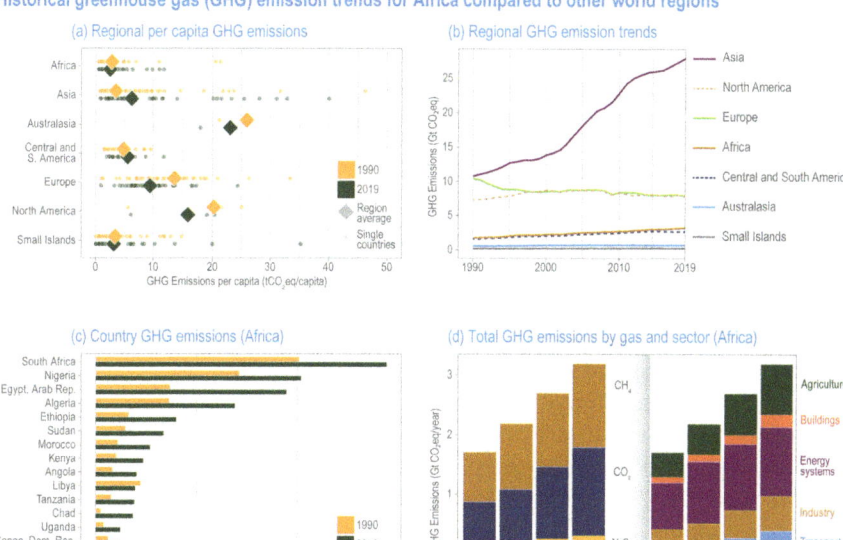

Fig. 4.6 Historical trends of greenhouse gas (GHG) emissions in Africa compared to other regions of the world (**a**) GHG emissions per person and by region, and their evolution from 1990 to 2019 (circles represent countries, diamonds represent the regional average); (**b**) Total GHG emissions by region since 1990; (**c**) Total GHG emissions in 1990 and 2019 for the fifteen highest emitting African countries; (**d**) Total emissions in Africa since 1990, broken down by GHG (on the left) and by sector (on the right). Methane and CO_2 emissions represent an almost equal share of GHG emissions in Africa, with the most significant emission sectors being energy and agriculture (Crippa et al. 2021). Agriculture emissions in panel (**d**) do not include land use, land-use change, and forestry (LULUCF CO_2). The global warming potentials over one hundred ans, in line with the estimates of Working Group I (WGI), are used. The emission data comes from Crippa et al. (2021), compiled in Chap. 2 of Working Group III (WGIII) of the IPCC's AR6

account the impact related to land use change, mainly due to deforestation, the GHG emission balance becomes much more significant and then represents 22% of global AFOLU sector GHG emissions (Brouziyne et al. 2023). The impact of land use change in Africa then accounts for half of GHG emissions. Similarly, the trend is towards an increase in emissions, rising from 1.8 Gt CO_2 eq in 2000 to 2.2 Gt CO_2 eq in 2018 (FAO 2020).

Box 4.3 The North Africa and Middle East region: a hotspot for climate change

The North Africa—Middle East (MENA) region is both a particularly high emitter of GHGs (in 2018, this region produced 8.7% of GHGs, while it only represented 6% of the world's population) and a region particularly vulnerable to the effects of climate change. It includes both high-income oil-exporting countries in the Persian Gulf area, middle-income countries (Morocco, Turkey, Egypt, etc.), and some of the least developed countries such as Sudan, Yemen, and Mauritania.

Climate change will indeed exert increasing pressure on the already scarce water and agricultural resources in the MENA region, thus threatening national security and the political stability of the entire region. The temperature increase observed in this region has been almost twice as fast as the global average. In addition to heatwaves, which are expected to be more intense and frequent, regions adjacent to the Mediterranean Sea should experience a decrease in rainfall ranging from 20% to 40%. Droughts are expected to last up to 90% longer. The region is also particularly sensitive to the phenomenon of desertification. Egypt, Jordan, and Palestine are already experiencing a decline in their vegetation, up to 80% of their land areas. In addition to increasing the aridity of the region, disrupting agricultural patterns, this phenomenon should also increase the number and intensity of sandstorms. In this part of the world, which already houses twelve of the seventeen countries most affected by water scarcity, and where the population is set to double by 2050, the consequences of the climate crisis on societies should be considerable. The consequences for agriculture will be major, as the IPCC estimates that nearly 45% of agricultural land should be exposed to salinity, soil nutrient depletion, and erosion. At the same time, this region heavily depends on imports of agricultural products: 40% of its food needs are currently covered by imports, and climate change should further accentuate this trend.

The continuation of the *status quo* will lead to an additional average regional warming that could reach 5 °C by the end of this century, thus risking making part of the MENA region uninhabitable, even though not all countries are equally affected by the impacts of climate change.

4.5.2 The Projected Climate Changes and Impacts in Southeast Asia

Alain Rival

4.5.2.1 The Climate Changes in Southeast Asia

Archipelagic by nature, Southeast Asian countries experience strong land-ocean-atmosphere interactions. The climate is predominantly tropical, warm and humid with abundant rainfall. The seasonal variability of precipitation in the region is primarily influenced by monsoon systems, the north-south migration of the intertropical convergence zone, and tropical cyclones (mainly for the Philippines and Indochina). Southeast Asia (SEA) proves to be one of the world's most vulnerable regions to climate change due to its extensive coastlines, heavy reliance on agriculture and forestry, and more generally on natural resources, including minerals. Each year, torrential monsoon rains and several cyclones cause hundreds of deaths in Southeast Asia (and affect millions of people); they also contribute to the rise in the price of rice and foodstuffs.

Southeast Asia is home to a significant concentration of countries heavily affected by climate change: four of the ten most affected countries in the world belong to this region, namely Myanmar, the Philippines, Thailand, and Vietnam. Indonesia has become the country with the highest GHG emission rate in the region in 2021, recording 2.05 billion tonnes. After Indonesia, Vietnam ranks second with 507.34 t, while Thailand ranks third with 452.12 Mt. Conversely, Timor-Leste has the lowest GHG emissions, at 9.38 Mt (Sonobe et al. 2024).

The region is facing increasing GHG emissions from agricultural sources: activities related to agriculture, forestry, and various land uses are essential for the region, for its food security, and for maintaining livelihoods in rural areas. These activities also significantly affect anthropogenic GHG emissions.

Deforestation is one of the main causes of GHG emissions from the agricultural sector. The massive conversion of land for agricultural purposes, especially for large-scale commercial agriculture including oil palm plantations, releases the carbon stored in trees and reduces the forests' ability to absorb CO_2.

The development of agricultural activities in Indonesia (Sonobe et al. 2024) has stimulated economic growth, but has also led to excessive ecological and economic costs. Between 2007 and 2018, this development generated about 48 billion dollars (5.7% of the gross domestic product, GDP), but air pollution and fires offset more than half of this amount. Measures to prevent fires and restore peatlands are now reducing these costs.

Highly exposed, being predominantly insular, and with low resilience, the region is vulnerable to the effects of climate change (IPCC 2022), with a foreseeable increase in the intensity and frequency of extreme climate-related events (floods and droughts). This radical change in seasonal precipitation patterns would have enormous repercussions on livelihoods, infrastructure, agricultural production, and the region's food security.

Temperature

The average temperature in Southeast Asia is expected to continue to rise during the xxi^e century. The projections of multi-model regional climate simulations given by the Coordinated Regional Climate Downscaling Experiment (Cordex-Sea) initiative show that the temperature increase on land would be between 3 °C and 5 °C by the end of the xxi^e century compared to the period before 1986–2005. For all levels of global warming (see Chap. 1), the land region is expected to warm slightly less than the global average.

Precipitation

The projections of future precipitation changes are very variable between the subregions of Southeast Asia and between the models (Fig. 4.7). Thus, the datasets provided by the Coupled Model Intercomparison Project (CMIP7) showed an increase in average annual precipitation over most land areas by the middle and end of the XXIe century, with a strong correlation to the model shown for higher warming levels only[3]. Based on the Cordex multi-model simulations for Southeast Asia, significant and robust increases in average precipitation over Indochina and the Philippines are projected, while a drying trend is noted over the maritime part for the early, middle, and end periods of the XXIe century.

By the end of the XXIe century, a 20% increase in average precipitation is expected in Myanmar, in north-central Thailand and northern Laos, and from 5% to 10% in the eastern Philippines and northern Vietnam. Significantly drier conditions are expected in other regions of Southeast Asia. In the Indonesian region, particularly on Java, Sumatra and Kalimantan, a decrease of 20–30% in average rainfall is expected by the end of the twenty-first century.

Fires

Nearly half of the world's tropical peatlands are located in Southeast Asia, where they constitute significant carbon reservoirs. Over the past three decades, most of the 25 Mha of tropical peatlands in the region have been deforested and drained. As a result, the lowering of the water table exposes the peat to oxidation, transforming the plant material accumulated over millennia into carbon dioxide and causing land subsidence. Nearly 80% of the region's natural peatlands have been deforested and drained, the majority of them are now occupied by perennial, forestry or agricultural plantations. This conversion increases the oxidation of the peat, which contributes to a rapid loss of carbon into the atmosphere in the form of greenhouse gas emissions, and increases their vulnerability to fires that regularly generate haze, with serious impacts on human health (Mishra et al. 2021).

Forest and vegetation fires, used as tools for agricultural expansion, are a major source of air pollution and can cause serious health problems in many regions of Southeast Asia. The economic cost of the forest fires that ravaged Indonesia during the El Niño episode of 2015 was estimated at over 16 billion US dollars, with over 100,000 premature deaths. For several days, the fires emitted more carbon dioxide in the region than the entire economy of the United States of America.

---

[4] Area of 462,840 km² for Papua New Guinea, and 20 km² for Nauru.

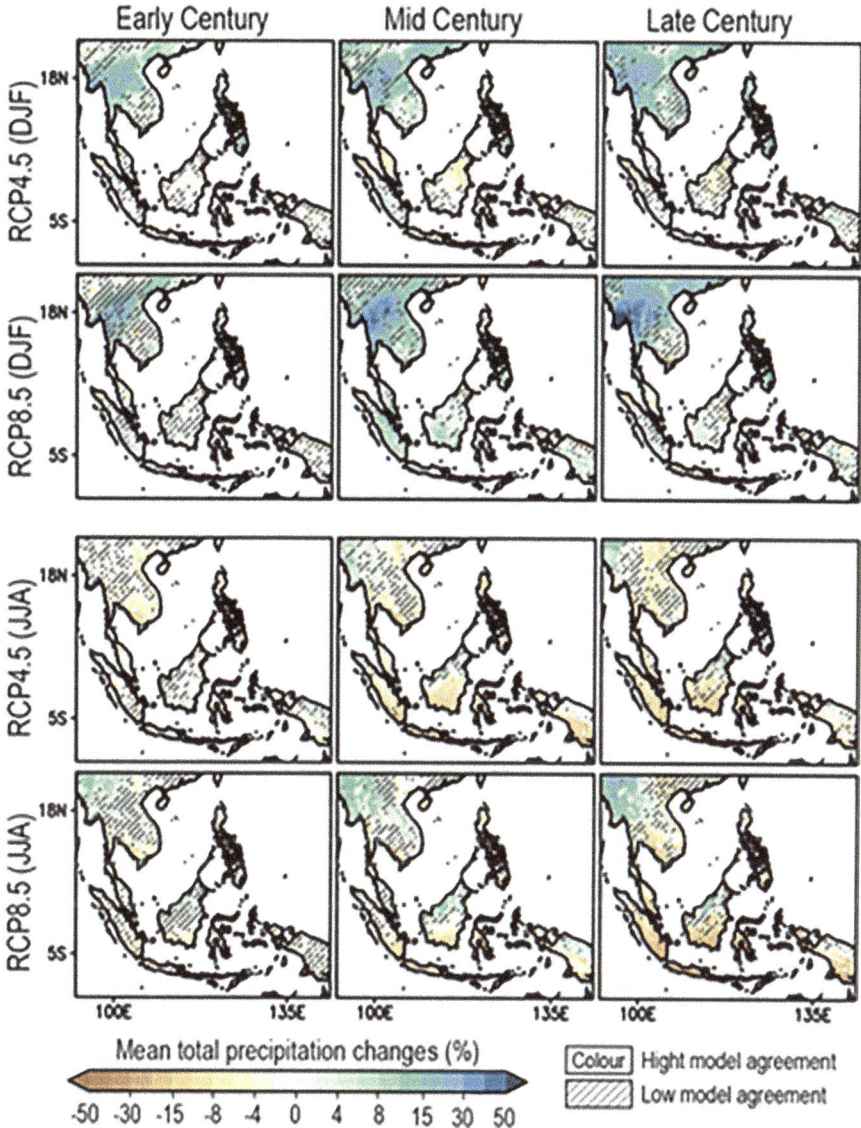

Fig. 4.7 Projection of expected changes through regional climate models (RCMs) for average rainfall between the beginning (2011–2040), middle (2041–2070) and end (2071–2099) of the twenty-first century and the historical period 1976–2005. (Source: Tangang et al. (2020); Fig. 10)) The data comes from Cordex-Sea downscaling simulations. The hatched lines indicate areas where model agreement is low (less than 80%). DJF: December-January-February; JJA: June-July-August.

Sea Levels

Changes in global sea level rise have undeniably begun to impact the island economies of Southeast Asia, which are already very vulnerable (Idris et al. 2023). Sea level rise can threaten the long-term sustainability of coastal communities and fragile ecosystems such as coral reefs, wetlands and mangroves. The Indo-Pacific region is home to most of the world's mangrove forests, although sediment input in this region is decreasing, due to the construction of dams on rivers.

Although the region tends to warm a little less than the global average, sea levels will rise faster elsewhere and shorelines will recede in coastal areas where 450 million people live. The rise in water levels is expected to cost billions of dollars in damage to major Asian cities during the current decade, and its impact is amplified by the tectonic movements of the fire belt and by the effects of massive groundwater pumping. Thus, nineteen of the twenty-five cities most exposed to a one-metre rise in sea level are located in Asia, seven of which are in the Philippines alone.

The coastal areas of Southeast Asia are densely populated and support a wide range of economic activities. The physical impact of sea level rise is manifested by coastal erosion and flooding of the lowest areas, salt intrusion, flooding due to storm surges and high tides as well as habitat loss. The economic impact of sea level rise results in the destruction of natural areas, fields and homes along the coasts, changes in land use, water management systems, navigation and waste management in coastal areas.

4.5.2.2 The Impacts of Climate Change on Different Sectors of Agriculture and Forestry in Southeast Asia

Agriculture and Forestry in Southeast Asia

The latest report from the Association of Southeast Asian Nations (ASEAN) on the state of climate change (ASEAN 2021) suggests that adaptation and mitigation actions should be synergistically implemented on the ground. Various examples include climate-smart agriculture and nature-based solutions, incorporating ecosystem-based adaptation such as agroforestry, protection of mangrove forests, and strengthening forest management through certification and reducing emissions from deforestation and forest degradation (REDD+) (see Chap. 6). Proper management of reservoirs for hydroelectricity will also protect local communities from river flooding and other extreme events, while contributing to climate change mitigation.

The IPCC report predicts particularly severe consequences for Southeast Asia, one of the most vulnerable regions of the planet to climate change. The regional bloc, primarily archipelagic, will be directly hit by rising sea levels, heatwaves, drought, and more intense and frequent rainfall episodes. Known as "rain bombs", episodes of heavy rainfall will intensify by 7% for each degree of global warming.

Biodiversity and ecosystem services play a crucial role in the socio-economic development as well as the cultural and spiritual fulfilment of the Asian population.

Species richness peaks in the Coral Triangle of Southeast Asia (Philippines and Indonesia), and the extent of mangrove forests in Asia accounts for about 38.7% of the global total. These coastal ecosystems provide multiple ecosystem services related to food production through fishing and aquaculture, carbon sequestration, coastal protection, and tourism and recreation.

Figure 4.8 schematically represents the expected impacts of climate change in some Southeast Asian countries. It clearly illustrates the spatio-temporal diversity of future impacts on food production, highlighting that there are "winners" and "losers" associated with climate change at different scales. The region's food security remains heavily dependent on rice production, which shows significant vulnerability, particularly in Vietnam and Thailand. Other scenarios, on the other hand, predict an increase in yields due to climate change, due to an increase in rice productivity per hectare.

Data on the observed effects of climate change on agriculture and food systems in Southeast Asia remain scarce. Most impacts have been associated with drought, monsoon, and oceanic oscillations, whose frequency and severity have been linked and associated by several authors with climate change. Generally, the major consequences on agricultural production, as observed by farmers in the Philippines and Indonesia, mainly concern delays in harvest dates, decrease in yields and product quality, increase in pest and disease incidence, growth delays, increased livestock mortality, and a drop in agricultural income.

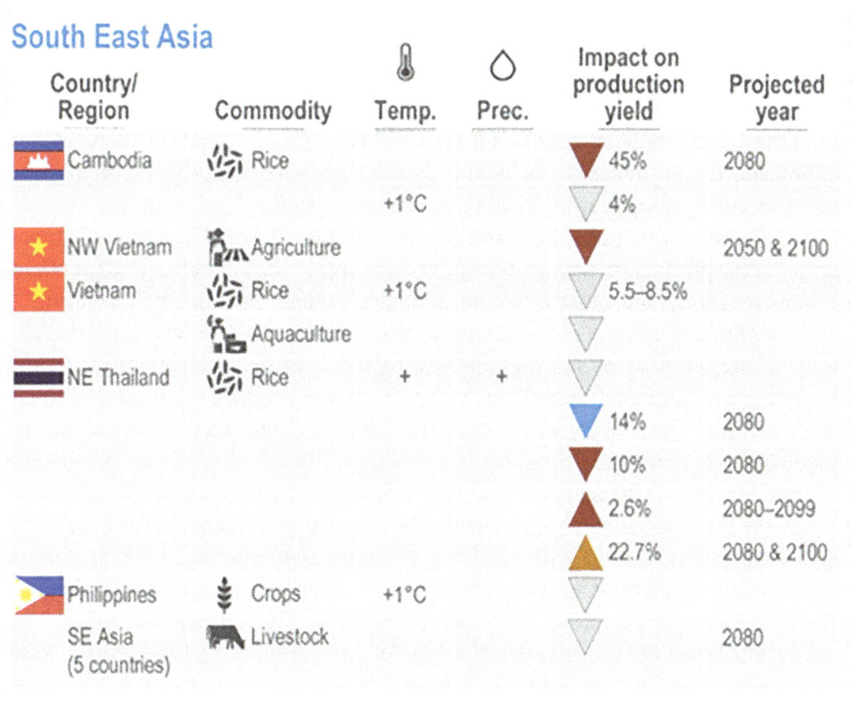

South East Asia

Country/Region	Commodity	Temp.	Prec.	Impact on production yield	Projected year
Cambodia	Rice			▼ 45%	2080
		+1°C		▼ 4%	
NW Vietnam	Agriculture			▼	2050 & 2100
Vietnam	Rice	+1°C		▼ 5.5–8.5%	
	Aquaculture			▽	
NE Thailand	Rice	+	+	▽	
				▼ 14%	2080
				▼ 10%	2080
				▲ 2.6%	2080–2099
				▲ 22.7%	2080 & 2100
Philippines	Crops	+1°C		▽	
SE Asia (5 countries)	Livestock			▽	2080

Fig. 4.8 Projection of the impacts of climate change on agriculture and food systems in the sub-regions of Asia, following the AR5 of the IPCC. (Source: Shaw et al. (2022; Fig. 10.6))

In most Southeast Asian countries, fisheries are small-scale activities, which are more vulnerable to the impacts of climate change than commercial sectors, with a general trend towards a decrease in the number of small units. The analysis of fishing-related activities shows a continuous decrease in catches, which will impact the future of the fisheries sector in the Philippines, Thailand, Malaysia, and Indonesia. Climate change is expected to reduce the overall production potential of fisheries due to a temperature increase of about 2 °C by 2050.

Just like fisheries, Asian aquaculture is highly vulnerable to climate change. The majority of shrimp farmers have observed that weather conditions have changed abruptly over the past five years and that high temperatures are the most damaging because they reduce growth rates, increase susceptibility to diseases, and affect farm productivity. Shrimp farms are also affected by changes in the variability and intensity of rainfall, perceived by the majority of farmers as part of the impacts of climate change. In Vietnam, small farmers are vulnerable to climate change: those who practice a type of extensive farming with few inputs are more fragile than those who follow a more intensive model, supported by more capital investments. Marine heatwaves pose a new threat to fishing and aquaculture, particularly in terms of disease spread. In the countries of Asia In Southeast Asia, it is predicted that over 30% of aquaculture zones will become unsuitable for production by 2050–2070, and that aquaculture production will decrease by 10–20% by 2050–2070 due to climate change.

Southeast Asia is home to nearly 15% of the world's tropical forests and includes at least four of the twenty-five global biodiversity hotspots. The region is also one of the main culprits for the disappearance of low-altitude and humid tropical forests. Thus, between 1990 and 2010, Southeast Asia recorded an average net loss of 1.6 Mha/year (0.6% per year), reducing the region's forest cover from 268 Mha to 236 Mha. Given these rates and the fact that over 90% of Southeast Asia's forests were still unprotected in the early 2010s, it is feared that more than 40% of the region's biodiversity may disappear by 2100 (Estoque et al. 2019).

A large number of published studies concern the predicted impacts of climate change on the production and economy of major crops: rice, maize, and wheat being the ones receiving the most attention. Climate change has, and will continue to have, a significant impact on agricultural production in various ways across Asia. An increasing number of sub-regional and regional studies using multiple modelling tools have provided significant evidence of the predicted impact of climate change on plant production, with clear indications of winners and losers depending on the countries concerned (Fig. 4.8).

While it is generally accepted that CO_2 promotes plant growth and productivity through the intensification of photosynthesis, uncertainty remains as to the extent to which carbon fertilisation will influence agricultural production in Asia. Indeed, plant nutrition is sensitive to rising temperatures, changes in water availability, and different adaptation measures employed in the field. The available literature indicates a clear trend towards the deterioration of grain quality and therefore a lower commercial value for rice grown in a high CO_2 environment. As global warming continues beyond 1.5 °C, the likelihood of negative impacts on agricultural and food security in many developing Asian regions will increase. There is a growing trend towards more integrated studies and modelling, which combine biophysical and socio-economic variables (including management practices) in a changing climate context, in order to reduce the uncertainty associated with future climate change impacts on the agricultural sector.

4.5.3 The Anticipated Climate Changes and Impacts in Central and South America

Lilian Blanc and Sylvain Schmitt

4.5.3.1 The Climate Changes in Central and South America

Central and South America are already experiencing the effects of global warming with an increase in cyclones, hurricanes, floods, droughts, sea-level rise, or the disappearance of glaciers. Beyond the social consequences (notably migratory movements), climate change also affects the supply of drinking water, food production (agriculture and livestock), and the natural ecosystems of this region, which contain 40% of the world's biodiversity and 25% of the world's forests. For this region, climate change amplify the net greenhouse gas emissions from the agriculture, forestry, and land-use change sectors, which are responsible for 25.3% and 20% of emissions respectively (Climate Watch 2022).

Temperature

In Central and South America, the increase in the greenhouse effect leads to a rise in average temperatures and average maximum temperatures of 2 °C to 6 °C depending on the regions and climate scenarios (Figs. 4.9 and 4.10), and a decrease in extreme low temperatures. The IPCC projects with certainty that these changes will affect all sub-regions of the continent with high to very high confidence. Moreover, the region currently has rates of temperature increase higher than the global average, an increase that is expected to continue in the future. This is a phenomenon with high confidence that affects (or will affect) all sub-regions of the continent.

Precipitation

The changes in precipitation in this region are less certain, but many studies predict increases in the northwest of South America and the Southeast, while decreases are expected in the Northeast and Southwest with very variable rates depending on the scenarios. These results are consistent for the end of the twenty-first century with the most intense increase scenarios (Fig. 4.9). Beyond trends, an increase in the

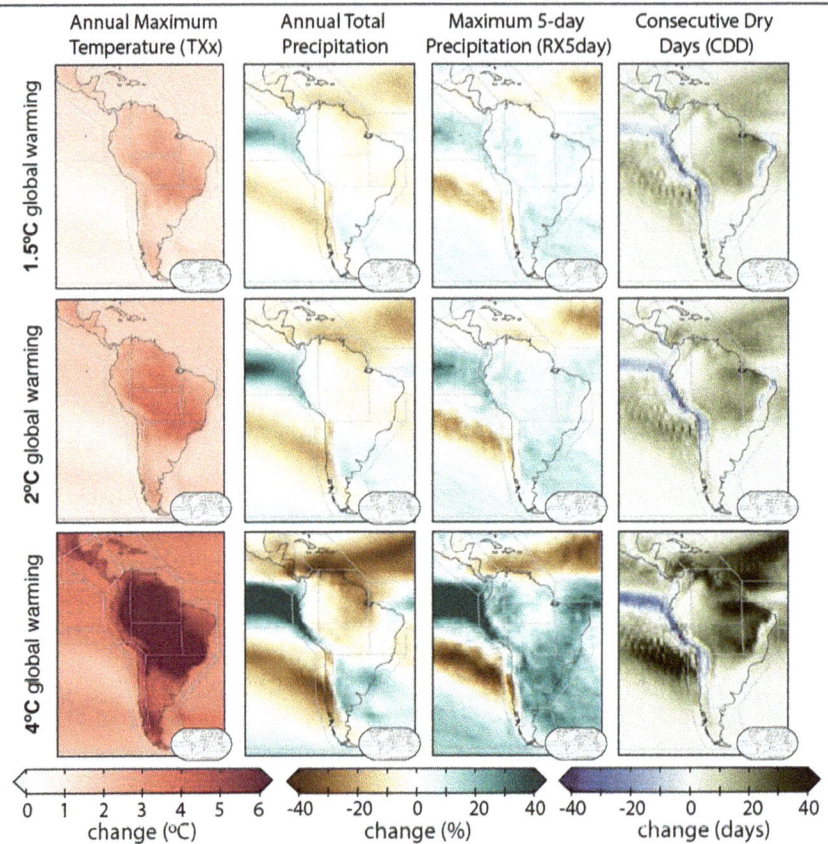

Fig. 4.9 Change in annual maximum temperature averages and precipitation in South and Central America under three global warming scenarios: +1.5 °C, +2 °C, and +4 °C compared to the 1850–1900 period. (Source: IPPC AR6 (2022))
The results are based on simulations from the CMIP6 multi-model ensemble (32 global models), using the SSP5–8.5 scenario to calculate warming levels.

intensity and frequency of extreme precipitation and floods is predicted in the north, northeast, southeast, and south of South America if the global temperature increase exceeds 2 °C (Fig. 4.9). These precipitations extremes lead to severe flooding and landslides. Conversely, in these regions, the number and frequency of dry days and droughts are also likely to increase. Southern Central America, and the south and southwest of South America are even expected to experience an increase in agricultural and ecological droughts by the mid twenty-first century . Finally, in the Andes, the loss of glacier volume and the thawing of permafrost will very likely continue under all IPCC scenarios, leading to large-scale glacial flooding and the overflow of glacial lakes.

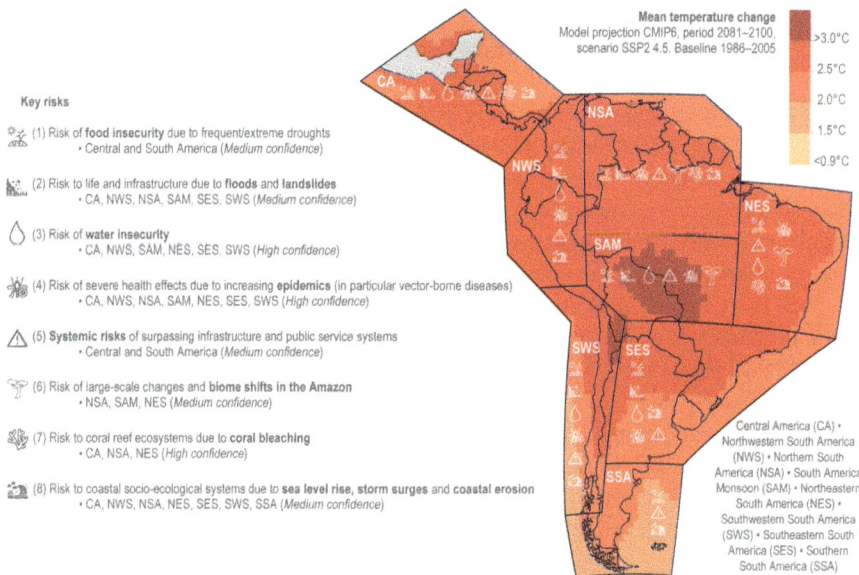

Fig. 4.10 Synthesis of the main risks related to climate change for South and Central America. (Source: IPPC AR6 (2022))
The base map indicates the average temperature change between the SSP2–4.5 scenario using the CMIP6 model projections for 2081–2100 and a reference period of 1986–2015.

Fires

The rise in temperatures, aridity, and more intense drought episodes create weather conditions conducive to fires, which are more frequent and intense, as illustrated by the recent mega-fires that have affected the Amazon. These fire-prone weather conditions are expected to increase in southern Central America and southwestern South America. On average, the inhabitants of the region have been more exposed to a high risk of fire, between 1 and 26 additional days depending on the sub-region, for the years 2017–2020 compared to 2001–2004 (Romanello et al. 2021). This increase in fires and aridity will potentially impact a wide range of sectors (including agriculture, forestry, health, and ecosystems).

Sea Level

In addition to the higher than average global temperature rise in South America, the relative sea level has risen more rapidly than the global average in the South Atlantic and subtropical North Atlantic, and less rapidly in the Eastern Pacific. This rise is expected to continue with a high degree of certainty (Fig. 4.10), contributing to the increase in coastal flooding in low-lying areas and the retreat of the coastline along most sandy coasts.

4.5.3.2 The Impacts of Climate Change on Various Sectors of Agriculture and Forestry in Central and South America

This region plays a major role globally both for biodiversity conservation (particularly with marine and terrestrial ecosystems in the Amazon region) and for agriculture with the production of agricultural commodities (livestock, coffee, bananas, sugar, soybeans, corn, sugarcane). This region is also characterised by two major factors that make it particularly vulnerable to climate change. On the one hand, significant land use changes (deforestation in the Amazon region) form a positive feedback loop with climate change, accentuating its effects. On the other hand, the region is characterised by strong social (access to land, for example) and geographical (urban *vs* rural) inequalities, with a high percentage of the population living below the poverty line.

Climate change has had a significant impact on agricultural production sectors in most regions with the magnitude of precipitation changes, extreme temperatures, and the changes in the agricultural calendar. Agriculture, livestock farming, fishing, and food systems in general have experienced medium to high impacts from climate change. In all regions, reductions in the growing season of crops have been observed for wheat (winter and summer), rice, maize, and soybeans, between 1981 and 2019. This reduction amounts to 10% for rice in South-West America and 5% for maize in Central America. The reduction in growing periods decreases yields.

For the three production sectors (annual and perennial crops, and livestock), however, strong regional variabilities are observed. In the northeast of South America, dry spells, with high temperatures and a decrease in precipitation, have been damaging to these three agricultural sectors. Other regions have also been affected such as Central America, the southeast and southwest of South America (annual crops), and the extreme south of the continent (livestock). In the future, these three production sectors will be greatly affected in the northeast and northwest regions, and to a lesser extent in the Cerrado region (centre of the continent). In this region, the yields of soybeans and maize will suffer one of the strongest negative impacts according to the estimates of the RCP4.5 and RCP8.5 scenarios and will require high levels of investment in adaptation if they continue to be grown in the same areas as currently (high confidence).

In Central America and the Andean foothills, annual crops will also be greatly affected. These effects are expected to worsen the situation for small and medium farmers and for indigenous populations in the mountainous areas of these regions. Under the A2 scenario (future with high GHG emissions leading to warming of 3.2 to 5.4 °C with regionalised economic growth and a strong increase in population), yield reductions in 2050 are estimated at 19% (beans) and between 4% to 21% (maize) for Central America. Food security in these two regions will therefore be largely compromised (high confidence).

The Amazon rainforest, one of the world's largest reservoirs of biodiversity and carbon, has already been and will continue to be highly vulnerable to drought (high degree of confidence). The Amazon rainforest has thus been heavily affected by the unprecedented droughts and high temperatures observed in 1998, 2005, 2010 and

2015/2016, which are partly attributed to climate change, causing an excess mortality of trees. The consequences of these strong climatic anomalies are an excess mortality of trees and a reduction in forest productivity at the basin scale. These strong changes in the functioning of forests temporarily transform forests from carbon sinks to carbon sources (high confidence). In addition, the combined effect of deforestation causing a change in land use, forest degradation (Bourgoin et al. 2024; Le Roux et al. 2022) and climate change increases the vulnerability of these terrestrial ecosystems to extreme climatic events (Phillips et al. 2009; Bennett et al. 2023). With these changes, the probability of transition, i.e. a transition from the tropical forest to other forest systems (seasonal forest) or savannah-like systems, is now assessed with a medium degree of confidence.

4.5.4 The Projected Climate Changes and Impacts in the Small States and Island Territories of Oceania

Séverine Bouard, Catherine Sabinot, Pierre-François Duyck, Philippe Birnbaum, Audrey Leopold, Julien Drouin, Fabian Carriconde, Laurent L'Huillier, and Christophe Menkès

4.5.4.1 An Uncertain Balance in the Face of Climate Change

The territories and countries of Oceania present a great diversity in terms of geography, geological history, size, topography, and local climatic conditions. However, this diversity, which gives rise to a unique cultural, biological, and food richness, also exposes them to a high vulnerability to climatic hazards. A single cyclone, a year of drought, or a biological invasion can wipe out the agricultural production of an entire country and eradicate numerous endemic species.

Very distant from each other, they have very variable sizes[4], leading Hau'ofa to name the region "our sea of islands" (Hau'ofa 2013) (Fig. 4.11). Located in a geologically very active region of the world, these archipelagos have complex geological histories, leading to a consequent diversity of their geological and pedological substrates. The Melanesian countries are rather large and mountainous, rich in mineral resources, while the Polynesian and Micronesian islands range from small volcanic islands to coral atolls. These islands have a great diversity of terrestrial and reef ecosystems, hosting a very rich biodiversity, a source of food for the communities.

If the island countries of the Pacific only represented 0.06% of the total carbon dioxide emissions in the world in 2019[5], they are particularly affected by climate change, even questioning their existence, as some are threatened with submersion (notably Kiribati, Tuvalu and the Marshall Islands).

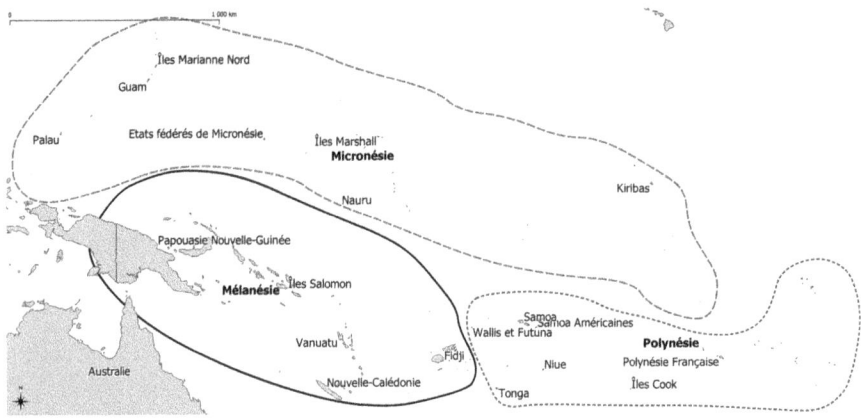

Fig. 4.11 The three major sub-regions of the South Pacific islands. (Source: map created by Jonas Brouillon, New Caledonian Agronomic Institute (IAC))

4.5.4.2 Insularity, a Geographical Challenge that Complicates Climate Projections

In this "ocean of islands", the most probable future scenarios are between SSP2–4.5 and SSP3–7.0 (IPCC 2022). However, the global climate models used in the Pacific and employed in the IPCC reports present numerous biases and uncertainties, particularly concerning the uncertain simulations of the South Pacific Convergence Zone (SPCZ) (Brown et al. 2020) regardless of the scenarios (Fig. 4.12).

These uncertainties have repercussions on the future evaluation of climate events such as El Niño-Southern Oscillation (ENSO), cyclones, heatwaves, heavy rainfall and droughts. In the South Pacific, studies predict a trend towards a decrease in the number of cyclones (with low confidence, however) in the scenarios considered realistic, but without consensus on the intensity of future cyclonic winds of these phenomena. However, the intensity of cyclonic rainfall will increase (Dutheil et al. 2020, 2019; Walsh et al. 2012, 2020; Knutson et al. 2020). Regardless of the scenarios, the expected increase in the number of heatwaves, the duration of heatwaves and temperature records is almost certain (Power and Delage 2019), but remains difficult to quantify due to model biases. Indeed, the IPCC models do not take into account local phenomena resulting from the interaction with the complex topography of the islands, as their grid is too coarse, on the order of a hundred kilometers (Evans et al. 2024). These gaps compromise the ability to reliably estimate the future evolution of precipitation, droughts and heatwaves.

---

[5] Data from the World Bank Wallis and Futuna which does not appear in this database. With Australia and New Zealand, this figure rises to 1.3% in 2023 (https://data.worldbank.org/indicator/EN.GHG.ALL.MT.CE.AR5). The figures show the significant contribution of the mining sector to emissions. With 6.614 Mt of CO_2 for 270,000 inhabitants in New Caledonia versus 3.401 Mt of CO_2 for 918,000 inhabitants in Fiji (latest available data from 2023).

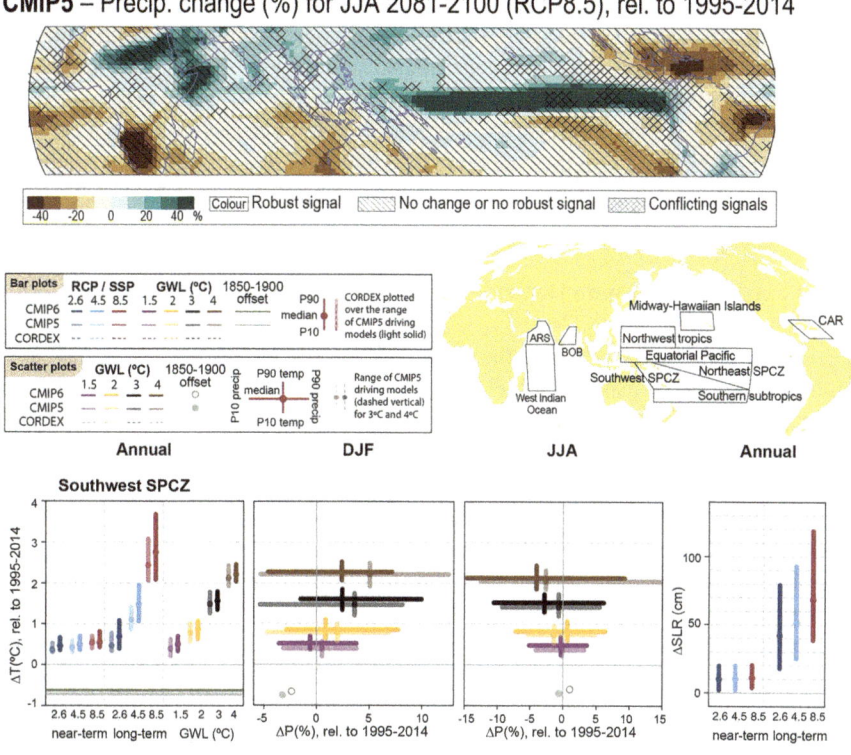

Fig. 4.12 Extract from the IPCC atlas for the SPCZ area
Top panel, expected average change (in %) of precipitation in the tropical Pacific, with hatched non-significant areas (including the SPCZ). Bottom panels: average changes in the annual average surface air temperature, precipitation and sea level rise relative to the 1995–2014 baseline in the Southwest SPCZ area. These changes are given annually or on a seasonal basis—December-January-February (DJF), June–July-August (JJA)—for the different scenarios (RCP/SSP), but also for warming levels (GWL: Global Warming Level), and for the CMIP6 and CMIP5 exercises (colours). The bars represent 90% of the distributions. (Source: IPCC (2021))

The few climate projections for the region, at a higher resolution, nevertheless indicate a general drying of the Central Pacific (Dutheil et al. 2020). Extreme rainfall events could increase in frequency from an occurrence every twenty years (period 1986–2005) to an event every ten, or even 4–6 years by 2090, according to certain greenhouse gas emission scenarios. Their intensity should increase. An increase of 0.5–1.3 °C in the average atmospheric temperature has been observed on the islands of the Western Pacific since the pre-industrial period (Whan et al. 2014). Depending on the emission scenarios, this increase in the average temperature could reach 2 °C to 3.7 °C by 2090 (Fig. 4.12). A sea-level rise of 20 cm to 120 cm by 2100 is projected according to scenarios, compared to the 1995–2014 averages (Fig. 4.12). The rise in sea level varies significantly between the archipelagos of the region. Importantly, this observed rise is not entirely attributable to changes in sea level due to climate change, as many Pacific islands also experience tectonic subsidence (Martínez-Asensio et al. 2019). Low-lying islands such as Kiribati, Tuvalu, and the Marshall Islands are already threatened with disappearance. The first bilateral treaty offering a migration path for Tuvaluans to Australia was signed in November 2023. In the Fiji Islands, a sea-level rise of nearly 5 mm per year is observed. This increase will thus heighten the risks of coastal flooding and intensify coastal erosion caused by cyclones, extreme distant swells, or tsunamis.

4.5.4.3 A Strong Climate Dependence of Ecosystems and Agrosystems

The Pacific encompasses five of the thirty-six "global biodiversity hotspots" and exhibits a high endemism of its biological resources. This is the result of the geographical isolation of these islands, the diversity of available habitats, and specific evolutionary processes. These territories host numerous remarkable plant and animal species found nowhere else (Payri and Vidal 2019). However, sea-level rise, extreme weather events, temperature changes, and changes in precipitation patterns threaten the specific habitats of these species.

A simulation based on nine climate change scenarios predicts that the distribution range of 87–96% of native and endemic tree species in New Caledonia will have decreased by 2070, that 52–84% of species will lose at least half of their current distribution range, and that 0–15% will disappear (Pouteau and Birnbaum 2016). Moreover, the intensive fragmentation of these ecosystems and the invasions of exotic species will disrupt the ecological balance in which these endemic species have developed. Climate change, in interaction with the increase in passenger flows and trade exchanges, changes the geographical distribution of pests and crop diseases, but also directly modifies their impacts (Bale et al. 2002; Hulme 2011).

The islands face multiple threats to their terrestrial resources (Nurse et al. 2014; Wong et al. 2014) and particularly to their soils. The effects of temperature increases are multiple and can affect their fertility and health (Filho et al. 2023). Soil organic matter (SOM) is crucial for soil fertility (see Chap. 17), but its stock can decrease with increasing temperatures which favour the mineralisation of SOM, affecting the water retention capacity of soils, their porosity, and can make them more susceptible to erosion. The intensification of rainfall events increases the disintegration, leaching (loss of nutrients) and acidification of soils, while droughts and sea-level rise increase salinity, reduce water availability and promote desertification. Climate change can also alter soil biodiversity. Thus, the soils of island territories, often naturally infertile (New Caledonia, Pillon et al. 2021) and with unique biodiversity (Carriconde et al. 2019), agricultural and forestry systems could be particularly transformed. While the soils of the region and their functioning remain largely unknown despite recent work on the subject (such as Carriconde et al. 2019; Demenois et al. 2020; Kulagowski et al. 2021; Léopold et al. 2021), limiting the effects of climate change on agroecosystems can be achieved by relying both on the knowledge already tested by the populations and on that brought by the scientific world. A dynamic of co-creation of knowledge will be essential to ensure sufficient resilience of agricultural productions.

4.5.4.4 Resilient Pacific Island Societies... But to What Extent?

The agricultural systems of the Pacific island states and territories are mainly based on family-type farming, with production primarily intended for local consumption. Staple food crops include yam, taro, sweet potato, cassava, bananas, tropical fruits, and vegetables. Cultivating these crops with contrasting water needs relies on techniques adapted to local conditions, different in Melanesia, Polynesia, and Micronesia. The indigenous knowledge associated with these practices is rich and diverse, but it is being undermined by transformations in lifestyles (migration and work reorganisation, etc.), with the reduction of available land and with the new climatic constraints and Environmental. The production systems in agriculture and livestock farming for the local market and export (kava, cocoa, spices, sugar cane, cattle, etc.) are also subject to the effects of climate change. Changes in the soil, the water demands of these systems, and pressures from pests and diseases jeopardise the financial resources of households that depend on them, as well as those of these countries with already unbalanced trade balances. Even capital investments that allow for a relative artificialisation of growing conditions (greenhouses, soil-less cultivation) are vulnerable to the most violent climatic events. As for livestock farming, efforts to tropicalise cattle herds, as in New Caledonia, could make these production systems relatively resilient, but the management of pastures, still unsophisticated, subject to the pressure of invasive plant species, and not agile in the face of the irregularity of the lengthening and intensification of droughts, will hardly achieve a production satisfying local needs.

Climate change will have significant repercussions on agricultural and food systems, with consequences for food security, health, and economic development (Barnett 2020; McCubbin et al. 2015; Klöck and Fink 2019). The intensification of floods and droughts has short term consequences—natural disasters can destroy entire crops and decimate livestock—as well as long-term effects on food production. These phenomena also accelerate the loss of arable land already occurring in some areas due to coastal erosion, sea-level rise, and saltwater intrusion. Combined with the processes described above, the cultivated and cultivable areas are likely to decrease, as is the diversity of plants adapted to these new climates and environments.

However, the inhabitants of the Pacific archipelagos have different local experiences of the scientific vision of climate change (Fache et al. 2019). Environmental and social changes are perceived as very closely linked and must be addressed simultaneously (Pascht 2019). The inhabitants have already developed practices and diversification strategies to limit damage to their crops, to cope with losses, and to adapt to environmental constraints. Subsistence strategies based on multi-activity are particularly called upon after natural disasters (see Chap. 5). Turning to the lagoon and its fishing resources is common following the destruction of crops and allows for a new production cycle to be launched immediately after the event. However, the frequency of devastating events and the concomitant transformation of terrestrial, marine, and coral ecosystems significantly weaken the security offered by multi-activity. This raises questions about the risks, challenges, and resilience of these inhabitants in the face of intense climatic hazards and prompts us to question the local and global policies to be implemented[6].

The small states and island territories of the Pacific, often overlooked by the major states, while they record the most significant negative impacts of climate change, face an urgent and vital need for more research and data.

4.5.5 The Anticipated Climate Changes and Impacts in Europe

Nicolas Viovy

4.5.5.1 The Climate Changes in Europe

Temperature

The first effect directly linked to the greenhouse effect resulting from the increase in CO_2 in the atmosphere is the rise in temperatures. This is a phenomenon on which we have a high degree of certainty and which will affect the entire continent. The increase in temperatures is faster on continents than on the ocean, and faster in temperate and boreal zones than in the tropics, so this temperature increase in Europe will be greater than the global average. Added to this phenomenon is the fact that air quality improvement policies since the 1980s, which have significantly reduced pollution, have reduced the amount of aerosols. However, these aerosols have a cooling power, so their reduction has contributed to the recent temperature increase. Thus, the temperature increase in France over the last decade has reached 1.5 °C since the pre-industrial era (1860) compared to 1.1 °C on a global average. If we look in more detail at the spatial and seasonal distribution of future temperature increase, we see a greater increase in the north and centre of Europe than in the south during winter, while this increase will be stronger around the Mediterranean during summer. This greater increase in the North and Centre in winter is linked to the snow feedback whose coverage will significantly decrease with the rise in temperatures. Surfaces will thus be darker (decrease in albedo) and absorb more energy. The increase in temperatures in the South during summer is linked to a decrease in evapotranspiration and therefore its cooling effect (the soils being drier). Climate change, accompanied by an increase in climate variability, the increase in average temperatures combined with a higher probability of extreme hot events will lead to a significant increase in heatwave periods, which we have already been observing for several years (for example 2019, 2020, 2022 and 2023).

---

[6] The Clipssa project, built on a partnership involving in particular AFD, IRD, Météo-France, IAC, Criobe and the communities of the overseas territories and Vanuatu, is interested in these issues: www.clipssa.org.

Precipitation

Even though the uncertainty related to changes in precipitation is greater than for temperatures, there is a consensus on a contrasting response between southern and northern Europe. During the winter, a large part of Europe should see an increase in its precipitation, while it will decrease around the Mediterranean. During the summer, a larger part of Southern Europe and a majority of Central Europe should see a decrease in precipitation, while from Germany to Scandinavia, precipitation should increase.

Extreme Events

Regarding extreme events, even though we should observe an increase in periods of summer droughts across a large part of Europe (except the extreme south and north) due to the combined effect of decreases in summer precipitation and higher temperatures, conversely, an increase in intense rainfall events is projected (Fig. 4.13). This phenomenon is explained by the increase in oceanic and atmospheric temperatures, which will lead to an increase in evaporation, and the increase in the capacity of the atmosphere to hold a larger amount of vapour. The amount of precipitable water will then be greater for given weather conditions. The indirect consequence will then be an increase in the frequency of floods.

Winds

There is no clear trend emerging in terms of the risk of storms. Even though their frequency should not increase, it is possible that an increase in their intensity may occur, particularly due to the accentuation of thermal contrasts between the ocean and the continents. This remains hypothetical and it is therefore difficult to conclude on the evolution of storm risk.

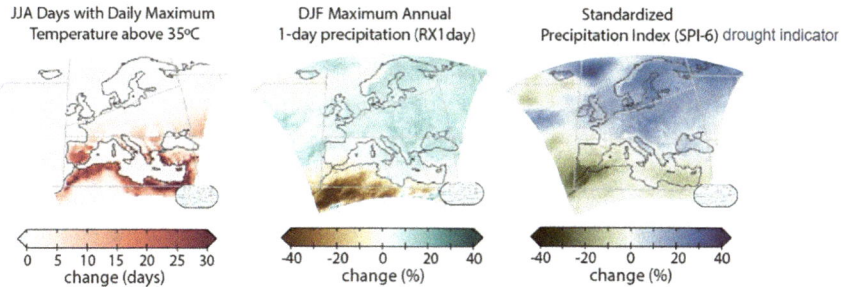

Fig. 4.13 Change in heatwave, heavy rain and drought indicators by 2050 for Europe. (Source: IPCC AR6 (2022))

Sea Level

The increase in temperatures has an impact on the rise in sea levels due to the melting of polar ice caps and the expansion of the oceans in connection with the increase in temperature (the proportion of these two factors being, at present, roughly equivalent). This concerns all European coastal areas. Due to the great inertia of the ocean, the rise in water levels will continue, even if we manage to stabilise atmospheric CO_2, long time after CO_2 begins to decrease. The sea level has risen by about 23 cm since 1880 and should gain another 30 cm by 2050 (IPCC 2013). The effect will be especially noticeable during episodes of storms combined with high tides which will increase the probabilities of submersion. It will also contribute to accentuating coastal erosion.

4.5.5.2 The Consequences of Climate Change on the Different Sectors of Agriculture in Europe

The effect of climate change will be contrasting, particularly between the south and the north of Europe. The combination of increasing temperatures and decreasing summer precipitation, particularly in southern Europe, will lead to an increase in periods of agricultural drought, especially noticeable around the Mediterranean. This will result in an increase in water stress and also thermal stress during the summer period, which will affect all agricultural sectors. Conversely, the increase in temperatures will allow for an earlier start to the vegetative cycle and an extension of this cycle. The impact will therefore be contrasting, both according to the type of crop and the regions (Fig. 4.14). With current agricultural practices, winter crops like wheat—which is relatively insensitive to summer drought, but can benefit from an earlier start—should see their yield increase in northern Europe, while they should slightly decrease in the Centre and the South. Similarly, soy, except in the most pessimistic scenario, should be able to be grown over a large part of Europe, where it is currently limited to a latitude of 50° N. Only Spain would see its yields decrease. However, maize, which is very water-demanding and grows during the summer, could only maintain its yields over a large part of Europe at the cost of a significant increase in irrigation which seems difficult to achieve and would lead to a strong usage tension on the water resource, already noticeable today. In terms of adaptation, new crops, like sorghum for animal feed, requiring less water and well adapted to high temperatures, could be developed particularly in the south and the centre of Europe. The increase in periods of heatwaves and droughts will also strongly affect the forestry sector, where an increase in diebacks is already being observed, particularly for species like Douglas fir or beech, which are very sensitive to water stress. Droughts and heatwaves also significantly increase the risk of fire which will then become possible over a large part of Europe. In 2022, a significant fire occurred in the Armorican forest, which had never happened before. For the livestock sector, droughts will decrease summer forage production. This

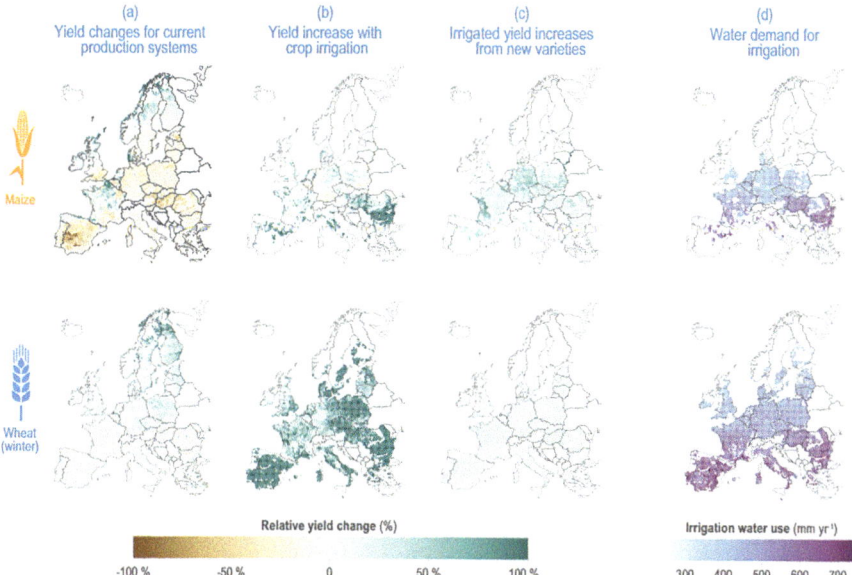

Fig. 4.14 Changes in wheat and maize yields in Europe for current practices and considering adaptations with associated water demands for a 2 °C temperature increase scenario. (Source: Bednar-Friedl et al. (2022))

phenomenon could however be partly compensated by earlier growth in spring and the possibility of grazing herds earlier and making summer forage stocks, which were usually intended for the winter period which on the contrary will shorten. But the increase in heatwaves will also pose a direct problem for livestock. The enteric fermentation of ruminants being exothermic, they are very sensitive to high temperatures, forcing them to reduce grazing in high heat, This reduces the production of meat or milk accordingly.

Another phenomenon linked to the increase in temperatures is paradoxically the increased risk of frost at the beginning of the vegetative cycle. Indeed, although the dates of the last frosts are becoming earlier and earlier, the advancement of the vegetative cycle linked to the increase in winter temperatures is faster than that of the period of the last frosts, thus increasing the risk that a frost occurs during the critical phases of bud burst and flowering. Vineyards and early fruit trees are particularly vulnerable in this regard.

4.5.5.3 Contribution of the Agricultural Sector to the Greenhouse Gas Balance

It is estimated that agriculture is responsible for 11.4% of greenhouse gas emissions at the European level, out of a total of 3.31 billion t eqCO_2. The share of greenhouse gas emissions from agricultural sources varies from one member state to another.

France is the country that emits the most in this sector, with 85 Mt of CO_2, followed by Germany (66 Mt) and Poland (57 Mt). These emissions are linked to various factors. There are of course CO_2 emissions related to fuel consumption across the farm and for the manufacture of inputs and phytosanitary products, but a large part of the emissions are also linked to methane (CH_4) produced by ruminants and nitrous oxide (N_2O) linked to fertilisers. The emissions of CH_4 are estimated at 5 Mt and those of N_2O at 660 Mt. As mentioned in the introduction, although these emissions are low, the global warming potential of CH_4 and N_2O is much greater than that of CO_2. But the lifespan of these different molecules also needs to be taken into account. Agricultural practices also affect the CO_2 balance by altering soil carbon. For example, in France (the situation is similar in the rest of Europe), areas of large-scale farming have already lost a large part of their soil carbon and continue to be a very slight source of carbon at 0.06 t eqCO_2/ha/year. But agriculture can also contribute to the sequestration of carbon in soils and thus to the mitigation of climate change. In France, grassland and forest soils have significant soil carbon contents (respectively 84 t/ha and 81 t/ha) and represent a carbon sink (from 0.37 to 0.80 t eqCO_2/ha/year for grasslands, and from 1.60 to 5.06 t eqCO_2/ha/year for forests depending on the studies) (EFESE Report 2019; INRAE Report 2020). Therefore, it is especially important to maintain existing stocks in grasslands and forests. Thus, reforestation, particularly linked to the abandonment of old agricultural lands over the last few decades (+ 9% in thirty years), has contributed to making the continent a carbon sink (excluding anthropogenic emissions). In general, large-scale crops, due to their low amount of carbon in the soil, offer significant storage potential. The most promising avenues are agroecology, the use of intermediate crops and the reconstitution of hedges (INRAE Report 2020). Even though concrete achievements are still very underdeveloped (compared to the United States for example), the European approach is much more systemic. Indeed, nature-based solutions must follow the recommendations of the IUCN by taking into account the impact of the proposed solutions on all ecosystem services.

4.5.6 The Anticipated Climate Changes and Impacts in North America

Nicolas Viovy

4.5.6.1 The Climate Changes in North America

Temperature

The increase in temperature will affect the entire North American continent (Lee et al. 2021) (Fig. 4.15). This temperature increase in North America will be greater than the global average. Upon closer examination of the spatial and seasonal

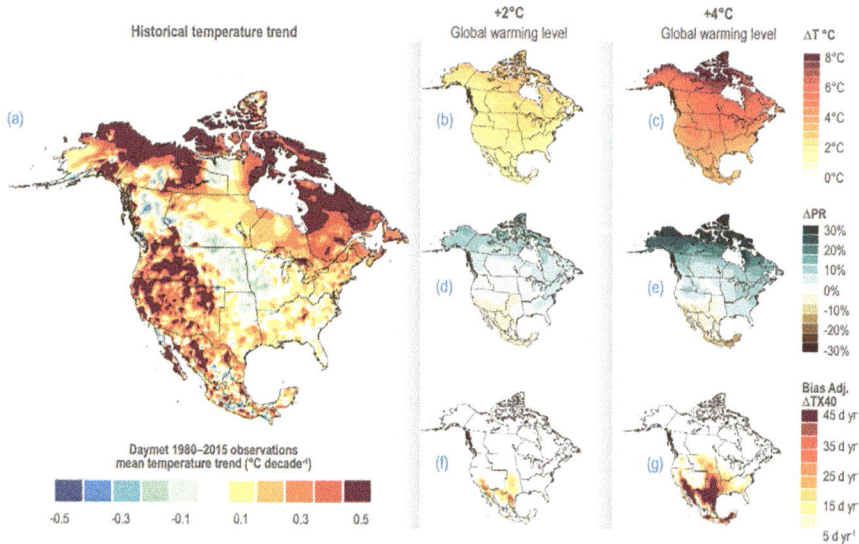

Fig. 4.15 Observed and projected climate change for North America. (Source: Hicke et al. (2022)) (**a**) Observed temperature trend from 1980 to 2015; (**b** and **c**) temperature change for scenarios +2 °C and +4 °C (ΔT); (**d** and **e**) change in precipitation (ΔPR); (**f** and **g**) change in the number of days with a temperature above 40 °C (ΔTX40) (Hsiang et al. 2017)

distribution of future temperature increase, an increase that grows with latitude and will also be more significant in the centre of the continent than on the coastal areas is observed. This stronger increase in the North and Centre in winter is linked to the snow feedback, whose coverage will significantly decrease with the rise in temperatures. The surfaces will thus be darker and absorb more energy. This also explains why the temperature increase will be particularly strong in winter in the northern part of the continent. In the scenario of a global average increase of 4 °C compared to the pre-industrial era, the increase in summer temperatures would also become very significant. Climate change, accompanied by an increase in climate variability, the increase in average temperatures combined with a higher probability of extreme hot events will lead to a significant increase in heatwave periods. This increase in temperatures and heatwaves will result in an increased risk of fire, particularly in the boreal zone. The 2023 fires in Canada, for example, destroyed 17 Mha. In the United States, the central west part should experience the greatest increases, both in the south (such as Arizona or Texas) and further north (in Montana and Dakota). Furthermore, the increase in the number of days where the temperature exceeds 40 °C in Texas or Arizona compared to the historical period could reach about thirty days per year (Lee et al. 2021). The average temperature increase being stronger in high latitudes than in the tropics, the latter will be even more pronounced in Canada than in the United States. In particular, a significant decrease in the freezing of soils, lakes and rivers can be expected, with consequences on soil stability (linked to the melting of permafrost) or on the modification of transport systems (the installation of camps or temporary roads on frozen lakes becoming impossible).

Precipitation

Even though the uncertainty related to precipitation change is greater than for temperatures, the emerging trends are an increase in winter precipitation, strong in the North, moderate in the Centre, but a significant decrease in precipitation over California, Arizona and New Mexico. During the summer, precipitation should increase in the North, but on the contrary decrease in the Centre. In Canada, precipitation should increase on average, particularly in the North.

Extreme Events

Regarding extreme events, an increase in periods of summer droughts should be observed, particularly across the entire central part to the west of the United States in line with heatwave areas, linked to the combined effect of decreases in summer precipitation and higher temperatures. Conversely, the IPCC modellers project an increase in intense rainfall events (Lee et al. 2021), which should be particularly marked in the north-west and north-east of the continent with the indirect consequence of an increase in flood frequencies. This phenomenon is explained both by the increase in oceanic and atmospheric temperatures, which will lead to an increase in evaporation, and by the increase in the atmosphere's capacity to hold a larger amount of vapour. The amount of precipitable water will then be greater for given weather conditions.

Winds

The United States is subject to frequent tornadoes. While it is difficult to estimate whether their number or intensity are likely to increase, it is very clear that there is a shift of the "tornado alley" towards the east of the continent, which should intensify in the future. This can be linked to the drying out of the central west mentioned above. Similarly, the southeast of the United States, particularly Florida, can be affected by hurricanes. However, while there is no emerging trend on a change in hurricane frequency, an increase in their intensity is projected. Similarly, the increase in ocean temperatures should lead to a further northward shift of cyclones, which could also affect the north of the east coast.

4.5.6.2 The Consequences of Climate Change on the Various Sectors of Agriculture in North America

The effect of climate change will be contrasted between the east and the west of the continent. In the East, climate change should lead to a notable decrease in yield and more generally in the South, while in the West some areas could see their yield increase (Fig. 4.16). However, it should be noted that it is precisely the large

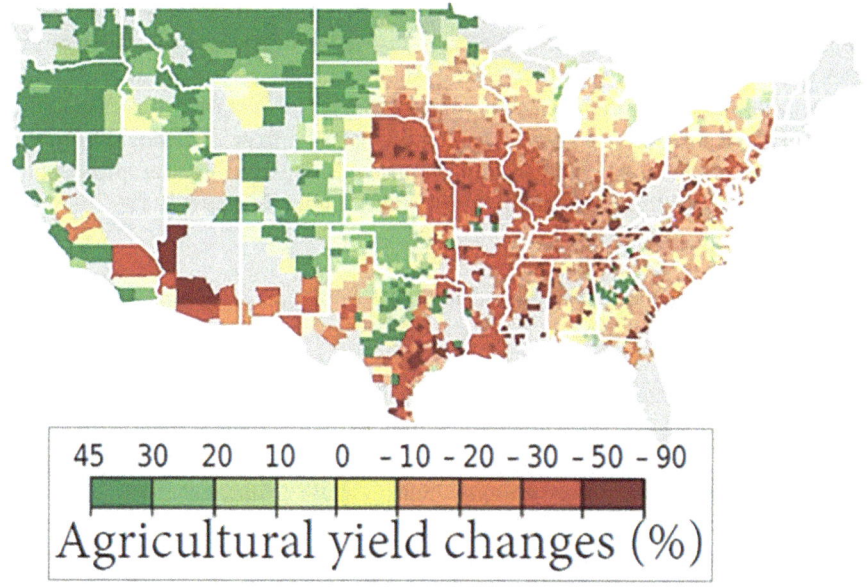

Fig. 4.16 Projected evolution of agricultural yield (all crops combined) between 2080 and 2099, under an RCP8 scenario. (Source: Hsiang et al. (2017))

agricultural plains of the Midwest, including the Corn Belt, which are expected to be heavily affected, while the areas where the yields could increase are not currently significant crop areas. This global view actually masks a wide disparity depending on the type of crop. Corn, particularly sensitive to summer drought due to high water demand and development in mid-summer, will be the most affected. Soybeans, which are also a summer crop, will also be negatively impacted although a little less than corn. On the other hand, wheat, which matures before the periods of drought, should benefit from an earlier emergence and thus see its yield increase. The maintenance of corn cultivation would require a significant increase in irrigation, but at the cost of an increase in available water volumes which would further accentuate the strong tensions on the already very constrained water resource. Climate change will also severely affect the forestry sector across the entire North American continent, both through increased fire risks and also through changes in species distribution ranges. In the centre, the increase in periods of heatwaves and droughts will heavily affect the forestry sector, where an increase in dieback is already being observed. Only the northern coastal areas, which are not limited in water and benefit from an extension of the growing period, should on the contrary benefit from climate change. For the livestock sector, droughts will decrease summer forage production. As in Europe, this phenomenon could however be partly compensated by earlier growth in spring and the possibility of grazing the herds earlier. But the increase in heatwaves will also pose a direct problem for livestock. The enteric fermentation of ruminants being exothermic, they are very sensitive to high temperatures, forcing them to reduce grazing in high heat, thereby reducing meat or milk production.

Another phenomenon related to the increase in temperatures is, paradoxically, the increase in the risk of frost in the early vegetative cycle. Indeed, although the dates of the last frosts are becoming earlier and earlier, the advancement of the vegetative cycle linked to the increase in winter temperatures is faster than that of the period of the last frosts, thereby increasing the risk of a frost occurring during the critical phases of budburst and flowering. Consequently, as on the European continent, early vine and fruit trees are the most vulnerable in this regard.

4.5.6.3 Contribution of the Agricultural Sector to the Greenhouse Gas Balance

The United States Department of Agriculture (USDA) estimates that agriculture is responsible for 11% of greenhouse gas emissions in the United States and 8% in Canada, out of a total of approximately 6.6 billion tonnes of CO_2 equivalent. This percentage of emissions linked to agriculture is close to the European fig (9%). Emissions per kilogram produced are, for example, around 30 kg CO_2 eq/kg for cattle (in carcass weight and not live weight) compared to 0.2 kg for cereals on average across the continent (FAOSTAT 2022 data[7]). Greenhouse gas emissions from the agricultural sector per capita are much higher in North America than in Europe (1.2 t CO_2 eq in the United States, 1.5 t CO_2 eq in Canada, compared to 0.87 t CO_2 eq in Europe) (Climate watch 2024 data[8]). This is due to both the ratio between plant and animal production (animal production being much more emitting), but also to the quantity produced per capita, which is higher in North America (around 2.9 t/inhab. Compared to about 2.3 t/inhab. in Europe). These emissions are linked to various factors. There are of course CO_2 emissions related to machinery, the manufacture of inputs and phytosanitary products, but a large part of the emissions are also linked to methane (CH_4) emissions produced by ruminants and rice fields, and nitrous oxide (N_2O), for the manufacture and use of chemical fertilisers. Annual emissions for the United States are estimated at 20 Mt for CH_4 and 930 Mt for N_2O. Although these emissions are low, the global warming potential of CH_4 and N_2O is much higher than that of CO_2 (28 times for CH_4, 273 times for N_2O), their effect on the radiative balance is very significant. It is also important to keep in mind the lifespan of these different molecules in the air. CH_4 has a lifespan of about 10 years; this duration is not negligible, but less problematic than that of N_2O which is 120 years. Agricultural practices also affect the CO_2 balance by modifying soil carbon. Therefore, it is especially important to maintain existing stocks in grasslands and forests. Even if agricultural practices are not to blame, the mega-fires of recent years in the west of the continent have destroyed large areas of forests. For example, the

---

[7] https://openknowledge.fao.org/server/api/core/bitstreams/cc09fbbc-eb1d-436b-a88a-bed42a1f12f3/content

[8] https://www.climatewatchdata.org/.

mega-fires of 2023 in Canada emitted 431 Mt of carbon (compared to 59 Mt on average each year), and significant reforestation programmes are planned in the United States which should help to rebuild some of the stocks. Unmanaged reforestation (i.e., linked to the abandonment of old agricultural lands over the last few decades), even though it is much weaker than in Europe for example, has partially contributed to making the continent a carbon sink (excluding anthropogenic emissions).

References

ASEAN State of Climate Change Report (ASCCR) (2021) Jakarta, ASEAN Secretariat. https://asean.org/wp-content/uploads/2021/10/ASCCR-e-publication-Correction_8-June.pdf

Baccini A, Goetz S, Walker W et al (2012) Estimated carbon dioxide emissions from tropical deforestation improved by carbon-density maps. Nat Clim Chang 2:182–185. https://doi.org/10.1038/nclimate1354

Bale JS, Masters GJ, Hodkinson ID, Awmack C, Bezemer TM, Brown VK et al (2002) Herbivory in global climate change research: direct effects of rising temperature on insect herbivores. Glob Chang Biol 8(1):1–16

Barnett J (2020) Climate change and food security in the Pacific Islands. In: Connell J, Lowitt K (eds) Food security in Small Island States. Springer Singapore, pp 25–38

Bednar-Friedl B, Biesbroek R, Schmidt DN, Alexander P, Børsheim KY, Carnicer J et al (2022) Europe. In: Climate Change 2022: impacts, adaptation and vulnerability. Cambridge University Press, Cambridge, UK/New York, pp 1817–1927. https://doi.org/10.1017/9781009325844.015

Bennett AC, Rodrigues de Sousa T, Monteagudo-Mendoza A et al (2023) Sensitivity of South American tropical forests to an extreme climate anomaly. Nat Clim Chang 13:967–974. https://doi.org/10.1038/s41558-023-01776-4

Bourgoin C, Ceccherini G, Girardello M et al (2024) Human degradation of tropical moist forests is greater than previously estimated. Nature 631:570–576. https://doi.org/10.1038/s41586-024-07629-0

Brouziyne Y, El Bilali A, Epule T, Ongoma V, Elbeltagi A, Hallam J et al (2023) Towards lower greenhouse gas emissions agriculture in north africa through climate-smart agriculture: a systematic review. Climate 11:139

Brown JR, Brierley CM, An S-I, Guarino M-V, Stevenson S, Williams CJR et al (2020) Comparison of past and future simulations of ENSO in CMIP5/PMIP3 and CMIP6/PMIP4 models. Clim Past 16(5):1777–1805

Carriconde F, Gardes M, Bellanger J-M, Letellier K, Gigante S, Gourmelon V et al (2019) Host effects in high ectomycorrhizal diversity tropical rainforests on ultramafic soils in New Caledonia. Fungal Ecol 39:201–212

Crippa M, Solazzo E, Guizzardi D, Monforti-Ferrario F, Tubiello FN, Leip A (2021) Food systems are responsible for a third of global anthropogenic GHG emissions. Nat Food 2(3):198–209

Demenois J, Merino-Martín L, Fernandez NN, Stokes A, Carriconde F (2020) Do diversity of plants, soil fungi and bacteria influence aggregate stability on ultramafic Ferralsols? A metagenomic approach in a tropical hotspot of biodiversity. Plant Soil 448(1–2):213–229. https://doi.org/10.1007/s11104-019-04364-8

Dutheil C, Bador M, Lengaigne M, Lefèvre J, Jourdain NC, Vialard J et al (2019) Impact of surface temperature biases on climate change projections of the South Pacific Convergence Zone. Clim Dyn 53(5–6):3197–3219

Dutheil C, Lengaigne M, Bador M, Vialard J, Lefèvre J, Jourdain NC et al (2020) Impact of projected sea surface temperature biases on tropical cyclones projections in the South Pacific. Sci Rep 10(1):4838

EFESE Report (2019) Carbon sequestration by ecosystems in France, Ministry of Ecological Transition and Territorial Cohesion. https://www.ecologie.gouv.fr/sites/default/files/publications/Th%C3%A9ma%20-%20La%20sequestration%20de%20carbone%20par%20les%20ecosysteme.pdf

Estoque RC, Ooba M, Avitabile V et al (2019) The future of Southeast Asia's forests. Nat Commun 10:1829. https://doi.org/10.1038/s41467-019-09646-4

Evans JP, Belmadani A, Menkes C et al (2024) Higher-resolution projections needed for small island climates. Nat Clim Chang 14:668–670. https://doi.org/10.1038/s41558-024-02028-9

Fache E, Dumas P, N'Yeurt ADR (2019) Introduction. Interdisciplinary synthesis of some discourses and responses related to climate in the Pacific. J Soc Oceanists 149:199–210

FAO (2020) Emissions due to agriculture. Global, regional and country trends 2000–2018, FAOSTAT Analytical Brief Series No 18. FAO, Rome. https://www.fao.org/3/cb3808en/cb3808en.pdf

FAO (2023) Pathways towards lower emissions—A global assessment of the greenhouse gas emissions and mitigation options from livestock agrifood systems. FAO, Rome. https://doi.org/10.4060/cc9029en

FAO (2024) Greenhouse gas emissions from agrifood systems—Global, regional and country trends, 2000–2022. FAOSTAT Analytical Brief Series, No. 94, Rome.

Fischer G, Shah MN, Tubiello F, Van Velhuizen H (2005) Socio-economic and climate change impacts on agriculture: an integrated assessment, 1990–2080. Philos Trans R Soc B: Biol Sci 360(1463):2067–2083

Hau'ofa E (2013) Our sea of islands. Pacific Islanders editions, p 40

Hicke JA, Lucatello S, Mortsch LD, Dawson J, Domínguez AM, Enquist CAF et al (2022) North America. In: Climate Change 2022: impacts, adaptation and vulnerability. Contribution of Working Group II to the Sixth Assessment Report of the Intergovernmental Panel on Climate Change. Cambridge University Press, Cambridge, UK/New York, pp 1929–2042. https://doi.org/10.1017/9781009325844.016

Hsiang S, Kopp R, Jina A, Rising J, Delgado M, Mohan S et al (2017) Estimating economic damage from climate change in the United States. Science 356(6345):1362–1369. https://doi.org/10.1126/science.aal4369

Hulme PE (2011) Biosecurity: the changing face of invasion biology. In: Fifty years of invasion ecology: the legacy of Charles Elton. Wiley Oxford, pp 73–88. https://doi.org/10.1002/9781444329988.ch23

Idris NH, Munadi MHF, Zheng YC, Lee BY, Vignudelli S (2023) Sea-level rise in Southeast Asia: a review of the factors, and the observed rates from tide gauge, satellite altimeters and assimilated data techniques. Int J Remote Sens:1–22. https://doi.org/10.1080/01431161.2023.2282408

INRAE Report (2020) Storing carbon in French soils: what potential in light of the 4 per 1000 goal and at what cost. https://www.inrae.fr/sites/default/files/pdf/Rapport%20Etude%204p1000.pdf

IPCC (1990) In: Houghton JT, Jenkins GJ, Ephraums JJ (eds) Climate change: the IPCC Scientific Assessment. Report prepared for IPCC by Working Group I. Cambridge University Press, p 364

IPCC (2013) Summary for policymakers. In: Climate change 2013: the physical science basis. Contribution of Working Group I to the Fifth Assessment Report of the Intergovernmental Panel on Climate Change. Cambridge University Press, Cambridge, UK/New York

IPCC (2014) Summary for policymakers. In: Climate change 2014: impacts, adaptation, and vulnerability. Part A: global and sectoral aspects. Contribution of Working Group II to the Fifth Assessment Report of the Intergovernmental Panel on Climate Change. Cambridge University Press, Cambridge, UK/New York, pp 1–32

IPCC (2021) Climate change 2021: the physical science basis. Contribution of Working Group I to the Sixth Assessment Report of the Intergovernmental Panel on Climate Change. Cambridge University Press, Cambridge, UK/New York

IPCC (2022) Climate change 2022: impacts, adaptation and vulnerability. Contribution of Working Group II to the Sixth Assessment Report of the Intergovernmental Panel on Climate Change. Cambridge University Press, Cambridge, UK/New York, p 3056. https://doi.org/10.1017/9781009325844

Klöck C, Fink M (2019) Dealing with climate change on small islands: toward effective and sustainable adaptation. Universitätsverlag Göttingen, p 337

Knutson T, Camargo SJ, Chan JC, Emanuel K, Ho C-H, Kossin J et al (2020) Tropical cyclones and climate change assessment: Part II: Projected response to anthropogenic warming. Bull Am Meteorol Soc 101(3):E303–E322

Kulagowski R, Thoumazeau A, Leopold A, Lienhard P, Boulakia S, Metay A et al (2021) Effects of conservation agriculture maize-based cropping systems on soil health and crop performance in New Caledonia. Soil Tillage Res 212:105079. https://doi.org/10.1016/j.still.2021.105079

Le Roux R, Wagner F, Blanc L et al (2022) How wildfires increase sensitivity of Amazon forests to droughts. Environ Res Lett. https://doi.org/10.1088/1748-9326/ac5b3d

Lee J-Y, Marotzke J, Bala G, Cao L, Corti S, Dunne JP, Engelbrecht F et al (2021) Future global climate: scenario-based projections and near-term information. In: Climate change 2021: the physical science basis. Contribution of Working Group I to the Sixth Assessment Report of the Intergovernmental Panel on Climate Change. Cambridge University Press, Cambridge, UK/New York, pp 553–672. https://doi.org/10.1017/9781009157896.006

Léopold A, Drouin J, Drohnu E, Kaplan H, Wamejonengo J, Bouard S (2021) Fire-fallow agriculture in Mare Loyalty Island: a sustainable cropping system for maintaining organic carbon in Gibbsic Ferralsol (New Caledonia, South West Pacific). Reg Environ Chang 21. https://doi.org/10.1007/s10113-021-01814-x

Martínez-Asensio A, Wöppelmann G, Ballu V, Becker M, Testut L, Magnan AK, Duvat VKE (2019) Relative sea-level rise and the influence of vertical land motion at Tropical Pacific Islands. Glob Planet Chang 176:132–143

McCubbin S, Smit B, Pearce T (2015) Where does climate fit? Vulnerability to climate change in the context of multiple stressors in Funafuti, Tuvalu. Glob Environ Chang 30:43–55

Mishra S, Page SE, Cobb AR et al (2021) Degradation of Southeast Asian tropical peatlands and integrated strategies for their better management and restoration. J Appl Ecol 58:1370–1387. https://doi.org/10.1111/1365-2664.13905

Nurse LA, McLean RF, Agard J, Briguglio LP, Duvat-Magnan V, Pelesikoti N et al (2014) Small islands. In: Climate change 2014: impacts, adaptation, and vulnerability. Part B: Regional aspects. Contribution of working group II to the fifth assessment report of the intergovernmental panel on climate change. Cambridge University Press, Cambridge, UK/New York, pp 1–32. 1613 p.

Panthou G, Lebel T, Vischel T, Quantin G, Sane Y, Ba A et al (2018) Rainfall intensification in tropical semi-arid regions: the Sahelian case. Environ Res Lett 13(6):064013. https://doi.org/10.1088/1748-9326/aac334

Pascht A (2019) Climate and knowledge worlds. Approaching climate change and knowledge creation in Vanuatu. Journal de la société des océanistes 149:235–244

Payri C, Vidal E (2019) Biodiversity, a pressing need for action in Oceania, Noumea. University Press of New Caledonia, p 64

Phillips OL, Aragão L, Fisher JB et al (2009) Drought sensitivity of the Amazon rainforest. Science 323:1344–1347. https://doi.org/10.1126/science.1164033

Pillon Y, Jaffré T, Birnbaum P, Bruy D, Cluzel D, Ducousso M et al (2021) Infertile landscapes on an old oceanic island: the biodiversity hotspot of New Caledonia. Biol J Linn Soc 133(2):317–341

Pouteau R, Birnbaum P (2016) Island biodiversity hotspots are getting hotter: vulnerability of tree species to climate change in New Caledonia. Biol Conserv 201:111–119

Power SB, Delage FPD (2019) Setting and smashing extreme temperature records over the coming century. Nat Clim Chang 9:529–534. https://doi.org/10.1038/s41558-019-0498-5

Romanello M, McGushin A, Di Napoli C et al (2021) The 2021 report of the lancet countdown on health and climate change. Lancet 398(10311):1619–1662. https://doi.org/10.1016/S0140-6736(21)01787-6

Rosenzweig C, Mbow C, Barioni LG, Benton TG, Herrero M, Krishnapillai M et al (2020) Climate change responses benefit from a global food system approach. Nat Food 1(2):94–97

Sanderman J, Hengl T, Fiske G (2017) The soil carbon debt of 12,000 years of human land use. PNAS 114(36):9575–9580. https://doi.org/10.1073/pnas.1706103114

Savadogo M, Somda J, Seynou O, Zabré S, Nianogo AJ (2011) Catalogue of good practices for adapting to climate risks in Burkina Faso. IUCN Burkina Faso, Ouagadougou. https://www.bretagne-solidaire.bzh/wp-content/uploads/sites/11/2020/03/catalogueriquesclimatiques_compressed-1.pdf

Shaw R, Luo Y, Cheong TS, Abdul HS, Chaturvedi S, Hashizume M, Insarov GE et al (2022) Asia. In: Climate change 2022: impacts, adaptation and vulnerability. Contribution of Working Group II to the Sixth Assessment Report of the Intergovernmental Panel on Climate Change. Cambridge University Press, Cambridge, UK/New York, pp 1457–1579. https://doi.org/10.1017/9781009325844.012

Sonobe T, Buchoud NJA, Akbar R, Altansukh B (2024) Transforming ASEAN strategies for achieving sustainable and inclusive growth. Asian Development Bank Institute, p 246. https://doi.org/10.56506/FYUK6909

Tangang F, Chung JX, Juneng L et al (2020) Projected future changes in rainfall in Southeast Asia based on CORDEX-SEA multi-model simulations. Clim Dyn 55:1247–1267. https://doi.org/10.1007/s00382-020-05322-2

Trisos CH, Adelekan IO, Totin E, Ayanlade A, Efitre J, Gemeda A, Kalaba K et al (2022) Africa. In: Climate change 2022: impacts, adaptation and vulnerability. Contribution of Working Group II to the Sixth Assessment Report of the Intergovernmental Panel on Climate Change. Cambridge University Press, Cambridge, UK/New York, pp 1285–1455. https://doi.org/10.1017/9781009325844.011

Tubiello FN, Karl K, Flammini A, Gütschow J, Obli-Laryea G, Conchedda G et al (2022) Pre- and post-production processes increasingly dominate greenhouse gas emissions from agri-food systems. Earth Syst Sci Data 14(4):1795–1809

UNEP Report (2005). https://reliefweb.int/map/world/africa-change-potential-cereal-output-2080

Walsh KJ, McInnes KL, McBride JL (2012) Climate change impacts on tropical cyclones and extreme sea levels in the South Pacific—A regional assessment. Glob Planet Chang 80:149–164

Walsh M, Backlund P, Buja L, DeGaetano A, Melnick R, Prokopy L, et al. (2020) Climate indicators for agriculture, USDA Technical Bulletin 1953. Washington, DC, 70 p. https://doi.org/10.25675/10217/210930

Whan K, Alexander L, Imielska A, McGree S, Jones D, Ene E et al (2014) Trends and variability of temperature extremes in the tropical Western Pacific. Int J Climatol 34(8):2585–2603

Wong PP, Losada IJ, Gattuso J-P, Hinkel J, Khattabi A, McInnes KL et al (2014) Coastal systems and low-lying areas. Clim Change 2104:361–409

Open Access This chapter is licensed under the terms of the Creative Commons Attribution-NonCommercial-NoDerivatives 4.0 International License (http://creativecommons.org/licenses/by-nc-nd/4.0/), which permits any noncommercial use, sharing, distribution and reproduction in any medium or format, as long as you give appropriate credit to the original author(s) and the source, provide a link to the Creative Commons license and indicate if you modified the licensed material. You do not have permission under this license to share adapted material derived from this chapter or parts of it.

The images or other third party material in this chapter are included in the chapter's Creative Commons license, unless indicated otherwise in a credit line to the material. If material is not included in the chapter's Creative Commons license and your intended use is not permitted by statutory regulation or exceeds the permitted use, you will need to obtain permission directly from the copyright holder.

Part II
Agricultural and Food Systems and the Land Sector: Contributors and Victims of Climate Change

Chapter 5
Family Farming in the Face of Climate Change: Potential for Adaptation Through Agroecology

Jean-Michel Sourisseau and Jean-François Le Coq

Abstract This chapter examines the role of family farming (FF) in climate change adaptation, a topic less studied despite FF being the most vulnerable agricultural system. It explores FF's potential to adapt to climate impacts compared to industrial farms, using the concept of family-based agroecology. Drawing on research from the Global South, it highlights the challenges and policy levers needed to support FF's resilience. The chapter aims to provide recommendations to enhance FF's capacity to cope with climate-related challenges and structural changes in agriculture.

While the implications of climate change for agriculture are relatively well understood, few studies have systematically analyzed the importance and consequences of agricultural production structures—from family farming to capital-intensive agribusiness—on climate-related issues. In particular, little is known about:

- The impacts of climate change on family farming (FF), despite the fact that these systems represent the most vulnerable category of agricultural producers;
- The potential of FF to adapt to climate change, especially in comparison to industrial-scale farms;
- The levers that could enable these millions of farms to cope with climate-related challenges.

This chapter aims to explore how climate issues in agriculture can be viewed through the lens of family farming, focusing on the adaptation of these systems to both extreme climate events and structural transformations. We draw on a non-exhaustive review of the academic literature and research conducted by CIRAD in various regions of the Global South. Our goal is to offer pathways for reflection and a number of recommendations to better understand the links between family farming and adaptation to climate change.

J.-M. Sourisseau (✉) · J.-F. Le Coq
Cirad, UMR Actors Resources Territories & Development (ART-Dev), Montpellier, France
e-mail: jean-michel.sourisseau@cirad.fr

We propose to address the relationship between family farming and climate through the concept of family-based agroecology, providing justification for this perspective. We then analyze the adaptive potential of family farming through the defining elements of agroecology, and the policy challenges involved in realizing this potential. We also present hypotheses regarding the comparative capacity of family farms (FFs) to adapt relative to other production systems. Finally, we conclude by identifying the necessary and feasible levers—particularly in terms of policy—to enable family farming to fully express its adaptive potential in the face of climate change.

5.1 How Should We Approach the Issue? The Case for Family-Based Agroecology

To clarify the relationship between family farming and climate issues, we begin by highlighting the importance, diversity, and definitions of family farming. According to statistical compilations by Lowder et al. (2021), 90% of farms worldwide are family-run, covering nearly 80% of agricultural land and producing over 75% of global food volume. However, small farms under 2 hectares—most of which are also family-run—account for 85% of all farms but only occupy 10% of agricultural land and produce 35% of food. These figures point to the immense diversity of family farming in terms of size and production capacity. Family farms are not exclusively small-scale, although most small farms are indeed family farms.

These figures also underscore that climate policies must explicitly target family farms, especially the smallest and most vulnerable ones. Doing so is a matter of climate justice and global solidarity. However, such targeting is not straightforward, as public policies at national and territorial levels often lack data that distinguish family farms—particularly due to insufficient information about the nature of agricultural labor (Bosc et al. 2014; Sourisseau 2014).

Identifying family farms is thus a necessary first step. But assuming there is political will, how can we concretely assess the links between the family-based nature of farming and climate change?

Let us begin with a definition. According to the FAO,[1] *"Family farming is a mode of organizing agricultural, forestry, fishery, pastoral, or aquaculture production that is managed and operated by a family and predominantly relies on family labor, both female and male. The family and the farm are linked, co-evolve, and fulfill economic, environmental, reproductive, social, and cultural functions."* While this is a broad definition, for the purposes of this chapter, we adopt a narrower one proposed by CIRAD researchers (Sourisseau 2014), based on the absence of permanent hired labor. This definition allows us to position "family businesses" or employer-operated farms (which may fall somewhere between purely family and purely corporate models) as part of the broader spectrum of family farming (Table 5.1).

---

[1] https://www.familyfarmingcampaign.org/fr/agricultura-familiar/

Table 5.1 Diversity of structure and organization of agricultural operations

	Entrepreneurial Agriculture	Family-based Agriculture	
	Firms	Owner-operated Farms	Family Dominance, no permanent employees
Labor	Exclusively salaried	Mixed, presence of permanent employees	Family Dominance, no permanent employees
Capital	Shareholders	Family or family association	Family
Management	Technical	Family and technical	Family
Consumption	Not applicable	Residual	Partial to dominant self-consumption
Legal status	Limited company or other corporate forms	Agriculture operator status, associative forms, rarely corporate forms	Informal or formal agriculture operator status
Land status	Ownership or formal indirect land tenure	Ownership or formal or informal land indirect tenure	

Source: adapted from Sourisseau (2014)

Crucially, the defining feature of family farms is not size, but the social and organizational relationships that govern agricultural production, characterized by the organic linkage between household and productive spheres.

We propose to examine how family farms adapt to climate change—both to extreme events and long-term changes—and how public policy might enable such adaptation. We do so by viewing these capacities through the lens of family-based organization, along with their access to production assets, capital, and resources.[2]

Beyond the diversity in farm size, the other essential characteristic of these agricultures is that they implement extremely diverse crop and livestock systems, from the most conventional to the most agroecological, from the most technological to the most artisanal (differential intensification in physical, financial and human capital).

This diversity makes it difficult to analyze their capacity to adapt to climate change. In fact, while there are analyses by production system, there is no analysis by type of production organization to assess the differentiated adaptation strategies specific to family farms. However, because of their sheer numbers, family farms (both large and small) are the ones most concerned by the need to adapt. They have a wide range of capacities to do so, even if the majority are suffering the effects of climate change with little room for manoeuvre, due to their weak production structure (FAO 2024). Those best endowed with capital, particularly natural and physical capital, and therefore also the largest and often the most profitable farms in our terminology, would have the resources to transform their cropping and production systems. However, they often have limited room for manoeuvre due to high levels

---

[2] The sustainable rural livelihood model is to measure the endowments of five main "capitals" (natural, physical, financial, social and human) and their dynamic changes to assess farm strategies and performance (Bosc et al. 2014).

of indebtedness, which pushes them to limit the technical changes needed to achieve a transformative agroecological transition.

To understand the link between family farming and climate change, we draw on the conclusions of the report by the Intergovernmental Panel on Climate Change (IPCC), which states that "ecosystem-based approaches, agroecology and other nature-based solutions in agriculture and fisheries can enhance resilience to climate change with multiple co-benefits" (IPCC 2022; p. 90). The IPCC proposes that agroecological forms of production should be favored, particularly on the criteria of cultivated biodiversity and resilience, over possible responses through productivist models: "Adaptation options based on conventional intensification of production have been widely adopted in agriculture for climate change adaptation, but with potential negative effects" (IPCC 2022).

The IPCC report refers to the definition of agroecology as constructed by the FAO (2018), around ten principles that should guide the transformation of agricultural systems (Fig. 5.1). The recommendations also refer, within this framework, to agroecological forms known as "transformative" in the terminology of Gliessman (2015). Gliessman distinguishes five forms of agroecological transition, ranging from the simple improvement of conventional inputs to the fundamental overhaul of agricultural and food systems. The transformative nature requires not marginal adjustments to the conventional model, but the redesign of agro-ecosystems and a move away from a focus on production to integrate environmental and social demands towards a different kind of food.

We therefore postulate, in line with the IPCC and with the conclusions of the 2019 report of the High-Level Panel of Experts on Food (HLPE 2019), that these transformative forms are the least destabilizing for the climate and the most adaptive to climate change. And we propose to formulate a series of hypotheses on the links between family farms and their potential to adapt to climate change through

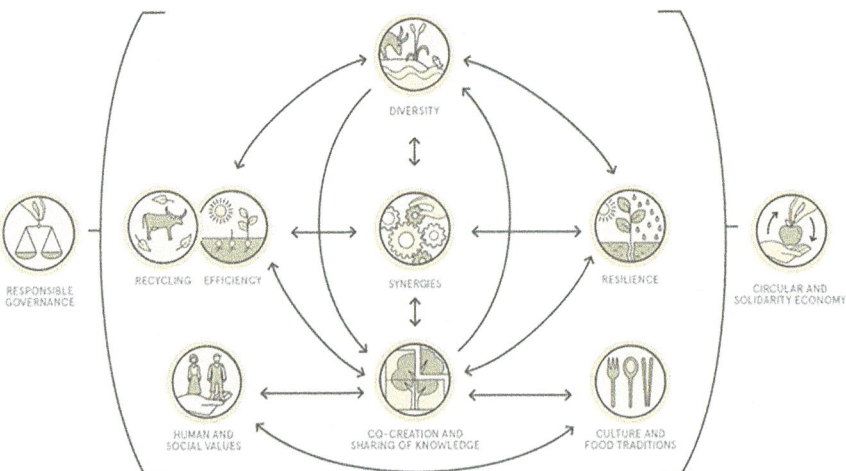

Fig. 5.1 Principles guiding the transformation of agricultural (and food) systems towards agroecology. (Source: FAO (2018))

their potential to adopt agroecological practices. In so doing, we acknowledge that family farms are not in themselves virtuous or climate-smart. But we hypothesize that they do have social, organizational and economic characteristics that could facilitate the transition to a transformative agroecology that would enable them to adapt to extreme events and more moderate but profound changes in climate. To systematize and organize our remarks, we will, like the IPCC, mobilize the principles that should guide agroecology proposed by the FAO (2024).

5.2 Family Agroecology and Its Potential for Climate Change Adaptation Through the Lens of Agroecological Principles

In this second part, we first review the FAO's ten guiding principles of agroecology to judge the capacity of family farms (more than firms) and their social community to adapt to climate change, by implementing agroecological practices. Then, we analyze in more detail the capacities of family farms for four of these principles guiding agroecology. Finally, we discuss the political dimensions of encouraging family farms to adapt to climate change.

5.2.1 The Agroecological Potential of Family Farming

The following table (Table 5.2) summarizes, item by item, the working hypotheses (yet to be debated) derived from our literature review. These hypotheses outline the characteristics and aptitudes of family farms (including their endowments of tangible and intangible capital), which are potential levers for achieving the agroecological transition.

5.2.2 Agroecology and Climate Adaptation in Family Farms

Building on these hypotheses, we examine four specific agroecological elements that are particularly well documented in the literature, using field-based evidence to explore their relationship to climate change adaptation in family farming.

5.2.2.1 Biodiversity

In a literature review of peasant (i.e., family-based) farming systems, Altieri et al. (2015) show that these systems mobilize multiple forms of diversification to adapt to climate variability. Other studies support these findings. For instance,

Table 5.2 Agroecological principles and family farming characteristics favoring agroecological transition in the context of climate challenges

Agroecological Principles	Potential Advantages of Family Farms for Agroecological Transition
Biodiversity and Landscape Diversity	Biodiversity is central to risk mitigation strategies in family and peasant farming (e.g., mixed cropping, agroforestry, livestock) → supports self-consumption, differentiated knowledge among household members. Small average size of family farms → need to optimize limited resources. Farm and landscape diversity → family farm landscapes are generally more heterogeneous.
Efficiency (reducing external inputs)	Autonomy → foundational element in definitions of peasant farming. Diverse of knowledge transmitted (gender- and generation-based) → mastering service plants for pest control in small-scale systems, and farmers' selection of plant material. Cultivated Biodiversity → system complexity, which calls for greater efficiency. Territorial anchoring → combination of parcel-level and landscape-level knowledge.
Food Culture and Traditions	Self-consumption → promotes food diversity and security. Intergenerational ties → preservation of culinary knowledge.
Resilience	Complex production systems → strategic flexibility and adaptive management. Multifunctional livelihoods → ability to shift the role of agriculture in household reproduction.
Human and Social Values	Role of women → contrasting dynamics: women are more exposed to crises (including climate), yet family-based work division can foster emancipation.
Recycling	Limited financial capital → propensity for resource recycling and natural resources uses optimization. Transmitted knowledge and territorial embedding → support recycling from parcel to territorial scale.
Co-creation and Sharing of Knowledge	Biodiversity → continuous learning in managing complex systems. Intra-household dynamics (gender and generations) → facilitated knowledge transmission. Community and territorial anchoring → support commons management and collective knowledge systems.
Synergy	Diversity and complexity of production systems → complementarity between farming and livestock at farm and landscape scales. Family reproduction and multiple activities → resource and objective synergies.
Circular and Solidarity Economies	Knowledge and territorial anchoring → circularity via resource flow management. Family and community ties → solidarity networks (formal/informal, material/immaterial).
Responsible Governance	Element exceeding the approach of family farming at the national, or even regional scale, but community anchoring and intra-family and territorial solidarity are foundations for responsible governance

Holt-Giménez (2002) highlights the resilience of peasant farms, showing that recovery was faster in complex, diversified systems than in monocultures after Hurricane Mitch hit Central America in 1998. Descheemaeker et al. (2016) show that in semi-humid Sahelian areas, cultivar choice, legume integration, and complex crop combinations are more feasible in small-scale family-managed systems. These practices not only aid in climate mitigation but also offer co-benefits. The integration of community knowledge and flexible division of labor within families enables the management of complex systems without mechanization. Moreover, cultivated biodiversity extends to the territorial level: agricultural mosaics are more complex where landscapes are occupied by diverse family farms (Bosc et al. 2014). This diversity can be compounded, generating new services and options, particularly when coordinated actions are implemented at the landscape level. The cumulative effect is especially significant in tropical settings and in peasant family farming systems (Harvey et al. 2014).

5.2.2.2 Efficiency

Efficiency—understood as reducing the use of external resources in agricultural and food systems (FAO 2018)—is closely linked to autonomy, a central theme in peasant farming (van der Ploeg 2008). Altieri and Nicholls (2017) document numerous examples of successful self-reliance among peasant farmers in the Global South, often grounded in high crop diversity. The combination of intergenerational and gendered knowledge, along with academic and normative inputs, supports more effective pest control in complex small-scale systems than in industrial farms. At the territorial level, the complexity of crop interactions is even greater due to system diversity. Provided there is adequate local knowledge, adaptation capacities can be enhanced by leveraging landscape-scale dynamics. Family farms' territorial anchoring further supports the accumulation and application of this knowledge.

5.2.2.3 Resilience

The propensity of family farms to implement resilient systems is well documented. Major reviews (e.g., Gliessman 2015; HLPE 2019) note that peasant (and thus family) farms are better at applying resilience-enhancing practices—such as managing biodiversity, water, soil fertility, or crop-livestock integration—because these are more compatible with micro-local, landscape-embedded management than with standardized industrial models. Pastoral practices in Africa—nearly all family-based—also illustrate resilience (Vayssières et al. 2017). Strategic mobility helps limit exposure to drought and other non-climate stressors. Additionally, resilience in family farming reflects flexibility in household objectives and performance. Families may adjust their consumption or diversify livelihoods through off-farm work and migration (Bosc et al. 2014). In this sense, the intertwining of domestic and productive spheres is a potential lever for adaptation to extreme or more moderate climatic events, especially when considering the possibilities of resorting to non-agricultural activities through pluriactivity and mobility.

However, the link between economic diversification and resilience is not always straightforward. Alfani et al. (2021) show that in Zambia, diversification—including livestock—does not consistently explain adaptation to El Niño-related droughts. Asfaw et al. (2019), in studies from Malawi, Niger, and Zambia, find that off-farm adaptive strategies are especially important for the poorest and often defensive in nature. Call et al. (2019) argue that long-term declines in family assets limit the effectiveness of diversification-based adaptation.

5.2.2.4 Family Farming and Human and Social Values

To simplify, and because this is central to intra-household dynamics in family farms, we address human and social values through the lens of gender in agroecology and climate adaptation. FAO's recent report on climate injustice (2024) reiterates long-standing findings about the critical role of women. Integrating women's knowledge and recognizing "reproductive" household activities enhance living conditions, support agroecological practices, and increase the flexibility of farming systems (Ume et al. 2022). Asfaw and Maggio (2018) suggest that matrilocality fosters more effective family-wide adaptation strategies. Even in patriarchal settings, positive outcomes for women are possible. Agamile et al. (2021), in Uganda, present a counterintuitive finding: extreme climate events sometimes offer women and youth emancipatory opportunities, as men disengage from cash crops, allowing women greater access to land for food crops. According to Ume et al. (2022), reproductive labor—especially by women—must be accounted for when assessing resilience and adaptation capacity. This lies at the heart of family farming: the intimate connection between productive and domestic spheres. Disaggregating these spheres obscures key adaptive strategies and levers for climate action.

Finally, Forest and Foreste (2022) propose broadening the reflection and action around gender and climate by documenting women's roles in collective resource management (commons). This suggests a promising extension of the family-based perspective and a complement to discussions on efficiency, biodiversity, and resilience through collective values and ecofeminist insights.

5.3 Public Policies to Support the Ecological Transition of Family Farming

While the family-based nature of agriculture presents favorable characteristics for an agroecological transition in the face of climate challenges, not all family farms adopt agroecological production systems. For those engaged in conventional farming, such a transition can entail additional costs. Meanwhile, family farms already engaged in agroecology—often as a response to limited access to financial or natural resources—face a range of economic challenges and risks.

Therefore, whether to facilitate the shift from conventional family farming toward agroecology in order to enhance climate resilience, or to reinforce the capacities of already agroecological farms, public policies are essential. Three major policy domains must be considered. Each currently integrates family farming to varying degrees—but overall, still insufficiently: Agricultural policies targeting family farming; Agroecology-oriented policies; Climate policies for the agricultural sector.

The overview of these three policy domains shows contrasting dynamics in terms of the concrete integration of the family issue. Although there are possible convergences between these dynamics, they deserve to be strengthened.

Indeed, Policies that explicitly target family farming have taken diverse forms, often justified by equity concerns: access to land and markets, poverty reduction, and/or food security. Despite the hopes raised by such targeted policies, a regional study in Latin America revealed their limitations and mixed outcomes (Sabourin et al. 2014). These policies have often struggled with effective implementation, limited funding, or inadequate institutional support.

Policies promoting agroecology have also diversified over the last decade, spurred by social movements—especially in Latin America and Brazil (Le Coq et al. 2020)—as well as by international cooperation, particularly in Africa (Milhorance et al. 2022). These policies draw on varying definitions of agroecology (with more or less transformative ambitions) and employ tools from multiple policy fields (e.g., agricultural extension, access to credit or land). Nonetheless, they face major constraints: funding remains limited compared to conventional or agribusiness-oriented models; implementation is often partial or discontinuous. While they often claim to target family farmers, in practice they rarely provide clear mechanisms for identifying or selecting beneficiaries, nor do they adequately address the structural nature of agroecological transitions. Indeed, successful transitions often require land, as well as social, human, financial, and even political capital. Without strengthened livelihoods, the most vulnerable family farms remain locked in conventional trajectories—without necessarily being able to deploy them effectively.

Climate policies for agriculture have also seen significant development over the past decade, particularly following the increasing integration of agriculture into the global climate agenda. These policies have integrated various response concepts, and can support different models - agroecology being just one of these models. Thus, in the case of Brazil, Milhorance et al. (2021) show that different concepts coexist in climate policies, some of which may be favourable to family farming, others more to agro-industrial models. Moreover, the need to consider combinations of policy instruments and their interactions to favourably reach family farmers in the face of climate challenges is precisely documented, particularly in the emblematic case of Brazil (Milhorance et al. 2020).

In short, while public policies (across their respective domains) offer levers to support the agroecological transition of family farming, these tools often remain inadequate to meet the scale of the challenge. Given the vast number of family farms and the urgency of climate action, more ambitious and better-coordinated efforts are required. Although this chapter focuses on supporting agroecological transitions among family farmers, it is equally important to maintain inclusive policies for those who are not (yet) practicing agroecology or do not wish—or are unable—to pursue that path. These farmers represent a significant share of both the agricultural population and agricultural land—and often include the most vulnerable segments. Finally, policy interventions must be conceived not only at the farm level but also at the scale of food systems. Only by addressing the broader political and economic environment can agroecological family farming truly scale up and fulfill its potential for climate adaptation.

5.4 Conclusion

Agroecology is built on a combination of techniques and social and economic dynamics of diverse origins, but its foundation lies in the maintenance and mobilization of local and micro-local knowledge. Such knowledge is essential for sustainably managing the complexity of agroecosystems and the interactions among their components. The literature reviewed in this chapter converges around the idea that these knowledge systems—and their operationalization at farm and territorial levels—are a comparative advantage of family-based farming systems over other forms of agriculture. Family farming, anchored in territories and communities, shaped by intra-family and intergenerational exchanges, governed by adaptable and responsive regulatory mechanisms, and mobilizing strong social capital connected to other livelihood assets, is potentially well aligned with the core principles of agroecology. As a result, family farms hold significant potential for sustainable adaptation to climate change.

However, this potential alone does not guarantee the sustainability of agroecological practices or the transition from conventional models. As with all agricultural transformations throughout history, realizing this potential requires a social and political compromise, accompanied by substantial means and public policy instruments. And not just a single tool: the complexity of the processes that shape farm structures and enable agroecology demands systemic approaches and combinations of collective and public actions that address all dimensions of sustainable development. Yet, only 2% of international public climate finance was allocated to small-scale family farmers and rural communities in 2021. Given that 95% of climate finance for smallholders comes from public sources, this amounts to just 0.3% of the USD 653 billion in total international climate finance from public and private sources (World Rural Forum 2023). Similarly, the political and financial weight of agroecology and family farming in agricultural policies remains far below that of conventional high-carbon intensification models. The road toward realizing the adaptive potential of family farming remains long and uncertain.

Research has a critical role to play in redirecting this trajectory—currently unfavorable to family farming and transformative agroecology. This chapter has explored existing findings, but also intuitions and hypotheses that deserve deeper examination and validation through broader evidence bases and new empirical work. A crucial task is to reduce uncertainties around greenhouse gas (GHG) metrics and mitigation (see Chap. 4). Our analysis calls for more precise metrics that account for the diversity of agrarian contexts and help explain scaling-up processes. Such data, along with more systematic measurement and better understanding of adaptation mechanisms, should enhance knowledge about how different forms of agriculture and livestock systems—varying in geography, technology, economy, politics, and society—interact with climate change. At the heart of this effort must be a renewed focus on the social and economic forms of production, with the goal of formulating more targeted and effective public policies.

References

Agamile P, Dimova R, Golan J (2021) Crop choice, drought and gender: new insights from smallholders' response to weather shocks in rural Uganda. J Agric Econ 72(3):829–856

Alfani F, Arslan A, McCarthy N, Cavatassi R, Sitko N (2021) Climate resilience in rural Zambia: evaluating farmers' response to El Niño-induced drought. Environ Dev Econ 26(5-6):582–604

Altieri M, Nicholls C (2017) The adaptation and mitigation potential of traditional agriculture in a changing climate. Clim Chang 140:33–45

Altieri M, Nicholls C, Henao A, Lana MA (2015) Agroecology and the design of climate change-resilient farming systems. Agron Sustain Dev 35:869–890

Asfaw S, Maggio G (2018) Gender, weather shocks and welfare: evidence from Malawi. J Dev Stud 54(2):271–291. https://doi.org/10.1080/00220388.2017.1283016

Asfaw S, Scognamillo A, Di Caprera G, Sitko N, Ignaciuk A (2019) Heterogeneous impact of livelihood diversification on household welfare: cross-country evidence from Sub-Saharan Africa. World Dev 117:278–295. https://doi.org/10.1016/j.worlddev.2019.01.017

Bosc P-M, Sourisseau JM, Bonnal P, Gasselin P, Valette E, Bélières JF (2014) Diversity of family farming: existence, transformation, becoming. Quæ Editions, 384 p

Call M, Gray C, Jagger P (2019) Smallholder responses to climate anomalies in rural Uganda. World Dev 115:132–144

Descheemaeker K, Oosting SJ, Homann-Kee TS, Masikati P, Falconnier G, Giller K (2016) Climate change adaptation and mitigation in smallholder crop-livestock systems in sub-Saharan Africa: a call for integrated impact assessments. Reg Environ Chang 16:2331–2343

FAO (2018) The 10 elements of agroecology. Guiding the transition towards sustainable food and agricultural systems. Rome, 15 p

FAO (2024) The unjust climate – Measuring the impacts of climate change on rural poor, women and youth. Rome, 120 p

Forest M, Foreste C (2022) What interpretive frameworks around the issues of Gender and Climate? Lessons from a bibliometric analysis. Research paper n°22. AFD, Paris, 107 p

Gliessman SR (2015) 3rd Agroecology: the ecology of sustainable food systems. CRC press, 405 p

Harvey CA, Chacon M, Donatti CI, Garen E, Hannah L, Andrade A et al (2014) Climate-smart landscapes: opportunities and challenges for integrating adaptation and mitigation in tropical agriculture. Conserv Lett 7(2):77–90

HLPE (2019) Agroecological and other innovative approaches for sustainable agriculture and food systems that enhance food security and nutrition. A report by the High Level Panel of Experts on Food Security and Nutrition of the Committee on World Food Security, Rome, 163 p

Holt-Giménez E (2002) Measuring farmers' agroecological resistance after Hurricane Mitch in Nicaragua: a case study in participatory, sustainable land management impact monitoring. Agric Ecosyst Environ 93:87–105

IPCC (2022) Climate change 2022: impacts, adaptation and vulnerability. Contribution of Working Group II to the Sixth Assessment Report of the Intergovernmental Panel on Climate Change. Cambridge University Press, Cambridge, UK/New York, 3056 p

Le Coq J-F, Sabourin E, Bonin M, Freguin-Gresh S, Marzin J, Niederle P, Vásquez L (2020) Public policy support for agroecology in Latin America: lessons and perspectives. Global J Ecol 5(1):129–138

Lowder S, Sanchez M, Bertini E (2021) Which farms feed the world and has farmland become more concentrated? World Dev 142:105455. https://doi.org/10.1016/j.worlddev.2021.105455

Milhorance C, Sabourin E, Le Coq J-F, Mendes P (2020) Unpacking the policy mix of adaptation to climate change in Brazil's semiarid region: enabling instruments and coordination mechanisms. Clim Pol 20(5):593–608

Milhorance C, Sabourin E, Chechi L, Mendes P (2021) The politics of climate change adaptation in Brazil: framings and policy outcomes for the rural sector. Environ Politics:1–23

Milhorance C, Camara AD, Sourisseau J-M, Piraux M, Assembène MC, Sirdey N, Sall MCA (2022) The integration of agroecology into public policies in Senegal. ISRA, Dakar, 54 p

Sabourin É, Marzin J, Le Coq J-F, Massardier G, Fréguin-Gresh S, Samper M, Sotomayor O (2014) Family farming in Latin America. Emergence, advances and limits of targeted policies. Revue Tiers Monde 4(4):23–41

Sourisseau JM (ed) (2014) Family farming and future worlds. Versailles, éditions Quæ, 360 p

Ume C, Nuppenau EA, Domptail SE (2022) A feminist economics perspective on the agroecology-food and nutrition security nexus. Environ Sustain Indic 16:100212

van der Ploeg JD (2008) The New Peasantries, struggles for autonomy and sustainability in an era of empire and globalization. Earthscan, 356 p

Vayssières J, Assouma MH, Lecomte P, Hiernaux P, Bourgoin J, Jankowski F et al (2017) Livestock at the heart of 'climate-smart' landscapes in West Africa. In: Caron P, Valette E, Wassenaar T, Coppens DEG, Vatché P (eds) Living territories to transform the world. Versailles, éditions Quæ, pp 111–117

World Rural Forum (2023) An untapped potential. An analysis of the international public financing flows for the fight against climate change destined for sustainable agriculture and family farmers. WRF, 14 p

Open Access This chapter is licensed under the terms of the Creative Commons Attribution-NonCommercial-NoDerivatives 4.0 International License (http://creativecommons.org/licenses/by-nc-nd/4.0/), which permits any noncommercial use, sharing, distribution and reproduction in any medium or format, as long as you give appropriate credit to the original author(s) and the source, provide a link to the Creative Commons license and indicate if you modified the licensed material. You do not have permission under this license to share adapted material derived from this chapter or parts of it.

The images or other third party material in this chapter are included in the chapter's Creative Commons license, unless indicated otherwise in a credit line to the material. If material is not included in the chapter's Creative Commons license and your intended use is not permitted by statutory regulation or exceeds the permitted use, you will need to obtain permission directly from the copyright holder.

Chapter 6
Climate Change, (Im)Mobility and Land Tenure: Challenges for Family Farming in the Global South

Sara Mercandalli, Hadrien Di Roberto, and Pierre Girard

Abstract By 2050, estimates indicate that there could be respectively 216 million internal and 200 million international climate migrants, especially impacting the Global South, notably Africa and Southeast Asia, where family farming dominates. Climate change affects land quality and tenure, influencing farmers' vulnerability and mobility. Mitigation and adaptation policies, such as fighting deforestation and recognizing climate migration, are crucial but face challenges in effectiveness and social impact. The chapter explores how climate change and climate policies affect land tenure and migration dynamics in rural areas, and their implications for family farming. It highlights land as a social relation and mobility as a complex coping strategy for family farmers under climate risks.

Recent estimates indicate that there could be respectively 216 and 200 million internal and international climate migrants[1] by 2050 (Clement et al. 2021; Burzyński

---

We extend our heartfelt thanks to Alain Karsenty (Cirad), Quentin Grislain (Cirad) and Jérémy Bourgoin (Cirad-ILC) for their valuable feedback and contributions to this chapter.

---

[1] Climate migration is the movement of a person or groups of people who, primarily for reasons of sudden or progressive change in the environment due to climate change are forced to leave their usual place of residence, or choose to do so, temporarily or permanently, within a state or across an international border. It is a subcategory of environmental migration (OIM 2019). Estimates of the number of environmental migrants are difficult and projections range from 25 million to 1 billion by 2050 (IOM 2009).

S. Mercandalli (✉)
Cirad, UMR ART-Dev, Montpellier, France
e-mail: sara.mercandalli@cirad.fr

H. Di Roberto
Cirad, UMR ART-Dev, Houphouet Bouany University, Abidjan, Ivory Coast

P. Girard
Cirad, UMR ART-Dev, Institute of Statistical Social and Economic Research (ISSER), Accra, Ghana

© The Author(s) 2026
V. Blanfort et al. (eds.), *Climate Impacts and Challenges in Agriculture, Forests and Food Systems*, https://doi.org/10.1007/978-3-032-04331-3_6

et al. 2022). Climate change also has impacts on the quality of arable land and its uses, which are closely linked to issues of land tenure transformations.

These land and migration issues related to climate change are particularly acute in the countries of the South; especially, Africa and Southeast Asia and the Pacific are both the regions that contribute the least to climate change and those that are most affected by its effects (Borderon et al. 2019; IPCC 2023). These two regions indeed have the highest number of environmentally displaced persons, respectively 7.5 and 22.6 million in 2023 (IDMC 2023). The stakes are high given the central role of the primary sector in the economy of these regions, and especially of family farming which is predominant there (Lowder et al. 2021). The diversity of family farming (see Chap. 5) implies varying levels of vulnerability and adaptation capacities to climate change, largely dependent on land tenure regimes and mobility (Zickgraf et al. 2016).

The countries of the South are also particularly concerned by climate change mitigation and adaptation policies. Firstly, the fight against deforestation in the tropical zone is a strategic lever in the negotiations of the United Nations Framework Convention on Climate Change (FAO 2022). Secondly, climate migration has been included as a mode of adaptation in several international frameworks[2]. However, the multiple causes and the uncertainty of the socio-ecological thresholds underlying climate migrations (Thalheimer et al. 2021) question the effectiveness and social effects of political action concerning migration, displacements and planned relocations (McDowell 2013).

Despite the centrality of land and mobility in the issues of agrarian justice and climate justice (Borras et al. 2023), the dialectic between climate change, land and mobility is under scrutinized (Zickgraf et al. 2016). The aim of this chapter is to highlight the links between climate change and land and migration dynamics in rural areas of the South, and their implications for family farming. What are the effects of climate change on land tenure regimes and the mobility of family farming? How do climate policies affect land and mobility of rural populations? To what extent can land and migration policies be levers in the fight against climate change?

To address these questions, we consider the implications of climate change both as a physical phenomenon and as a social and political one, i.e., by examining the implications of measures to combat climate change. Land is defined as "the set of social relations between people about the possession and use of land, as well as the control of this use" (Colin and Daoudi 2022). Mobility, on the other hand, is part of a complex decision-making system of rural families to cope with both socio-economic and climate risks; it can be voluntary or involuntary (forced) and can cover a diversity of temporal and spatial scales - temporary or permanent mobility, circular, national or international (Gemenne and Blocher 2017).

---

[2] Cancún Adaptation Framework (2010), Paris Agreement of 2015, Global Compact for Safe, Orderly and Regular Migration (A/RES/73/195) (UN 2018), Sendai Framework for Disaster Risk Reduction (UN 2015).

The chapter first probes the relationships between the biophysical dimensions of climate change, land and mobility and their implications for family farming, then explores the land and migration dimensions associated with public policies of climate action, as well as their effects.

6.1 The Relationships Between Climate Change and Land and Migration Dynamics at the Heart of the Livelihoods of Family Farming

Family farming in the South is confronted with increasingly frequent and intense disasters, or with progressive changes in temperatures and precipitation. These two situations, with different temporalities, involve critical social relations over land and other resources, between rural populations and other actors. These social relations transform family farming through forms of endogenous adaptation, around local resources (Agrawal 2010), or involving mobility when the former are insufficient (Gemenne and Blocher 2017).

6.1.1 Climate Change and Family Farming: What Implications for Land Tenure?

By disappearing or degrading arable land, forests or grazing areas, climate change affects the very basis of the production processes of family farming. Its effects on land use changes are relatively well documented (see Chap. 4), but the changes produced on land tenure regimes are less so. It can be expected that the scarcity of land and its commodification will produce, depending on local contexts, differentiated effects in terms of equity, either by balancing land endowments or by reinforcing inequalities (Anseeuw and Baldinelli 2021). Climate change can also be a source of conflicts over land use and control, as shown by Vesco et al. (2020). Indeed, the scarcity or degradation of land can lead to new competitions and conflicts over land access, within family farming for agricultural or pastoral uses, or with other actors for other uses (energy, carbon sequestration). Climate change sometimes opens opportunities for land access for certain actors at the expense of family farming. Land acquisitions by new investors (international or national elites) can be linked to exogenous political mechanism in the context of mitigation practices (see Sect. 6.2.1) or to more endogenous dynamics related to the effects of climate change on the scarcity or degradation of land. The decline in the livelihoods of family farming due to climate change can force some families to sell their land to these local or foreign investors. Moreover, large-scale investments create local uncertainty in terms of land security, which can deter family farms from devoting labour or resources to managing climate risks on their own land

(Nyantakyi-Frimpong 2020). The implications of climate change on land tenure systems can therefore be multiple, but observations are still insufficient to really isolate it from other factors of evolution of land systems and their implications on family farming (Murken and Gornott 2022).

6.1.2 Climate Change and Migration Dynamics of Family Farming

Mobility has always been a response of rural populations to historical and contemporary climate changes (Gemenne and Blocher 2017; Cattaneo et al. 2019). The effects of climate change on mobility act in combination with other factors of mobility, such as demographic, socio-economic or political ones (Black et al. 2011). Thus, dissociating climate change-related mobility from mobility resulting from other factors remains a methodological challenge (Meze-Hausken 2008). Depending on the context, families' responses take various forms of chosen or forced (im) mobility, at different spatial and temporal scales, accompanied by arrangements between the migrant, his family and the group, on the resources of the places of origin and destination (Agrawal 2010).

Mobility can first be a punctual response to intra or interannual disruptions of the agricultural production system (Lalou and Delaunay 2015). The temporary economic mobility, often national, of a family member then ensures the continuity of livelihoods without further modifying the intra-family rules on resources.

Mobility can also be a structural response of families to persistent degradation of the agricultural production system, by a multi local social and residential organisation based on national or international circular mobility of one or several family members, between a rural origin space and one or more destinations (Potts 2010; Brüning and Piguet 2018; Mercandalli et al. 2019). In contexts even more constrained by resources, permanent migration to the city is frequent, contributing to the urban transitions at work in sub-Saharan Africa (Barrios et al. 2006). This type of migration is no longer a family adaptation system to an ecological crisis, it is an abandonment of the territory and a dispersion of the group (Mounkaïla 2002).

Whether it is temporary, circular or permanent mobility, climate change is most often an amplifier of more or less old migration dynamics, being part of the functioning of rural communities (Morrissey 2014). Climate change can, however, also contribute to initiating and structuring new mobilities, related to favourable socio-economic or ecological dimensions in the arrival spaces. Finally, in Africa as in Asia, the vast majority of mobility in a context of climate change follows the dynamics of rural mobility which take place first within countries (Mercandalli et al. 2019). International migrations, on the other hand, take place more within sub-regions (Leal et al. 2022).

Despite the attention given to mobility, those who leave are generally the minority compared to those who stay. Like mobility, immobility in a situation of climate change is more or less constrained. Individual decisions or consequences in terms of immobility also take various degrees of vulnerability and/or resilience (Zickgraf 2021; p. 127–28). Boas et al. (2022) note that mobility and immobility are potential "acts of resistance" in a changing climate. The recent report from the IPCC (2023) underlines that, in the decades to come, some people will not be able or will not want to leave places where they could nevertheless be vulnerable to the impacts of climate change. According to Benveniste et al. (2022), the climate change would result in a decrease in emigration of the lowest-income individuals by more than 10% in 2100 for average development and climate scenarios, and up to 35% for more pessimistic scenarios. However, how climate immobility should be managed and the standards that should underpin this management are rarely addressed (Thornton et al. 2023).

6.1.3 Interactions Between Land Dynamics and Migration in the Context of Climate Change

Mobility and access to land are two interdependent dynamics, reinforced by the effects of climate change. On the one hand, land inequalities, exacerbated by climatic factors, largely determine the mobility of rural people (Obeng-Odoom 2017). On the other hand, in contexts marked by a scarcity of arable land due to climate change, mobility, through the resources it provides, can ensure a complementary form of access to land (Rakotomalala et al. 2022). All these dynamics involve increased competition for land with a propensity to create changes in land regimes (Quan and Dyer 2008).

Depending on whether the effects of climate change allow for the continuity of agricultural activity or not, different sociopolitical configurations produce a plurality of mobilities and land effects in terms of access and land security. In the case of slow degradation of the quality and quantity of cultivated land, one scenario is where the quest for additional land takes place in new spaces, through circular or permanent migrations. The initial choice of destination spaces sometimes takes the form of opening or expanding a pioneer front. But often, these mobilities occur within communities presenting opportunities for activities and where the nature of local land regimes plays a determining role for access and securing the rights of migrants. A second scenario is where support policies for commercial agriculture create or reinforce flows of migrant workers affected by climate change, and in search of land and income. The populations of spaces receiving migrants can here find themselves in competition with the new arrivals for access to work and land, creating land conflicts. Finally, when there is no more margin on cultivated lands and the possibilities of pioneer fronts are exhausted, a third configuration can be identified in which mobilities are oriented towards protected spaces. In this case, illegal access to land followed by deforestation by migrants counteracts environmental policies while creating land conflicts in these spaces (Moizo 2000).

In the case of effects of climate change making lands unfit for agriculture or causing them to disappear, other problems tend to appear in the affected areas and the installation zones, such as competition for land between new arrivals, occupation of local properties or the overload of public land administration systems (Jacobs and Almeida 2020).

6.2 Land and Mobility in Mitigation and Adaptation Policies

Climate change as a field of political action translates into mitigation and adaptation policies that can have impacts on land and on the mobility of family farmers.

6.2.1 Contested Effects of Mitigation Policies on Land and Migration

The main mitigation pathways concern the land sector (IPCC 2019). They take various forms such as support for energy transitions (promotion of biofuels, wind or photovoltaic energy), protection of wetlands and forests, reforestation initiatives, afforestation, or forms of more sustainable agriculture. These mitigation policies affect land regimes (Hunsberger et al. 2017; Borras et al. 2020; Le Meur and Rodary 2022) and mobilities (Malkamäki et al. 2018).

Firstly, the promotion of biofuels has fuelled a "land rush" triggered by the rise in oil and agricultural commodity prices from the second half of the 2000s (German and Schoneveld 2012). Today, 17% of large-scale land acquisitions are linked to the production of biofuels according to the Land Matrix. These land acquisitions can be accompanied by the exclusion of local populations from access to land, particularly when land rights are not formally recognised (Cotula 2012). They sometimes undermine land reforms that aim to strengthen local rights, by encouraging the state to reaffirm its ownership in view of the economic interest of these projects (Burnod 2022). The anticipated decline of fossil fuels is also contributing to land pressure through the expansion of spaces reserved for production of wind or photovoltaic energy (Scheidel and Sorman 2012). The energy transition also drive to the growing demand for rare minerals. These mines particularly affect the lands of vulnerable populations according to Owen et al. (2023), who estimate that half of them are located on or near the lands of indigenous peoples or small family farmers.

Secondly, the conservation measures of mitigation policies have land and migratory effects through the regulation of access and authorised land uses (Le Meur and Rodary 2022). National parks, protected areas, classified forests are all institutional tools that are remobilised and extended in the fight against climate change, particularly in connection with the financing opportunities offered by carbon offset mechanisms (Cavanagh and Benjaminsen 2014). These mechanisms have direct impacts on land tenure, as they often aim to regulate or withdraw land rights from populations. Afforestation initiatives—like the "Great Green Wall" projects in the Sahel or in northern China—have also led to forms of enclosure with the physical exclusion and migration of former resource users (Turner et al. 2023).

Finally, market instruments and economic incentives aim to guide land use and agricultural practices. They do not directly modify land rights, but can have indirect effects. The definition of land rights and the identification of rights holders are central issues in the implementation of these instruments; this is the case for payments for environmental services (PES) and REDD+ projects (reducing emissions from deforestation and forest degradation). PES compensate actors for a certain use of land in order to create incentives for the provision of an ecosystem service. According to Karsenty (2019; p. 89), "the issue of property rights is important because the possibility of fulfilling the contract implies that the provider of the environmental service has management and exclusion rights over the land or natural resources concerned[3,4]". REDD+ projects rely on carbon finance to subsidise initiatives to reduce emissions linked to deforestation. They allow the creation of specific property rights for carbon credits, often disconnected from the recognition of land ownership. These projects have raised concerns from family farming coalitions and indigenous populations about who will decide on the authorised uses of forests and how the benefits from the sale of carbon credits will be shared (Larson et al. 2013). In this sense, a green grabbing (Fairhead et al. 2012) can occur without excluding populations from the land, but through their marginalisation in decision-making processes concerning resources and the unequal distribution of benefits derived from carbon credits (He and Wang 2023). The certification of carbon rights ensures certain benefits to populations, but in return requires the production of complex compliance rules and technocratic processes producing forms of exclusion. Land conflicts associated with REDD+ projects are particularly concentrated in protected areas, when carbon finance provides the means for their managers (state services or the private sector) to regain control over these spaces by excluding the populations who had been able to settle there for agriculture. The clarification of the land regime, the securing of the rights of populations and their involvement at different stages of projects contribute to better equity and viability of REDD+ projects (Larson et al. 2013).

When land rights are withdrawn from local populations, mitigation policies consequently have effects on mobilities. Former users may be forced to migrate (or to modify grazing routes) and these migrations can in turn give rise to new tensions over rural land. Moreover, the production of biofuels or large reforestation projects encourage the migration of labour to carry out strenuous agricultural work for low wages (Richardson 2010).

---

[3] https://landmatrix.org/

[4] For example, in the case of sharecropping, a PES aiming at restricting usage rights for the sharecropper could have an impact on the owner's income. In this case, land ownership will count in establishing a sharing of payments between the sharecropper and the owner (Karsenty 2019).

6.2.2 Land and Migration: Levers and Limits for Mitigation and Adaptation to Climate Change

Land security appears as an important lever for adaptation to climate changes for family farming (Castro and Kuntz 2022; Murken and Gornott 2022). Firstly, land tenure security can encourage investments for adapting cropping systems to new climatic conditions, even if the positive effects of land security on agricultural investments are not always verified in the litterature (Colin and Daoudi 2022). The anticipation of a yield decrease linked to climate change can lead to a reduction in investments regardless of the security of land rights. Thus, depending on the importance of the response to be provided by producers, a policy of securing land rights could have differentiated effects on investments and on adaptation strategies. Secondly, the securing of land transactions opens the opportunity for adaptation strategies through land markets. Indeed, transfers and acquisitions (permanent or temporary) can allow for diversification of cultivated plots and associated climatic risks. Transfers can also finance non-agricultural activities and migrations (Castro and Kuntz 2022). Furthermore, Improved land security can facilitate the implementation of policies to reduce CO_2 emissions (Djenontin Ida Nadia et al. 2018). However, when the formalisation of land rights is a prerequisite for the implementation of mitigation and adaptation policies, there is a risk that it may further marginalise the most vulnerable (Almeida and Jacobs 2022). The formalisation of private property does not necessarily mean an improvement in land security for populations (Colin et al. 2009). As for the effects of recognising community land rights on deforestation, the empirical literature is not unanimous. Local land regimes are diverse and embedded in broader institutional contexts and therefore have differentiated effects on deforestation, to be analysed in light of specific local contexts (Robinson et al. 2014).

Migration is also a central element of climate change mitigation and adaptation policies, the effects of which on family farming are questioned. A firts aspect of existing initiatives involves disaster risk reduction (DRR) policies. The aim is first to prevent disasters through infrastructure (dikes, drainage channels, etc.) and to improve warning systems (Giry 2023). Then, it is about mitigating the effects of the disaster by relocating and resettling populations living in high-risk areas (Almeida and Jacobs 2022). The construction of infrastructure or the resettlement of people often depends on land acquisition, affecting the rights of those who are expropriated and those who welcome new arrivals. Finally, when disasters occur, displacement[5] is the major adaptation mechanism. These displacements are temporary or permanent with resettlement (or planned relocation) and occur primarily within a country's borders (Naser 2012). They can lead to land insecurity and promote conflicts if not implemented carefully (Brzoska 2019). Indeed, these displacements inherently

---

[5] In the context of natural disasters, displacement is then a forced movement of people due to a public measure (CAF 2010). In 2020, of more than 30 million displacements related to natural disasters, 98% were due to rapid-onset events such as floods, storms or fires (IOM 2022).

bring about changes in land access, both in departure and arrival areas, making land rights a dual issue in responding to climate change (Quan and Dyer 2008; Jacobs and Almeida 2020). The choice of relocation sites is crucial. Prioritising public land status for these displacements can exclude options where the people concerned prefer family or group-based mobility, less prone to land conflicts than movements outside the group. This was the case with the successful resettlement within the island of Samoan families following the 2009 tsunami (Charan et al. 2017). Indeed, customary land systems deserve more recognition in responding to displacement and facilitating resettlement (Fitzpatrick 2022), including considering that some state-owned spaces are not exempt from customary rights and users without formal rights. Experience shows that voluntary programmes can be more effective and that planned resettlement over short distances is better suited (Correa et al. 2011). Finally, displacements have differentiated effects on access to assistance resources: while some families are better able to capture aid thanks to their social capital, displacement can also be favourable to more vulnerable populations who become autonomous thanks to new land rights (Giry 2023).

The second aspect of public action involves medium-term development policies in which authorities have begun to include mobility as a lever for adaptation, particularly in nationally determined contributions[6] (NDCs) and national adaptation plans (NAPs)[7] (Oakes et al. 2022). NDCs and NAPs can be significant vectors for integrating human mobility into national policy frameworks (Mombauer et al. 2023). Indeed, NDCs and NAPs can provide answers to questions of human (im)mobility related to climate change, firstly by mitigating migration pressures by avoiding displacements and reducing the need for planned relocation, and secondly by supporting migration as adaptation (Warner et al. 2014). While mobility is increasingly present in NDCs, few countries propose interventions to remedy its adverse effects or to promote adaptive aspects. Mombauer et al. (2023) shows that the dominant vision in NAPs is that of mobility as a risk or problem. However, NDCs and NAPs can improve the integration of human mobility as an opportunity, in a range of policy sectors that are priorities for adaptation and for loss and damage. This highlights the need for adequate funding and institutional capacities to strengthen the integration of human mobility into these instruments.

---

[6] NDCs are vehicles articulating each country's commitments to reduce national emissions and adapt to the impacts of climate change in accordance with the objectives of the Paris Agreement (UNFCCC 2015).

[7] The NAPs aim to identify medium and long-term adaptation needs and to develop and implement strategies, programmes and plans to address them. The NAP process was established at the 16th Conference of the Parties (COP) in 2010 within the Cancún Adaptation Framework, in order to identify medium and long-term adaptation needs and to develop and implement strategies, programmes and plans to address them (UNFCCC 2011).

References

Agrawal A (2010) Local institutions and adaptation to climate change. In: Mears and Norton (ed) Social dimensions of climate change: equity and vulnerability in a warming world, pp 173–198

Almeida B, Jacobs C (2022) Land expropriation – the hidden danger of climate change response in Mozambique. Land Use Policy 123:106408. https://doi.org/10.1016/j.landusepol.2022.106408

Anseeuw W, Baldinelli GM (2021) Uneven ground – Land inequalities at the heart of societal inequalities. ILC, OXFAM

Barrios S, Bertinelli L, Strobl EE (2006) Climatic change and rural–urban migration: the case of Sub-Saharan Africa. J Urban Econ 60(3):357–371. https://doi.org/10.1016/j.jue.2006.04.005

Benveniste H, Oppenheimer M, Fleurbaey M (2022) Climate change increases resource-constrained international immobility. Nat Clim Chang 12(7):634–641. https://doi.org/10.1038/s41558-022-01401-w

Black R, Adger WN, Arnell NW, Dercon S, Geddes A, Thomas D (2011) Migration and global environmental change. Glob Environ Chang 21:S1–S2. https://doi.org/10.1016/j.gloenvcha.2011.10.005

Boas I, Wiegel H, Farbotko C, Warner J, Sheller M (2022) Climate mobilities: migration, Im/mobilities and mobility regimes in a changing climate. J Ethn Migr Stud 48(14):3365–3379. https://doi.org/10.1080/1369183X.2022.2066264

Borderon M, Sakdapolrak P, Muttarak R, Kebede E, Raffaella Pagogna R, Sporer E (2019) Migration influenced by environmental change in Africa: a systematic review of empirical evidence. Demogr Res 41:491–544. https://doi.org/10.4054/DemRes.2019.41.18

Borras SM, Franco JC, Nam Z (2020) Climate change and land: insights from Myanmar. World Dev 129:104864. https://doi.org/10.1016/j.worlddev.2019.104864

Borras S, Scoones I, Baviska A, Edelman M, Peluso LN, Wolorf W (2023) Climate change and critical agrarian studies. In: Climate change and critical agrarian studies. Routledge, London/New York (N.Y.). https://pure.eur.nl/en/publications/climate-change-and-agrarian-struggles

Brüning L, Piguet E (2018) Environmental changes and migration in West Africa. A review of case studies. Belgeo 1. https://doi.org/10.4000/belgeo.28836

Brzoska M (2019) Understanding the disaster–migration–violent conflict nexus in a warming world: the importance of international policy interventions. Soc Sci 8(6):167. https://doi.org/10.3390/socsci8060167

Burnod P (2022) Large land acquisitions: realities, challenges and trajectories. In: Colin J-P, Lavigne DP, Léonard É (eds) Rural land in the South: Issues and keys to analysis, Objectifs Suds. IRD Éditions, Marseille, pp 633–716. https://doi.org/10.4000/books.irdeditions.45410

Burzyński M, Deuster C, Docquier F, De Melo J (2022) Climate change, inequality, and human migration. J Eur Econ Assoc 20(3):1145–1197. https://doi.org/10.1093/jeea/jvab054

Castro B, Kuntz C (2022) Land tenure insecurity and climate adaptation: socio-environmental realities in Colombia and implications for integrated environmental rights and participatory policy. In: Holland MB, Masuda YJ, Robinson BE (eds) Land tenure security and sustainable development. Springer International Publishing, Cham, pp 177–199. https://doi.org/10.1007/978-3-030-81881-4_9

Cattaneo C, Beine M, Fröhlich CJ, Kniveton D, Martinez-Zarzoso I, Mastrorillo M et al (2019) Human migration in the era of climate change. Rev Environ Econ Policy 13(2):189–206. https://doi.org/10.1093/reep/rez008

Cavanagh C, Benjaminsen TA (2014) Virtual nature, violent accumulation: the 'spectacular failure' of carbon offsetting at a Ugandan National Park. Geoforum 56:55–65. https://doi.org/10.1016/j.geoforum.2014.06.013

Charan D, Kaur M, Singh P (2017) Customary land and climate change induced relocation—a case study of Vunidogoloa Village, Vanua Levu, Fiji. In: Filho WL (ed) Climate change adaptation in pacific countries, Climate Change Management. Springer International Publishing, Cham, pp 19–33. https://doi.org/10.1007/978-3-319-50094-2_2

Clement V, Rigaud KK, de Sherbinin A, Jones B, Adamo S, Schewe J, et al (2021) Groundswell Part 2: acting on internal climate migration. http://hdl.handle.net/10986/36248

Colin J-P, Daoudi A (2022) Land dynamics, agrarian dynamics. In: Lavigne DP, Léonard É (eds) Rural land in the South: issues and keys to analysis, Objectifs Suds. IRD Éditions, Marseille, pp 399–471. https://doi.org/10.4000/books.irdeditions.45233

Colin J-P, Le Meur P-Y, Léonard E (2009) Land rights registration policies: from legal framework to local practices. Karthala, Paris, p 538

Correa E, Ramírez H, Sanahuja H (2011) Populations at Risk of Disaster: a resettlement guide. World Bank, Washington D.C. 157 p

Cotula L (2012) The international political economy of the global land rush: a critical appraisal of trends, scale, geography and drivers. J Peasant Stud 39(3–4):649–680. https://www.tandfonline.com/doi/abs/10.1080/03066150.2012.674940

Djenontin Ida Nadia S, Foli S, Zulu LC (2018) Revisiting the factors shaping outcomes for forest and landscape restoration in Sub-Saharan Africa: a way forward for policy, practice and research. Sustainability 10(4):906. https://doi.org/10.3390/su10040906

Fairhead J, Leach M, Scoones I (2012) Green Grabbing: a new appropriation of nature? J Peasant Stud 39(2):237–261. https://doi.org/10.1080/03066150.2012.671770

FAO (2022) The state of the world's forests 2022. Food and Agriculture Organisation. https://doi.org/10.4060/cb9360fr

Fitzpatrick D (2022) Research brief on land tenure and climate mobility in the Pacific Region. Pacific Islands Forum Secretariat, PRP

Gemenne F, Blocher J (2017) How can migration serve adaptation to climate change? Challenges to fleshing out a policy ideal. Geogr J 183(4):336–347. https://doi.org/10.1111/geoj.12205

German L, Schoneveld G (2012) A review of social sustainability considerations among EU-approved voluntary schemes for biofuels, with implications for rural livelihoods. Energy Policy (Renewable Energy in China) 51:765–778. https://doi.org/10.1016/j.enpol.2012.09.022

Giry B (2023) Sociology of disasters. Repères 813. La Découverte, Paris. 128 p

He J, Wang J (2023) Certificated exclusion: forest carbon sequestration project in Southwest China. J Peasant Stud 50(6):2165–2186. https://doi.org/10.1080/03066150.2022.2163163

Hunsberger C, Corbera E, Borras SM, Franco JC, Woods K, Work C et al (2017) Climate change mitigation, land grabbing and conflict: towards a landscape-based and collaborative action research agenda. Can J Dev Stud/ Rev Can Dev Stud 38(3):305–324. https://doi.org/10.1080/02255189.2016.1250617

IDMC, Internal Displacement Monitoring Centre (2023) 2023 Global Report on Internal Displacement (GRID). https://www.internal-displacement.org/publications/2023-global-report-on-internal-displacement-grid/

IPCC (2019) Climate Change and Land: an IPCC special report on climate change, desertification, land degradation, sustainable land management, food security, and greenhouse gas fluxes in terrestrial ecosystems

IPCC (2023) Climate Change 2022 – Impacts, Adaptation and Vulnerability: Working Group II Contribution to the Sixth Assessment Report of the Intergovernmental Panel on Climate Change. Cambridge University Press. https://doi.org/10.1017/9781009325844

Jacobs C, Almeida B (2020) Land and climate change: rights and environmental displacement in Mozambique. Research report. Van Vollenhoven Institute for Law, Governance and Society (VVI)

Karsenty A (2019) PES in developing countries: to compensate or reward? In: Laglais A (Coord) Agriculture and payments for environmental services. What legal questions?, PUR, 447 p. 978-2-7535-7601-8. https://hal.science/hal-02080034

Lalou R, Delaunay V (2015) Chapter 14. Seasonal migrations and climate change in rural Senegal. In: Sultan B, Lalou R, Sanni MA, Oumarou A, Soumaré MA (eds) Rural societies facing climate and environmental changes in West Africa. IRD Éditions, pp 287–313. https://doi.org/10.4000/books.irdeditions.9830

Larson AM, Brockhaus M, Sunderlin WD, Duchelle A, Babon A, Dokken T et al (2013) Land tenure and REDD+: the good, the bad and the ugly. Glob Environ Chang 23(3):678–689. https://doi.org/10.1016/j.gloenvcha.2013.02.014

Leal Filho W, Olaniyan OF, Nagle Alverio G (2022) Where to Go? Migration and climate change response in West Africa. Geoforum 137:83–87. https://doi.org/10.1016/j.geoforum.2022.10.011

Le Meur P-Y, Rodary E (2022) Land and environmental measures. In: Lavigne DP, Léonard É (eds) Rural land in the South: issues and keys to analysis, Objectifs Suds. IRD Éditions, Marseille, pp 863–940. https://doi.org/10.4000/books.irdeditions.45513

Lowder SK, Sánchez MV, Bertini R (2021) Which farms feed the world and has farmland become more concentrated? World Dev 142:105455. https://doi.org/10.1016/j.worlddev.2021.105455

Malkamäki A, D'Amato D, Hogarth NJ, Kanninen M, Pirard R, Toppinen A, Zhou W (2018) A systematic review of the Socio-economic impacts of large-scale tree plantations, worldwide. Glob Environ Chang 53:90–103. https://doi.org/10.1016/j.gloenvcha.2018.09.001

McDowell C (2013) Climate-change adaptation and mitigation: implications for land acquisition and population relocation. Dev Policy Rev 31(6):677–695. https://doi.org/10.1111/dpr.12030

Mercandalli S, Losch B, Belebema MN, Bélières J-F, Bourgeois R, Dinbabo MF et al (2019) Rural migration in sub-Saharan Africa: patterns, drivers and relation to structural transformation. FAO, Cirad, Rome. https://doi.org/10.4060/ca7404en

Meze-Hausken E (2008) On the (Im-)possibilities of defining human climate thresholds. Clim Chang 89(3–4):299–324. https://doi.org/10.1007/s10584-007-9392-7

Moizo B (2000) Deforestation and migration dynamics (Madagascar). In: Gillon Y, Chaboud C, Boutrais J, Mullon C, Weber J (eds) Good use of renewable resources. IRD Éditions, pp 169–185. https://doi.org/10.4000/books.irdeditions.25394

Mombauer D, Link A-C, Van Der Geest K (2023) Addressing climate-related human mobility through NDCs and NAPs: state of play, good practices, and the ways forward. Front Clim 5:1125936. https://doi.org/10.3389/fclim.2023.1125936

Morrissey J (2014) Environmental change and human migration in Sub-Saharan Africa. In: Piguet E, Laczko F (eds) People on the move in a changing climate, Global Migration Issues, vol 2. Springer Netherlands, Dordrecht, pp 81–109. https://doi.org/10.1007/978-94-007-6985-4_4

Mounkaïla H (2002) From circular migration to the abandonment of local territory in Zarmaganda (Niger). Eur J Int Migr 18(2):161–187. https://doi.org/10.4000/remi.1662

Murken L, Gornott C (2022) The importance of different land tenure systems for farmers' response to climate change: a systematic review. Clim Risk Manag 35:100419. https://doi.org/10.1016/j.crm.2022.100419

Naser MM (2012) Climate change, environmental degradation, and migration: a complex nexus. William Mary Environ Law Policy Rev 36(3)

Nyantakyi-Frimpong H (2020) What lies beneath: climate change, land expropriation, and Zaï Agroecological innovations by smallholder farmers in Northern Ghana. Land Use Policy 92:104469. https://doi.org/10.1016/j.landusepol.2020.104469

Oakes R, Van Der Geest K, Corendea C (2022) Any Port in a Storm? Climate, mobility, and choice in Pacific Small Island Developing States. In: Behrman S, Kent A (eds) Climate refugees. Cambridge University Press, pp 249–260

Obeng-Odoom F (2017) Unequal access to land and the current migration crisis. Land Use Policy 62:159–171. https://doi.org/10.1016/j.landusepol.2016.12.024

OIM (2019) Glossary on Migration. International migration law, n°34, 248 pp. ISSN 1813-2278. https://environmentalmigration.iom.int/sites/g/files/tmzbdl1411/files/iml_34_glossary.pdf

OIM (2009) Migration, Environment and Climate Change: Assessing the evidence. 448 pp, ISBN 978-92-9068-454-1https://publications.iom.int/system/files/pdf/migration_and_environment.pdf

Owen JR, Kemp D, Lechner AM, Harris J, Zhang R, Lèbre É (2023) Energy transition minerals and their intersection with land-connected peoples. Nat Sustain 6(2):203–211. https://doi.org/10.1038/s41893-022-00994-6

Potts D (2010) Circular migration in Zimbabwe & contemporary sub-Saharan Africa. James Currey, Woodbridge, Suffolk/Rochester. 312 p

Quan J, Dyer N (2008) Climate change and land tenure. The implications of climate change for land tenure and policy. FAO

Rakotomalala H, Bouquet E, Burnod P (2022) Land markets and access to land for migrants in Western Madagascar: opportunities and constraints. Rural Econ 381:79–93. https://doi.org/10.4000/economierurale.10375

Richardson B (2010) Big Sugar in Southern Africa: rural development and the perverted potential of sugar/ethanol exports. J Peasant Stud 37(4):917–938. https://doi.org/10.1080/03066150.2010.512464

Robinson BE, Holland MB, Naughton-Treves L (2014) Does secure land tenure save forests? A meta-analysis of the relationship between land tenure and tropical deforestation. Glob Environ Chang 29:281–293. https://doi.org/10.1016/j.gloenvcha.2013.05.012

Scheidel A, Sorman AH (2012) Energy transitions and the global land rush: ultimate drivers and persistent consequences. Glob Environ Chang 22(3):588–595. https://doi.org/10.1016/j.gloenvcha.2011.12.005

Thalheimer L, Williams DS, Van Der Geest K, Otto FEL (2021) Advancing the evidence base of future warming impacts on human mobility in African Drylands. Earth's Future 9(10):e2020EF001958. https://doi.org/10.1029/2020EF001958

Thornton F, Andreolla Serraglio D, Thornton A (2023) Trapped or staying put: governing immobility in the context of climate change. Front Clim 5:1092264. https://doi.org/10.3389/fclim.2023.1092264

Turner MD, Davis DK, Yeh ET, Hiernaux P, Loizeaux ER, Fornof EM et al (2023) Great green walls: hype, myth, and science. Annu Rev Environ Resour 48(1):263–287. https://doi.org/10.1146/annurev-environ-112321-111102

Vesco P, Dasgupta S, De Cian E, Carraro C (2020) Natural resources and conflict: a meta-analysis of the empirical literature. Ecol Econ 172:106633. https://doi.org/10.1016/j.ecolecon.2020.106633

Warner K, Kälin W, Martin SF et al (2014) Integrating human mobility issues within National Adaptation Plans. Policy brief n°9. UNU-EHS, Bonn. https://collections.unu.edu/eserv/UNU:1838/pdf11800.pdf

Zickgraf C (2021) Theorizing (Im)mobility in the face of environmental change. Reg Environ Chang 21(4):126. https://doi.org/10.1007/s10113-021-01839-2

Zickgraf C, Vigil S, Longueville F, Ozer P, Gemenne F (2016) The impact of vulnerability and resilience to environmental changes on mobility patterns in West Africa. World Bank

Open Access This chapter is licensed under the terms of the Creative Commons Attribution-NonCommercial-NoDerivatives 4.0 International License (http://creativecommons.org/licenses/by-nc-nd/4.0/), which permits any noncommercial use, sharing, distribution and reproduction in any medium or format, as long as you give appropriate credit to the original author(s) and the source, provide a link to the Creative Commons license and indicate if you modified the licensed material. You do not have permission under this license to share adapted material derived from this chapter or parts of it.

The images or other third party material in this chapter are included in the chapter's Creative Commons license, unless indicated otherwise in a credit line to the material. If material is not included in the chapter's Creative Commons license and your intended use is not permitted by statutory regulation or exceeds the permitted use, you will need to obtain permission directly from the copyright holder.

Chapter 7
Water, Agriculture and Climate Change: Global Perspectives

Magalie Bourblanc, Caroline Lejars, and Pierre-Louis Mayaux

Abstract One third of the global population is projected to face severe water shortages, largely due to climate change exacerbating existing unsustainable water use. The FAO's 2021 report highlights unprecedented pressure on water and soil resources, pushing them to their limits. Water is framed as a global public issue linked to security and potential conflicts, driving supply-focused policies and large infrastructure projects. Climate change adds urgency, promoting techno-solutions, but there is a risk of maladaptation if actions are rushed. Water is crucial for climate adaptation, requiring careful, balanced approaches.

According to current projections, one third of the world's population is expected to face severe water shortages for part of the year in the coming years, particularly as a result of climate change. Will climate disruption lead to equally dramatic hydrological disruption? In reality, climate change is more likely to reinforce and exacerbate—rather than solely initiate—an already unsustainable trajectory of water resource use, a pattern that has been unfolding in many parts of the world for several decades. The latest FAO report, which assesses the state of water and soil resources worldwide, is unequivocal. Entitled "Systems at breaking point" (FAO 2021), it leaves little room for doubt: "unprecedented pressure" is being exerted on natural resources, which are being "pushed to the limits of their production capacities".

In the first section, we will share a number of observations regarding the increase in water uses, outlining the main pressures exerted on water resources. We will then examine how water has been defined for many years as a global public problem, underlining—among other things—the security dimension that this global framing often emphasizes, and the role that unequal distribution of water could play in the outbreak of conflicts. We will subsequently analyse how this framing at the international level has supported the resurgence of supply-oriented policies focused on the

---

M. Bourblanc (✉) · C. Lejars · P.-L. Mayaux
Cirad, joint research unit Water Management, Actors, Uses (G-EAU),
Université de Montpellier, Montpellier, France
e-mail: magalie.bourblanc@cirad.fr

construction of large hydraulic infrastructures, particularly evident in various development aid policies. We will also consider how the challenge of combating climate change now provides a new rationale for techno-solutionism in general. Indeed, the issue of water is of paramount importance in the broader context of climate change adaptation. As the World Bank states, "water is to adaptation what energy is to mitigation, and the challenges the planet will face in adapting to water-related issues are immense".[1] It is essential not be blinded by the "climate emergency" and to avoid rushing into the implementation of measures that amount to maladaption (Boutroue et al. 2022).

7.1 Increasing Pressure on Water Resources: Systems at Breaking Point

7.1.1 Alarming Findings

The current state of global freshwater resources' availability is deeply alarming. For instance, 47% of the world's population currently lives in areas suffering from water scarcity at least 1 month a year. This figure is expected to rise to 57% by 2050, according to the United Nations global report.

Water stress[2] obviously varies depending on geographical areas. North Africa, Southern Africa and West Africa have less than 1700 m^3 per capita, a level considered critical for a nation's ability to sustainably meet food and water demands across sectors.

The trajectory of available freshwater resources is equally concerning (Fig. 7.1). Groundwater levels, primarily tapped by agriculture, are collapsing (Jasechko et al. 2024). Between 2000 and 2018, despite a decrease in per capita withdrawals, global renewable water resources per capita declined by approximately 20%.

The evidence is unequivocal. Over the past 50 years, i.e., prior even to the first observable impacts of climate change, global water demand has increased at twice the rate of population: a 600% increase over the past hundred years, more than double the pace of demographic expansion (Wada et al. 2016). Both consumption and withdrawals are skyrocketing. By 2050, global water demand is projected to increase by 55%, with industrial water demand expected to rise almost everywhere except possibly in North America and Western Europe. In Africa, where industrial use remains marginal, demand could increase by 800%, and by 250% in Asia.

---

[1] Global Water Expertise Centre, 2016. Soon to be dry? Climate change, water and economy, analytical summary, World Bank.

[2] Water stress is the ratio between water demand and available resources. It can be measured in percentage (and reach or exceed 100%) or from the amount of water available per person per year (m^3/person/year) (Falkenmark indicator). Areas with water availability per year and per person less than 1700 m^3/person/year are considered in a situation of "water stress".

7 Water, Agriculture and Climate Change: Global Perspectives

With every increment of global warming, regional changes in mean climate and extremes become more widespread and pronounced

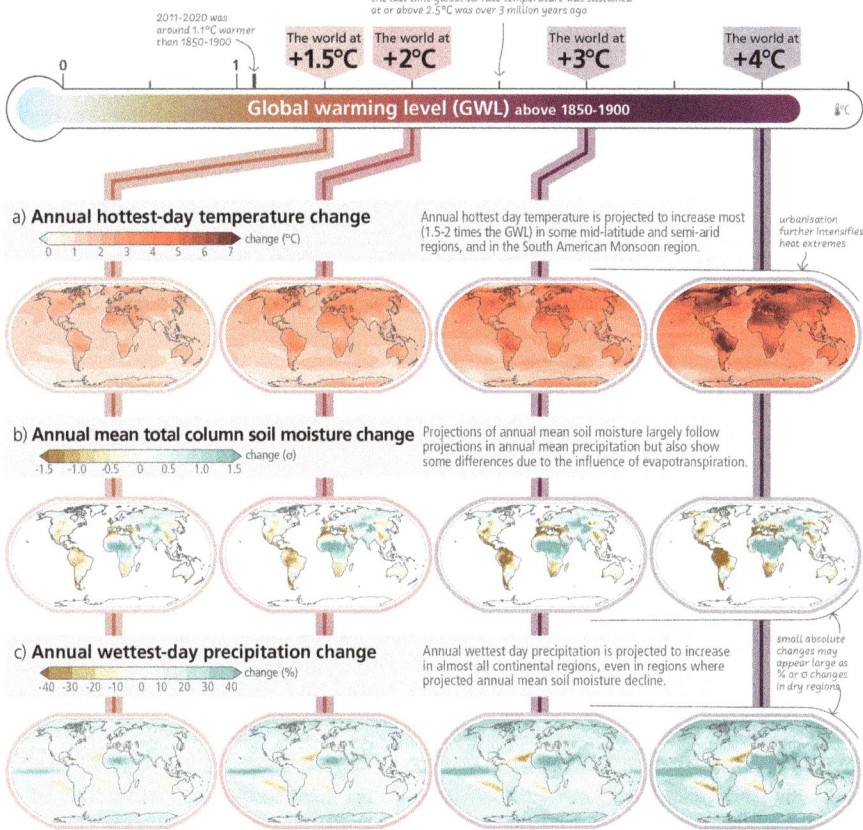

Fig. 7.1 Assessment of the impact of climate change on soil moisture and annual rainfall on a global scale for different levels of temperature increase. (Source: IPCC (2023))

Global water demand from the manufacturing sector alone is projected to rise by 400% (Boretti and Rosa 2019).

As for agriculture, which remains the largest water consumer across the globe, with an average consumption[3] of 60% of "blue" water (i.e., water flowing through surface and groundwater systems), the area equipped for irrigation has more than doubled in recent decades -from 139 million hectares in 1961 to 320 million hectares in 2012 (FAO 2014). Between 1970 and 2015, while the world population doubled, cereal production nearly tripled, vegetable production quadrupled, tomato production quintupled, and soybean production increased eightfold (FAO 2020).

---

[3] Consumption (or net withdrawal) corresponds to the part of the water withdrawn and not returned to aquatic environments.

Despite uncertainties in global projections, food demand is expected to rise by another 60% by 2050 (WWAP 2018).

Map

Global warming leads to an increase in average temperature as well as more pronounced extreme events, with significant regional variability. (a) The number of days with extreme temperatures is mainly increasing in semi-arid regions. (b) Soil moisture, as well as evapotranspiration, are affected. (c) Annual rainfall increases in continental regions, even in areas where soil moisture is decreasing.

7.1.2 Pressures on the Resource: The Specific Impact of Climate Change Remains Difficult to Determine

The main pressures on resources stem from population growth and the adoption of increasingly meat-based diets. Population growth alone is estimated to account for 54% of the increase in demand, while dietary shifts contribute approximately 25% (FAO 2020).

Global population growth drives an overall rise in water demand, even though per capita demand tends to decline (FAO 2020). According to FAO projections, the global population is expected to rise from 7.7 billion in 2019 to 9.7 billion by 2050—an increase of 26%.

Changes in dietary patterns represent the second key factor influencing increased water demand. Higher incomes and urban lifestyles are pushing food consumption towards higher intake of animal protein, fruits, and vegetables—all of which require greater resource inputs. In this context, the 58% global increase in meat consumption over the past two decades is likely to exert considerable pressure on water demand. These assumptions are based on average calculations suggesting that 1000 liters of water are needed to produce 1 kg of cereals, compared to 15,000 liters for 1 kg of beef. However, this generalised estimate overlooks the distinction between "green" and "blue" water. The livestock sector is indeed a major water consumer, but the vast majority (93%) of that use pertains to green water (rainwater) linked to crop production, and is not directly withdrawn from freshwater reserves. Only intensive, industrial livestock production systems consume significant amounts of blue water—i.e., water drawn from rivers, lakes, and aquifers.

Other growing pressures on resources include urbanization and the decline of land suitable for agriculture. Agricultural production has increased over the past 70 years despite a decrease in available farmland, largely due to intensified agricultural practices. This trend has led to a 20% expansion in irrigated land over the last two decades.

Water demand is also strongly correlated with energy demand. On the one hand, the development of energy crops in rainfed or irrigated systems can enhance energy supply, but it may also trigger increased competition over land ands water resources (Giovanetti and Ticci 2016). On the other hand, the expansion of hydropower intensifies river withdrawals, which can disrupt natural flow regimes and negatively impact local communities and ecosystems.

Global water scarcity is not solely attributable to physical shortages, but also to the progressive degradation of water quality in many countries (FAO 2017). This deterioration stems from insufficient sanitation (90% of wastewater is discharged untreated; WHO and UNICEF 2015), industrial pollution (300–400 million tons of waste are discharged annually into water bodies; WWAP 2018), and agricultural runoff. Agriculture is the predominant source of nitrogen contamination in water and an important source of phosphorus (UNEP 2016). In OECD countries, current nitrogen and phosphorus pollution from agriculture already exceeds the planet's absorption capacity. Yet, nitrogen and phosphorus discharges are projected to increase by 180% and 150%, respectively, by 2050 (OECD 2012).

In this context, the impact of climate change on water resources remains difficult to quantify, given the high degree of variability across and within countries. Climate simulations from the IPCC report reveal highly divergent outcomes, with some regions expected to experience increased water availability (e.g., Northern Europe, Russia, Central Asia), while others (e.g., Sub-Saharan and Saharan Africa) are already facing, or will face, significant declines. Additionally, scenarios indicate an increase in extreme weather events such as droughts and floods.

Finally, it is particularly challenging to disentangle the relative contributions of resource withdrawals and climate change to declining water availability. These two dynamics—resource depletion and rising demand—act synergistically, as global warming and reduced precipitation compel farmers to irrigate more. As a result, without adequate adaptation measures, more than 4 billion people could face water scarcity by 2050—a shortage that would also jeopardize global food security (FAO 2020). These raw data feed a narrative around a looming "water crisis", one that could exacerbate existing tensions, and potentially even fuel actual water-related conflicts, as we will see in the paragraph below.

7.2 From Crisis to Conflicts, and Ultimately to Water Wars?

The discourse surrounding water scarcity as a potential driver of full-scale crises is widely propagated by international institutions and amplified through media channels. Within the extensive body of international literature situated at the intersection of security, defence, and international development, a strong conceptual proximity exists between the themes of climate-induced conflict and water-related conflict (Homer-Dixon 1999; Koubi 2019). This literature often conceptualises scarcity as a destabilising force for states and societies. It is already well established that limited access to water can trigger large-scale migratory movements, primarily due to the food crises it engenders. When rural areas suffer, urban centres inevitably bear the consequences. The Middle East has long served as a case study for analyses linking political unrest to water shortages, and such analyses have since been extrapolated to the broader developing world (Gleick 2014). The hypothesis of a causal link between drought and violent conflict has, more recently, been posited for numerous countries ranging from Egypt and Indonesia to sub-Saharan Africa (Werrell et al. 2015; Caruso et al. 2016; Buhaug et al. 2015).

The dominant focus of this literature targets the "poor countries", which according to Baechler (1998; p. 25) include: Africa, Latin America, Central and South Asia. In doing so, it helps to reproduce an imaginary division between a conflict-ridden, anarchic "South", and a peaceful, civilised "North". Violent conflicts over water are generally portrayed as phenomena occurring only beyond the Mediterranean or the Rio Grande. This framing may come as a surprise to French readers aware of the fierce confrontations over "mega-basins"[4] in Deux-Sèvres region, or to American readers familiar with the credible threats made by Oregon's far-right militias to seize local dams in order to prioritise water flows into their own state.[5]

Indeed, the so-called "North" is by no means immune to water-related conflict, as illustrated by the resurgence of tensions over the Tagus River, a transboundary watercourse between Spain and Portugal. Portuguese environmental organisations now contest the Albufeira Convention signed in 2000, which permits the diversion of water towards the intensive agricultural zones of southern Andalusia.[6] These groups accuse Spain of excessive water extraction to the detriment of downstream Portugal. In this sense, water is poised to become a critical geopolitical issue not only for the Global South, but also for the North.

For a long time, academic scholarship sought to temper this growing alarmism and to challenge the supposedly inescapable logic of water-induced conflict. Scholars often highlighted the cooperative dimensions of water management and its role in cementing peace between rival states (Lasserre and Brun 2018), even between those sharing water-stressed basins or those otherwise politically hostile to one another.[7] A most commonly cited example is the existing agreement between India and Pakistan for the sharing of the Indus, in effect since 1960, which has endured despite three wars between the two nations. At the domestic level, numerous studies have also concluded that isolating water scarcity as a specific driver of social unrest remains methodologically challenging.[8]

---

[4] Barroux R., 2023. Megabasins: at Sainte-Soline, a massive mobilisation marked by violent confrontations, *Le Monde*.

[5] Wilson J., 2021. Amid mega-drought, rightwing militia stokes water rebellion in US West, *The Guardian*.

[6] N. Hervé-Fournereau quoted in the article by: Cailloce L., 2023. Will there be a water war?, *Le Journal du CNRS*.

[7] The database on international treaties concerning freshwater resources, maintained by the University of Oregon, estimates that out of 1831 "significant interactions" recorded since 1945, 1228 were "cooperative". It also lists over 400 active international treaties, most of which are "very durable".

[8] For a review of this empirical literature, see: Fröhlich C.J., 2012. Water: Reason for Conflict or Catalyst for Peace? The Case of the Middle East, *L'Europe en Formation*, 365(3), 139–161.

Another critical limitation of the "water wars" literature lies in its tendency to naturalise scarcity. While it is true that resource depletion is partly driven by "natural" causes—such as demographic growth, declining precipitation, and increased evapotranspiration due to climate change—scarcity is equally the outcome of a range of social decisions operating across multiple temporal scales. These include choices around agricultural production, integration into global trade, the condition of water storage infrastructure, energy systems, spatial planning policies, and consumption practices. The complex entanglement of these socio-political processes and their evolving nature over time are often overlooked by positivist scholars influenced by natural sciences (Jabri 1996; Magrin 2023).

Sociologist Lyla Mehta (2001) has been among the most prominent voices in articulating this critique.[9] Her aim is not to reject the validity of the raw data generated by this globalising framework, but to interrogate its underlying framing effects. By promoting "an absolute conception of environmental scarcity", such a perspective risks masking how scarcity affects territories and social groups in profoundly unequal ways. She contends that "real scarcity is, in most cases, not caused by the physical absence of water, but by a lack of financial resources and economic or political power". This aligns with other scholars who emphasise the capacity to mobilise water as a more decisive factor than raw availability (Blanchon 2024). In a similar vein, Fernandez (2014) has critiqued the depoliticising effects, because "averaging", of the widely cited "Falkenmark" water stress indicator, commonly used in media and public policy. Such an indicator obscure social inequalities in water access, both by relying on national averages and by implicitly endorsing the Malthusian premise that population growth is the principal driver of water stress (population increase mechanically degrading the indicator), thereby neglecting key social and political dynamics.[10]

Despite these nuances and appeals for analytical caution, the idea of heightened tensions around "blue gold" (akin to the past scramble for oil or "black gold") regularly resurfaces. The prevailing consensus now holds that water resources are rarely the primary cause of conflict, yet the changing climate may exacerbate pre-existing tensions. This elevates water to a matter of high political salience and global significance, as the following section will further explore. Indeed, by invoking the spectre of impending water-related conflicts, the international community may find justification for the standardisation of public policy solutions at a global scale.

---

[9] Mehta (2001). See also: Huff A., Mehta L., 2019. Untangling Scarcity, in Brewer J., Fromer N., Jonsson F.A., Trentmann F., *Scarcity in the Modern World History, Politics, Society and Sustainability, 1800–2075*, Bloomsbury Publishing, chap. 3, p. 27–46.

[10] This indicator also sidesteps the very political issue of division, between what should be produced on national territory with available water resources, and what should instead be imported.

7.3 Standardised and Widespread Policies at the International Level

7.3.1 The Resurgence of Resource Augmentation Projects

Large-scale hydraulic infrastructure projects for water storage had been declining since the 1990s. Indeed, NGOs had called for a moratorium, given their impact on biodiversity, the massive population displacements they caused and their mixed record in poverty reduction in developing countries. They were instrumental in the establishment of the World Commission on Dams under the auspices of the World Bank.

However, after a decline in investments at the turn of the twenty-first century, large hydraulic works made a strong comeback as early as 2003, supported by World Bank financing (Lynch 2013), with an increasing share of funding now coming directly from states themselves through projects in China, India, Iran, Turkey, Ethiopia, and others. The development of desalination projects constitutes another example of investment in techno-solutionism. By 2013, more than 17,000 seawater desalination plants operated worldwide, serving over 300 million people across 150 countries.

This techno-solutionism is often reinforced by the issue of climate change, both in its mitigation and adaptation dimensions: water indeed sits at the crossroads of these two facets of the fight against climate change. Today, this declared climate motivation increasingly dominates the justification for the continuation of these large infrastructure projects (Barone and Mayaux 2019). Limiting greenhouse gas (GHG) emissions, as committed under the Paris Agreement, requires clean energy production technologies. Hence, the resurgence of large hydropower projects, which on average produce only a tenth of the GHG emissions of gas or coal.

Yet, it is primarily in its contribution to climate change adaptation that the water sector assumes a crucial role today, as will be explored in the next section.

7.3.2 Water, a Core Public Policy Sector in the Climate Change Adaptation Strategy

Generally speaking, the number of adaptation measures has increased considerably since 2014. Most of these measures focus on the water sector, which currently accounts for 60% of all recorded actions. Moreover, the majority concentrate on the agricultural sector (IPCC report 2023), where the expansion of irrigation and its corollary, water storage, are often presented as a panacea. Indeed, global food production heavily depends on irrigation. Advocates of irrigation thus emphasise that: "Although irrigated lands only account for 20% of the global total, they alone provide 40% of global agricultural production. Without the 300 million hectares of irrigated land, we would need to mobilise an additional 600 million hectares from forests and pastures,"[11] which would further increase greenhouse gas emissions.

In this context, a clear disparity in treatment has long been observed in development aid policies, with funding primarily allocated to large-scale infrastructure projects, which are more visible and more valorised in the communication strategies of both national and international donors. This is also partly explained by the greater ease of disbursement for major projects. Conversely, the proliferation of small-scale projects involving new plot-level practices, cover crops, rainwater harvesting, the selection of more resilient crops, or even precision agriculture is relatively inexpensive, but always more time-consuming to manage. Donor investments thus tend to come at the expense of adapting production systems based on rainfed agriculture, which nevertheless accounts for 80% of cultivated lands and represents 60% of global food production.

The narrative constructed around climate disruption further accentuates this trend and now extends to so-called developed countries. In France, for example, the prospect of climate change has opportunistically been seized upon by interest groups clustered around irrigation to advocate for this singular solution: irrigation. This practice, which has been on the rise since the 1980s, emerging alongside the expansion of maize cultivation, is more topical than ever, especially with the promotion of water storage reservoirs. This lobbying effort has likely spurred a strong response from public authorities, who speak of a "declared emergency,"[12] especially given the new geopolitical context shaped by the Russo-Ukrainian conflict: water and agriculture have now acquired a particularly important geostrategic dimension in terms of defending food sovereignty.

7.3.3 The Silent Revolution of Groundwater

Alongside investment policies in large-scale irrigation infrastructure, highly problematic dynamics are observed in semi-arid regions, where the expansion of groundwater pumping irrigation is underway. An agricultural boom based on groundwater, which began in the 1960s, was triggered in many semi-arid regions by the combination of now easily accessible pumping and irrigation technologies and private individual initiatives. Molle et al. (2003) referred to this agricultural boom centred on groundwater as a "silent revolution," driven by small private investments often operating on the margins of public policies and associated with a high degree of informal arrangements (Shah 2009; López-Gunn et al. 2012; Kuper et al. 2016; Lejars et al. 2017).

---

[11] See p. 7 of the following publication: Benoît G., 2020. Agriculture, land, water and climate, *Futuribles*, 438(5), 5–27.

[12] Tandonnet H., Lozach J.-J., 2016. Water: emergency declared, Senate information report, no 616.

The laissez-faire approach of public authorities toward small private investments linked to groundwater exploitation is currently resulting in resource depletion, water quality degradation, land subsidence, biodiversity loss, and the deepening of social inequalities. These global trends have been documented in numerous arid and semi-arid countries such as Algeria, Australia, China, India, Mexico, Morocco, Spain, Tunisia, and the United States of America, where the irregular nature of surface water availability makes groundwater a strategic resource for irrigation (Scott and Shah 2004; Konikow and Kendy 2005; Llamas and Martínez-Santos 2005; Shah 2009; Ross and Martinez-Santos 2010; Kuper et al. 2016; IWMI 2020).

In many countries, the expansion of groundwater exploitation for irrigation is supported by public policies through direct subsidies to irrigation equipment, as well as through energy and land policies that facilitate the development of new irrigated areas. Access to land and water is eased by laws that allow farmers without formal property rights over land or water to obtain access rights—such as by legalizing existing wells—and enable new farmers employing capital-intensive production methods to settle in local production zones. By granting subsidies for agriculture and irrigation equipment, states often encourage and finance the development of new technologies such as drip irrigation and, ultimately, the expansion of irrigated lands. Finally, a recent development is the subsidy policy for photovoltaic panels observed in certain countries like India, which significantly reduces the energy costs typically associated with groundwater pumping, thereby incentivizing irrigators to exploit the resource without limits and threatening to rapidly deplete groundwater aquifers.[13]

7.4 A Necessary Shift in Policies Towards Less Intensive and Less Water-Demanding Models

Over the past decade, countries have significantly intensified their efforts to modernise their water and agriculture sectors. The primary justification for public support of large infrastructure projects, as well as the promotion of irrigation, rested on their presumed connection to the developmental agenda and the poverty reduction it was supposed to foster in countries of the Global South. These trends are currently being reinforced and legitimised by climate change. Financial support for large hydraulic infrastructures, irrigation equipment, and drip irrigation has increased across all areas, with a growing share directly funded by national governments. Simultaneously, public subsidies for irrigation equipment and energy have generated and sustained a groundwater boom in many regions worldwide.

These massive investments have contributed to feeding the 800 million people added to the global population between 2010 and 2020. They have also helped accommodate evolving dietary patterns and global meat consumption.

---

[13] https://e360.yale.edu/features/solar-water-pumps-groundwater-crops

However, research indicates that although investments in irrigation development have recently contributed to overall wealth increases, this has come at the cost of rising inequalities, increased migration, and greater pressure on water resources (Mayaux and Lejars 2022). In India, for example, the large-scale population displacements caused by the construction of dams and reservoirs for irrigation disproportionately affect poor communities compared to other societal groups (Duflo and Pande 2005). The overall economic benefits themselves remain questionable. Studies concluding on positive contributions on economies generally focus on national or regional scales, as data are often unavailable at sub-national levels (Narayanamoorthy 2018). Even at these levels, other authors highlight the increased energy consumption (Bazilian et al. 2012). In Morocco, Doukkali and Lejars (2015) demonstrated the economic burden of this new energy dependence, concluding that the Moroccan economy would have benefited more from investments in rainfed agriculture. Not all farmers engaged in this groundwater-based economy have been able to equitably reap the benefits, which are primarily captured by well-endowed entrepreneurs and wealthier agricultural operators. Capital has supplanted land as the main production factor in this new economy, marginalizing family farmers or exposing them to considerable risks given the necessary investments, sometimes leading to bankruptcy (Ameur et al. 2017; Mayaux 2021). Other studies reveal that this dependency renders societies particularly vulnerable, with some even questioning their social and economic collapse following an initial period of environmental decline (Petit et al. 2017).

Although these pressures vary considerably between and within countries, they tend to intensify, particularly under the effect of increased climate variability. In the years to come, governments will need to reassess their current levels of political and financial commitment to their water and agriculture sectors, while undertaking significant policy shifts toward less resource-intensive agricultural models.

7.5 Conclusion

The consequences of climate change on water resources remain uncertain, insofar as model projections have primarily focused on the global level, without refining forecasts through more regionalised scenarios. However, it is highly likely that these consequences will exacerbate already problematic dynamics in water management, which currently affect the hydrological cycle and have led to the near disappearance of groundwater in certain areas. In response, numerous reports and analyses—not solely linked to climate change—have, for several years, warned of escalating risks associated with water issues. This has generated a degree of apprehension, particularly within international organisations, which have amplified alerts and alarmist discourse: for instance, a 2016 World Bank report predicts new civil wars linked to water scarcity.

These tensions, which could intensify both internationally and locally, in countries of the Global South as well as the Global North, should necessarily prompt us to reconsider how we use, manage and share water resources. This is all the more pressing given that water management currently stands at the intersection of several major challenges: the human right to water; biodiversity and the preservation of aquatic ecosystems; and new agricultural demands. In other words, the water sector is not only strategic in the fight against climate change, but is also central to debates on various issues, some of which lie at the heart of our societal projects. Consequently, water has become increasingly politicised.

In the face of these new challenges, the primary response emerging appears to be a continuation of the same model. The inevitability of climate change is conveniently presented as the main cause of present and future water shortages, despite not currently being the predominant pressure factor, as noted earlier. This seems to foster an environment conducive to the reactivation of former public policy solutions that were already leading water policies into a dead end. Following the major shift toward demand-side management, supply-side policies have regained legitimacy through the lens of climate change: we are yet again looking to exploit hypothetical new water resources, even though this often amounts more to a mirage than a reliable promise.

In a way, this observation is not surprising. The solutions put forward in the fight against the impact of climate change on water resources are dominated by the agricultural sector, which is particularly mobilised on these issues, and for whom the prospect of climate change serves as an opportunity to better advocate for a single solution, resorting to irrigation. In contrast, water sector actors have thus far appeared more reticent on these issues, partly for reasons previously discussed in this chapter: the trajectory of the resource, even without climate change impacts, was already extremely worrying. It is also worth noting that, while the greening or "ecologisation" of water policies has been evident at least since the 2000s and the adoption of the Water Framework Directive at the European level, their "climatisation" has yet to be effectively realised. It therefore seems essential today to more fully engage in this debate in order to propose alternative responses and advocate for greater sobriety alongside mechanisms aimed at better controlling water demand.

References

Ameur F, Amichi H, Kuper M, Hammani A (2017) Specifying the differentiated contribution of farmers to groundwater depletion in two irrigated areas in North Africa. Hydrogeol J 25:1579–1591

Baechler G (1998) Why environmental transformation causes violence: a synthesis, vol 24(4). Environmental Change and Security Project Report, pp 24–44

Barone S, Mayaux P-L (2019) Water policies. LGDJ. 160 p

Bazilian M, Rogner H, Howells M, Hermann S, Arent D, Gielen D, Yumkella KK (2012) Considering the energy, water and food nexus: towards an integrated modelling approach. Energy Policy 39(12):7896–7906

Blanchon D (2024) Geopolitics of water. Between conflicts and cooperations. Le cavalier bleu. 144 p

Boretti A, Rosa L (2019) Reassessing the projections of the world water development report. NPJ Clean Water 2:15. https://doi.org/10.1038/s41545-019-0039-9

Boutroue B, Bourblanc M, Mayaux P-L, Ghiotti S, Hrabanski M (2022) The politics of defining maladaptation: enduring contestations over three (mal)adaptive water projects in France, Spain and South Africa. Int J Agric Sustain 20(5):892–910

Buhaug H, Benjaminsen TA, Sjaastad E, Theisen MO (2015) Climate variability, food production shocks, and violent conflict in Sub-Saharan Africa. Environ Res Lett 10:125015

Caruso R, Petrarca I, Ricciuti R (2016) Climate change, rice crops, and violence: evidence from Indonesia. J Peace Res 53:66–83

Doukkali MR, Lejars C (2015) Energy cost of irrigation policy in Morocco: a social accounting matrix assessment. Int J Water Res Dev 31(3):422–435

Duflo E, Pande R (2005) Dams. Social Science Research Network. 56 p. https://ssrn.com/abstract=796170

FAO (2014) AQUASTAT: area equipped for irrigation. In: FAO [online]. Rome. Cited 09 June 2016. https://www.fao.org/aquastat/en/geospatial-information/global-maps-irrigated-areas

FAO (2017) Water pollution from agriculture: a global review. Colombo, FAO & IWMI

FAO (2020) FAOSTAT. https://www.fao.org/faostat

FAO (2021) The state of the world's land and water resources for food and agriculture: systems at breaking point. Rome

Fernandez S (2014) Governing water since 1945. Internationalisation and intensification of capital flows, techniques and models. In: Pestre D (ed) The governance of technosciences. Governing progress and its damages since 1945. Paris, La Découverte, Recherches, 6, pp 203–230

IPCC (2023) 6th assessment report of the intergovernmental panel on climate change. Geneva

Giovanetti G, Ticci E (2016) Determinants of biofuel-oriented land acquisitions in sub-Saharan Africa. Renew Sustain Energy Rev 54(C):678–687

Gleick PH (2014) Water, drought, climate change, and conflict in Syria. WCAS 6:331–340

Homer-Dixon T (1999) Environment, scarcity, and violence. Princeton University Press. 272 p

IWMI (2020) Groundwater governance in the Arab World. Cited 6 August 2021 http://gw-mena.iwmi.org/outputs/

Jabri V (1996) Discourses on violence. Conflict analysis reconsidered. Manchester University Press. 204 p

Jasechko S, Seybold H, Perrone D (2024) Rapid groundwater decline and some cases of recovery in aquifers globally. Nature 625:715–721. https://doi.org/10.1038/s41586-023-06879-8

Konikow LF, Kendy E (2005) Groundwater depletion: a global problem. Hydrogeol J 13(1):317–320

Koubi V (2019) Climate change and conflict. Annu Rev Polit Sci 22(1):343–360

Kuper M, Faysse N, Hammani A, Hartani T, Marlet S, Hamamouche MF, Ameur F (2016) Liberation or anarchy? The Janus nature of groundwater use on North Africa's new irrigation frontiers. In: Jakeman T, Barreteau O, Hunt R, Rinaudo JD, Ross A (eds) Integrated groundwater management. Springer, the Netherlands, pp 583–615

Lasserre F, Brun A (2018) The sharing of water. A geopolitical reflection. Odile Jacob, Paris. 208 p

Lejars C, Daoudi A, Amichi H (2017) The key role of supply chains actors in the development of groundwater irrigation in North Africa. Hydrogeol J 25:1595–1603. https://doi.org/10.1007/s10040-017-1571-7

Llamas MR, Martínez-Santos P (2005) Intensive groundwater use: silent revolution and potential source of social conflicts. J Water Resour Plan Manag 131(5):337–341

López-Gunn E, Rica M, Van Cauwenbergh N (2012) Taming the groundwater chaos. In: de Stefano L, Llamas MR (eds) Water, agriculture and the environment in Spain: can we square the circle? CRC Press, pp 227–240

Lynch B (2013) Rivers of contention: scarcity discourse and water competition in highland Peru. Geo J Int Comp L 42:69–92

Magrin G (2023) Agriculture, climate crisis and violence: anything new under the Sahel sun? Bull Assoc Fr Geogr 100(1):6–19

Mayaux P-L (2021) Layering and perpetuating: the logics of conservative reforms in Morocco's irrigation policies. J North Afr Stud. https://doi.org/10.1080/13629387.2021.195034

Mayaux P-L, Lejars C (2022) Enabling institutional environments conducive to livelihood improvement and adapted investments in sustainable land and water uses. FAO., 58 p, Rome. https://doi.org/10.4060/cc0950en

Mehta L (2001) The manufacture of popular perceptions of water scarcity: dams and water-related narratives in Gujarat, India. World Dev 29(12):2025–2041

Molle F, Shah T, Barker R (2003) The groundswell of pumps: multilevel impacts of a silent revolution. Conference paper for the International Commission on Irrigation and Drainage/Asia Regional Workshop: Management and Operation of Participatory Irrigation Organizations. 10–12 November, 2003. Taipei (CHN)

Narayanamoorthy A (2018) Financial performance of India's irrigation sector: an historical analysis. Int J Water Res Dev 34(1):116–131

OECD (2012) OECD environmental outlook to 2050. Paris. https://doi.org/10.1787/9789264122246-en

Petit O, Lopez GA, Kuper M, Rinaudeau JD, Lejars C, Douadi A (2017) Resilience or collapse? Looking at agricultural groundwater economies under stress. Hydrogeol J 25:1549–1564. https://doi.org/10.1007/s10040-017-1567-3

Ross A, Martinez-Santos P (2010) The challenge of groundwater governance: case studies from Spain and Australia. Reg Environ Change 10(4):299–310. https://doi.org/10.1007/s10113-009-0086-8

Scott C, Shah T (2004) Groundwater overdraft reduction through agricultural energy policy: insights from India and Mexico. Water Res Dev 20(2):149–164

Shah T (2009) Taming the anarchy: groundwater governance in South Asia. Routledge. 324 p

UNEP (2016) A snapshot of the world's water quality: towards a global assessment

Wada Y, Flörke M, Hanasaki N, Eisner S, Fischer G, Tramberend S, Satoh Y et al (2016) Modelling global water use for the 21st century: the water futures and solutions (WFaS) initiative and its approaches. Geosci Model Dev 9:175–222

Werrell CE, Femia F, Sternberg T (2015) Did we see it coming? State fragility, climate vulnerability, and the uprisings in Syria and Egypt. SAIS Rev Int Aff 35:29–46

WHO, UNICEF (2015) Progress on sanitation and drinking water: 2015 update and MDG assessment. New York

WWAP (2018) The United Nations world water development report 2018. UNESCO, Paris. www.unwater.org/publications/world-water-development-report-2018/

Open Access This chapter is licensed under the terms of the Creative Commons Attribution-NonCommercial-NoDerivatives 4.0 International License (http://creativecommons.org/licenses/by-nc-nd/4.0/), which permits any noncommercial use, sharing, distribution and reproduction in any medium or format, as long as you give appropriate credit to the original author(s) and the source, provide a link to the Creative Commons license and indicate if you modified the licensed material. You do not have permission under this license to share adapted material derived from this chapter or parts of it.

The images or other third party material in this chapter are included in the chapter's Creative Commons license, unless indicated otherwise in a credit line to the material. If material is not included in the chapter's Creative Commons license and your intended use is not permitted by statutory regulation or exceeds the permitted use, you will need to obtain permission directly from the copyright holder.

Chapter 8
Food Systems: Both Responsible for and Victims of Climate Change

Hélène David-Benz, Arlène Alpha, Victoria Bancal, Carine Barbier, Damien Beillouin, Yannick Biard, Daniel Fonceka, Franck Galtier, Sandra Payen, Ninon Sirdey, and Mathieu Weil

Abstract Climate change and food systems are now key priorities on global and national agendas, with resilience emphasized at the 2021 UN Food Systems Summit and COP28 in 2023. Food systems include all stages from input production to consumption and waste, contributing about one-third of global greenhouse gas emissions. Industrialized countries are major emitters due to their production and consumption patterns, while developing countries, though less responsible, face greater climate impacts. This chapter analyzes both emission sources within food systems and climate change's effects, especially on post-production stages. It highlights increasing food security risks, particularly in low-income countries.

H. David-Benz (✉)
Cirad, UMR MoISA, Saint-Pierre, La Réunion, France

MoISA, Univ Montpellier, CIHEAM-IAMM, Cirad, INRAE, Institut Agro, IRD, Montpellier, France
e-mail: helene.david-benz@cirad.fr

A. Alpha · F. Galtier · N. Sirdey
MoISA, Univ Montpellier, CIHEAM-IAMM, Cirad, INRAE, Institut Agro, IRD, Montpellier, France

Cirad, UMR MoISA, Montpellier, France

V. Bancal
Cirad, UMR Qualisud, Abidjan, Ivory Coast

UFR Sciences and Technologies of Food, University Nangui Abrogoua, Abidjan, Ivory Coast

Qualisud, Univ Montpellier, Avignon University, Cirad, Institut Agro, University of La Réunion, Montpellier, France

C. Barbier
UMR International Research Centre on Environment and Development, CNRS, ENPC, Cirad, AgroParisTech, EHESS, Montpellier, France

D. Beillouin
Cirad, UPR HortSys, Caribbean Agro-Environmental Campus (CAEC), Martinique, France

HortSys, Cirad, University of Montpellier, Montpellier, France

© The Author(s) 2026
V. Blanfort et al. (eds.), *Climate Impacts and Challenges in Agriculture, Forests and Food Systems*, https://doi.org/10.1007/978-3-032-04331-3_8

The interactions between climate change and food systems are now prominently featured on both national and international agendas. The 2021 United Nations Food Systems Summit identified resilience to climate change as a key priority. Likewise, at COP28 in 2023, 159 countries endorsed a declaration advocating the inclusion of agriculture and food in their national climate commitments (see Chap. 2). Indeed, food systems represent a critical lever for mitigation—particularly in industrialised countries—and pose significant challenges for adaptation, as they are directly impacted by the effects of climate change.

Food systems encompass all the activities required to feed a population, including pre-production (such as the manufacture of inputs), primary production, and post-production stages (transport, processing, distribution, consumption) as well as waste management throughout the chain. The term agri-food systems is also used, including in several chapters of this book, and similarly includes all activities from upstream to downstream. The systemic approach, implied by the concept of *food system*, highlights the interconnections among various activities and their socio-economic, environmental (notably greenhouse gas emissions, GHGs) and nutritional impacts (David-Benz et al. 2022; FAO 2018; HLPE 2017).

When considering all stages, food systems account for approximately one-third of global anthropogenic GHG emissions (Crippa et al. 2021; Rosenzweig et al. 2020; Tubiello et al. 2022). Simultaneously, they are also affected by effects of climate change.

This chapter explores both dimensions of the relationship between food systems and climate changes, with a particular emphasis on the disparities between country categories. Indeed, industrialised countries, due to the characteristics of their production systems, supply chains and consumption patterns, are major contributors to GHG emissions. In contrast, developing countries are both the least contributors to emissions and the most affected by climate change-related impacts.

Y. Biard
Cirad, UPR HortSys, Montpellier, France

HortSys, Cirad, University of Montpellier, Montpellier, France

ELSA (Environmental Life cycle and Sustainability Assessment), Montpellier, France

D. Fonceka
Cirad, UMR AGAP, Montpellier, France

Cirad, INRAE, AGAP, University of Montpellier, Institut Agro, Montpellier, France

S. Payen
Cirad, UMR ABSys, Montpellier, France

UMR ABSys, University of Montpellier, Cirad, INRAE, Institut Agro, Montpellier, France

M. Weil
Qualisud, Univ Montpellier, Avignon University, Cirad, Institut Agro, University of La Réunion, Montpellier, France

Cirad, UMR Qualisud, Montpellier, France

The first section examines the various GHG emission sources within food systems. The second section addresses the impact of climate change on these systems, with a particular focus on downstream stage—an area that has received little scholarly attention compared to the impacts on primary production. This analysis highlights the growing risks to food security, particularly in low-income countries.

8.1 Heterogeneous Food Systems, Responsible for Nearly One-Third of GHG Emissions

8.1.1 Pre- and Post-harvest Stages as Key Contributors to the Rise in GHG Emissions from Food Systems

Recent studies on GHG emissions from food systems differentiate three main emission sources: (1) land use changes, (2) production, (3) pre- and post-production, which includes the production of inputs and all downstream activities (Tubiello et al. 2022). This approach, by incorporating land use changes, sheds light on the major climate impact of deforestation, driven by agricultural expansion.

GhG emissions originate from a variety of sources and are influenced by agricultural practices, the types of food produced and consumed, and waste management strategies. Drawing on FAOSTAT data, Tubiello et al. (2022) demonstrate that over three decades (1990–2019) total emissions from the food system increased by 17%, primarily due to a doubling of emissions from pre- and post-production activities. Emissions from land use changes decreased by 25%, reflecting a deceleration in deforestation), while farm-level emissions increased by 9% (Fig. 8.1).

Emissions associated to pre-production, encompassing all emissions generated by the manufacture and transport of inputs, vary widely across food system, and their precise share remains difficult to quantify (Tubiello et al. 2022). The five stages of post-production (transport, transformation, distribution, consumption, waste management) account for approximately 20–30% of global food system emissions. Food transport, often over long distances, contributes significantly to emissions, though estimates vary. Crippa et al. (2021) estimate this contribution at 4.8% of total food system emissions, while Li et al. (2022) suggest it could be as high as 20.0%. These emissions are higher in industrialised countries than in developing countries due to significant reliance on road transport in the former. The processing and packaging stage accounts for approximately 5–10% of total emissions. These processes are energy-intensive and often rely on fossil fuels. Distribution and retail are also GHG emitters, notably due to the refrigeration of food in supermarkets; they represent about 2–4% of total emissions. Finally, food waste—when disposed of landfill, produces methane during decomposition. Waste management thus contributes around 3–4% of total GHG emissions.

While the preparation of meals at home or in restaurants accounts for only 3–5% of total emissions, dietary choices themselves significantly influence the food

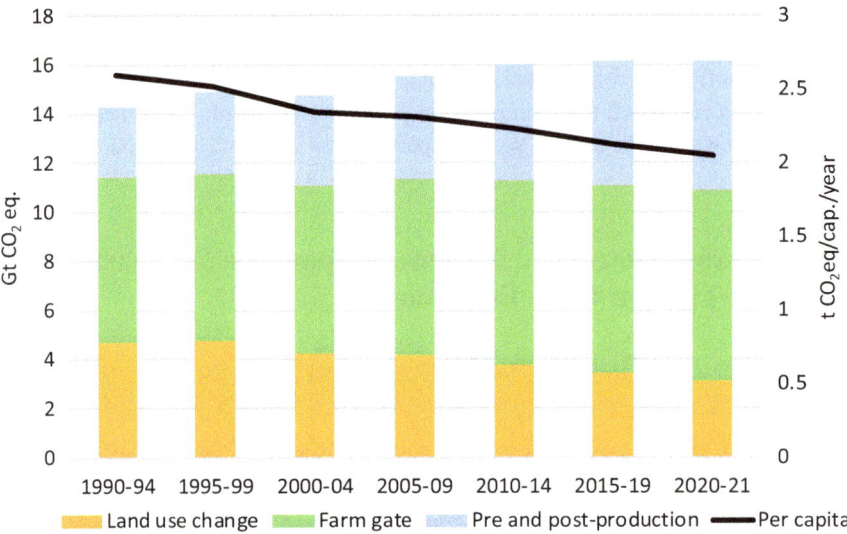

Fig. 8.1 Evolution of GHG emissions by food systems by major source, 1990–2021. (Source: FAOSTAT data. https://www.fao.org/faostat/en/#data/GT and https://www.fao.org/faostat/en/#data/OA). Total GHG emissions from food systems rose from 14.3 to 16.1 gigatonnes CO_2 equivalent between 1990–94 and 2020–21. This increase is primarily related to the "Pre- and post-production" item, which grew from 20% to 32% of the total during this period. The "Farm Gate" category remained the largest contributor, though relatively but stable (47–48% of the total), while emissions from "Land use change" declined substantially (33–19%). Per capita emissions from food systems have declined over the same period, from 2.6 to 2.0 tonnes CO_2 equivalent

system's overall carbon footprint. Indeed, emission intensities vary greatly between food types: producing 1 kg of beef emits approximately 14 kg CO_2 eq/kg, compared to 3 kg for pork, 2 kg for poultry,[1] and less than 1 kg for plant-based food (often around 0.1 kg). As a result, both the consumption of meat and the cultivation of feed crops for livestock—which is often traded internationally—significantly increase the carbon footprint of food systems. Animal production accounts for nearly 75% of agricultural production emissions (WHO 2023). Methane emissions from enteric fermentation of ruminants and manure management are the most significant sources, followed by nitrous oxide emissions resulting from nitrogen fertiliser use—primarily for feed crops in high-income countries. Depending on dietary patterns, carbon footprints can vary substantially: between vegetarian or vegan diets and heavily meaty-based diets (100–170 g of meat/day), emissions may increase by a factor of 3–6 (Barbier et al. 2019; Scarborough et al. 2023). Nevertheless, consumption of animal products varies significantly across and within countries. In 2021, per capita meat consumption was 92 kg/capita/year in high-income countries, compared to the global average of 43 kg and just 12 kg in low-income countries (Ritchie et al. 2024). Dairy consumption, which also contributes to GHG emissions (primarily from

---

[1] These emissions are estimated per kilogram of live weight of the animals.

ruminants), shows a similar disparity. Global average milk consumption reached 88 g/day/capita in 2018—double the amount in 1990 (Miller et al. 2022). The highest consumption levels are recorded in Mexico, in the UK, the USA, and France (188–206 g/day), while the lowest are observed in China, Bangladesh, and the Democratic Republic of Congo (31–37 g/day).

Beyond dietary patterns, GHG emissions from the same product can vary considerably depending on production methods and post-production processes. These differences can be captured through life cycle assessment analysis (Box 8.1).

8.1.2 Significant Disparities Between Industrialised and Other Food Systems

Numerous studies have demonstrate that GHG emissions from food systems vary considerably across countries (Fig. 8.2).

In Africa and Asia, per capita GHG emissions are significantly lower than in other regions (Fig. 8.2). However, due to population growth and shifting dietary

> **Box 8.1 Why adopt a holistic approach to the food system? The life cycle assessment (LCA) approach**
>
> In both French and international literature, the Life Cycle Assessment (LCA) approach is considered the standard for evaluating environmental impacts. Based on a multicriteria framework, it offers a comprehensive view of all sources of impact on ecosystems, human health, and non-renewable resources, using either a product-based perspective or, increasingly, territorial scales (Cornelus et al. 2021). LCA is particularly effective in quantifying GHG emissions across entire value chains, enabling a more detailed understanding than that offered by sectoral approaches.
>
> **LCA in the Analysis of Value Chains.**
>
> The **Value Chain Analysis for Development (VCA4D)** project, funded by the European Commission and implemented by Agrinatura, employs a systematic methodological framework to analyse agricultural value chains, providing evidence-based insights to inform decisions aimed at enhancing their sustainability (Fabre et al. 2021). The LCA approach identifies and quantifies the resources consumed and emissions generated across entire value chains, enabling a holistic evaluation of environmental impacts. Since 2016, more than 45 value chains in low-income countries have been examined, revealing, among other findings, the critical role of infrastructure in shaping the carbon footprint of various food systems. These studies also demonstrate that inadequate infrastructure—such as limited access to electricity, poor road networks, unreliable hydrocarbon supply, lack of clean water, and inefficient transport and logistics—indirectly exacerbates environmental impacts by increasing product losses, particularly for perishable goods (Parrot et al. 2018).

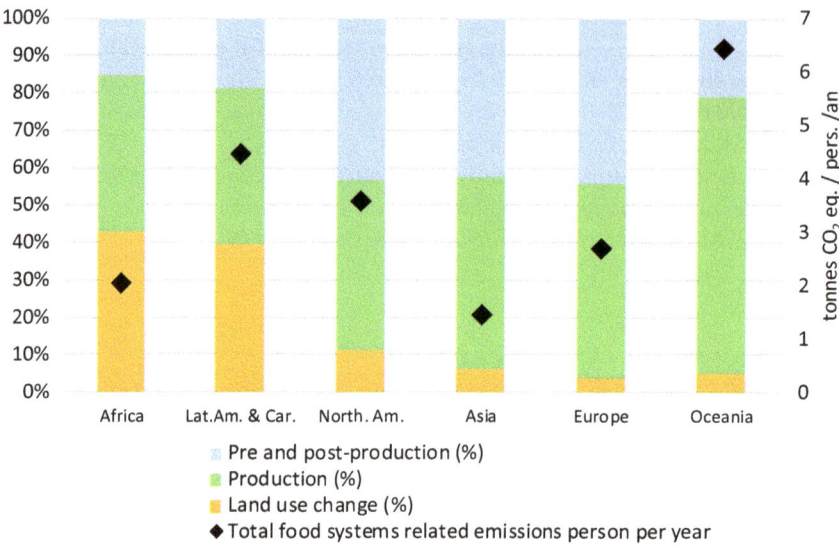

Fig. 8.2 GHG emissions from food systems by major regions (2021). (Source: FAOSTAT data). Per capita GHG emissions from food systems, expressed in kilograms per person per year, are as follows (in ascending order): 1.5 in Asia, 2 in Africa, 2.7 in Europe, 3.6 in North America, 4.4 in Latin America/Central America/Caribbean and 6.4 in Oceania. Latin America/Central America/Caribbean and Africa are notable for the high proportion of emissions attributable to "Land Use Change" (40–43%). Asia, North America and Europe are characterized by a significant share of "Pre- and Post-production" (42–44%). Oceania stands out for the predominant share of "Production" (74%)

habits—particularly toward increased consumption of meat and processed foods—total emissions from food systems are rising rapidly, with a notable increase in pre- and post-production stages in Asia. In Europe and North America, per capita emissions are considerably higher, primarily due to more meat-intensive diets and the increasing contribution of post-production processes. Emissions are also very high in Latin America, largely as a result of continued deforestation

8.2 Food Systems Affected by Climate Change

Climate change is characterised by shifts in temperature and precipitation patterns, as well as an increased frequency and intensity of climate-related shocks. These various manifestations of climate change impact the entire food system, from production to consumption. The effects on agricultural production have been extensively documented in the scientific literature, which shows that global agricultural output is expected to experience overall reductions in yield, a decline in product quality, and decreased livestock productivity. Retrospective studies report actual yield losses of 4–5% for major crops such as maize, wheat, rice, and soybean (Iizumi et al. 2018; Moore et al. 2015). Projections further suggest an average yield reduction of 11% by 2050 in the absence of adaptation measures (Hasegawa et al. 2022). These negative effects are expected to be particularly severe in vulnerable regions, including sub-Saharan Africa and Southeast Asia.

In contrast, the literature on the broader impacts of climate change on food systems—particularly on post-harvest stages and especially in low-income countries—is considerably less developed. Nevertheless, the associated risks are substantial, with serious implications for food security. This is particularly concerning as nutrient-rich foods such as fruits, vegetables, meat, and fish are among the most perishable. Even less perishable products, such as cereals and tubers, are increasingly susceptible to losses due to rising temperatures, which promote the spread of pests and toxic microorganisms.

This section first examines the impacts of climate change on food losses, marketing, and product quality. It then addresses the broader implications for food security, with a particular focus on the most vulnerable countries.

8.2.1 Impacts on Losses, Product Quality, and Trade

Climate change affects all actors within food systems. Rising temperatures and increasing frequency of climate shocks heighten the risk of losses and raise storage and transport costs, thereby impacting the incomes of value chain actors and prices for consumers. In high-income countries, this primarily results in increased costs for processing, packaging, storage, and transport. In low-income countries, losses are exacerbated by inadequate infrastructure and limited preservation methods, particularly for highly perishable products (Box 8.2). The literature identifies three major post-harvest impacts: degradation of the nutritional and visual quality of products, heightened risks to food safety, and increased market instability.

8.2.1.1 Degradation of Nutritional and Visual Quality of Products

According to Christopoulos and Ouzounidou (2020), climate change exerts varied effects on the quality of fruits and vegetables. While it can enhance the synthesis and accumulation of carbohydrates and antioxidants and stimulate plant defence mechanisms, it may also reduce the content of proteins, minerals, and amino acids, while degrading visual appeal. Elevated temperatures lead to discoloration, "sunburn," and changes in texture. Accelerated production cycles may reduce fruit size and increase the accumulation of undesirable compounds. For example, tomatoes cultivated under high-temperature conditions have demonstrated lower concentrations of essential micronutrients (K, Mg, Ca) as well as lycopene, carotene, and antioxidants (Rosales et al. 2011).

> **Box 8.2 Perception of risks and adaptation strategies among fruit and vegetable traders**
>
> A survey conducted under the SAFOODS project among 796 tomato, mango, and leafy vegetable retailers and wholesalers in Côte d'Ivoire and Senegal found that 78% of respondents identified high temperatures and 48% heavy rainfall as the most severe manifestations of climate change. These events adversely affect both the quantity and quality of production (and thus the traders' supply capacity), and damage transportation infrastructure. Reported effects include leaf desiccation, accelerated ripening, fruit senescence, rot, insect damage, and disruption of commercial activity—ultimately resulting in economic and/or nutritional losses.
>
> Adaptation strategies employed by traders include increasing the frequency and reducing the volume of purchases to ensure same-day sales, sourcing directly from producers, sorting products to sell at differentiated prices, and identifying outlets prior to purchase. Despite these efforts, traders report average losses of 2.4% for tomatoes, 6.5% for mangoes, and 3.8% for leafy vegetables. In addition, 8.3%, 10.5%, and 11.6%, respectively, are either sold at a loss or given away—resulting in considerable economic losses that are likely to intensify with the increasing frequency of extreme climate events. Increased risks to the food safety.

Higher temperatures and humidity levels promote the proliferation of foodborne pathogens and contaminants that compromise the safety (and potentially the nutritional and sensory quality) of food products. This includes the growth of pathogenic bacteria such as *Salmonella* and *Campylobacter*, which thrive under elevated temperatures (Akil et al. 2014). Similarly, mycotoxigenic fungi—such as those producing aflatoxins—pose serious threats to human and animal health, including carcinogenic risks (Battilani et al. 2016; Watson et al. 2016; Wild et al. 2015; WHO and Joint FAO/WHO Expert Committee on Food Additives 2017). For instance, certain strains of *Aspergillus flavus*, which affect a variety of crops, produce aflatoxins whose production is significantly stimulated by increased CO_2 levels and drought stress (Medina et al. 2015).

In Europe, a temperature increase of just 2 °C could cause 40% of maize production to exceed legal aflatoxin limits (Battilani et al. 2016). The presence of toxin-producing fungi is also expected to expand into new geographic areas, thereby exacerbating food safety risks, particularly in regions lacking sufficient monitoring and risk management infrastructure (Miller 2016).

8.2.1.2 Increasing Market Instability, to Which International Regulatory Instruments Are Ill-Suited

Climate change is expected to disrupt production zones, affect volumes and stability of supply, and hinder transport—thus destabilising both domestic and international markets (IPCC 2023). The estimated impacts vary depending on the climate scenario, region, commodity, and methodology. By 2050, food prices could rise by 5–20% relative to a no-climate-change scenario, and price volatility could increase by 10% (Chen and Villoria 2019; Nelson et al. 2014; Wiebe et al. 2015).

Moreover, climate change interacts with other structural factors contributing to higher and more volatile food prices: population growth, rising meat consumption in emerging economies, increased use of crops for non-food purposes (e.g., biofuels), and promotion of more sustainable but less productive farming systems (Brunelle and Dumas 2019; Galtier 2019). In this context, Daviron (2020) and Galtier (2021) note that biofuel production absorbs 15% of global maize and vegetable oil output (Chap. 19), while China has restructured its agricultural policy to reduce soil pollution, offsetting lower domestic production by securing imports via the "new silk roads."

These dynamics have triggered successive price crises on international markets (2008, 2010–2011, 2021–2023), with severe repercussions for global food security—a situation projected to worsen (Galtier 2019). In this evolving context, international trade and stock management will play a crucial role in mitigating risks (Chen and Villoria 2019; Wiebe et al. 2015). However, the current rules of the World Trade Organization (WTO)—designed during a period of agricultural surplus—are poorly adapted to contemporary challenges of scarcity. While they restrict export subsidies, they do not prevent export restrictions or bans, and they severely constrain the ability of countries to build public stock, viewing such measures as market distortions that support producers (Galtier 2023). Several reform attempts have failed due to opposition from key exporting nations. An additional threat to food security.

Extreme weather events are now among the principal drivers of food insecurity and malnutrition, alongside armed conflict, economic slowdowns, and widening inequalities (Bezner et al. 2022; FAO et al. 2023). Low-income countries are particularly affected, as their food supply largely depends on small-scale family farming, which primarily relies on rainfed crops and pastoral livestock systems. These systems often lack adequate early warning mechanisms, rendering them especially vulnerable to climate shocks. Declining incomes among smallholder farmers—and among small-scale coastal fishers affected by the depletion of fishery resources—will reduce their capacity to secure sufficient and nutritious food (FAO 2018; Mbow et al. 2019). Indigenous populations also rank among the most climate-vulnerable groups. Often poor and marginalised, they face heightened risks due to the close ties between their traditional food systems and increasingly threatened ecosystems (Jantarasami et al. 2018; Smith and Rhiney 2016).

The volatility and upward trajectory of food prices will compel low-income households to reduce the diversity of their diets. These trends are likely to disproportionately affect landlocked, low-income countries that are heavily dependent on food imports, as well as poor households located in flood-prone regions. Furthermore, declining water quality and rising temperatures are expected to exacerbate food insecurity. Children in rural areas of low-income countries will be particularly at risk due to diminished availability and diversity of food, greater exposure to extreme heat, and the increased incidence of diarrhoeal and vector-borne diseases (Oppenheimer and Anttila-Hughes 2016).

8.3 Conclusion

Food systems contribute significantly to GHG emissions, particularly in high-income countries and in regions where deforestation remains extensive. Consequently, the transformation of food systems represents a major lever for GHG mitigation.

At the same time, food systems are also profoundly affected by climate change, across both production and post-production stages. Rising temperatures and the increasing frequency of climate shocks have cumulative effects on food production—impacting its quantity, stability, and quality—as well as on food preservation and processing. These changes exacerbate market instability and contribute to rising food prices. As a result, they affect the incomes of producers and downstream actors in the supply chains, while also impacting consumers through reduced product availability and/or increased prices—particularly for the most perishable yet nutritionally essential products, such as fruits and vegetables.

However, these impacts are highly heterogeneous. They disproportionately affect the most vulnerable regions and populations in low-income countries, which are already heavily impacted by multifactorial food insecurity. This is due to their greater exposure to climate change impacts, reliance on vulnerable production systems, limited access to equipment, weaker infrastructure, and low purchasing power.

In a context of multiplying crises and shocks of various kinds, climate change thus represents an additional constraint to achieving the goal of feeding the nine billion people expected by 2050.

References

Akil L, Ahmad HA, Reddy RS (2014) Effects of climate change on salmonella infections. Foodborne Pathog Dis 11(12):974–980

Barbier C, Couturier C, Pourouchottamin P, Cayla J-M, Silvestre M, Pharabod I (2019) The energy and carbon footprint of food in France—from production to consumption. CIRED. https://www.centre-cired.fr/wp-content/uploads/2021/05/empreinte_carbone_alimentation_en_france_fr_052019.pdf

Battilani P, Toscano P, Van Der Fels-Klerx HJ, Moretti A, Camardo LM, Brera C et al (2016) Aflatoxin B1 contamination in maize in Europe increases due to climate change. Sci Rep 6(1):24328

Bezner KR, Hasegawa R, Lasco I, Bhatt D, Deryng A, Farrell A et al (2022) 2022: Food, fibre, and other ecosystem products. In: Intergovernmental Panel On Climate Change (Ipcc) (ed) Climate change 2022: impacts, adaptation and vulnerability. Contribution of Working Group II to the Sixth Assessment Report of the Intergovernmental Panel on Climate Change, Cambridge University Press, Cambridge, UK and New York, NY, pp 713–906. https://doi.org/10.1017/9781009325844.007

Brunelle T, Dumas P (2019) Risks of higher food prices on international markets. In: Dury S, Bendjebbar P, Hainzelin E, Giordano T, Bricas N (eds) Food systems at risk: new trends and challenges. FAO, Cirad and European Commission, pp 103–105. https://openknowledge.fao.org/handle/20.500.14283/ca5724en

Chen B, Villoria NB (2019) Climate shocks, food price stability and international trade: evidence from 76 maize markets in 27 net-importing countries. Environ Res Lett 14(1):014007

Christopoulos M, Ouzounidou G (2020) Climate change effects on the perceived and nutritional quality of fruit and vegetables. J Innov Econ Manag 34(1):79–99

Cornelus M, Pradinaud C, Villevieille A, Roux P (2021) Overview of environmental assessment methods. Methodological Guide Version 1.1, ELSA-PACT. https://www.elsa-pact.fr/content/download/3775/37021?version=1

Crippa M, Solazzo E, Guizzardi D, Monforti-Ferrario F, Tubiello FN, Leip A (2021) Food systems are responsible for a third of global anthropogenic GHG emissions. Nat Food 2(3):198–209

David-Benz H, Sirdey N, Deshons A, Orbell C, Herlant P (2022) Conceptual framework and method for national and territorial diagnostics—activating the sustainable and inclusive transition of our food systems. FAO, Cirad, European Union., 70 p. https://doi.org/10.4060/cb8603fr

Daviron B (2020) Biomass. A history of wealth and power. Quae editions, Versailles, p 392

Fabre P, Dabat M-H, Orlandoni O (2021) Methodological note for the analysis of agricultural value chains. Frameworks and tools—key elements. Version 2, Technical and Research Document, Paris, France, Agrinatura EEIG. 43 p

FAO (2018) The future of food and agriculture—alternative pathways to 2050. Summary version. FAO, Rome, Italy. 60 p

FAO, IFAD, WHO, WFP, UNICEF (2023) The State of Food Security and Nutrition in the World 2023. Urbanisation, transformation of Agri-food systems and access to healthy food along the rural-urban continuum. FAO, UNICEF, IFAD, WFP, WHO. 343 p

Galtier F (2019) Why food prices are likely to become more unstable. In: Dury S, Bendjebbar P, Hainzelin E, Giordano T, Bricas N (eds) Food systems at risk: new trends and challenges. FAO, Cirad, European Commission, pp 107–110. https://doi.org/10.19182/agritrop/00105

Galtier F (2021) Reading note on "biomass. A history of wealth and power". Nat Sci Soc 29(2):242–244

Galtier F (2023) Take an inch for a mile. About an error of metrics in WTO rules and its impact on the ability of countries to build public stocks for food security. Food Pol 116:102400

Hasegawa T, Wakatsuki H, Ju H, Vyas S, Nelson GC, Farrell A et al (2022) A global dataset for the projected impacts of climate change on four major crops. Sci Data 9(1):58

HLPE (2017) HLPE Report # 12—Nutrition and food systems. https://openknowledge.fao.org/server/api/core/bitstreams/4ac1286e-eef3-4f1d-b5bd-d92f5d1ce738/content

Iizumi T, Shiogama H, Imada Y, Hanasaki N, Takikawa H, Nishimori M (2018) Crop production losses associated with anthropogenic climate change for 1981-2010 compared with preindustrial levels. Int J Climatol 38(14):5405–5417

IPCC (2023) Climate change 2023: synthesis report. Contribution of Working Groups I, II and III to the Sixth Assessment Report of the Intergovernmental Panel on Climate Change

Jantarasami L, Novak R, Delgado R, Narducci C, Marino E, McNeeley S et al (2018) Chapter 15: tribal and indigenous communities. In: Impacts, risks, and adaptation in the United States: the Fourth National Climate Assessment, vol II. U.S. Global Change Research Program

Li M, Jia N, Lenzen M, Malik A, Wei L, Jin Y, Raubenheimer D (2022) Global food-miles account for nearly 20% of total food-systems emissions. Nat Food 3(6):445–453

Mbow C, Rosenzweig C, Barioni LG, Benton TG, Herrero M, Krishnapillai M (2019) Food security. In: Climate change and land: an IPCC special report on climate change, desertification, land degradation, sustainable land management, food security, and greenhouse gas fluxes in terrestrial ecosystems

Medina Á, Rodríguez A, Sultan Y, Magan N (2015) Climate change factors and *Aspergillus flavus*: effects on gene expression, growth and aflatoxin production. World Mycotoxin J 8(2):171–180

Miller JD (2016) Mycotoxins in food and feed: a challenge for the twenty-first century. In: Li D-W (ed) Biology of microfungi. Springer, pp 469–493

Miller V, Reedy J, Cudhea F, Zhang J, Shi P, Erndt-Marino J et al (2022) Global, regional, and national consumption of animal-source foods between 1990 and 2018: findings from the Global Dietary Database. Lancet Planet Health 6(3):e243–e256

Moore M-L, Riddell D, Vocisano D (2015) Scaling out, scaling up, scaling deep: strategies of non-profits in advancing systemic social innovation. JCC 58:67–84

Nelson GC, Valin H, Sands RD, Havlík P, Ahammad H, Deryng D et al (2014) Climate change effects on agriculture: economic responses to biophysical shocks. Proc Natl Acad Sci 111(9):3274–3279

Oppenheimer M, Anttila-Hughes J (2016) The science of climate change. Future Child 26(1):11–30

Parrot L, Biard Y, Kabré E, Klaver D, Vannière H (2018) Analysis of the mango value chain in Burkina Faso—final report, expertise report. Cirad, Montpellier, France. 230 p

Ritchie H, Rosado P, Roser M (2024) Meat and dairy production. Our World in Data. https://ourworldindata.org/meat-production

Rosales M, Cervilla L, Sánchez-Rodríguez E, Rubio-wilhelmi M d M, Blasco Leon MB, Rios J et al (2011) The effect of environmental conditions on nutritional quality of cherry tomato fruits: evaluation of two experimental Mediterranean greenhouses. J Sci Food Agric 91(1):152–162

Rosenzweig C, Mbow C, Barioni LG, Benton TG, Herrero M, Krishnapillai M et al (2020) Climate change responses benefit from a global food system approach. Nat Food 1(2):94–97

Scarborough P, Clark M, Cobiac L, Papier K, Knuppel A, Lynch J et al (2023) Vegans, vegetarians, fish-eaters and meat-eaters in the UK show discrepant environmental impacts. Nat Food 4:565–574. https://doi.org/10.1038/s43016-023-00795-w

Smith R-AJ, Rhiney K (2016) Climate (in)justice, vulnerability and livelihoods in the Caribbean: the case of the indigenous Caribs in northeastern St. Vincent. Geoforum 73:22–31

Tubiello FN, Karl K, Flammini A, Gütschow J, Obli-Laryea G, Conchedda G et al (2022) Pre- and post-production processes increasingly dominate greenhouse gas emissions from Agri-food systems. Earth Syst Sci Data 14(4):1795–1809

Watson S, Chen G, Sylla A, Routledge MN, Gong YY (2016) Dietary exposure to aflatoxin and micronutrient status among young children from Guinea. Mol Nutr Food Res 60(3):511–518

WHO, Joint FAO/WHO Expert Committee on Food Additives (2017) Evaluation of certain contaminants in food: eighty-third report of the Joint FAO/WHO Expert Committee on Food Additives. World Health Organization, Geneva. 182 p. https://iris.who.int/handle/10665/254893

WHO (2023) Red and processed meat in the context of health and the environment: many shades of red and green: information brief. https://www.who.int/publications/i/item/9789240074828

Wiebe K, Lotze-Campen H, Sands R, Tabeau A, Van Der Mensbrugghe D, Biewald A et al (2015) Climate change impacts on agriculture in 2050 under a range of plausible socioeconomic and emissions scenarios. Environ Res Lett 10(8):085010

Wild CP, Miller JD, Groopman JD (2015) Mycotoxin control in low- and middle-income countries, vol 9. IARC Working Group Report. 66 p

Open Access This chapter is licensed under the terms of the Creative Commons Attribution-NonCommercial-NoDerivatives 4.0 International License (http://creativecommons.org/licenses/by-nc-nd/4.0/), which permits any noncommercial use, sharing, distribution and reproduction in any medium or format, as long as you give appropriate credit to the original author(s) and the source, provide a link to the Creative Commons license and indicate if you modified the licensed material. You do not have permission under this license to share adapted material derived from this chapter or parts of it.

The images or other third party material in this chapter are included in the chapter's Creative Commons license, unless indicated otherwise in a credit line to the material. If material is not included in the chapter's Creative Commons license and your intended use is not permitted by statutory regulation or exceeds the permitted use, you will need to obtain permission directly from the copyright holder.

Chapter 9
Forests and Climate Change

Jacques Tassin, Alexandre Caron, Vincent Freycon, Bruno Hérault,
Bruno Locatelli, Marie Ange Ngo Bieng, Régis Peltier, and Camille Piponiot

Abstract A forest is defined by the FAO as an area with over 10% tree cover and trees taller than 5 m. Forests vary greatly in structure and their integration with agricultural landscapes. Although agriculture causes 80% of deforestation—about 50,000 km² annually since 1990—the climate change impacts on forest-agriculture interactions are little studied. Multi-sectoral territorial projects like TerrAmaz in the Brazilian Amazon are rare. This chapter focuses on how forests influence and are affected by climate change, proposing actions to help human-transformed forests adapt.

A forest, according to the FAO definition, is an area characterised by a tree cover rate of more than 10% and a tree height exceeding 5 m (Keenan et al. 2015). It exhibits a wide diversity of structural attributes, compositions and extents, as well as various degrees of integration within agricultural landscapes. The impact of climate change on the interactions between forests and agriculture, beyond the deforestation induced by 80% of agricultural fronts, amounting to approximately 50,000 km² per year since 1990 (MacDicken et al. 2016), are however scarcely addressed in the scientific literature. Projects dedicated to a territorial and multi-sectoral approach, such as TerrAmaz in Brazilian Amazonia (Poccard-Chapuis 2022), remain rare. Therefore, this chapter focuses on forests *per se*, and briefly

---

J. Tassin (✉) · R. Peltier · C. Piponiot
Cirad, UPR Forests and Societies, Montpellier, France
e-mail: jacques.tassin@cirad.fr

A. Caron
UPR Forests and Societies, UMR ASTRE, Cirad, Montpellier, France

V. Freycon · B. Hérault · B. Locatelli
UPR Forests and Societies, Cirad, Montpellier, France

M. A. N. Bieng
UPR Forests and Societies, Cirad, Montpellier, France

Unit Forests and Biodiversity in Productive Landscapes, Guatemala City, Guatemala

© The Author(s) 2026
V. Blanfort et al. (eds.), *Climate Impacts and Challenges in Agriculture, Forests and Food Systems*, https://doi.org/10.1007/978-3-032-04331-3_9

explains how forests govern and are affected by climate change. It also proposes some possible levers for action to enable forests, as spaces transformed by humans, to adapt to climate change.

9.1 The Forest as a Fundamental Link in the Global Climate

Forests cover the planet over 3900 Mha divided into three major geographical domains: the tropical domain accounting for approximately 50% (2000 Mha), the boreal (1100 Mha) and the temperate (750 Mha) domains. The annual monitoring of forests, as is set up in French Guiana, helps to clarify their role in the global carbon cycle. Naturally regenerated secondary forests represent 57% of the total forest area, and only 7% of global forests are plantations (FAO 2020).

The carbon stocks of the planet's forests are of the order of 860 GtC, of which 380 Gt (44%) is in the soil (up to about 1 m deep), 360 GtC (42%) in living biomass (above and below ground), 75 GtC (9%) in dead wood, and 45 GtC (5%) in litter. Tropical forests represent more than half of the stocks (470 GtC, 55%) and the rest is distributed in boreal (270 GtC, 31%) and temperate forests (120 GtC, 14%) (Pan et al. 2011). These forests play a crucial role in climate change. They represent the main carbon sink in continental environments and the second sink after the oceans. They would capture, according to estimates, between 70% and 100% of the 1.8 GtC/year absorbed on continents (Wigneron and Ciais 2021). Forests could store 226 GtC more than currently if they were allowed to regenerate in areas often on the fringes of large tropical forest fragments, which are no longer used for agriculture and remain free from urbanisation (Mo et al. 2023).

Forests play a role in mitigating global climate change and in adapting to its effects, at regional and local scales. Their role in mitigation, through photosynthesis which absorbs atmospheric CO_2 and induces carbon storage, is recognised in international negotiations and in major forest policy instruments, such as REDD+ (reducing emissions from deforestation and forest degradation). This is the first sector to be covered by these instruments. However, their role in adaptation is less well-known. This role is provided by ecosystem services that reduce the impacts of climate change or help human societies to adapt (Locatelli et al. 2015a). For example, through evapotranspiration, forests regulate continental or regional rainfall and local temperature (De Frenne et al. 2013),and help to maintain a climate favourable to agriculture and other human activities. In watersheds, they absorb excess water and then release it into the atmosphere during droughts and promote infiltration into the soil along root systems (Bradshaw et al. 2007). In regions exposed to extreme weather events, they protect soils from erosion (Elliot et al. 2018). They also contribute to the conservation of biodiversity by sheltering numerous animal and plant species (Putz et al. 2001). Forests and their 60,000 tree species indeed host 80% of amphibian species, 75% of birds and 65% of mammals. This biodiversity, a driver of ecosystem resilience, is a valuable asset for adaptation to climate disturbances. It is a matter of no longer seeing forests only as carbon sinks (Locatelli et al. 2015b).

9.2 Direct and Feedback Links Between Climate Change and Forests

With the increase in atmospheric CO_2 levels, the above-ground biomass of mature trees theoretically increases (Körner 2017), but the tree mortality rate rises during extreme drought episodes either due to photosynthesis deficit or embolisms in the raw sap of trunks (McDowell et al. 2018). Water deficit dries out the upper part of the tree crowns, thereby making them fragile. Leaf phenology is also altered, with rising temperatures extending the vegetation season and increasing unmet water needs. This sometimes results in mega-fires generating peaks of greenhouse gas emissions (Bowman et al. 2021).

Extreme dry seasons induced by climate change have variable effects on tropical rainforests (TRF), which are exposed to an average annual rainfall of 2200 mm and a dry season of 3.7 months on average. However, these averages vary significantly on a global (Amazon, Congo Basin, Borneo-Mekong Basin), regional, or local scale. For example, the length of the dry season ranges from 0.6 to 4.5 months, respectively in the northwest and southeast of the Amazon TRF. In Southeast Asia, in particular, El Niño corresponds to extreme drought, linked to the reversal of the South Equatorial Current (Glantz and Ramirez 2020). Its impact varies by site, with a low impact on tree mortality in sites with a marked dry season, but high impact in sites without a marked dry season, such as in Malaysia where this rate increased from 0.9% to 6.4% during El Niño 1997–1998 (Clark 2004). In sites with a marked dry season, species are indeed adapted to drought. Over the past 30 years, several El Niño events have led to a floristic change in newly recruited trees in the Amazon, with an increase in drought-adapted genera and a decrease in genera associated with a wetter climate.

Water exchanges between the atmosphere and soils, as well as cloud mass circulation and the triggering of of precipitation influenced by forest dynamics, are in turn modified by climate change through feedback mechanisms that remain to be clarified and quantified. By ensuring 80% of carbon exchanges between terrestrial ecosystems and the atmosphere and by dissipating some of the solar energy through evapotranspiration, forests act as highly efficient thermostabilisers. However, due to rising temperatures and recurrent droughts, they feed feedback loops that amplify climate change.

The upcoming climatic conditions, some of which have never been experienced in human history, and which now involve forests in a runaway process, will alter the role of these forests in sequestering atmospheric carbon. Warming indeed increases plant respiration, and soil drying increases the mortality of young trees, thereby increasing carbon release. Certainly, the increase in CO_2 from human activities enhances the photosynthetic efficiency of plants, and thus the absorption of carbon in forests. But this same increase in atmospheric CO_2 concentration leads to a rise in global temperatures. The effect of this rise on carbon sequestration in forests therefore depends on latitude. In boreal zones, which are little fragmented by human activities, the lengthening of the growing season allows the forest to advance towards the poles. At lower latitudes, extreme temperatures can lead to a high mortality rate of trees, resulting in a massive release of carbon. The organic matter in forest soils could itself be affected by warming, which increases biological activity and thus soil respiration, releasing the carbon it contains. Climate change will therefore likely increase tree mortality and gaseous carbon emissions from soil organic matter degradation, seriously reducing, or even reversing, their function as carbon sinks.

Climate change also induces cascading effects on the autecology of forest species, risking their extinction or proliferation. It interacts with land transformation and the global connectivity of living beings where new interfaces between humans, animals, and pathogens are created. The Ebola virus epidemic in West Africa in 2014 started in the Guinean forest, where the tropical forest landscape has been transformed into an agrosystem dotted with forest islands. In these reconfigured territories, animal communities, such as bats, and their pathogens, adapt their autecology and modify the risks of pathogen transmission between species. The epidemic, or even pandemic, risk associated with a "forest exit" of a pathogen, results from human transformations carried out in the forest.

9.3 Factors of Variability of Forest Vulnerability to Climate Change

At the scale of the forest plot, heterogeneity and diversity of species and age classes condition the vulnerability of tree populations. With their specificities related to phenology, rooting, and autecology, these species present a diversity of vulnerability to climate change. Mixed forests appear to be less vulnerable to drought and pests than monospecific forests (Bauhus et al. 2017), but not all mixtures are equivalent. Older trees are more resistant to drought, but they are also less resilient (Au et al. 2022).

At the landscape scale, viewed as a complex of ecosystems, the spatial and structural heterogeneity of forests determines their permeability to fire and their resistance to wind. Natural forests are therefore less vulnerable than planted forests (Liu et al. 2022). Natural forests that have been shaped over time by fire benefit from a mosaic spatial heterogeneity, which reduces their overall conductivity to fire, as revealed in Venezuela and Brazil (Mistry et al. 2016).

At the scale of the forest region and over the past two decades, the rise in temperature and the scarcity of precipitation have affected the resistance of forests to pest attacks, fires, or storms, in high-latitude temperate zones, but also in the intertropical zone. Modelling allows us to predict the future of forest structure and composition among others. It appears, for example, that in the northern and southern forest margins of the Congo basin, the Atlantic forests and most of those in the Democratic Republic of Congo are among the most vulnerable (Réjou-Méchain et al. 2021).

9.4 Forestry Practices and Other Adaptive Measures

Climate change, which alters thermal conditions, photosynthetic efficiency, access to water and nutrient resources, calls for adaptive management. Paradoxically, the forestry profession has only been aware of this recently. The main difficulty lies in a triple uncertainty represented by the low predictability of the effects of climate change, the variability of forest responses to this change, and the unknown timing of these responses (Jandl et al. 2019). Therefore, historical analyses, investigating past evolutions spread over at least several centuries, remain of little use for designing appropriate adaptive measures. So-called ecological niche or vegetation succession models remain informative, but insufficient to predict forest futures.

Furthermore, the increase in such knowledge will never measure up to the extreme variability of forests and their dynamics. Recommendations for forest management remain, therefore, of a generic, even empirical, order, and generally favour (1) the diversity and heterogeneity of species and structures as factors of resilience, (2) the extension of the exploitation age which allows for more carbon capture and protection of young trees, and (3) the accompaniment and real-time observation by technicians and practitioners, with regard to the forms of natural adaptation that manifest themselves, even beyond interventionist silvicultural practices (Boxes 9.1 and 9.2). To a large extent, these measures remain of a prospective nature, the practices conducive to the expression of forest resilience still being of an exploratory nature (Chuine et al. 2023).

Box 9.1 Increasing forest resilience in Ivory Coast

In the forest plantations of Ivory Coast, diversifying species and tree ages is a fundamental strategy for strengthening the adaptation of restored forests to climate change (Fig. 9.1). Firstly, species diversity increases the ecosystem's adaptation to the changing climatic conditions of West Africa. The plantations will face changes in rainfall patterns, extreme weather events, and temperature variations. By introducing various species, the probability of including trees adapted to specific conditions is increased. Some species may be more resistant to drought, and others may thrive in wetter conditions. This diversity provides a solid foundation for future natural regeneration and resilience to various climate scenarios (Messier et al. 2022). Moreover, diversifying tree ages plays a key role in forest resilience. Young trees are often more inclined to adaptive growth, while mature trees have developed long-term resistance mechanisms. By integrating a range of ages, plantations can better adapt to disturbances such as fires or seasonal droughts, and increase ecosystem stability and the longevity of forest cover. Finally, diversification creates a natural shield that increases resistance to diseases and pests. On the other hand, monospecific plantations are more vulnerable to epidemics, which spread rapidly among trees of the same species.

Box 9.2 An adaptive forestry practice accessible to all

Another adaptive practice, in the many countries where shifting cultivation on forest slash-and-burn is still implemented, is assisted natural regeneration (ANR). Particularly used in the Central African Republic (Kpolita et al. 2022) and in the Democratic Republic of Congo (Peltier et al. 2014), it involves preserving, during the cultivation period, seedlings, suckers or shoots of pre-existing forest species and promoting their growth through selective weeding, thinning and pruning (Fig. 9.2). These species, perceived as not very disruptive to food crops, prove to be productive (in charcoal, fruits, caterpillars, fibres and honey, for example) during the fallow period or in a perennial agroforestry association. ANR avoids clearing new forest lands and emitting large quantities of CO_2. It also increases the volume of wood available in forest fallows and sequesters as much carbon. By valuing a higher volume of wood energy over an equivalent fallow period, it reduces the possible use of fossil energy. It also preserves a higher species diversity by preventing the invasion of pyrophilic introduced species and the savannisation of landscapes. Most of the species chosen in ANR are said to be heliophilous, resistant to heat and atmospheric drought and are therefore better adapted to climate change. Some farmers extend ANR by creating a perennial agroforestry garden combining palm trees, fruit and forest trees, and mixed food crops, renewed as they are harvested.

9.5 International Bodies, Economic Incentives and Science-Society Interactions

While understanding the relationships between forests and microclimates is rapidly advancing, transposing this to an international scale remains more complex. Sharing concerns among stakeholders requires a long time that does not easily align with the urgency of a global response. Moreover, indispensable equity is required, as it is the so-called Global South countries that suffer most from climate changes, with rising sea levels, increasing climatic extremes, floods and forest fires.

The United Nations Framework Convention on Climate Change (UNFCCC) and its 1997 Kyoto Protocol led to a Clean Development Mechanism (CDM) that does not directly constrain emissions, even though it finances projects to reduce GHG emissions. International discussions highlight the importance of reducing emissions from deforestation and forest degradation (REDD+, the "+" corresponding to the consideration of increasing carbon stocks, for example through adapted silvicultural practices or plantations). The aim of the REDD+ programme, launched in 2008, is to encourage countries to take measures to protect their forests. The difficulty is in measuring the deforestation avoided by this mechanism, as it is based on projections and unverifiable predictive hypotheses. No model can predict the evolution of

Fig. 9.1 Plantations of local forest species mixed with plantain in a taungya type system in the classified forest of Téné, Ivory Coast (Credit: photo B. Hérault). The photo shows how agricultural crops (here plantain) can be used to accompany the development of local species (here Bété *Mansonia altissima*) during the first years of the plantation. The photo illustrates a Taungya agroforestry system implemented in the classified forest of Téné, in Ivory Coast. It shows plantations of local forest species associated with agricultural crops. Here, plantain seedlings, a food crop, coexist with young Bété trees (*Mansonia altissima*). This system allows agricultural crops to go with the development of forest species during their early years of growth. The Taungya system thus promotes forest regeneration while ensuring agricultural production, thereby optimising the use of space and available resources

economic and climatic variables controlling deforestation rates (agricultural product prices, droughts and precipitation, fires), leaving the door open to "optimised" scenarios, with variables chosen according to the strategic interests of states or private actors proposing them (Karsenty 2023). These financial mechanisms struggle to intertwine societal, environmental, and development aspects.

Fig. 9.2 During the harvest of maize and the cutting of cassava, a farmer from the Batèkè plateau in the Democratic Republic of Congo marks with a stake the natural seedling of a tree (*Millettia laurentii*, or wenge) that he wishes to preserve (Credit: photo R. Peltier). The photo shows how, in a practical way, a farmer can plant a painted stake to indicate where is a tree that he wishes to protect is. The photo is taken in a field, at the edge of a valley that begins the Batéké plateau. In the foreground, the farmer, accompanied by his young son, is painting in white a stake about 1 m high, which he has planted next to a young Wengé seedling, about 30 cm high. This is in a field where maize has just been harvested, the dry stalks of which can be seen, and where the farmer is going to cut cassava. The stake will indicate to the weeding staff that this young tree should not be cut. In the background, in the valley, remnants of forest can be seen, which are being cleared. In the third plane, on the other side of the valley, the grassy savannah, typical on the sandy deposits of the Batéké plateau, can be seen

Climate change and forests suffer from being considered in their own, as if they existed independently of the socio-economic determinants that shape their future. Their evolutions and degradations are primarily the result of human injustices and inequities within societies, with neoliberal logics playing a leading role (Stephens 2022). The sciences of forest ecology and silviculture are therefore ill-equipped to confront changes whose political causes are placed outside their field of investigation. Societal discourses advocate adaptive approaches to increase forest resilience to climate change and to seek solutions drawn from technology, but they refrain from investing in the necessary transformative social changes. In fact, they contribute to the disengagement of populations, who are the only ones capable of implementing the necessary social shifts. As the social dimensions of climate change and their intersections with forest realities remain too little understood, the policies implemented exclude economically precarious populations, women, and non-white communities (Reames 2016).

New forms of social facilitation remain to be explored, including so-called serious games (Garcia et al. 2016). ClimateRush is a game project aimed at alerting our societies to the vulnerability of tropical forests. It involves the variability of climate vulnerability among different tropical forest tree species (Veintimilla et al. 2019), facing temperature increases calibrated on the eight climate warming scenarios produced by the IPCC. This variability of vulnerability is materialised by specific temperature optima defining a thermal niche unique to each species and represented by the average temperature where the species is present (Hernández Gordillo et al. 2021). The final version will be a board game aimed at environmental education and consultation. It aims to strengthen societal action power through reduction, adaptation, and compensation strategies proposed by the IPCC, in order to limit the climate vulnerability of forest ecosystems.

9.6 Conclusion

This overview reveals the complexity of the mechanisms coupling forests and climate change, and the uncertainties they cover, in perspectives of runaway scenarios for which scientific research remains ill-equipped. It is confronted with a range of trends that sometimes oppose each other, making it particularly difficult, if not risky, to specify in which direction their resultant is heading.

The future of the forest appears to be more determined by agricultural dynamics and social aspirations than by climate warming itself or the way the forest is managed. Once again, climate warming appears uncontrollable, if we persist in considering it as a problem whose causes we should not confront. The forest cannot be considered as an autonomous entity, outside of socio-ecological continuities, whose evolutions remain determined by processes of environmental and social justice, tragically disregarded.

References

Au TF, Maxwell JT, Robeson SM, Li J, Siani SM, Novick KA et al (2022) Younger trees in the upper canopy are more sensitive but also more resilient to drought. Nat Clim Chang 12(12):1168–1174

Bauhus J, Forrester DI, Gardiner B, Jactel H, Vallejo R, Pretzsch H (2017) Ecological stability of mixed-species forests. In: Mixed-species forests: ecology and management. Springer, pp 337–382

Bowman DM, Williamson GJ, Price OF, Ndalila MN, Bradstock RA (2021) Australian forests, megafires and the risk of dwindling carbon stocks. Plant Cell Environ 44(2):347–355

Bradshaw CJA, Sodhi NS, Peh KSH, Brook BW (2007) Global evidence that deforestation amplifies flood risk and severity in the developing world. Glob Chang Biol 13:2379–2395

Chuine I, Ciais P, Cramer W, Laskar J (2023) Report: French forests facing climate change. Academy of Sciences, Paris

Clark DA (2004) Sources or sinks? The responses of tropical forests to current and future climate and atmospheric composition. Philos Trans R Soc Lond B Biol Sci 359(1443):477–491

De Frenne P, Rodríguez-Sánchez F, Coomes DA, Baeten L, Verstraeten G, Vellend M et al (2013) Microclimate moderates plant responses to macroclimate warming. Proc Natl Acad Sci 110(46):18561–18565

Elliot WJ, Page-Dumroese D, Robichaud PR (2018) The effects of forest management on erosion and soil productivity. In: Soil quality and soil erosion. CRC Press, pp 195–208

FAO (2020) Global forest resources assessment 2020: key findings. FAO, Rome

Garcia C, Dray A, Waeber P (2016) Learning begins when the game is over: using games to embrace complexity in natural resources management. GAIA-Ecol Perspect Sci Soc 25(4):289–291

Glantz MH, Ramirez IJ (2020) Reviewing the Oceanic Niño Index (ONI) to enhance societal readiness for El Niño's impacts. Int J Disaster Risk Sci 11:394–403

Hernández Gordillo AL, Vilchez MS, Ngo Bieng MA, Delgado D, Finegan B (2021) Altitude and community traits explain rain forest stand dynamics over a 2370-m altitudinal gradient in Costa Rica. Ecosphere 12(12):e03867. https://doi.org/10.1002/ecs2.3867

Jandl R, Spathelf P, Bolte A, Prescott CE (2019) Forest adaptation to climate change—is non-management an option? Ann For Sci 76(2):1–13

Karsenty A (2023) Inadequacy of international mechanisms for the environment: can we end deforestation through carbon offsetting? Int Strat Rev 3:95–105

Keenan RJ, Reams GA, Achard F, de Freitas JV, Grainger A, Lindquist E (2015) Dynamics of global forest area: results from the FAO Global Forest Resources Assessment 2015. For Ecol Manag 352:9–20

Körner C (2017) A matter of tree longevity. Science 355(6321):130–131

Kpolita A, Dubiez E, Yongo OD, Peltier R (2022) First evaluation of the use of Assisted Natural Regeneration by Central African farmers to restore their landscapes. Trees Forest People 7:100165

Liu D, Wang T, Peñuelas J, Piao S (2022) Drought resistance enhanced by tree species diversity in global forests. Nat Geosci 15(10):800–804

Locatelli B, Catterall CP, Imbach P, Kumar C, Lasco R, Marín-Spiotta E et al (2015a) Tropical reforestation and climate change: beyond carbon. Restor Ecol 23(4):337–343. https://doi.org/10.1111/rec.12209

Locatelli B, Pavageau C, Pramova E, Di Gregorio M (2015b) Integrating climate change mitigation and adaptation in agriculture and forestry: opportunities and trade-offs. WIREs Clim Change 6(6):585–598. https://doi.org/10.1002/wcc.357

MacDicken K, Jonsson Ö, Piña L, Maulo S, Contessa V, Adikari Y et al (2016) Global forest resources assessment 2015: how are the planet's forests changing? FAO, Rome. 54 p

McDowell N, Allen CD, Anderson-Teixeira K, Brando P, Brienen R, Chambers J et al (2018) Drivers and mechanisms of tree mortality in moist tropical forests. New Phytol 219(3):851–869

Messier C, Bauhus J, Sousa-Silva R, Auge H, Baeten L, Barsoum N et al (2022) For the sake of resilience and multifunctionality, let's diversify planted forests! Conserv Lett 15(1):e12829

Mistry J, Bilbao BA, Berardi A (2016) Community owned solutions for fire management in tropical ecosystems: case studies from Indigenous communities of South America. Phil Trans R Soc Lond B Biol Sci 371(1696). https://doi.org/10.1098/rstb.2015.0174

Mo L, Zohner CM, Reich PB, Liang J, de Miguel S, Nabuurs G-J et al (2023) Integrated global assessment of the natural forest carbon potential. Nature 624:92–101. https://doi.org/10.1038/s41586-023-06723-z

Pan YD, Pan Y, Birdsey RA, Fang J, Houghton R, Kauppi PE, Kurz WA et al (2011) A large and persistent carbon sink in the world's forests. Science 333:988–993. https://doi.org/10.1126/science.1201609

Peltier R, Dubiez E, Diowo S, Gigaud M, Marien J-N, Marquant B et al (2014) Assisted Natural Regeneration in slash-and-burn agriculture: results in The Democratic Republic of the Congo. Trop Woods For 321(3):67–79

Poccard-Chapuis R (2022) Opportunities for forest restoration in Paragominas. TerrAmaz Project, Paragominas, Brazil. 120 p

Putz FE, Blate GM, Redford KH, Fimbel R, Robinson J (2001) Tropical forest management and conservation of biodiversity: an overview. Conserv Biol 15(1):7–20

Reames TG (2016) Targeting energy justice: exploring spatial, racial/ethnic and socioeconomic disparities in urban residential heating energy efficiency. Energy Policy 97:549–558

Réjou-Méchain M, Mortier F, Bastin JF, Cornu G, Barbier N, Bayol N et al (2021) Unveiling African rainforest composition and vulnerability to global change. Nature 593(7857):90–94

Stephens JC (2022) Beyond climate isolationism: a necessary shift for climate justice. Curr Clim Chang Rep 8(4):83–90

Veintimilla D, Ngo Bieng MA, Delgado D, Vilchez Mendoza SJ, Zamora N, Finegan B (2019) Drivers of tropical rain forest composition and alpha-diversity patterns over a 2520 m altitudinal gradient. Ecol Evol 9(10):5720–5730

Wigneron J, Ciais P (2021) The role of forests in the planet's carbon balance. Planet-Vie website, ENS, EDUSCOL. https://planet-vie.ens.fr/thematiques/ecologie/cycles-biogeochimiques/role-des-forets-dans-le-bilan-de-carbone-de-la-planete

Open Access This chapter is licensed under the terms of the Creative Commons Attribution-NonCommercial-NoDerivatives 4.0 International License (http://creativecommons.org/licenses/by-nc-nd/4.0/), which permits any noncommercial use, sharing, distribution and reproduction in any medium or format, as long as you give appropriate credit to the original author(s) and the source, provide a link to the Creative Commons license and indicate if you modified the licensed material. You do not have permission under this license to share adapted material derived from this chapter or parts of it.

The images or other third party material in this chapter are included in the chapter's Creative Commons license, unless indicated otherwise in a credit line to the material. If material is not included in the chapter's Creative Commons license and your intended use is not permitted by statutory regulation or exceeds the permitted use, you will need to obtain permission directly from the copyright holder.

Chapter 10
Agriculture and Climate Change Debates: The Case of Livestock Production

Christian Corniaux, Vincent Blanfort, Mathieu Vigne, Jonathan Vayssières, and Guillaume Duteurtre

Abstract Animal production is a major contributor to greenhouse gas emissions, accounting for 12% of global emissions and about 40% of agri-food system emissions, mainly due to methane. It also faces criticism for its roles in deforestation, land competition, pollution, and health concerns, sparking debates and calls for reduced meat consumption, especially in the Global North. However, livestock systems vary widely worldwide, with many in the Global South playing vital social, economic, and ecological roles, such as supporting livelihoods, maintaining soil fertility, and preserving ecosystems. Pastoral and mixed farming systems contribute to biodiversity and carbon storage, highlighting the complexity of livestock's impacts. This chapter reviews livestock's climate impact globally and addresses controversies surrounding its environmental, social, and economic functions.

10.1 Livestock Farming, a Controversial Activity

Animal production is among the most controversial sectors in major environmental issues, particularly in debates on climate change. A recent study (FAO 2023) estimates that livestock farming accounts for 12% of total anthropogenic greenhouse gas (GHG) emissions, which represents about 40% of the total emissions from agri-food systems. Methane (CH_4) is the main cause, accounting for more than half of the emissions linked to animal production, while carbon dioxide CO_2 accounts for just under a third, and nitrous oxide (N_2O), for 15%. This role of the livestock sector in greenhouse gas emissions, and therefore in climate change, is often highlighted within the scientific community, leading to significant research on potential mitigation levers.

C. Corniaux (✉) · V. Blanfort · G. Duteurtre
Cirad joint research unit Selmet "Tropical and Mediterranean Animal Production Systems",
Montpellier Univ, Inrae, Institut Agro, Montpellier, France
e-mail: christian.corniaux@cirad.fr

M. Vigne · J. Vayssières
Cirad, UMR SELMET, Univ. Montpellier, INRAE, Agro Institute, Montpellier, France

In addition to these criticisms, there are controversies over the role of livestock farming in deforestation, competition for land use between animal feed and human food in land degradation, or pollution of groundwater. The scientific community is also regularly questioning the impact of meat consumption on human health, and on animal welfare in intensive farming. In response to these debates, many civil society organisations in the North advocate vegetarianism or flexitarianism, or develop anti-dairy discourses or campaigns for reducing or stopping meat consumption of meat consumption (particularly red meat) and processed meats (see Chap. 23). Livestock farming is thus at the heart of many societal debates on agriculture, food and climate change. To support radical advocacy, some civil society organisations unjustly accuse livestock farming of being responsible for more than 50% of anthropogenic GHG emissions[1], which highlights the crystallisation of societal debates on the issue of the links between livestock farming and climate change.

However, these controversies must be examined considering the diversity of contexts. Globally, there is a great diversity of production systems, species raised, modes of transformation and marketing, and consumption patterns. Animal feeding systems are extremely varied, from the exploitation of grazing lands by pastoral farming to feed-lots[2]. These feeding systems play a crucial role in the levels of greenhouse gas emissions, but also in the ability of grasslands and grazing lands to store carbon in soils and vegetation, in groundwater pollution, in product quality, etc. Denouncing the impacts of livestock farming as a homogeneous whole therefore makes little sense, if this diversity of situations and local contexts is not considered (see Chap. 16).

In the Global South, in particular, livestock farming methods can be very different from those found in the Global North, and livestock activities play a structuring social and economic role in many countries (Blanfort et al. 2015). The livestock sector represents about 40% of the agricultural wealth produced globally, and 1.3 billion jobs. It is estimated that livestock farming provides or contributes to the livelihoods and food security of nearly one billion people worldwide, including 430 million poor farmers (FAO 2023).

Livestock activities also fulfil multiple functions in peasant systems, where they are strongly integrated into agricultural systems (see Chaps. 12 and 16). These mixed systems are predominant in the Global South (Richard et al. 2013; Herrero et al. 2016). In many vulnerable communities, the herd often constitutes long-term insurance and savings. Animals also play a role in the production of organic manure essential for maintaining soil fertility in countries where farmers have little access to synthetic fertilizers. In mixed systems, animals also provide animal traction for crops (Alary et al. 2011). Pastoral systems also cover 30% of the terrestrial space across all latitudes, particularly in marginal areas for agriculture (Herrero et al. 2016). Their importance is crucial in Africa and Central Asia, where ruminant farming contributes to the balance and sustainability of ecosystems by limiting bush encroachment. It contributes to the maintenance of biodiversity, stimulates plant growth, participates in nutrient cycles, seed dispersal, and improves rainwater infiltration, over vast territories often devoid of any other economic activity.

Revisiting the multiple roles of livestock farming in the Global South highlights functions it also plays in developed economies. In many countries of the Global North, livestock farming also has environmental, social, or economic functions that justify not considering it, mechanically, only as an adjustment factor to reduce greenhouse gas emissions (Ryschawy et al. 2017; Dumont et al. 2016). To shed light on the necessary trade-offs between these environmental, economic and social functions, the contribution of different types of livestock farming to climate change needs to be better understood.

Given the significant changes that the various livestock sectors have undergone in recent decades, particularly their growth (see Chap. 16), this chapter presents a global overview of the impact of the livestock sector on climate change. Then we propose to shed light on some of the controversies associated with this issue, which now make livestock farming a particularly controversial subject.

10.2 What Are the Contributions of Livestock Farming to Climate Change?

The strong dynamics characterizing the livestock sector, particularly its strong growth, have major repercussions on the environment. Intensive animal production has developed worldwide, accompanied by industrialized value chains (Steinfeld et al. 2006; de Haan et al. 1997). This "livestock revolution", associated with the expansion of land cultivated for animal and human food, has exacerbated human pressure on land and natural resources (see Chap. 1).

According to the latest estimates from the FAO (2023), the livestock sector is responsible for 12% of global anthropogenic GHG emissions (6.2 Gt CO_2 eq/year), which are divided into direct and indirect emissions (Fig. 10.1).

Direct emissions account for more than half of the sector's total emissions (3.7 Gt eqCO_2). These include methane emissions (CH_4) from enteric fermentation, CH_4 and N_2O emissions related to effluent management, as well as CO_2 emissions related to energy consumption directly on the farm.

Indirect CO_2 emissions include energy consumption upstream of production systems, for the manufacture of fertilizers and pesticides for production, the processing and transport of animal feed, live animals and livestock products. Indirect emissions of CH_4 and N_2O come from the manufacture of feed and fertilizers. Finally, specific indirect CO_2 emissions are particularly associated with the conversion of forests into pastures and the expansion of oil palm and soy plantations to produce animal feed.

Beyond these global figures, GHG emissions from the livestock sector vary greatly depending on the regions of the world, without being completely correlated with animal production volumes (Fig. 10.2). They are also closely related to species (large ruminants: cattle 65%, buffaloes 9%, small ruminants 7%, poultry 10% and

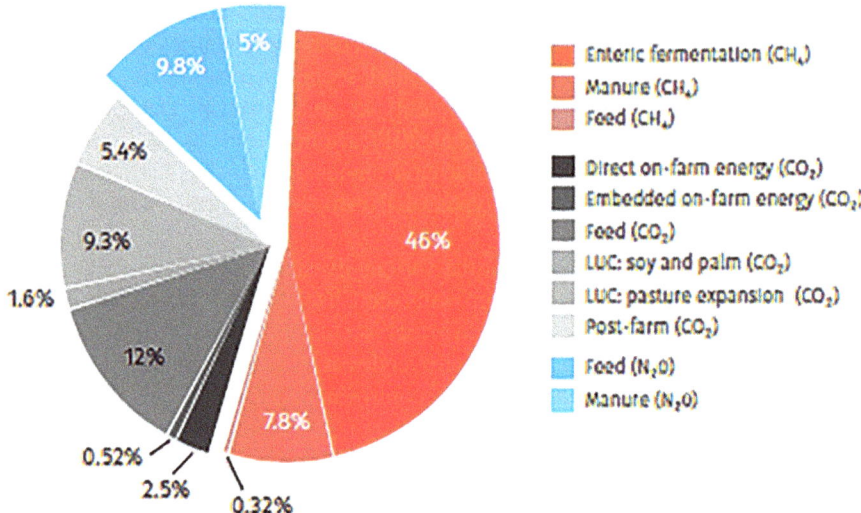

Fig. 10.1 Distribution and type of emissions from the livestock sector. (Source: based on FAO (2023))

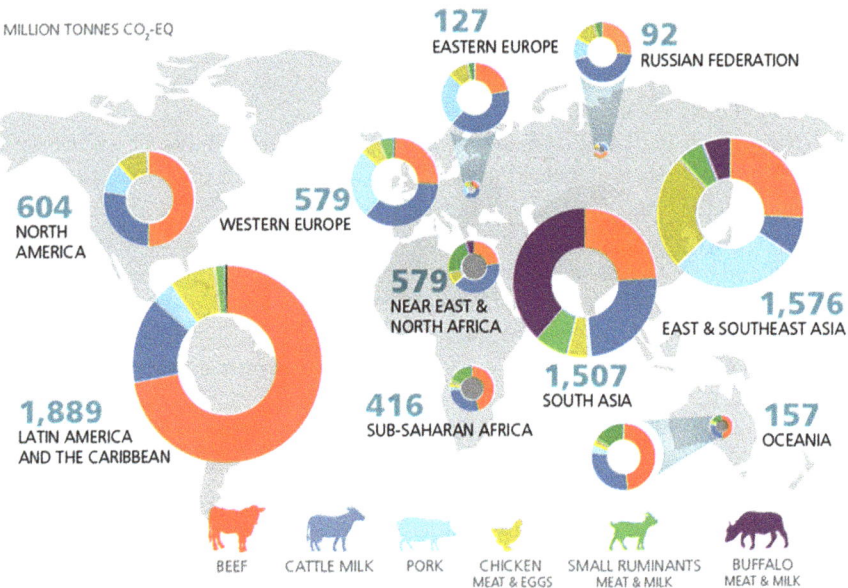

Fig. 10.2 Regional and species distribution of emissions from livestock. (Credit: Food and Agriculture Organization of the United Nations CC BY-NC-SA 3.0 IGO: FAO (2023))

pigs 9%). Ruminants as a whole are therefore the largest contributors to livestock emissions (over 80%), due to methane emissions accounting for 54% (31% for carbon dioxide CO_2, 15% for nitrous oxide N_2O) (Gerber et al. 2013).

This is why emissions per unit of protein produced are higher for meat and milk from ruminants than for pork or poultry (Duteurtre et al. 2019). However, it is important to consider the great diversity of livestock sub-systems, some of which appear to be less carbon-emitting. Pasture-based livestock systems, based on the use of natural pastures and cultivated meadows, as well as integrated crop-livestock systems, generally generate fewer emissions: they would be responsible for only 20% of total livestock emissions (Gerber et al. 2013).

These figures are still the subject of debate within the scientific community. In particular, the contribution of livestock to emissions related to land use change and carbon sequestration is not yet correctly assessed. The analysis by major world region (Gerber et al. 2013) shows that Sub-Saharan Africa, due to its low productivity, and Latin America and the Caribbean, due to the conversion of primary forests into pasture and crops for animal feed, are the regions with the highest emissions per kilogram of carcass produced (70 kg $eqCO_2$/kg). However, references on the specific contribution of Southern countries' livestock systems to GHG emissions are insufficient and these figures hide a diversity highlighted by assessments such as those carried out by Cirad on contrasting livestock systems (Vigne et al. 2015).

Furthermore, the assessment of livestock systems is highly dependent on the methods and metrics used (Blanfort et al. 2022), where the standards are sometimes unsuitable for this diversity (see Chap. 16).

10.3 What Options for Sustainable Food Systems?

Over the past 40 years, the rapid growth of animal production (see Chap. 16) has led to a proportional increase in GHG emissions. This trend is made more problematic as it is the most intensive, and the most specialized, farming systems that have contributed to this surge in animal production. This cycle requires a rethinking of the transformation prospects of the sector. However, it is also essential to refine this global vision by nuancing the intensity of these emissions in the diversity of contexts that have just been quickly mentioned in the previous paragraph. Considering global options could endanger rural communities and agro-pastoral territories that contribute in many regions to the sustainable development of the planet. In particular, it is better to think about the links between the production and consumption of animal products.

10.3.1 Reinventing Local Food Systems

Population growth is the powerful driver of the dynamics observed on the different continents (see Chap. 16). To feed nearly ten billion people by 2050, according to United Nations projections, it will be necessary to continue to increase animal origin production. This is all the truer as the areas where the population continues to grow rapidly are precisely those where individuals, mostly poor, currently consume very few animal products and aspire to consume more meat, milk and eggs (Sub-Saharan Africa, India, Southeast Asia, etc.). To meet this increase in demand for animal products, countries have four options available: increase the number of animals raised without modifying the systems; intensify their production by conventional methods; import animal products; or intensify the systems in an ecological way. These four options do not have the same impact on greenhouse gas emissions and on the carbon balance of the livestock sector (Table 10.1).

Table 10.1 Options for development strategies of animal productions to meet the increase in domestic demand, and associated impacts on climate change

Options	Characteristics	Impacts on climate change
Import animal products	Logistics (sea transport of meat, powdered milk). Intensified production systems and transformation processes (milk drying towers, refrigeration or freezing of meat).	Increase in the carbon footprint linked to transport and imported deforestation.
Increase the number of animals on the national territory without changing the systems	Unmodified farming systems. For grazing ruminants, possible mitigation through carbon storage, or in soils benefiting from animal manure.	Increase in GHG emissions (see Chap. 24). Soil carbon (see Chap. 17), increases or decreases depending on practices.
Intensify farming systems in a conventional way	Increased use of concentrated feeds, mechanisation or robotisation, standardisation (genetics...), production of manure or slurry. Specialisation of farming systems, disconnection of exchanges between livestock and agriculture. ↓ grazed areas, ↑ use of forages and cultivated feeds.	GHG reduction per kg of product is linked to the increase in productivity. Reduction of carbon storage in livestock feeding systems.
Intensify farming systems in an ecological way by using agroecology	Development of agroecological farming systems. Ecological intensification of mixed farming and livestock systems. Reconnection of exchanges between livestock and agriculture in the territories through the valorisation of manure, the valorisation of crop residues, etc. ↑ Grazed areas/↑ use of forages and cultivated feeds.	Reduction of GHG emissions per kg of product, linked to the increase in carbon sequestration by mixed systems. Increase in carbon storage in livestock feeding systems (pasture and meadows) (Blanfort et al. 2022). Increase in carbon storage in cropping systems.

10.3.2 Encourage Substitutions and Loss Reduction to Promote Sustainable Sectors

In fine, it appears difficult, if not impossible, to reconcile both the increase in the supply of animal products and the limitation of GHG emissions on a global scale. Some solutions aim to drastically reduce the place of meat and milk in human consumption. The development of protein crop sectors (soy, peas, beans, etc.) is currently supported by aggressive public policies. This is the case, for example, with the European Green Deal[3]. However, substituting meat or milk with soy is not always an acceptable solution from an environmental standpoint. More confidentially, new sectors are trying to establish themselves sustainably (insect farming, in vitro meat production), while they themselves are subject to controversy: consumer acceptance, environmental cost, intensive production mode, large-scale transition. Ultimately, these so-called "alternative" solutions remain based on an intensified, industrialized model, relying on international trade of products upstream and downstream.

Another often proposed option is to primarily limit the consumption of red meat by substituting it with white meat or fish, productions a priori less emitting of GHGs. Moreover, excessive consumption of red meat is now described as problematic for human health. But again, the increase in white meat production does not challenge the intensified and industrialized model, criticized for its socio-economic externalities ("feed-food" competition, animal/human food) and negative environmental impacts (importation of deforestation).

To avoid this pitfall, another path is possible, that of agroecology. It also seems opportune to limit losses and waste throughout the animal product sectors. Globally, they would be around 15% for milk and meat. For more than half, this waste would come from the consumption link. This value is considerable, but the saving of these losses, complex to implement, would remain modest in the face of the scale of the challenge.

Box 10.1 Livestock in the circular bioeconomy: a lever to reduce food losses and contribute to environmental protection

The circular bioeconomy offers significant potential to support sustainable economic development integrated with major global challenges (food security, climate change, management of natural resources within planetary boundaries or biodiversity preservation). The levers to promote the circular bioeconomy are multiple (Muscat et al. 2021). They can consist of process optimisation or the reorganisation of human activities and the emergence of new forms of cooperation between local actors.

(continued)

Box 10.1 (continued)

As a major driver of biomass flows at different scales, the contribution of the livestock sector to the circular bioeconomy is significant both as a beneficiary and as a contributor. A major lever involving livestock is based on the use of resources that do not compete with human food (natural resources, agricultural by-products, by-products from food or biofuel processing, food waste, etc.) for feeding or laying animals. These biomasses are then converted into fertilizing matter, rich in carbon, organic matter and nutrients to be used in plant production or as food for humans. Moreover, the by-products of livestock, whether they come from production systems or processing, can also be used as biomaterials (leather, biopolymers, etc.) or bioenergy (combustion of manure or biogas from manure).

By promoting the recycling of biomasses, these levers allow to improve the overall efficiency of food systems: firstly because they help to reduce food losses by finding other ways of valorizing these biomasses; then because they reduce losses of carbon and nutrients to the environment, major contributions of food systems to global warming and ecosystem pollution. In Reunion Island for example, the overall nitrogen efficiency of the agri-food system is low (35% according to Alvanitakis et al. 2024). Although the livestock sector is already involved in 31% of biomass flows at the island scale (Kleinpeter et al. 2023), encouraging the circularity of biomass around livestock could improve this efficiency by +42% (Alvanitakis et al. 2024).

Ultimately, would eating less meat or dairy to reduce the number of farmed animals, particularly ruminants, be the only effective and acceptable solution? International organisations such as the FAO or WHO no longer publish recommendations on individual consumption of milk or meat. However, figures of around 100 kg/year/person for milk and 30 kg/year/person for meat are still circulating. These figures are significantly exceeded in Northern countries (respectively around 200 and 100 kg/year/person). Reducing individual consumption of milk and meat in areas where it is excessive would significantly reduce herd sizes and thus greenhouse gas emissions. On the other hand, consumption of animal products is generally below these thresholds in the majority of Southern countries. The scope for encouraging a reduction in consumption is therefore very narrow, if not non-existent, and unacceptable from a social and nutritional standpoint.

10.4 Conclusion: Livestock Farming, a Problem, But Also a Solution for the Planet

Animal industries are booming in the South. These developments are resulting in a rapid increase in greenhouse gas emissions, and in the development of specialised and intensive production systems, the carbon impact of which is problematic. In developed countries, these developments are older, and the consumption of animal products is no longer increasing. But the systems continue to specialise and industrialise, creating other controversies over the unsustainable use of resources such as water (see Chaps. 7 and 18) or energy (see Chap. 19), and over the reduction of the multiple functions of livestock farming.

Methane emissions are a major stumbling block to societal acceptance of the development of livestock farming. A new global agreement on reducing methane emissions was enacted at COP26 (Global Methane Pledge) to reduce CH_4 emissions by 30% by 2030 compared to 2020 levels; agriculture and livestock farming are affected, but only as secondary sectors compared to those of energy and waste (see Chap. 24). Carbon storage dynamics must be encouraged by reinventing new territorial synergies or by promoting agroecological systems. The impact of the rise in animal production on climate change must be analysed in light of the great diversity of the systems involved and the associated solutions (see Chap. 16). More than a variable for adjusting greenhouse gas emissions, livestock farming must be considered in its multiple functions in order to enhance the potential of agriculture-livestock integration (Blanfort et al. 2022).

References

Alary V, Duteurtre G, Faye B (2011) Livestock and societies: the multiple roles of livestock in tropical countries. INRAE Anim Prod 24(1):145–156. https://doi.org/10.20870/productions-animales.2011.24.1.3246

Alvanitakis M, Kleinpeter V, Vigne M, Benoist A, Vayssieres J (2024) A substance flow analysis to assess the potential benefits of livestock based circularity for nutrient use efficiency and carbon return to soils in the agro-food-waste system of a tropical island. Agric Syst 219:104046. 15 p. https://doi.org/10.1016/j.agsy.2024.104046

Blanfort V, Vigne M, Vayssières J, Lasseur J, Ickowicz A, Lecomte P (2015) The agronomic roles of livestock in contributing to adaptation and mitigation of climate change in the North and South. Agron Environ Soc 5(1):107–115. https://publications.cirad.fr/une_notice.php?dk=592032

Blanfort V, Assouma MH, Bois B, Edouard-Rambaut LA, Vayssières J, Vigne M (2022) Efficiency to account for the complexity of the contributions of grazing livestock systems to climate change. In: Ickowicz A, Moulin C-H (eds) Grazing livestock and sustainable development of Mediterranean and tropical territories. Recent knowledge on their strengths and weaknesses. Quæ editions, Versailles, pp 86–104

Dumont B, Dupraz P (2016) Roles, impacts and services from livestock in Europe. Synthesis of collective scientific expertise. INRA, France, p 1048

Duteurtre G, Assouma MH, Poccard-Chapuis R, Dumas P, Touré I, Corniaux C et al (2019) Climate change, animal product consumption and the future of food systems. In: Dury S, Bendjebbar P, Hainzelin E, Giordano T, Bricas N (eds) Food systems at risk. New trends and challenges. Cirad, FAO, pp 39–42. https://doi.org/10.19182/agritrop/00088

FAO (2023) Pathways towards lower emissions—A global assessment of greenhouse gas emissions and mitigation options from livestock agrifood systems. FAO, Rome. https://doi.org/10.4060/cc9029en

Gerber PJ, Steinfeld H, Henderson B, Mottet A, Opio C, Dijkman J, Falcucci A, Tempio G (2013) Tackling climate change through livestock—A global assessment of emissions and mitigation opportunities. FAO, Rome, p 115 de Haan C, Steinfeld H, Blackburn H (1997) Livestock and the environment: finding a balance. Report of the Commission of the European Communities, the World Bank and the governments of Denmark, France, Germany, The Netherlands, UK and USA.

Herrero M, Henderson B, Havlík P, Thornton PK, Conant RT, Smith P et al (2016) Greenhouse gas mitigation potentials in the livestock sector. Nat Clim Chang 6:452–461. https://doi.org/10.1038/nclimate2925

Kleinpeter V, Alvanitakis M, Vigne M, Wassenaar T, Lo SD, Vayssières J (2023) Assessing the roles of crops and livestock in nutrient circularity and use efficiency in the agri-food-waste system: a set of indicators applied to an isolated tropical island. Resour Conserv Recycl 188:106663. 13 p. https://doi.org/10.1016/j.resconrec.2022.106663

Muscat A, de Olde EM, Ripoll-Bosch R et al (2021) Principles, drivers and opportunities of a circular bioeconomy. Nat Food 2:561–566. https://doi.org/10.1038/s43016-021-00340-7

Richard D, Dourmad J-Y, Coulon JB, Picon-Cochard C (2013) Livestock and animal sectors. In: Soussana J-F (ed) Adapting to climate change. Agriculture, ecosystems and territories. Quæ editions, Versailles, 296 p

Ryschawy J, Disenhaus C, Bertrand S, Allaire G, Aznar O, Plantureux S et al (2017) Assessing multiple goods and services derived from livestock farming on a nation-wide gradient. Animal 11(10):1861–1872

Steinfeld H, Gerber P, Wassenaar T, Castel V, Rosales M, De Haan C (2006) Livestock's long shadow: environmental issues and options. FAO, Rome, p 414

Vigne M, Blanfort V, Vayssières J, Lecomte P, Steinmetz P (2015) Constraints on livestock farming in the South: ruminants between adaptation and mitigation. In: Torquebiau E (ed) Climate change and world agriculture. Quæ editions, Versailles, pp 123–135

Open Access This chapter is licensed under the terms of the Creative Commons Attribution-NonCommercial-NoDerivatives 4.0 International License (http://creativecommons.org/licenses/by-nc-nd/4.0/), which permits any noncommercial use, sharing, distribution and reproduction in any medium or format, as long as you give appropriate credit to the original author(s) and the source, provide a link to the Creative Commons license and indicate if you modified the licensed material. You do not have permission under this license to share adapted material derived from this chapter or parts of it.

The images or other third party material in this chapter are included in the chapter's Creative Commons license, unless indicated otherwise in a credit line to the material. If material is not included in the chapter's Creative Commons license and your intended use is not permitted by statutory regulation or exceeds the permitted use, you will need to obtain permission directly from the copyright holder.

Chapter 11
Agriculture, Health and Climate Change: Towards a "One Health" Vision

Marisa Peyre, Didier Lesueur, Flavie L. Goutard, Alexandre Hobeika, Alexis Delabouglise, Maxime Tesch, Daan Vink, and François Roger

Abstract Human-induced climate change increases the incidence of infectious and vector-borne diseases by altering ecosystems and expanding pathogen transmission. It also worsens food security and nutrition through impacts on agriculture, while extreme weather and human activities raise risks of respiratory and waterborne diseases. Vulnerable populations suffer most from these health and environmental disruptions. The IPCC recommends a holistic "One Health" approach to address interconnected impacts on human, animal, and environmental health. This chapter explores these links and advocates sustainable agricultural methods that protect health and the environment.

The links between human-induced climate change and the increase in the incidence of infectious diseases have been demonstrated (Romanello et al. 2021; Van de Vuurst and Escobar 2023). The publications of the Intergovernmental Panel on Climate Change (IPCC) highlight the growing dangers related to climate change, including impacts on the health of humans, animals and the environment. The impact of other human activities such as the transformation and fragmentation of

---

M. Peyre (✉) · A. Delabouglise · M. Tesch · D. Vink
Cirad, UMR ASTRE, Montpellier, France
e-mail: marisa.peyre@cirad.fr

D. Lesueur
Cirad, UMR Eco&Sols, Montpellier, France

Alliance of Biodiversity International and International Center for Tropical Agriculture (Ciat), Asia hub, Common Microbial Biotechnology Platform (CMBP), Hanoi, Vietnam

F. L. Goutard
Cirad, UMR ASTRE, National Institute of Animal Science (NIAS), Hanoi, Vietnam

A. Hobeika
Cirad, UMR MoISA, Montpellier, France

MoISA, Univ Montpellier, Cirad, CIHEAM-IAMM, INRAE, Institut Agro, IRD, Montpellier, France

F. Roger
Cirad, DGD-RS, Van Phuc Diplomatic Compound, Hanoi, Vietnam

natural habitats also alters ecosystems and promotes the spread of vectors such as mosquitoes that spread dengue, malaria and the Zika virus, or ticks that cause Lyme disease (Calvin et al. 2023). The increase in temperatures and variations in rainfall extend the transmission periods of these pathogens and introduce diseases, particularly vector-borne ones, into new areas (Caminade et al. 2019; de Souza and Weaver 2024). The intensification of interactions between humans, domestic animals and wildlife, linked to this degradation of ecosystems, also leads to an increase in the circulation and emergence of zoonotic diseases such as Ebola, SARS-CoV (1 and 2), influenza or the Nipah virus (Leroy et al. 2009). The increase in heatwaves induces thermal stress that increases both human and animal mortality, and accentuates these risks of emergence (Ebi et al. 2021).

It is also documented that climate change negatively affects agriculture and indirectly human health (Wheeler and von Braun 2013). It jeopardises food security (more severe and prolonged droughts, increase in floods) and consequently contributes to malnutrition in populations. Fires caused by droughts, but also the very frequent slash-and-burn practices in many countries, combined with rampant urbanisation grouping increasingly large populations in large cities, increase air pollution leading to respiratory conditions (Tran et al. 2023). The lack of access to clean water and the numerous floods lead to an increase in the spread of waterborne diseases such as cholera or leptospirosis (Levy et al. 2018). All these extreme weather conditions, combined with the impact of human activities on nature, have a significant effect on agriculture and biodiversity, thereby increasing the risks to human and animal health (Owino et al. 2022). This also causes psychological trauma and significant stress, increasing the risk of mental illnesses (Hayes et al. 2018). As is often the case, the most vulnerable populations are the first victims of all these disruptions and changes. The IPCC has highlighted in its recommendations the need to promote holistic "One Health" approaches to better account for the multifactorial effects of climate change on agriculture and health (animal, human, and environmental) to implement appropriate actions to prevent and limit global health impacts (Mahon et al. 2024).

The aim of this chapter is to deepen the understanding and analysis of the relationships between climate change, agriculture, and "healths" to promote sustainable agricultural methods that respect human health, as well as the environment, which has been largely overlooked until now (Fig. 11.1). Its objective is also to examine how a comprehensive approach that considers the complex interactions between agriculture and animal, human, and environmental health could effectively implement sustainable production systems without negative consequences for the climate, but also for humans, biodiversity, and the environment (Banerjee and van der Heijden 2023).

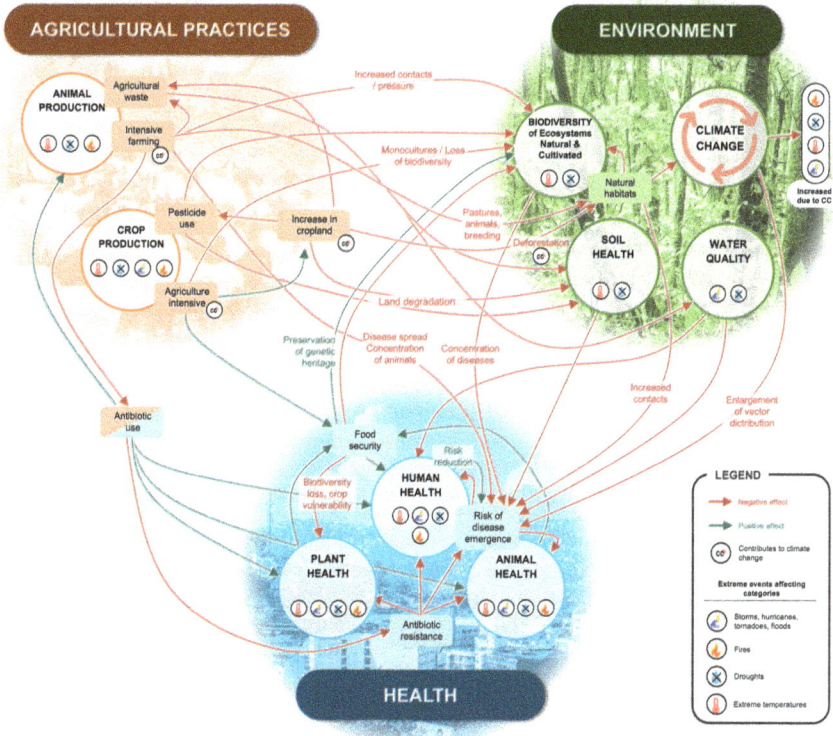

Fig. 11.1 Interconnections between agricultural practices, environment, and healths. The figure presents the relationships between agricultural practices for animal and plant productions (use of pesticides, intensive agriculture, agricultural waste...) the environment (biodiversity, climate change, soil health, water quality) and healths (human, animal, and ecosystem health). The diagram identifies interactions between these three elements, which are classified according to their positive or negative impact and also their links with climate change. Many negative processes are identified such as soil degradation, loss of biodiversity, antibiotic resistance, expansion of vector zones, etc. Some positive interactions are also highlighted, notably food security and preservation of genetic heritage

11.1 Agriculture, Climate Change, and Health

Healthy and sustainable agriculture is essential for human health. It must provide healthy ingredients with high nutritional value for the population's diet, which guarantees them better health. Otherwise, it can provide food contaminated with harmful chemical residues or with low nutritional value or exposed to polluting and harmful chemical inputs for humans and animals, thus endangering their health. In 2020, the FAO showed how agriculture directly contributes to climate change through greenhouse gas (GHG) emissions, mainly linked to the excessive application of nitrogen fertilisers (emission of N_2O for example), but also through intensive livestock farming which produces significant methane CH_4 emissions (FAO 2022). All these phenomena contribute significantly to global warming, promoting the spread of zoonotic and vector-borne diseases (Fig. 11.1).

In addition to the direct effects on climate change, intensive agriculture mainly based on the use of chemical inputs to combat crop diseases, but also on antibiotics given to livestock, also promotes the development of pathogen resistance, making their management more complex and their impacts more significant in a context further aggravated by the effects of climate change (negative spiral) (Gilbert et al. 2017). Traditionally, agricultural practices can be classified into three types: conventional practices based on non-renewable resources (E), those integrating renewable resources (S), and those rethinking agroecosystems according to the principles of agroecology (R) (Gliessman 2014; Hill and MacRae 1996). In plant health, agroecological strategies such as agroecological crop protection (Deguine et al. 2023) stand out for their low carbon footprint, avoiding the "waste" of high GHG emission inputs (Lamichhane et al. 2015; Wyckhuys et al. 2022). These approaches, particularly the S and R strategies, are also effective in limiting infectious risks, notably zoonoses, thanks to the biological control by conservation of habitats (Ratnadass and Deguine 2021). Conversely, conventional practices based on agrochemistry exacerbate pesticide resistance, increasing risks related to vectors and disease reservoirs (malaria, leptospirosis) and resistant fungal pathogens (Ratnadass et al. 2023). Rodenticides and Triazole fungicides, widely used in agriculture, also negatively impact public health and environmental health. Sustainable and resilient practices, although imperfect, generally reduce risks to ecosystems and human health, while increasing resilience to climate change. They provide an appropriate response to limit antimicrobial resistance, reduce greenhouse gas emissions, and strengthen the sustainability of agricultural systems in the face of current global challenges. It is by emphasising their adoption that we can hope for an improvement in the current situation.

The impact of climate change on agriculture and therefore on health differs between women and men, due to their positions within agri-food, health, and social systems. Women are more vulnerable to extreme climate events, as they are often responsible for managing natural resources and family food security. They also have limited access to resources, training, or technologies to cope with these challenges (Lecoutere et al. 2023). If we want suitable and inclusive solutions, it is therefore essential to address the issue of gender in all global initiatives, including the "One Health" approach (Galiè et al. 2024).

11.2 Degradation of Natural Ecosystems and Emergence of Diseases

11.2.1 Agricultural Practices and the Consequences of Climate Change

In their quest to find new arable land, humans encroach on existing natural ecosystems. This contributes to intensifying their interactions with wild animals, whose natural habitat is disrupted by human actions, but also by the effects of climate change (droughts, fires, floods), with the direct consequence of an increased risk of disease spread. Deforestation and the transformation of forests into agricultural areas (slash-and-burn agriculture, needs for land for intensive agriculture, especially soy, etc.) contribute to climate change by limiting the capture of carbon emitted into the atmosphere and harm the habitats of many wild species (Seymour et al. 2022). These disturbances bring humans, domestic animals, and wildlife closer together, creating interface zones allowing interspecies transmission and the emergence of diseases such as Ebola, Nipah virus infection, and coronavirus diseases (Magouras et al. 2020).

Extreme weather conditions, such as periods of drought or flooding, alter the distribution and availability of agricultural and natural resources and also intensify interactions between different species (Bartlett et al. 2022; FAO et al. 2020). Food security is affected by these losses in agricultural production, which in some parts of the world promotes an increase in meat consumption and intensive farming of domestic animals (Bonwitt et al. 2018). This consumption constitutes an additional risk of interspecies disease transmission through the handling of contaminated carcasses or animal products or through their ingestion under limited biosecurity conditions (little or no protective equipment, low-temperature cooking) (Cornélis et al. 2022). On the other hand, intensive farming, with limited biosecurity conditions, promotes the transmission and multiplication of pathogens and increases the risks of spread and emergence of animal or zoonotic diseases with pandemic potential (such as avian flu) (Delabouglise et al. 2022; Gilbert et al. 2017).

These disturbances exerted on natural ecosystems also allow the introduction and proliferation of parasites that vector these diseases, in connection with global warming. Higher temperatures increase the periods of activity and extend the geographical areas of mosquitoes and ticks, which are vectors of many diseases such as dengue, malaria, Lyme disease, etc. New animal, zoonotic and human diseases, such as Rift Valley fever or West Nile fever, Crimean-Congo haemorrhagic fever (CCHF) or Zika and Oropouche viruses, etc. (Anikeeva et al. 2024; Morand and Lajaunie 2021), thus emerge in new geographical areas.

11.2.2 Climate Change Exacerbates Soil Deterioration: A Health Hazard Too Long Neglected

Agroecology is a practice that aims to make agricultural production sustainable while preserving the environment. Most often, this involves better soil health, which is characterised by significant natural biodiversity. Soil health is essential for the sustainable sequestration of carbon, thus helping to limit global warming while ensuring sustainable yields. From a "One Health" perspective, soil health is central, as it positively interacts with the health of plants, but also with that of animals and humans in a balanced and healthy environment. It is therefore crucial to implement environmentally friendly agricultural practices to maintain a balance between the Food security, ecosystem health, and general well-being in a changing climate context. For many years, soil health was not considered in the scientific literature related to human or livestock health, but increasingly studies show that it is central and that preserving the health of soils equates to preserving human health (Romano and Zelikoff 2024; Singh et al. 2023; Sun et al. 2023).

11.3 The Urgency of an Integrated "One Health" Approach to Limit Climate Change and Reduce Pandemic Risks

Following the Covid-19 outbreak, which highlighted the links between global warming and the acceleration of disease emergence, both linked to human activities including agriculture, it has become clear that priority should be given to a holistic approach called "One Health" that would take into account all health aspects, their interconnections and the determinants that affect them. A definition of the "One Health" approach, inclusive of the environment, has been adopted internationally (One Health High-Level Expert Panel, Food and Agriculture Organisation of the United Nations, World Health Organisation of the United Nations, World Organisation for Animal Health and the United Nations Environment Programme) (One Health High-Level Expert Panel et al. 2022). In 2023, the COP28 for the first time included in its programme a day specifically dedicated to the health issues of global warming. The Covid-19 pandemic has proven that cooperation between different sectors (agriculture, animal and human health, environment, but also public policy, economy, social, education) is essential to prevent, prepare for and respond effectively to future crises (FAO et al. 2022; Sanga et al. 2024), the frequency and number of which are increasing exponentially, like extreme weather events. However, consideration of prevention and risk reduction is still too limited and it is important to act quickly for things to change positively (Peyre et al. 2021). One Health and prevention of emerging risks have been included in the legally binding International Pandemic Accord adopted by WHO member states in May 2025.

Although the importance of these issues is recognised globally, the lack of action persists mainly due to the delay in financial and political involvement.[1] Given that the origins of climate change and the occurrence of pandemics are intimately linked by common determinants (human impact on nature), it is imperative to re-examine political and economic priorities to establish concrete and effective measures against these global threats. Despite promising progress made in the international climate agenda, with increasingly precise goals and national responsibilities, tangible and operational measures are still too limited. This inertia is mainly due to a lack of political will and dominant economic logics. Moreover, climate policies are too often fragmented and divided by sector, which hinders a global and inclusive approach. To try to harmonise climate and public health management strategies, it is imperative to implement a "One Health" approach, which will aim to align initiatives to combat climate change and prevent pandemics.

In this context, the approach to territorial health, particularly through the Socio-Ecological System Health (SESH) and the One Health in Social-Ecological Systems (OHSES) approaches, allows us to respond to these challenges by providing an operational dimension to these integrated approaches, in connection with the resilience of agricultural and production systems (De Garine-Wichatitsky et al. 2021; Zinsstag et al. 2024). The concept of territory is a complex concept encompassing physical, social, economic and political dimensions, defined by its boundaries, whether physical or symbolic, and by the interactions between its components (see Chap. 21). These interactions are all the more important in the context of climate change, which increases pressures on territorial ecosystems and modifies socio-ecological dynamics (Wilcox et al. 2019). By emphasising interdisciplinarity, resilience and community participation, these frameworks integrate the impacts of climate into disease prevention and management (De Garine-Wichatitsky et al. 2021; Zinsstag et al. 2024). They also use tools such as systems thinking and participatory modelling to analyse the dynamic interactions between climate, ecosystems and human systems (Binot et al. 2015; Duboz et al. 2018; Étienne 2010). The Health and Territories project illustrates this approach by promoting the agroecological transition in *living laboratories* in Senegal, Benin, Laos and Cambodia, combining climate resilience, soil health and sustainable food systems (Duboz et al. 2018). This approach is essential to address the increasing impacts of climate change, while strengthening the health and sustainability of territories.

---

[1] Pandemic: "Funding prevention actions is necessary to avoid future epidemics and limit climate risks", 2023. *The World*. https://www.lemonde.fr/sciences/article/2023/06/22/pandemie-financer-les-actions-de-prevention-est-necessaire-pour-eviter-les-prochaines-epidemies-et-limiter-les-risques-climatiques_6178682_1650684.html

11.4 Conclusion

The "One Health" approach is a crucial tool for better adaptation to climate change, offering a global perspective on the health of humans, animals, and the environment. It is necessary to confront emerging infectious diseases, whether zoonotic or not, by also addressing the management of socio-ecosystems (agricultural practices, urbanisation, etc.), the preservation of the environment, including biodiversity and soil health. Although the "One Health" approach is gaining recognition, it is still insufficiently integrated into national and local policies. It is hindered by inadequate coordination between the health, agriculture, environment and public policy sectors on the ground, which prevents effective management of climate and health threats. Moreover, funding and management mechanisms, often fragmented and ill-suited to global challenges, hinder the practical implementation of these prevention strategies. The nations most exposed to climate change, particularly those in development, lack the essential resources to establish effective prevention mechanisms. The absence of globally coordinated responses exacerbates the situation, largely dominated by reactive measures rather than preventive ones. The "One Health" approach aims to overcome these obstacles by promoting global and interdisciplinary solutions (Olive et al. 2022). It promotes the establishment of synergies between public health, ecosystem preservation and sustainable agriculture, taking into account the social and environmental aspects of global issues. The implementation of sustainable and resilient agricultural methods in the face of climate change is essential to maintain a balance between food security, the health of socio-ecosystems and overall well-being. Adopting the "One Health" approach in global strategies to combat global warming and health risks would both reduce these risks and improve our resilience to these multiple and simultaneous crises.

References

Anikeeva O, Hansen A, Varghese B, Borg M, Zhang Y, Xiang J, Bi P (2024) The impact of increasing temperatures due to climate change on infectious diseases. BMJ 387:e079343

Banerjee S, van der Heijden MGA (2023) Soil microbiomes and one health. Nat Rev Microbiol 21(1):6–20

Bartlett H, Holmes MA, Petrovan SO, Williams DR, Wood JLN, Balmford A (2022) Understanding the relative risks of zoonosis emergence under contrasting approaches to meeting livestock product demand. R Soc Open Sci 9(6):211573

Binot A, Duboz R, Promburom P, Phimpraphai W, Cappelle J, Lajaunie C et al (2015) A framework to promote collective action within the One Health community of practice: using participatory modelling to enable interdisciplinary, cross-sectoral and multi-level integration. One Health 1:144–148

Bonwitt J, Dawson M, Kandeh M, Ansumana R, Sahr F, Brown H, Kelly AH (2018) Unintended consequences of the 'bushmeat ban' in West Africa during the 2013-2016 Ebola virus disease epidemic. Soc Sci Med 200:166–173

Calvin K, Dasgupta D, Krinner G, Mukherji A, Thorne PW, Trisos C et al (2023) IPCC, 2023: climate change 2023: synthesis report. Contribution of Working Groups I, II and III to the Sixth Assessment Report of the Intergovernmental Panel on Climate Change, Geneva, Switzerland

Caminade C, McIntyre KM, Jones AE (2019) Impact of recent and future climate change on vector-borne diseases. Ann N Y Acad Sci 1436(1):157–173

Cornélis D, Vigneron P, Vanthomme H (coord) (2022) Gabon—towards sustainable village hunting management. In-depth diagnosis of the Mulundu department and strategic recommendations. SWM Programme. Rome, FAO, Cirad, CIFOR and WCS. https://doi.org/10.4060/cb9765fr

De Garine-Wichatitsky M, Binot A, Ward J, Caron A, Perrotton A, Ross H et al (2021) "Health in" and "Health of" social-ecological systems: a practical framework for the management of healthy and resilient agricultural and natural ecosystems. Front Public Health 8. https://doi.org/10.3389/fpubh.2020.616328

Deguine J-P, Aubertot J-N, Bellon S, Côte F, Lauri P-E, Lescourret F et al (2023) Chapter One—Agroecological crop protection for sustainable agriculture. In: Sparks DL (ed) Advances in agronomy, vol 178. Academic Press, pp 1–59

Delabouglise A, Guerin J-L, Lury A, Binot A, Paul M, Peyre M et al (2022) Intensification of livestock systems and pandemic risks. Cah Agric 3116

de Souza WM, Weaver SC (2024) Effects of climate change and human activities on vector-borne diseases. Nat Rev Microbiol 22(8):476–491

Duboz R, Echaubard P, Promburom P, Kilvington M, Ross H, Allen W et al (2018) Systems thinking in practice: participatory modelling as a foundation for integrated approaches to health. Front Vet Sci 5

Ebi KL, Vanos J, Baldwin JW, Bell JE, Hondula DM, Errett NA et al (2021) Extreme weather and climate change: population health and health system implications. Annu Rev Public Health 42:293–315

Étienne M (coord) (2010) Companion modelling. A participatory approach in support of sustainable development. Versailles, Quæ editions, 384 p

FAO (2022) Greenhouse gas emissions from agrifood systems. Global, regional and country trends, 2000–2020, Rome. 12 p

FAO, Cirad, CIFOR, WCS (2020) Build back better in a post-COVID-19 world – Reducing future wildlife-borne spillover of disease to humans: Sustainable Wildlife Management (SWM) Programme, Rome, 48 p

FAO, UNEP, WHO, WOAH (2022) One health joint plan of action, 2022–2026, Rome, 86 p

Galiè A, McLeod A, Campbell ZA, Ngwili N, Terfa ZG, Thomas LF (2024) Gender considerations in One Health: a framework for researchers. Front Public Health 12

Gilbert M, Xiao X, Robinson TP (2017) Intensifying poultry production systems and the emergence of avian influenza in China: a 'One Health/Ecohealth' epitome. Arch Public Health 75(1):48

Gliessman S (2014) Agroecology: the ecology of sustainable food systems, 3rd edn. CRC Press. 405 p

Hayes K, Blashki G, Wiseman J, Burke S, Reifels L (2018) Climate change and mental health: risks, impacts and priority actions. Int J Ment Heal Syst 12(1):28

Hill SB, MacRae RJ (1996) Conceptual framework for the transition from conventional to sustainable agriculture. J Sustain Agric 7(1):81–87

Lamichhane JR, Barzman M, Booij K, Boonekamp P, Desneux N, Huber L et al (2015) Robust cropping systems to tackle pests under climate change. A review. Agron Sustain Dev 35(2):443–459

Lecoutere E, Mishra A, Singaraju N, Koo J, Azzarri C, Chanana N et al (2023) Where women in agri-food systems are at highest climate risk: a methodology for mapping climate–agriculture–gender inequality hotspots. Front Sustain Food Syst 7

Leroy EM, Epelboin A, Mondonge V, Pourrut X, Gonzalez J-P, Muyembe-Tamfum J-J, Formenty P (2009) Human Ebola outbreak resulting from direct exposure to fruit bats in Luebo, Democratic Republic of Congo, 2007. Vector Borne Zoonotic Dis 9(6):723–728

Levy K, Smith SM, Carlton EJ (2018) Impacts of climate change on waterborne diseases: progressing towards designing interventions. Curr Environ Health Rep 5(2):272–282

Magouras I, Brookes VJ, Jori F, Martin A, Pfeiffer DU, Dürr S (2020) Emerging zoonotic diseases: should we reconsider the animal-human interface? Front Vet Sci 7

Mahon MB, Sack A, Aleuy OA, Barbera C, Brown E, Buelow H et al (2024) A meta-analysis on global change drivers and the risk of infectious disease. Nature 629(8013):830–836

Morand S, Lajaunie C (2021) Outbreaks of vector-borne and zoonotic diseases are associated with changes in forest cover and oil palm expansion on a global scale. Front Vet Sci 8

OHHLEP, Adisasmito WB, Almuhairi S, Behravesh CB, Bilivogui P, Bukachi SA et al (2022) One health: a new definition for a sustainable and healthy future. PLoS Pathog 18(6):e1010537

Olive M-M, Angot J-L, Binot A, Desclaux A, Dombreval L, Lefrançois T et al (2022) One health approaches to address emergencies: a necessary dialogue between state, science, and society. Nat Sci Soc 30(1):72–81

Owino V, Kumwenda C, Ekesa B, Parker ME, Ewoldt L, Roos N et al (2022) The impact of climate change on food systems, diet quality, nutrition, and health outcomes: a narrative review. Front Clim 4

Peyre M, Vourc'h G, Lefrançois T, Martin-Prevel Y, Soussana J-F, Roche B (2021) PREZODE: preventing zoonotic disease emergence. Lancet 397(10276):792–793

Ratnadass A, Deberdt P, Martin T, Sester M, Deguine J-P (2023) Impacts of crop protection practices on human infectious diseases. In: Vithanage M, Prasad MNV (eds) One health: human, animal, and environment triad. Wiley, pp 287–308

Ratnadass A, Deguine J-P (2021) Crop protection practices and viral zoonotic risks within a one health framework. Sci Total Environ 774:145172

Romanello M, McGushin A, Di Napoli C, Drummond P, Hughes N, Jamart L et al (2021) The 2021 report of the *lancet* countdown on health and climate change: code red for a healthy future. Lancet 398(10311):1619–1662

Romano I, Zelikoff JT (2024) Soil health is human health. Explore 20(6):103047

Sanga VT, Karimuribo ED, Hoza AS (2024) One health in practice: benefits and challenges of multisectoral coordination and collaboration in managing public health risks: a meta-analysis. Int J One Health:26–36

Seymour F, Wolosin M, Gray E (2022) Not just carbon: capturing all the benefits of forests for stabilising the climate from local to global scales. World Resources Institute, Washington

Singh BK, Yan Z-Z, Whittaker M, Vargas R, Abdelfattah A (2023) Soil microbiomes must be explicitly included in one health policy. Nat Microbiol 8(8):1367–1372

Sun X, Liddicoat C, Tiunov A, Wang B, Zhang Y, Lu C et al (2023) Harnessing soil biodiversity to promote human health in cities. Npj Urban Sustain 3(1):1–8

Tran HM, Tsai F-J, Lee Y-L, Chang J-H, Chang L-T, Chang T-Y et al (2023) The impact of air pollution on respiratory diseases in an era of climate change: a review of the current evidence. Sci Total Environ 898:166340

Van de Vuurst P, Escobar LE (2023) Climate change and infectious disease: a review of evidence and research trends. Infect Dis Poverty 12(1):51

Wheeler T, von Braun J (2013) Climate change impacts on global food security. Science 341(6145):508–513

Wilcox BA, Aguirre AA, De Paula N, Siriaroonrat B, Echaubard P (2019) Operationalising one health employing social-ecological systems theory: lessons from the greater Mekong subregion. Front Public Health 7

Wyckhuys KAG, Furlong MJ, Zhang W, Gc YD (2022) Carbon benefits of enlisting nature for crop protection. Nat Food 3(5):299–301

Zinsstag J, Meyer JM, Bonfoh B, Fink G, Dimov A (2024) One Health in human-environment systems: linking health and the sustainable use of natural resources. CABI One Health 3(1). https://doi.org/10.1079/cabionehealth.2024.00

Open Access This chapter is licensed under the terms of the Creative Commons Attribution-NonCommercial-NoDerivatives 4.0 International License (http://creativecommons.org/licenses/by-nc-nd/4.0/), which permits any noncommercial use, sharing, distribution and reproduction in any medium or format, as long as you give appropriate credit to the original author(s) and the source, provide a link to the Creative Commons license and indicate if you modified the licensed material. You do not have permission under this license to share adapted material derived from this chapter or parts of it.

The images or other third party material in this chapter are included in the chapter's Creative Commons license, unless indicated otherwise in a credit line to the material. If material is not included in the chapter's Creative Commons license and your intended use is not permitted by statutory regulation or exceeds the permitted use, you will need to obtain permission directly from the copyright holder.

Chapter 12
What Pastoralism Tells Us About Climate Change

Saverio Krätli, Véronique Ancey, and Der Dabire

Abstract Definitions of tropical pastoralism have been shaped by colonial and economic interests, often based on outdated categories. Recent research urges a critical reassessment of these classifications and their ideological roots. Pastoralism is viewed as a strategy that adapts to and benefits from environmental variability through mobility. Climate change is not just a natural issue but a socio-economic one, tied to global industrial practices. The language used to describe climate risks plays a role in shaping political responsibility and action.

12.1 Introduction

12.1.1 Definitions Are Political

Definitions of tropical pastoralism have been produced in a context of economic interests and colonial administrative needs: distinctions according to gradients of mobility (nomads, transhumants, sedentary), agricultural activity (herders, agropastoralists, peasants) or market orientation. The evolution of scientific knowledge about pastoralism challenges this portfolio of categories still used in rural development and calls for their decolonisation (Krätli 2016; FAO 2021) or, less

---

S. Krätli
International Union of Anthropological and Ethnological Sciences (IUAES), Commission on Nomadic Peoples (CNP), Lewes, UK

V. Ancey (✉)
Cirad, UMR Actors Resources Territories & Development (ART-Dev), Montpellier, France
e-mail: veronique.ancey@cirad.fr

D. Dabire
International Research-Development Centre on Livestock in the Subhumid Zone (CIRDES), Bobo-Dioulasso, Burkina Faso

metaphorically,[1] to critically examine and make explicit the ideological assumptions that link them to economic and environmental ideas.

In this spirit, the distinction that matters here refers to how pastoral systems interact with the environment. Through the prism of social-ecological,[2] are pastoral "functions" optimised through ecological integration, by "working with" nature, or by distancing from it, and creating artificially stable environments to shelter themselves from the uncertainty of natural conditions?

Under this light, pastoralism is a livelihood strategy based on rearing livestock and specialised in managing the largely unpredictable variability of rangeland conditions, even benefitting from it, while often engaging in productive or commercial relationships with other livelihood systems (FAO 2021). Pastoral systems in cold, temperate or tropical environments use mobility as an essential strategy, operable at different scales of time (daily, seasonal, multi-year) and space. This chapter concerns pastoral systems including or not including cultivation practices at the scale of the family economy.[3]

The term "climate change" euphemises the risk factors of climate disruption. Vocabulary matters to define the problem before measuring the risks: it is part of the political framing that distributes—or erases—responsibilities. Designating everyone—or no one—as responsible for global warming contributes to framing it as inevitable and insoluble.

In the literature on pastoralism and resilience, or vulnerability, "climate" challenges are usually distinguished from "socio-economic" challenges such as lack of basic services, poor governance or adverse policies. However, insofar as global warming is caused by a particular economic model associated with industrialisation and the race for growth, the term *climate* in "climate change" no longer means "natural", and consequently, the distinction between climatic and socio-economic factors is misleading. The climatic manifestations of global warming (droughts, floods, extreme weather events) are just as socio-economic as the bursting of a poorly constructed dam, leaks polluting water from a chemical plant, or a landslide hitting a city following deforestation on the slopes of a hill. The only difference is the length of the chain of causes and effects that we examine.

---

[1] "Decolonisation is not a metaphor" (Tuck and Wayne 2022).

[2] We understand by socio-ecosystem (with Berkes and Folke 1998) a category created to recognise that separating the social and ecological is never really possible in practice and that, consequently, the analysis must take into account this inevitable integration.

[3] The category "agropastoralism" introduced in the 1970s in postcolonial research on livestock farming in so-called "less advanced" countries, refers to the integration of pastoralism and crop farming. This integration—often associated with high levels of sustainability—can occur at multiple scales, including at the territorial level between specialised groups of herders and farmers. However, the term *agropastoralism* has most often been used to describe an alternative to pastoralism, in which livestock is integrated solely at the scale of a single sedentary household and its farm. This narrow usage overlooks the historical depth, the actual possibilities and the broader realities of agropastoral integration, particularly in the Sahel.

12.1.2 Context and Scope

This chapter contributes to the dialogue between science and society on climate disruption, using pastoralism in the Sahel as a case study. This case is emblematic for three main reasons: (1) it highlights the historical tension between the central role of pastoral systems in shaping territories and the persistent technopolitical push toward sedentary intensification—first under colonial rule and later through development policy (Carougeau 1930; Bocco 1990); (2) it exemplifies a social-ecological system where the political economy of land and mobility rights determine whether people can harness the environment's intrinsic variability—increasingly exacerbated with climate change (Krätli 2015); (3) it reflects a broader trajectory in which Sahelian states have been integrated into the global economy primarily as recipients of externally driven climate policy recommendations (Losch 2023).[4]

By identifying the core functional traits[5] of pastoralism, we can explore what this livelihood strategy can teach us about engaging with climate disruption—across all environments, whether cold, temperate, or tropical. Our discussion draws on three key insights from the specialized literature. Firstly, pastoral systems are characterised by their deep ecological integration. They are not a leftover from a bygone era or a stage on the path toward intensified livestock production. In environments where forage resources are highly variable and unpredictable—such as arid and semi-arid zones, as well as mountains and high plateau regions (Ellis and Swift 1988; Behnke et al. 2011)—pastoralists are able to secure relatively stable outcomes by interfacing environmental variability with equally flexible and adaptive practices embedded in their ways of operating (Krätli and Schareika 2010; Kammili et al. 2011; Meuret and Provenza 2014; Roe 2020; Sharifian et al. 2022; Scoones 2023). Secondly, the global presence, sustainability, and relative performance of pastoralism remind us that environmental variability is not inherently an anomaly or a threat. Pastoral systems show that climatic variability can be a resource. They also serve as a warning: there can be no meaningful protection against global warming through strategies that continue to force local stability—attempting to control fluctuations and uncertainties in production factors through intensification and environmental simplification—at the cost of further climate disruption (FAO 2021; GIZ 2022). On the other hand, they remind us that there can be no meaningful protection against global warming through strategies that continue to force local stability at the cost of further climate disruption—strategies that attempt to control fluctuations and uncertainties in production factors through intensification sustained by artificially stabilizing the environment (FAO 2021; GIZ 2022). Thirdly, countries of the Global South are not historically responsible for triggering global warming and, even today,

---

[4] And its successive visions: climate governance through shared but equitable collective effort (Article 2 of the Paris Agreement), combining adaptation and mitigation, differs from the previous dogma assigning adaptation to the South and mitigation to the North.

[5] Unlike descriptive attributes, which vary with contexts and times: animal breeds, distance and duration of transhumance, grazing rates, etc.

most remain negligible contributors to artificial greenhouse gas emissions (GHG) from burning fossil fuel. Yet they are confronted not only with climate challenges but also with incomplete demographic transitions and limited economic diversification. Moreover, the carbon balance of pastoral systems in arid and semi-arid areas—*when measured in ways that account for their ecological integration*—is potentially neutral or even negative (Assouma et al. 2019); an approach that contrasts sharply with dominant assessment models, which focus on the animal in isolation from the environment (see Chaps. 17 and 21). This reality calls into question both the relevance and the appropriateness of the climate mitigation demands placed on pastoral systems (and taken up by states), which circulate widely in the worlds of agriculture and development aid (Fig. 12.1).

A common question about pastoralism in the face of climate change is whether these systems are still viable in light of increasing environmental variability—in other words, what climate change tells us about pastoral development. This chapter proposes to reverse the question: what can we learn from pastoralism about climate change and related policies?

The first part of the chapter broadly outlines the evolution of institutional representations of pastoralism/rangeland integration. The second part addresses the key factors enabling or hindering the resilience of pastoral systems in the face of global warming. In conclusion, we suggest that by working with, rather than against,

Fig. 12.1 A Turkana girl milking in the morning, at Lotere, located 2 days' walk from Lokiriama towards the Ugandan border, in the Turkana district, Kenya. (Credit: photo by S. Krätli 2000)

environmental variability, pastoralism invites us to imagine alternative ways of organising human life within dynamic living systems. It also offers a valuable lens through which to rethink the resilience of production systems that depend on controlling and stabilizing the environment.

12.2 The History of Livestock Policies Applied to Pastoralism, in the Light of Climate Change

12.2.1 A History of Numbers

Historically, pastoralism has been treated as having secondary importance in livestock sector policies and agricultural priorities, which have focused on so-called 'useful' areas and on intensifying production. For administrations shaped by sedentary and agrarian reference points, the pastoral way of life—dispersed, mobile, and closely tied to natural resources—was difficult to control and transform. Unsurprisingly, global estimates of livestock numbers and pastoral populations vary depending on how pastoralism is defined and which sources are used (Pic-Ciamarra et al. 2014; Johnsen et al. 2019). At a minimum, more than 180 million people were living from 'pastoral and agropastoral' livestock systems in 2016, across 75% of countries (Kieta et al. 2016). Official figures are often reproduced uncritically or mechanically[6]; but in the absence of better data—and given both population growth over the past decade and the fact that roughly half of the Earth's land surface is classified as rangeland (ILRI et al. 2021),[7] it is reasonable to assume that pastoralism supports the livelihoods of hundreds of millions of people.

12.2.2 Successive Narratives on the Links Between Pastoralism, Environment and Climate

Over the past decades, the relationship between pastoralism and its environment has been a subject of debate across multiple institutional arenas—intersecting international development, sub-Saharan geopolitics, and agricultural and livestock sector policies.

---

[6] The figures of 200–500 million repeated by relatively recent sources extrapolate those of 2014 (WISP/UNEP 2014) including ranchers and conservation areas. Moreover, it is necessary to distinguish the sources on pastoralism and on all livestock systems, concerning animal productions and land areas, cultivated or not, which are concerned.

[7] Rangelands are characterised by indigenous vegetation of grasses, forbs and shrubs including spontaneous ecosystems. They feed domestic and wild animal species grazing and browsing and mainly host populations of pastoralists, agropastoralists and ranchers. About half of the earth's surface is classified as rangelands (ILRI 2021).

From the colonial era through the early decades of development interventions in the 1960s, pastoralism was widely associated with the phenomenon of desertification (Davis 2016). This perception persisted until research demonstrated that, in arid environments, pastoralism is not the problem but part of the solution (Swift 1996). Today, the capacity of pastoral systems to accommodate imbalance and manage variability is increasingly recognised (Hubert 2012; Scoones 2023).

Yet pastoral mobility is still sometimes framed narrowly—as a coping strategy in a hostile environment—implicitly suggesting that the ideal would be intensive, sedentary livestock production in a temperate, artificially stabilised setting. This perspective inherits a view of nature's variability as an anomaly, or even a threat. Similarly, defining pastoralism by its use of arid lands—places "where no crop grows"—implicitly prioritises crop cultivation wherever it is possible. In practice, however, more humid cultivated areas—where pastoral herds often remain for several months each year, sometimes longer than in arid zones—are sites of mutual benefit. Here, livestock manure and crop residues are exchanged, creating synergies between farmers and herders and contributing to soil fertility. Coexistence and complementarity are made possible where herders, farmers, and both customary and formal authorities recognise overlapping rights and shared access to land and resources.

In the field of international security, a growing body of literature since the 2010s has focused on the relations between pastoralism and conflicts. Much of this work has postulated a causal chain linking environmental degradation, vulnerability to terrorist recruitment, and the escalation of latent (agro)pastoral tensions into violent conflicts at national or regional scale. Yet, empirical studies have not confirmed this correlation (Benjaminsen et al. 2012; Krätli and Toulmin 2020). Instead, they show that the ethnicisation of the multifaceted crisis in the Sahel tends to polarise society: on one side, those who see pastoralists—particularly the Fulani—as the main contingent of armed jihadist groups; on the other, those same pastoralists, who reappropriate the stigma by framing their involvement as a response to the social and political marginalization they have faced since the emergence of nation-states. Field-based analyses converge in identifying the political and security crisis in the Sahel as a crisis of rural governance with a strong pastoral component (Rangé et al. 2020).

History has shown the limitations of policy approaches imposed on pastoral systems that rely on sedentarisation: a disconnection from the environment, scaling up of standardised practices, spatial specialisation, and market-driven technical "rationalization" (Box 12.1). Today, climate change reveals the risks of such approaches for the future—not only for pastoral systems, but for society more broadly.

Box 12.1 The land issue: resources, social relations and the role of the State

Pastoral spaces are generally distinguished between aggregation points—often near infrastructure or urban centres—where mobile herders tend to temporarily gather, governed by authorised, exclusive, collective, or individual access rights; and rangeland areas with open, regulated, or collectively exclusive access. These spatial boundaries and associated rights are not fixed—they evolve over time (Lavigne et al. 2022; Moritz 2016). A decisive factor in securing pastoral systems lies in maintaining a balance between their structured ties to territory (through recognised bundles of rights) and their capacity for mobility (Ancey et al. 2020). The recognition of pastoral practices and customary rights of shared access to pastures and water points—their reality on the ground, their potential effects on resources and the social systems that activate them, and their sustainability—has given rise to a substantial body of literature in both the Global North and South, and to countless controversies (Dell'Angelo et al. 2017; Eggertsson 2009). These issues are also addressed in institutional recommendations (FAO 2012; Davies et al. 2018). However, experience shows that recognising commons merely as a type of land, rather than recognising them as rights (e.g. grazing, gathering), is not enough to protect pastoral livelihoods. Once reduced to a category of land, "commons" can be titled and sold, often becoming simply synonymous with land waiting to be privatised. Pastoralists do not categorise land for the market, but as a functional space—integrating resources, practices and rights into people's livelihood. In Senegal's Ferlo region, for instance, a *good place* (*modji jofde*) is defined by pastoralists according to a combination of qualities: diversity, nutritional qualities, abundance of the grass, soil characteristics, air quality, absence of parasites, secure access, and peacefulness (Ancey and Diao 2004). What is at stake is recognizing that pastoral commons are usually linked to *in rem* rights—e.g. to access grazing opportunities and water—that are inalienable from the territory. In Europe, the "commons" (The term can strictly refer to uncultivated lands—forests, pastures, heaths—belonging to a village community or the lord and subject to shared uses; and rights of use on private lands, fields or forests (gleaning, common grazing, etc.) (Béaur 2006)) were considered in the eighteenth century an obstacle to the modernization. Their suppression despite widespread rural resistance, was never fully accomplished (Vern 2021; Vivier 2003). By contrast, in most regions of the global south, the customary rights associated with pastoral rangelands were simply erased or overlooked as colonial administrations introduced their land tenure systems. In sub-Saharan Africa, the expansion of cultivated areas at the expense of rangelands and fallows has led to increasing fragmentation of grazing areas making access more difficult. This has required more careful negotiation by pastoralists, tighter herd supervision, and greater precautions (Sambo et al. 2022). Since the 1990s and 2000s, some countries have introduced institutional responses in the form of pastoral codes, but these have largely failed to secure agropastoral spaces (Bonnet 2013).

12.2.3 How Do Environmental Policies and Global Warming Affect Pastoral Systems?

12.2.3.1 Distinguishing Variability, Global Warming and Its Manifestations in Pastoral Contexts

Pastoralism has co-evolved with its environment; particularly by actively engaging with the spatio-temporal variability of rainfall that defines arid and semi-arid zones. In these ecosystems, fluctuations in rainfall—and therefore in the availability of grazing opportunities—are constitutive features. The coefficient of annual rainfall variation (standard deviation/mean) has consistently hovered around 30%. A notably wet period was recorded between 1950 and 1968, followed by a prolonged dry phase from 1972 to 1995. Since 1996, overall rainfall levels have increased but remain highly irregular, resembling the patterns observed before the 1930s. In addition to this temporal variability, the storm-driven nature of monsoon rains generates strong spatial heterogeneity: within a single year, cumulative rainfall can differ dramatically across distances of just 20–30 km (Ancey and Ickowicz 2015).

Since 2004, the increase in rainfall relative to the 1971–2000 average has been depicted on maps as a "northward shift in isohyets," though these trends remain uncertain. This apparent shift, along with observed increases in biomass production, has led some authors to refer to a "greening of the Sahel" (Bégué et al. 2011). However, climate projections for the region remain inconclusive, due to the complexity of the African monsoon system. Existing models continue to produce contradictory results (Hiernaux and Soussana 2011). By the end of the 2000s, some projections suggested a rise in temperature ranging from 1.8 °C to 4 °C, while also emphasising that local variability may obscure broader trends, and that the most significant impacts may arise from extreme events—such as heat stress, droughts, or floods (Thornton et al. 2009).

12.2.3.2 Late and Limited Recognition of Pastoralism in International Climate Negotiations: Are Mitigation Challenges for Pastoralism a Non-issue?

In the history of international climate negotiations, livestock farming first appeared on the negotiators' agenda in 2017, framed under the broader category of agriculture (see Chap. 2).[8] Initially excluded from the text of Decision 4/CP.23, pastoralism was later incorporated—under the term *agropastoral systems*—in May 2018, during the first session of the Subsidiary Body for Scientific and Technological Advice (SBSTA) of the United Nations Framework Convention on Climate Change (UNFCCC). In the substance of these international decisions, pastoralism has been

---

[8] Through decisions 4/CP.23 of COP23 and decision-/CP.27 of COP27, respectively establishing the Koronivia Joint Work on Agriculture (SP/UNFCCC 2017) and the Charm el-Cheikh quadrennial joint initiative on the implementation of climate action for agriculture and food security (FCCC/CP/2022/L.4).

addressed from two main angles: improving livestock systems and enhancing their contribution to the sustainability of agricultural systems—particularly in terms of soil carbon storage, soil health and fertility, nutrient recycling and effluent management) (SP/UNFCCC 2017, p21). This framing reflects the dual objectives of climate policies, namely combining mitigation and adaptation.

For context, the contribution of livestock to agricultural greenhouse gas (GHG) emissions varies considerably by regions, production systems and practices, but it is widely recognized as substantial, accounting for an estimated 14.50% of global emissions (Gerber et al. 2013) (Box 12.2).

Box 12.2 The case of Burkina Faso: sectoral policies to promote "environmentally respectful pastoralism"

In Burkina Faso, 90.6% of greenhouse gas (GHG) emissions originate from the AFOLU sector (agriculture, forestry, and other land uses). An estimate 15.57% of AFOLU emissions are attributed specifically to livestock, while 75% of this livestock is estimated to be in pastoral systems (https://bulletin.woah.org/?panorama=praps-burkinafaso). Despite these apparently impressing figures, Burkina Faso's total contribution to global GHG emissions is estimated at just 0.02% (Our World in Data 2022) or 0.07% (IEA-EDGAR 2022), depending on the source. A draconian 50% cut in GHG emissions from pastoralism—nested within livestock, itself nested within agriculture—would reduce global GHG emissions by 0.000021–0.000074%, a vanishingly small change. To put this in perspective: if total global emissions were represented as a full year, even the entire annual emissions from pastoralism in Burkina Faso would last less than the blink of an eye.

Livestock sector policy in Burkina Faso prioritises production, reflecting the centrality of food security in the national agenda, while greenhouse gas emissions are addressed only marginally. Nonetheless, environmental concerns are not absent from national programmes supporting pastoralism. The principle of sustainable pastoralism has been recognized in the country's legal and policy framework for some time—explicitly in the guiding law on pastoralism (law n° 034–2002/an) since 2002, and in the agro-silvopastoral, fisheries and wildlife orientation law (law no. 070–2015/CNT) since 2015. Since 2018, the sectoral policy for agro-silvopastoral, fisheries and wildlife production has also incorporated this concern, referring to the goal of developing an "environmentally responsible" (*respectueux de l'environnement*) form of livestock farming. In practical terms, the actions focus on: routine awareness-raising on rotational grazing to prevent overgrazing and improve regrowth quality; the rehabilitation and enhancement of degraded lands for pastoral purposes; the promotion of agro—fodders reserves (including annual and shrub fodder crops), cutting and storing straw to reduce pressure on the environment; the promotion of biodigesters to recycle livestock waste into cooking energy and manure for fields (since 2010) and the mandatory requirement for environmental impact assessments of pastoral infrastructures such as livestock farms, vaccination yards, water points and slaughterhouses (since 2017).

In implementing these actions, agropastoral populations are encouraged to reduce herd size; carry out reforestation using seedlings provided by environmental services; cultivate fodder crops using seeds distributed by livestock services; build up fodder reserves—through the cutting and conservation of straw and the collection of crop residues—using equipment distributed by livestock services; and construct manure pits.

These measures reflect a certain awareness of the constraints and challenges facing pastoralism. However, when it comes to the environment, the prevailing narrative within technical services continues to depict pastoralism as a driver of environmental degradation and greenhouse gas emissions. Within this framing, the construction of an "environmentally responsible pastoralism" becomes a matter of changing herders' practices. These initiatives are in fact led by the Ministry of the Environment, through the National Office for Environmental Assessments. Yet, given the minimal contribution that agriculture, livestock, and pastoralism in Burkina Faso are expected to make in terms of global GHG reductions, it is worth questioning what the actual drivers of the displayed efforts.[9]

12.3 The Key Factors Enabling or Hindering the Resilience of Pastoral Systems in the Face of Climate Change

To explain how climate change affects the populations of pastoral systems, we distinguish three dimensions, two of which are already visible and one is increasingly likely.

Firstly, pastoralist populations are bearing the consequences of measures taken by more powerful groups to protect themselves from the threat of climate disruption—what might be called a geopolitical rush for potential resources. In some cases, this occurs under the guise of adaptation and mitigation efforts, as rangelands are closed off to make way for mining or energy production projects intended to "green" national economies. Other rural populations beyond pastoralists are facing similar pressures.

Secondly, pastoralist populations are directly experiencing the consequences of climatic disruption through new forms of climate variability. In the past, although unpredictable from 1 year to the next, the variability of natural conditions was expressed within familiar spectra; now, these spectra themselves are changing. The challenge of these changes is compounded by growing restrictions on mobility: pastoralists' main strategy for managing climate risk while taking advantage of the space-temporal variability of grazing resources. The less herders are able to move with their herds, the more their exposure to climate risk increases and the more

---

[9] For more information, see the Cassecs project, a research and development project supporting innovation for the resilience of pastoral and agropastoral livestock farming in Sahelian countries, funded by the European Desira programme.

likely they are to experience what they describe as "drought"—which, in pastoral terms refers to a lack of access to grazing resources.

Thirdly, as global average warming exceeds +1.5 °C, the likelihood of triggering one or more tipping points in the Earth system increases significantly (Lenton et al. 2023). In the event of such runaway climate change, the issue would shift from resilience to survival. Yet, thanks to the characteristics outlined above, even in the face of catastrophic scenarios, pastoralist populations may not be in the worst position to face the biophysical challenges of global warming.

To explore the key factors underpinning the resilience of pastoral systems in the face of global warming, this section focuses on the first two dimensions mentioned earlier. Six factors are briefly outlined below.

1. *Specialisation anchored in adaptive institutions.* Faced with increasing climate variability, the core source of resilience in pastoral systems lies in their specialization, grounded in specific institutional arrangements. Rather than shielding production from variability by creating artificially stable environments, pastoralism works with natural dynamics—adapting food production systems to variability This defining feature of humans– animal– environment relations in pastoral systems aligns closely with the principles of agroecology, which many see as essential for both climate adaptation and mitigation (FAO 2021).

2. *Operational variability to match ecological variability.* Pasture growth and the nutritional quality of forage resources are shaped by inherent environmental variability—daily, seasonal, multi-year, and spatial (across ecosystems, landscape features, plant species, and even parts of the same plant). By preserving the overall adaptability of the production system, allowing for strategies and day-to-day operations that can adjust to fluctuations in natural inputs, pastoralists are 'prepared to be surprised'—an increasingly vital strategy as climate change disrupts familiar patterns of variability.

3. *Strategic mobility as systemic plasticity.* Mobility is the clearest expression of the systemic plasticity of pastoralism. It allows herders to reduce their exposure to drought and other stressors by actively seeking out better conditions. Beyond its importance for herd survival, mobility underpins the ecological and economic sustainability of pastoral systems—and contributes to the resilience of broader socio-economic systems, including crop-livestock interactions, settlements, and markets (Corniaux et al. 2016).

4. *Exceptional herd biodiversity and ecological mimicry.* Pastoral herds maintain exceptional levels of biodiversity, including both inter and intra species genetic diversity as well as epigenetic variation.[10] Combined with herd management strategies—particularly mobility—this complex biodiversity allows pastoral systems to mimic the ecological interactions between wild ruminants and their

---

[10] Gene expression, without DNA modification, combining inherited and rapidly adaptable behaviours and performances related to the environment: abilities to feed, to shelter from heat, knowledge of the territory, attachment to the breeder; complex behaviours composing the animal culture (Krätli 2023; Provenza 2008; Landau and Provenza 2020).

environments. These features are also keys to the estimated carbon neutrality of such systems (feeding selectivity, adaptation to resources variations) (see Chap. 24).

5. *Extended social networks as enablers of mobility.* Rural and urban networks of relatives and friends, whether close or distant, strengthen both intra and inter community ties and facilitate routine and exceptional forms of mobility. This highlights the importance of integrating pastoralism into national policies—through youth support, decentralised service infrastructures, and recognition of customary pastoral land rights—as well as regional frameworks supporting cross border mobility and trade (Thébaud 2017; Davies et al. 2018; Sourisseau and Ancey 2021).

6. *Inclusive customary institutions and resource governance.* In arid zones, customary social organisation and institutions—when based on inclusive practices and a balance between individual and collective decision-making—help maintain flexible access to pastures and water (Thébaud 2017; Davies et al. 2018) (Fig. 12.2).

Throughout the history of pastoral development, pastoral systems have often been portrayed as struggling against a hostile natural environment. With the expected increase in climate variability, this environment is now seen as becoming even more hostile. As a result, pastoralist populations are frequently listed among the groups most vulnerable to climate disruption—especially in contrast to intensive

Fig. 12.2 The herd leaves the camp at dawn for the day, in the Moroto district, of the Karamoja region, in northeastern Uganda. (Credit: photo by S. Krätli 2000)

agriculture. Yet the resilience of intensive agriculture to climate variability depends on creating and maintaining artificially stable environments—at the cost of accelerating global warming. By contrast, the resilience of pastoralism, maintained in symbiosis with a variable environment, may appear fragile—but it does not expose humanity to the risk of runaway climate change (GIZ 2022).

12.4 Conclusions: Pastoralism's Lessons in the Face of Climate Change

The central lesson pastoralism offers in the context of climate disruption is that unpredictable variability in natural conditions is not inherently problematic. Placing agricultural production within artificially stabilized environments is not a necessity: it is a historical choice made by part of humanity. It is only from this perspective that environmental variability becomes a problem in itself. Pastoralism demonstrates that a different choice,—anchored in distinct production strategies—can lead to use climate variability as a resource for food production—but in this case 'reducing climate risk' calls for a more sophisticated approach than increasing efforts to introduce stability.

Observing pastoralism helps us rethink what resilience to climate change really entails, particularly as the conventional "modern" approach based on introducing stability to shelter certain human systems from nature, has been rendered obsolete by its own role in driving climate disruption for all. This also exposes the anachronism of the resilience strategies currently favoured in climate policy and based on the principle of externalising climate variability by building artificially stable production environments.

Understanding how ecosystems function in pastoral areas depends on recognising the spatial dynamics that sustain them—particularly the movement of herds, seasonal human migrations, and flows of economic exchange (Corniaux et al. 2016). Going beyond systems integration, a *territorial lens*—as developed in the *approche territoriale*—adds a political dimension to the analysis, for instance by examining how decentralisation affects governance and access to resources (Sourisseau and Ancey 2021).

Unless these insights are taken seriously, pastoralists' climate resilience will continue to be undermined by stress factors such as sedentarisation policies, shrinking and fragmented rangelands, and inappropriate land governance. More broadly, when it comes to food systems and the policies shaping global food security, the functioning of pastoralism offers valuable lessons for envisioning the systemic transformations required to face climate disruptions.

References

Ancey V, Diao CA (2004) Information on natural resources in pastoral environments: the secret of the "modji jofde", the good places. Communication at the XI World Congress of Rural Sociology, Trondheim, Norway

Ancey V, Ickowicz A (2015) Pastoral water points in the Sahel: at the heart of pastoral practices and territorial development. In: Lepart J (ed) (coord.), Water for livestock in alpine pastures and routes: a resource to manage, develop, share. French Association of Pastoralism, Cardère ed, pp 61–72

Ancey V, Rangé C, Magnani S, Patat C (2020) Young shepherds in the city—final synthesis. Supporting the economic and social integration of young shepherds—Chad and Burkina Faso. FAO, Rome

Assouma MH, Hiernaux P, Lecomte P, Ickowicz A, Bernoux M, Vayssières J (2019) Contrasted seasonal balances in a Sahelian pastoral ecosystem result in a neutral annual carbon balance. J Arid Environ 162:62–73

Béaur G (2006) In a dubious debate. The commons, what stakes in France of the 18th–19th centuries? Rev Mod Contemp Hist 3(1):89–114

Bégué A, Vintrou E, Ruelland D, Claden M, Dessay N (2011) Can a 25-year trend in Soudano-Sahelian vegetation dynamics be interpreted in terms of land use change? A remote sensing approach. Global Environ Change 21(2):413–420. https://doi.org/10.1016/j.gloenvcha.2011.02.002

Behnke R, Fernandez-Gimenez ME, Turner M, Stammler F (2011) Pastoral migration: mobile systems of livestock husbandry. In: Milner-Gulland EJ, Fryxell JM, Sinclair ARE (eds) Animal migration: a synthesis. Oxford University Press

Benjaminsen TA, Alinon K, Buhaug H, Tove Buseth J (2012) Does climate change drive land-use conflicts in the Sahel? J Peace Res 49(1):97–111. https://doi.org/10.1177/0022343311427343

Berkes F, Folke C (eds) (1998) Linking social and ecological systems: management practices and social mechanisms for building resilience. Cambridge Univ. Press. 476 p

Bocco R (1990) The sedentarisation of nomadic shepherds: international experts facing the Bedouin issue in the Arab Middle East (1950–1970). Cah Sci Hum 26(1–2):97–117

Bonnet B (2013) Pastoral vulnerability and public policies for securing pastoral mobility in the Sahel. Worlds Dev 164(4):71–91. https://doi.org/10.3917/med.164.0071

Carougeau J (1930) Methods to recommend for increasing and improving French colonial livestock. Col Agron 19(150):161–169

Corniaux C, Ancey V, Touré I, Camara A, Cesaro J-D (2016) Pastoral mobility, a Sahelian issue that has become sub-regional. In: Pesche D, Losch B, Imbernon J (eds) (dir.), Atlas for the rural futures Programme of NEPAD, second revised and expanded edn, Montpellier, Cirad, NEPAD, pp 60–61

Davies J, Ogali C, Slobodian L, Roba G, Ouedraogo R (2018) Crossing boundaries: legal and policy arrangements for cross-border pastoralism. FAO

Davis DK (2016) The arid lands. History, power, knowledge. The MIT Press. 296 p

Dell'Angelo J, D'Odorico P, Rulli MC, Marchand P (2017) The tragedy of the grabbed commons: coercion and dispossession in the global land rush. World Dev 92:1–12

Eggertsson T (2009) Hardin's brilliant tragedy and a non-sequitur response. J Nat Resour Policy Res 1(3):265–268

Ellis JE, Swift DM (1988) Stability of African ecosystems: alternate paradigms and implications for development. J Range Manag 41(6):450–459

FAO (2012) Voluntary guidelines on the responsible governance of tenure of land. Fisheries and Forests

FAO (2021) Pastoralism—making variability work. FAO Animal Production and Health Paper No 185. https://doi.org/10.4060/cb5855en

Gerber PJ, Steinfeld H, Henderson B, Mottet A, Opio C, Dijkman J et al (2013) Tackling climate change through livestock—a global assessment of emissions and mitigation opportunities. FAO, Rome

GIZ (2022) Pastoralism and resilience of food production in the face of climate change. Technical background paper, Saverio Krätli et al.. 20 p

Hiernaux P, Soussana JF (2011) Climate changes and their expected impacts in warm regions. In: Proceedings of the seminar Livestock and Environment in Warm Regions. Overview of the controversies, Study Methodologies, Research Directions, Inra-Cirad, pp 8–12

Hubert B (2012) Preface. In: Toutain B, Marty A, Bourgeot A, Ickowicz A, Lhoste P (eds) Pastoralism in dry zones. The case of sub-Saharan Africa. CSFD/Agropolis International, Montpellier. 60 p

ILRI, IUCN, FAO, WWF, UNEP and ILC (2021) Rangelands Atlas. ILRI, Nairobi

IEA-EDGAR (2022). https://edgar.jrc.ec.europa.eu/report_2023#:~:text=all%20world%20countries-, Main%20findings,61.6%25%20of%20global%20GHG%20emissions

Johnsen KI, Niamir-Fuller M, Bensada A, Waters-Bayer A (2019) A case of benign neglect: knowledge gaps about sustainability in pastoralism and rangelands. UNEP and GRID-Arendal

Kammili T, Hubert B, Tourrand JF (2011) A paradigm shift in livestock management: from resource sufficiency to functional integrity. Hohhot, China. Ed. de la Cardère. 272 p

Kieta N, Kvinikadze G, Pica-Ciamarra U, Bourn D, Honhold N, Georgieva N, Bako D (2016) Guidelines for the enumeration of nomadic and semi-nomadic (transhumant) livestock. FAO, Rome

Krätli S (2015) Valuing variability: new perspectives on climate resilient drylands development. IIED. https://www.iied.org/10128iied

Krätli S (2016) Discontinuity in pastoral development: time to update the method. OIE Sci Tech Rev 35(2):485–497

Krätli S (2023) Issues of declining livestock breeds: revisiting domestic animal diversity in pastoral systems. Past Pastoral 1:1–14

Krätli S, Schareika N (2010) Living off uncertainty: the intelligent animal production of dryland pastoralists. Eur J Dev Res 22:605–622. https://doi.org/10.1057/ejdr.2010.41

Krätli S, Toulmin C (2020) Farmer-herder conflict in sub-Saharan Africa?. London, UK: International Institute for Environment and Development (IIED)

Landau SY, Provenza FD (2020) Of browse, goats, and men: contribution to the debate on animal traditions and cultures. Appl Anim Behav Sci 232:105127

Lavigne DP, Ancey V, Fache E (2022) Commons and governance of shared resources. In: Colin J-P, Lavigne DP, Léonard E (eds) Rural land in the countries of the south: issues and keys to analysis. IRD, Quæ editions, pp 177–255

Lenton TM, Armstrong McKay DI, Loriani S, Abrams JF, Lade SJ, Donges JF et al (eds) (2023) The global tipping points report 2023. University of Exeter

Losch B (2023) Chapter 71. Trajectories of sub-Saharan Africa: plurality, specificities and asynchronies in globalisation. In: Boyer R, Chanteau JP, Labrousse A, Lamarche T (eds) (coord.), Regulation theory: A new state of knowledge. Dunod, pp 585–591. https://doi.org/10.3917/dunod.boyer.2023.01.0584

Meuret M, Provenza F (eds) (2014) The art and science of shepherding. Tapping the Wisdom of French Herders, Austin, Acres. 432 p

Moritz M (2016) Open property regimes. Int J Commons 10(2):688–708. https://doi.org/10.18352/ijc.719

Our World in Data (2022). https://ourworldindata.org/co2-and-greenhouse-gas-emissions; https://ourworldindata.org/co2-and-greenhouse-gas-emissions?insight=there-are-large-differences-in-emissions-across-the-world#key-insights

Pic-Ciamarra U, Baker D, Morgan N, Zezza A, Azzarri C, Ly C et al (2014) Investing in the livestock sector. Why good numbers matter. A sourcebook for decision makers on how to improve livestock data., World Bank Report Number 85732-GLB. The World Bank, Washington

Provenza FD (2008) What does it mean to be locally adapted and who cares anyway? J Anim Sci 86:E271–E284

Rangé C, Magnani SD, Ancey V (2020) "Pastoralism" and "Insecurity" in West Africa: from reifying narrative to political dispossession. Int Rev Dev Stud 243:115–150

Roe E (2020) A new policy narrative for pastoralism? Pastoralists as reliability professionals and pastoralist systems as infrastructure. STEPS Working Paper 113, IDS, Brighton

Sambo N, Sayadi AC, Clochard O (2022) Transborder transhumance in the ECOWAS space. In: Migreurop CCS (ed) (dir), Atlas of migrations in the world. Freedom of movement, borders and inequalities. Armand colin, pp 64–65

UNFCCC Secretariat (2017) Report of the conference of the parties on its twenty-third session. UNFCCC Secretariat, Bonn, p 38

Scoones I (ed) (2023) Pastoralism, uncertainty and development. Practical Action Publishing., 180 p. https://doi.org/10.3362/9781788532457

Sharifian A, Gantuya B, Wario HT, Kotowski MA, Barani H, Manzano P et al (2022) Global principles in local traditional knowledge: a review of forage plant-livestock-herder interactions. J Environ Manag 328:116966

Sourisseau JM, Ancey V (2021) A territorial and anticipatory approach for a peaceful transhumance at the border between Togo and Burkina Faso. Synthesis. (Vol. 17). FAO, 44 p. https://doi.org/10.4060/cb7925fr

Swift J (1996) Desertification narratives, winners & losers. In: Leach M, Mearns R (eds) The lie of the land, Challenging received wisdom on the African environment. The International African Institute with James Currey Ltd, London and Oxford

Thébaud B (2017) Pastoral and agropastoral resilience in the Sahel: portrait of the 2014–2015 and 2015–2016 transhumance (Senegal, Mauritania, Mali, Burkina Faso, Niger). Nordic Consulting Group, with ISRA-Bame and Cirad, for AFL, UK Aid

Thornton PK, van de Steeg J, Notenbaert A, Herrero M (2009) The impacts of climate change on livestock and livestock systems in developing countries: a review of what we know and what we need to know. Agric Syst 101:113–127

Tuck E, Wayne YK (2022) Decolonisation is not a metaphor. Ed. Rot.Bo.Krik. 128 p

Vern F (2021) The form of commons in civil property law. In: Joye J-F (ed) (coord.), The commons in the 21st century: a collective property between history and modernity. Savoie Mont-Blanc University Press, pp 295–316

Vivier N (2003) Collective property in Western Europe. Introduction. In: Démélas M-V, Vivier N (eds) (coord.), Collective properties facing liberal attacks (1750–1914): Western Europe and Latin America. Rennes University Press, pp 15–38

WISP/UNEP (2014) Pastoralism and the green economy—a natural nexus? IUCN-World Initiative for Sustainable Pastoralism (WISP and UNEP), Nairobi

Open Access This chapter is licensed under the terms of the Creative Commons Attribution-NonCommercial-NoDerivatives 4.0 International License (http://creativecommons.org/licenses/by-nc-nd/4.0/), which permits any noncommercial use, sharing, distribution and reproduction in any medium or format, as long as you give appropriate credit to the original author(s) and the source, provide a link to the Creative Commons license and indicate if you modified the licensed material. You do not have permission under this license to share adapted material derived from this chapter or parts of it.

The images or other third party material in this chapter are included in the chapter's Creative Commons license, unless indicated otherwise in a credit line to the material. If material is not included in the chapter's Creative Commons license and your intended use is not permitted by statutory regulation or exceeds the permitted use, you will need to obtain permission directly from the copyright holder.

Chapter 13
Major Crops and Climate Change: The Cases of Rice, Sorghum, Sugarcane and Cotton

Alexia Prades, Patricio Mendez del Villar, Didier Tharreau,
Edward Gérardeaux, Raphaëlle Ducrot, Aude Ripoche, David Pot,
Mohamed Lamine Tékété, Cyril Diatta, Laurent Laplaze, Boris Parent,
Isabelle Basile-Doelsch, Christine Granier, Myriam Adam, Julie Dusserre,
Michel Vaksmann, Mathias Christina, Christophe Poser, Bruno Bachelier,
and Romain Loison

Abstract This chapter discusses how climate change poses major challenges for tropical crops like rice, sorghum, sugarcane, and cotton, which are mostly grown on small family farms. While large-scale farming is a major contributor to greenhouse gas emissions, these smallholder systems also have an environmental impact. Crops like flooded rice and burned sugarcane fields emit GHGs, and all four crops are vulnerable to climate change. Cotton, though less directly affected, faces risks due to water dependence and pest pressure. The chapter outlines production systems, climate impacts, and proposes adaptation and mitigation strategies.

A. Prades
Cirad, DGD-RS, Montpellier, France
e-mail: alexia.prades@cirad.fr

P. M. del Villar (✉)
Cirad, UMR TETIS, Univ Montpellier, AgroParisTech, Cirad, CNRS, INRAE, Montpellier, France
e-mail: patricio.mendez@cirad.fr

D. Tharreau
Cirad, UMR PHIM, Montpellier, France

E. Gérardeaux · J. Dusserre · M. Christina · C. Poser · B. Bachelier · R. Loison
Cirad, UPR Agroecology and Sustainable Intensification of Annual Crops (Aïda), Montpellier, France

Aïda, Univ. Montpellier, Cirad, Montpellier, France

R. Ducrot
Cirad, UMR G-EAU, Royal University of Agriculture, Phnom Penh, Cambodia

A. Ripoche
Cirad, UR GECO, Station de La Bretagne, Saint-Denis, La Réunion, France

13.1 Introduction: Tropical Crops Facing Climate Change, Significant Challenges for Predominantly Family Farming

Alexia Prades

Cirad, DGD-RS, Montpellier, France

In certain parts of the world, it is now accepted that large-scale farming significantly contributes to climate change, as they have been conducted since the 1945–1950 period, through massive intensification processes, consuming chemical inputs, energy from fossil resources, and generating greenhouse gases (GHGs). In all OECD countries, total GHG emissions from agriculture have increased on average by 3.8% between 2005 and 2021.

The term *large-scale farming*, used in this chapter and applied to developing countries in tropical and subtropical areas, might appear inappropriate, as the cereals (rice, sorghum) or commercial crops (sugarcane, cotton) presented here are mostly extensively grown on small family farms. They contribute either directly as

---

D. Pot
Cirad, UMR AGAP Institute, Montpellier, France

UMR AGAP Institute, Univ Montpellier, Cirad, INRAE, Institut Agro, Montpellier, France

M. L. Tékété
Rural Economy Institute, Bamako, Mali

C. Diatta
Regional Study Centre for Drought Adaptation Improvement, Senegalese Institute of Agricultural Research, Thiès, Senegal

L. Laplaze
UMR DIADE, University of Montpellier, IRD (Research Institute for Development), Montpellier, France

B. Parent
UMR LEPSE, University of Montpellier, INRAE, Institut Agro, Montpellier, France

I. Basile-Doelsch
Aix-Marseille University, CNRS, IRD, INRAE, CEREGE, Aix-en-Provence, France

C. Granier
INRAE, UMR AGAP Institute, Montpellier, France

M. Adam
Cirad, UMR AGAP Institute, Montpellier, France

UMR AGAP Institute, Univ Montpellier, Cirad, INRAE, Institut Agro, Montpellier, France

Faculty of Agriculture and Food Processing, National University of Battambang, Battambang, Cambodia

M. Vaksmann
Cirad, UMR AGAP Institute, Montpellier, France

UMR AGAP Institute, Univ Montpellier, Cirad, INRAE, Agro Institute, Montpellier, France

food or indirectly as a source of income to the food security of millions of families and take part to the food and energy sovereignty of these countries (Sourisseau 2014). Indeed, rice, sorghum, sugarcane, and cotton crops also contribute, to varying degrees, to climate change. But while it has been proven that flooded rice fields emit methane and that the practice of burning for sugarcane produces GHGs, it has also been proven that these crops will be greatly affected by climate change. Given the existing knowledge, this seems a little less true for cotton species. However, the dependence of this crop on water resources and its sensitivity to the pressure of pests are factors those farmers will have to take into account.

This chapter presents key facts on the production of four major crops cultivated in tropical areas: rice, sorghum, sugarcane, and cotton. For each of them, it includes a description of the production systems and their international context. It describes the impacts of climate change on the production and ends with proposals of solutions for adaptation and mitigation.

13.2 Towards Rice Farming Adapted to Global Changes

Patricio Mendez del Villar, Didier Tharreau, Edward Gérardeaux, Raphaëlle Ducrot, and Aude Ripoche

Cirad, UMR TETIS, Univ Montpellier, AgroParisTech, Cirad, CNRS, INRAE, Montpellier, France

Cirad, UMR PHIM, Montpellier, France

Cirad, UPR Agroecology and Sustainable Intensification of Annual Crops (Aïda), Montpellier, France

Aïda, Univ. Montpellier, Cirad, Montpellier, France

Cirad, UMR G-EAU, Royal University of Agriculture, Phnom Penh, Cambodia

Cirad, UMR Aïda, Station de La Bretagne, Saint-Denis, La Réunion, France

13.2.1 Rice Feeds Half of Humanity

Rice is the staple food for more than half of the world's population, particularly the poorest. In many Southern countries, food security and social and political stability depend on the availability of rice at "affordable" prices. Rice is also a source of household income in many countries. It is also an essential commodity for emergency aid when it comes to meeting the needs of populations after extreme weather events, the frequency and amplitude of which are increasing.

The more than a thousand-year-old exploitation of many rice plains continuously (without crop rotation and sometimes in double or triple annual culture) testifies to the resilience of aquatic rice ecosystems that provide more than three-quarters of the world's production. In many regions of the world, including Europe and France, rice farming also plays an important role in sustainable development and in the conservation of biodiversity in delta areas. Rice is also grown in more fragile mangrove and non-flooded ecosystems (strict rainfed rice farming), particularly in Africa and Latin America.

13.2.2 Diversified Rice Systems

One of the characteristics of the rice sector is probably the great diversity of existing agro-socio-economic situations. This diversity includes the integration of rice farming into complex agricultural systems (for example, rice-fish farming), including livestock, and interaction with other sectors (for example, market gardening). While keeping this diversity in mind, it is nevertheless possible to identify quite distinct characteristics of the sector: three main agrosystems (irrigated rice farming, lowland and strict rainfed), subsistence farming *vs* Marketed production, including for export, input-intensive *vs* labour-intensive, small producers *vs* large producers. Certain combinations of these characteristics correspond to very representative agro-socio-economic situations on a global scale: small but intensive farming producers in Asia, small producers with low input use in Sub-Saharan Africa, large intensive farming producers in South America, particularly in southern Brazil, in Uruguay and Argentina. These situations can be likened to different value chains, each facing different challenges and defining a different initial context to be taken into account when implementing changes.

13.2.3 A Global Rice Market Facing Global Shocks

By 2023, according to the FAO (2025), global paddy rice production had reached 805 Mt. (or 535 Mt. in milled rice basis), of which nearly 90% are produced and consumed in Asia. The Asian continent is also the main surplus hub and rice supplier for the rest of the world, particularly Sub-Saharan Africa, the main rice deficit area and the leading import hub with a third of global imports. Despite a high production potential in arable lands, Sub-Saharan Africa thus has to import between 30% and 50% of the rice consumed. This external dependence strongly exposes it to the instability of international markets and increases food insecurity (availability and access), especially in urban areas.

The 2008 crisis, illustrated by the surge in global prices, was a significant shock, especially for the most impoverished populations. To respond to this crisis, African governments, with their international partners, have implemented new policies to support local sectors with the aim of achieving rice self-sufficiency. However, support programmes are not achieving the expected goals of significantly reducing rice dependence on the international market. Crises of all kinds, like the Covid-19 pandemic or the war in Ukraine, combined with climate change, underline the need to develop more resilient rice systems. However, on certain national markets in the South, local production sectors are facing the globalisation of trade and are subject to the competition of import sectors. The low added value of sectors promoting quality for export (organic or *sustainable rice*) is a contextual element to be taken into account.

Global rice sectors must face multiple challenges of different kinds (biophysical and environmental, technical, socio-economic and political), which add additional constraints to climate change management. The first of these is to generate value, distribute it equitably among actors, and improve the living standards of 400 million poor rice farmers to reduce inequalities, without prejudice to the 500 million consumers living below the poverty line for the majority of them.

13.2.4 Meeting the Needs of Consumers

Another significant challenge is to respond to the diversification of demands in terms of organoleptic and industrial quality of rice, and to increased requirements for nutritional and health quality. Quality is a challenge throughout the value chain, from the production to the consumption of rice. This involves not only addressing the technological issue (research on plant functioning, organoleptic qualities, etc.), but also leveraging this quality to better value rice commercially, which simultaneously requires organisational innovations.

13.2.5 Rice Systems, Living Spaces

Beyond considerations related to the marketing of rice, irrigated rice-growing areas are territories in which populations are settled and live. The permanent presence of water gives these areas a great specificity with, in particular, consequences on human health, issues on water uses, a particular vulnerability to climate change. Irrigated rice farming is also a significant source of GHGs (methane, nitrous oxide) in agriculture and therefore at the heart of issues related to climate change (IPCC 2021a; Sinha et al. 2020).

13.2.6 Meeting Future Food Security in the Face of Global Changes

On a global scale, the challenges to be met in terms of production can be summarised as follows: for each additional billion people in the world population, an additional 100 Mt. of paddy rice (13% of total production in 2024) per year would need to be produced (Trébuil and Hossain 2004); and this with less land, less water and less labour, with more environmentally friendly production systems, more resilient to climate change and emitting less GHGs. By 2025, 10–15% of irrigated rice fields will be subject, to varying degrees, to a water constraint resulting from increased urban and industrial demands and the effects of climate change (Tuong and Bouman 2003).

Rice farming must therefore, more than any other crop, face the instability of the climate and its consequences: the increase in daytime and nighttime temperatures can harm flowering and therefore production; the increase in CO_2 atmospheric increase in biomass and if nutrient availability cannot meet the increased demand, the biomass gain will not translate into grain yield gains. This situation can even impair quality. Drought, soil salinity and flooding are increasingly frequent abiotic constraints that reduce yields (Lafarge et al. 2015; Ahmadi et al. 2015). Climate change can also promote the emergence or re-emergence of rice diseases that strongly affect yields and quality.

In a context of climate change, whose impacts are intensifying, water, necessary for the majority of rice systems, is becoming scarce and the multiplication of its uses reinforces competition between users in the same territory. The rice sector will thus have to manage the high consumption of a water resource under stress in irrigated systems in Asia, Africa and Latin America. It will be necessary to produce with less water and co-develop (up to sustainable appropriation) organisational and institutional innovations for better conflict management around this precious resource. Better water management can also contribute to promoting efficient systems emitting less GHGs. Rice cultivation is indeed responsible for a significant part of anthropogenic GHG emissions (22% of global agricultural methane emissions) (IPCC 2021a, b).

13.2.7 The Agroecological Transition in Rice Cultivation, a Technical Challenge

Global rice cultivation will also have to face technical challenges related to climate change. How to stabilise or increase production while limiting negative impacts on humans and the environment? The demand for rice will continue to increase due to the growth of the world population by an additional two billion by 2050, including one billion in Africa.[1] In input-intensive systems, particularly in Asia, the objective

---

[1] UN, 2024. Population. https://www.un.org/en/global-issues/population

will be to produce more by reducing the use of chemical inputs (fertilizers, pesticides) and accompanying the potential reduction of labour (mechanization). Reducing the use of certain synthetic inputs improves the carbon balance, but it also reduces the dependence of rice farmers on often imported products, not always available or accessible, and dependent on global geopolitical stability. Finally, reducing inputs contributes to reducing production costs given the increase in prices. In Africa, the context is quite different as the quantities of chemical fertilizers are relatively reduced (between 100 and 200 kg/ha), this being largely due to financial constraint and their reduced availability; a situation that tends to worsen with the quadrupling of fertilizer prices between 2020 and 2022, due to the pandemic, and with new increases since the war in Ukraine in 2022 and its repercussions on international trade.

Certain rice agrosystems (deltas, protected areas, rainfed rice cultivation) are particularly fragile and must receive special attention. Agroecological solutions are a way to protect them. Increasing plant diversity in rice systems is necessary to limit soil fertility loss and to increase resilience to climate change, but it remains a significant technical challenge in irrigated or flooded rice cultivation.

13.3 Sorghum and Climate Change: Balancing Adaptation Needs with Mitigation Potential

David Pot, Mohamed Lamine Tékété, Cyril Diatta, Laurent Laplaze, Boris Parent, Isabelle Basile-Doelsch, Christine Granier, Myriam Adam, Julie Dusserre, and Michel Vaksmann

Cirad, UMR AGAP Institute, Montpellier, France

UMR AGAP Institute, Univ Montpellier, Cirad, INRAE, Institut Agro, Montpellier, France

Rural Economy Institute, Bamako, Mali

Regional Study Centre for Drought Adaptation Improvement, Senegalese Institute of Agricultural Research, Thiès, Senegal

UMR DIADE, University of Montpellier, IRD (Research Institute for Development), Montpellier, France

UMR LEPSE, University of Montpellier, INRAE, Institut Agro, Montpellier, France

Aix-Marseille University, CNRS, IRD, INRAE, CEREGE, Aix-en-Provence, France

INRAE, UMR AGAP Institute, Montpellier, France

Faculty of Agriculture and Food Processing, National University of Battambang, Battambang, Cambodia

Cirad, UPR Agroecology and Sustainable Intensification of Annual Crops (Aïda), Montpellier, France

Aïda, Univ. Montpellier, Cirad, Montpellier, France

UMR AGAP Institute, Univ Montpellier, Cirad, INRAE, Agro Institute, Montpellier, France

Sorghum, a species native to northeastern Africa, has a dual identity. It is, on one hand, a cornerstone of food security in semi-arid regions around the world; on the other, it serves as a driver of agroecological transition in the production systems of Northern countries. In both contexts, sorghum must adapt to the challenges posed by climate change, while also contributing to its mitigation.

13.3.1 Sorghum: Ensuring Food Security and Driving the Agroecological Transition

Sorghum ranks as the fifth most produced cereal worldwide and serves as the staple food for over 500 million people living in semi-arid regions. It is extensively cultivated—covering more than 5000 hectares per country—in 72 countries, occupying a total global area of 40.2 Mha, with an average annual production of 59.1 Mt. (between 2018 and 2022).

The largest sorghum cultivation areas are found in sub-Saharan Africa—particularly in Sudan and South Sudan (7.6 Mha), Nigeria (5.69 Mha), and Niger (3.72 Mha)—as well as in India (4.42 Mha) and in the United States (2.09 Mha). In Europe, on average 277,000 ha of sorghum were cultivated annually between 2018 and 2022. On the same period, the five largest sorghum-producing countries were the United States (8.71 Mt./year), Nigeria (6.71 Mt./year), Ethiopia (4.79 Mt./year), Mexico (4.54 Mt./year), and India (4.40 Mt./year).

In West Africa, sorghum yields have changed little between 1998–2002 and 2018–2022, with only a modest increase of 6.4%. As a result, production growth in the region has largely been driven by the expansion of cultivated areas.

In Africa, human consumption—which encompasses a wide variety of food products—accounts for nearly three-quarters of total sorghum use. In contrast, in Northern and emerging countries, sorghum is primarily used as animal feed. Additionally, it is emerging as a key crop for bioenergy production (Thomas et al. 2021) and shows promising potential for the development of biomaterials.

The development of sorghum-based value chains is supported by the crop's distinctive biological characteristics. As a C4 photosynthetic species, sorghum exhibits efficient carbon assimilation under high-temperature conditions. Its strong tolerance to abiotic stresses—particularly water scarcity and low soil fertility—has been well documented (Schlegel et al. 2018), contributing to its relatively low environmental footprint. Finally, although it can be affected by various pests and diseases, sorghum remains a robust species with a relatively low sensitivity to biotic stress, thereby requiring minimal plant protection treatments.

13.3.2 How Climate Change Is Shaping Sorghum Production

Climate change is already impacting West Africa. Between 2000 and 2009, a 1 °C increase compared to pre-industrial levels resulted in sorghum yield reductions ranging from 5% to 15% (Sultan et al. 2019).

Future projections indicate varying degrees of yield reductions. In West Africa, climate scenarios of +1.5 °C and + 2 °C predict yield declines (excluding Niger) of approximately 2% and 5%, respectively, regardless of fertilization levels (Faye et al. 2018). More recent projections from 2020, covering the five main sorghum-producing countries in West Africa, estimate losses between 15% and 28% (Defrance et al. 2020). Notably, Niger appears to benefit from a favorable impact of climate change.

Yield reductions are therefore to be expected, but these data alone cannot predict whether the needs of the populations will be met. Considering the evolving needs of the populations in West Africa (+300% based on demographic changes) there is a substantial anticipated gap between demand and future sorghum production capacity (Defrance et al. 2020).

Changing environmental conditions are also expected to impact sorghum quality. Under water deficit conditions, increases in grain hardness and protein content have been observed, alongside a decrease in protein digestibility and alterations in micronutrient balance (Impa et al. 2019). In contrast, heat stress leads to reductions in protein content and digestibility, coupled with an increase in grain hardness. Both stresses are associated with an overall decline in micronutrient levels (Impa et al. 2019).

The quality of sorghum biomass, which is crucial for both animal feed and bioenergy production, will also be impacted by environmental stress. Water deficit has been shown to increase soluble sugar content while reducing lignin and cellulose levels in stems (Luquet et al. 2019).

However, it is important to emphasize that, for both grain and biomass, the impacts on the final products consumed or utilized have yet to be thoroughly analyzed.

13.3.3 Sorghum: A Contributor to Climate Change Mitigation

The potential of crops to mitigate climate change primarily relies on two mechanisms: their capacity to sequester atmospheric carbon in the soil and their ability to reduce greenhouse gas emissions, including CO_2, CH_4, and N_2O.

Regarding soil carbon sequestration, sorghum benefits from a deep root system. A recent comprehensive study of biomass contributing to carbon storage—particularly through the incorporation of crop residues into the soil—across cereals and legumes in Africa and Asia was provided by Kuyah et al. (2023). This study reveals that sorghum ranks among the cereals with the highest potential for carbon storage in both above-ground biomass and root systems, emphasizing the significant role of carbon released through rhizodeposition.

The most comprehensive study to date—considering both geographical scope and duration—estimates CO_2 emissions at 250 g CO_2 per kilogram of sorghum grain produced.[2] Although these figures should be interpreted cautiously, they compare favorably to maize grain emissions in the same region, which are estimated at 390 g CO_2/kg (Adom et al. 2012). In Sub-Saharan Africa, the use of sorghum straw to supplement the poor-quality forage available at the end of the dry season has been shown to reduce enteric methane emissions from ruminants (Fulani zebu) by 21% (Gbenou et al. 2024).

Regarding N_2O emissions, sorghum's root exudates have been shown to inhibit soil nitrification, which may lead to lower emissions compared to other major crops (Subbarao et al. 2013).

13.3.4 Sorghum in a Changing Climate: Strategies for Adaptation and Climate Mitigation

Despite sorghum's inherent resilience, its cultivation systems must evolve to meet the future demands of producers and consumers. Notably, farmers are already adapting to climate change by drawing on traditional knowledge—adjusting sowing dates and varieties, relocating crops to more favorable environments, and adopting water-saving techniques (Amadou et al. 2022) as well as crop diversification strategies (Traore et al. 2023). This highlights the critical importance of fostering close collaboration between farmers and researchers to develop effective solutions that mitigate the anticipated adverse impacts of climate change.

Optimizing cropping systems that integrate sorghum-legume associations (such as cowpea and others; see Chap. 21) will be a crucial strategy for adaptation. Access to fertilizers will also play a key role in maintaining or even increasing yields, although it may increase the system's vulnerability to climate change (Adam et al. 2020). The development of improved varieties is equally essential. Beyond adjusting phenology and selecting for the stay-green trait—already key targets of breeding programs (see Chap. 21)—additional traits must be considered. Although sorghum is generally regarded as drought-tolerant, significant variability exists within its genetic diversity, and the underlying physiological and genetic mechanisms remain largely unexplored. It is therefore critical to deepen our understanding of water use efficiency by examining the plant's hydraulic traits, including the aerial apparatus (e.g., transpiration efficiency, xylem cavitation) and the root system (anatomy, architecture, and rhizodeposition). A detailed analysis of root anatomy, exudate production, and the plant's ability to recruit beneficial microbiome partners will also enhance soil carbon sequestration potential. Finally, the development of resilient varieties and sustainable cropping systems must address the combined and recurrent stresses—such as drought and heat—that plants will increasingly face in the future.

---

[2] SGS North America, 2015. The-Carbon-Footprint-of-Sorghum.pdf.

13.3.5 Evolution of Sorghum Production Systems: Towards a Multi-Species Approach Embracing Diverse Production Contexts

The evolution of sorghum-based production systems will need to acknowledge the roles and benefits of associated crops, inevitably leading to changes in both cultivation practices and the traits targeted in new varieties. Consumer needs must be systematically and more thoroughly integrated to guide the desired development of cropping systems and variety selection. Identifying relevant ideotypes—optimal trait combinations—will undoubtedly benefit from advances in comparative and translational biology. Rather than framing sorghum and maize as simple opposites—with sorghum traditionally viewed as more drought-tolerant, a notion now being challenged (Rotundo et al. 2024)—synergies between research on these two crops should be fostered for the advantage of both producers and consumers. Finally, it is essential to recognize that adapting cropping systems to future conditions will depend not only on agronomic factors such as yield and product quality, but also on socio-economic considerations, including labor availability and the potential for intensification (e.g., input quality and availability, biomass flow management).

13.4 Sugarcane in the Face of Climate Change

Mathias Christina and Christophe Poser

Cirad, UPR Agroecology and Sustainable Intensification of Annual Crops (Aïda), Montpellier, France

Aïda, Univ. Montpellier, Cirad, Montpellier, France

13.4.1 Impacts and Challenges Related to Climate Change for Sugarcane Cultivation

Sugarcane cultivation is an economic pillar for many tropical and subtropical countries, providing not only sugar for human consumption, but also biofuels, animal feed and bedding, and industrial by-products. Sugarcane is currently grown on about 26 million hectares of agricultural land worldwide, employing millions of people in the cultivation, transport, processing and distribution sectors. The largest global producers are located in Brazil, Southeast Asia (India, China, Thailand), but sugarcane is also present in all tropical and subtropical areas (United States, South and West Africa, Australia). Faced with increasing demand, cultivation areas are expanding.

Like other major crops such as corn or wheat, sugarcane cultivation significantly contributes to GHG emissions (Tongwane et al. 2016). The cultivation process represents the majority of emissions compared to the industrial transformation process (Macedo et al. 2008), due to mechanisation, the use of chemical inputs or the practice of burning, which is still common in some countries, although its cessation is widely promoted to limit GHGs. For example, a study, conducted in Brazil between 2005 and 2006, showed that the transport of sugarcanes after harvesting accounted for only 7% of GHG emissions, compared to 11% for the production of fertilisers and 19% resulting from the practice of burning (Macedo et al. 2008).

Climate change also poses a serious threat to this industry, but the expected consequences vary greatly depending on the areas concerned (Linnenluecke et al. 2018). The challenges are numerous: variations in precipitation patterns, rising temperatures, rising sea levels in coastal areas and the increased frequency of extreme weather events such as cyclones and floods (Warren et al. 2024).

Studies have shown contrasting projections depending on the climate zones, due to the combination of favourable conditions for C4 crops[3] (increase in temperature and CO_2) and unfavourable precipitation patterns (Fig. 13.1). Indeed, sugarcane requires specific climate conditions to grow optimally. It thrives in regions where rainfall is abundant, ideally over 1500 mm per year, and where average daily temperatures are between 22 °C and 30 °C.[4] Well-drained and fertile soils are also essential for high yields. However, the increase in average temperatures and the increased variability of precipitation make crop management more difficult and compromise yield stability.

The impacts of climate change on sugarcane are multiple and varied. Changes in precipitation patterns can lead to more frequent and severe drought periods (Carvalho et al. 2015), affecting the growth of sugarcane which requires regular water inputs to maximise yields (Jones et al. 2015). Good irrigation management could indeed increase yields in many regions (Linnenluecke et al. 2018). However, the decrease in precipitation and the increase in evapotranspiration due to higher temperatures reduce the availability of fresh water, exacerbating competition for this resource between agriculture, domestic and industrial uses and even more in the context of a demand for improved water quality.

The increase in temperatures also poses a significant challenge. The maturation of the sugarcane (accumulation of sugar in the stems) is disrupted by excessive heat, particularly by the increase in night-time minimums. In addition, higher temperatures and humidity promote the proliferation of pests and diseases, such as brown rust of sugarcane or weeds, which can harm production (Goebel and Sallam 2011). However, the increase in temperatures is seen as more beneficial for sugarcane cultivation in many regions, provided there is satisfactory water availability.

---

[3] C4 crop: a crop whose photosynthesis is of the C4 type, meaning that the first carbohydrate formed has four carbon atoms. C4 plants are characterised by a higher photosynthetic yield and better water use than C3 plants.

[4] https://www.fao.org/land-water/databases-and-software/crop-information/sugarcane/en/

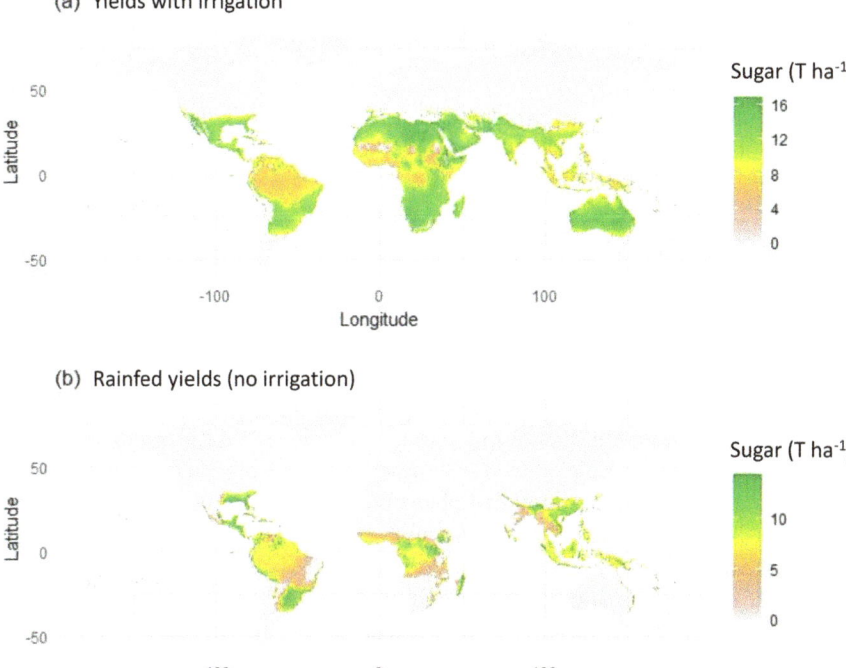

Fig. 13.1 Example of a gradient of potential sugarcane yields in the context of climate change: period 2041–2070. (Source: data from gaez-services.fao.org). Sugar yield (t/ha) under irrigated (a) or rainfed (b) conditions, GFDL-ESM2M climate model, RCP4.5 scenario. The maps show a colour gradient from brown for low sugar yields to green for high sugar yields worldwide. Two situations are represented, with irrigated yields at the top and non-irrigated yields at the bottom. They illustrate the diversity of situations that will be encountered by the main sugarcane producing countries in the future between 2040 and 2070

The rise in sea level could also exert increasing pressure on sugarcane production areas (Warren et al. 2024). Coastal areas, where sugarcane is often grown due to their fertile soils and favourable climate, are particularly vulnerable to floods favoured by rising waters. Sugarcane is a crop sensitive to flooding, leading to root rot and a reduction in the quality and quantity of crops (Gomathi et al. 2015). A rise in sea level can also result in increased soil salinisation, making agricultural land less productive or even infertile. Therefore, some of the main producing regions, such as those located in deltas and coastal plains, could see their production decrease significantly.

13.4.2 Adaptation Pathways and Mitigation Potential of Sugarcane Cultivation

To cope with climate change, farmers will need to adopt effective technical adaptation and mitigation strategies but also review the geographical distribution of cultivated areas. One of the main technical adaptation pathways is the selection and cultivation of sugarcane varieties that are more resistant to drought, heat and diseases (Goebel and Sallam 2011; Grandis et al. 2024). In particular, drought resistance is seen as an essential characteristic to develop future varieties. An increase in the rooting depth and water use efficiency of varieties is a sought-after trait for these drought-tolerant varieties. The long (10–15 years) and costly varietal creation and selection programs, coupled with the improvement of agronomic practices on these improved varieties, play a crucial role.

Optimising irrigation is also essential in many regions of the world. Drip irrigation techniques, for example, allow for more efficient use of water, reducing losses through evaporation and ensuring that plants receive the necessary moisture during periods of drought. Moreover, practices limiting evaporation such as mulching can improve water retention, thus increasing the resilience of crops to climate variations.

Another adaptation pathway, more specific to certain regions, involves shifting crops to regions that have become more favourable for sugarcane cultivation in connection with the increase in temperatures or the adaptation of varieties (Poser et al. 2020). For example, high-altitude areas in the tropical islands of the Indian Ocean will be more favourable for sugarcane cultivation even under non-irrigated conditions (Christina et al. 2024). Also, a shift in the harvest period in certain regions like the north of South Africa has shown potential to improve yield in the future (Park et al. 2007). Consequently, there is already an increased interest in sugarcane in historically low-producing regions, such as southern Europe for example.

Crop diversification is another adaptation strategy, particularly for small-scale farmers. By integrating complementary crops, farmers can reduce their dependence on a single crop, thus diversifying their sources of income and improving the resilience of their farms to climate hazards (Aurand et al. 2022). Furthermore, the development of sustainable agroecological practices, such as crop associations, can help maintain long-term land productivity by improving soil structure, increasing biodiversity and reducing erosion, while decreasing pesticide use (Soulé et al. 2024), thus contributing to the reduction of GHG emissions.

Moreover, sugarcane crops can play a significant role in mitigating climate change through several mechanisms, including soil carbon storage, but also through a direct impact on the local climate (Loarie et al. 2011). Sugarcane has a dense and deep root system, which allows for effective soil carbon sequestration (see Chap. 17), compared to other crops (La Scala Junior et al. 2012). For example, a study in Brazil estimated that 2.4 t/ha of CO_2 was emitted during a one-year sugarcane cultivation cycle (de Figueiredo et al. 2010), which is less than the amount of carbon present in the sugarcane roots at harvest (2–6 t/ha of organic carbon) (Chevalier et al. 2023). In addition, sugarcane is often used to produce biomass and bioethanol (see Chap. 19), alternatives to fossil fuels that help reduce GHG emissions. Initiatives aimed at reducing GHG emissions from sugarcane production can also contribute to climate change mitigation. In particular, the use of sugarcane residues for bioenergy production and improving energy efficiency in the transformation processes are examples of measures that can reduce the carbon footprint of the sector (Cherubin et al. 2021). The same applies to water savings made in the factory and in the fields.

In conclusion, climate change poses significant challenges for sugarcane cultivation, but adaptation strategies exist, and this crop has a strong potential for mitigation through carbon storage *via* its deep root system. By combining agronomic innovation, sustainable farming practices, and the necessary political support for the transition, it is possible to strengthen the resilience of sugarcane to climate impacts and ensure the sustainability of this essential sector for many national economies. However, small-scale farmers, who account for a large part of global sugarcane production, are particularly vulnerable to climate change. Therefore, government policies and investments in agricultural infrastructure, such as water management systems and support programs for small-scale farmers, are essential to strengthen the resilience of the crop.

13.5 Cotton in the Face of Climate Change

Edward Gérardeaux, Bruno Bachelier, and Romain Loison

Cirad, UPR Agroecology and Sustainable Intensification of Annual Crops (Aïda), Montpellier, France

Aïda, Univ. Montpellier, Cirad, Montpellier, France

13.5.1 The Global Cotton Sector: A Distribution Across Five Continents with Significant Disparities

With an annual production of over 24 Mt., cotton has been the leading natural fibre and the second textile fibre in volume, behind synthetic fibres, for several decades. In 2022–2023, two-thirds of the world's cotton fibre production came from the Asia-Oceania region, a quarter was produced on the American continent and less than a tenth in Africa.

The average global yield of 800 kg of fibre per hectare hides large disparities between production areas. Thus, while Australia obtains more than 2000 kg/ha, Africa only produces an average of 400 kg/ha. This gap reflects the diversity of abiotic, biotic, technical and organisational conditions of cotton production. In practical terms, the spectrum ranges from small-scale, low-mechanisation family farming in the South to intensive farming in developed countries.

In cotton-growing countries where cultivation is not intensive, cotton is mainly grown in small family farms, on a few hectares, almost exclusively rainfed, with small-scale mechanisation, rarely motorised. Harvesting is still largely manual, which is a constraint due to labour shortages. Food crops (legumes, cereals, etc.) are rotated in the cropping system and some benefit from the residual effect of cotton fertilisation. Depending on the production areas, either the producers source inputs from the local market and sell their harvest to collectors (Asia) or the producers contract with a cotton company, public or private, which provides them with inputs on credit at the beginning of the season and guarantees the purchase of their harvest at a price set at the beginning of the season (Africa). Although providing a low margin to producers under these conditions, cotton cultivation is considered a cash crop, contributing significantly to the monetary income of several million families. In addition, there are tens of millions of people involved in the various stages of the sector, from production to processing.

In the most advanced countries (United States, Australia), cotton cultivation is practised by large farms, motorised from sowing to harvest, irrigated and intensive according to the principles of precision agriculture, and conducted using high-performance varieties, often transgenic. When they are not themselves ginners,[5] the producers subcontract the primary processing of their harvest to private companies and remain owners of the fibre and seeds obtained, which they then sell to commercial intermediaries.

The secondary processing of cotton fibre into textiles has great difficulty developing locally in Africa, which exports nearly 90% of its production and thus loses almost all of the added value associated with textile products (yarn, fabric, clothing, etc.). In a century, it has significantly declined on other continents, mainly in favour of a few Asian countries, whose industries process more than 80% of global production. This strong concentration in countries with low labour costs allows the global market to be supplied with textiles, some of which feed the fast-fashion industry.

While the fibre represents three-quarters of the value of the harvest, the seed also contributes to the economy of the sector. It is an important oilseed source, most often valued locally in human (oil) and animal (cake) food, but also in cosmetics.

13.5.2 Impacts of Climate Change on Major Cotton Production Areas

As with all crops, climate change will have an effect on growth conditions, and therefore on yields and the sustainability of cotton industries. Thanks to several of its traits, cotton will respond differently compared to other crops. It is a plant resistant to drought and high temperatures, mainly due to its physiological characteristics.

---

[5] A ginner is a public or private entrepreneur responsible for ginning (or primary processing) of cottonseed (the product of cotton harvesting, composed of seeds bearing fibres) to separate the fibre from the seed.

- Firstly, its base temperature[6] of 12 °C makes it a plant with high growth and functioning ranges: the optimum temperatures are between 27 °C and 32 °C for vegetative growth and between 24 °C and 27 °C for fruit development.
- Secondly, the indeterminate growth of cotton, which produces fruiting and vegetative organs concurrently throughout its cycle and has the ability to shed underdeveloped fruiting organs, gives it a certain plasticity. Thus, a period of high heat or moderate duration drought does not irreversibly affect production, as the plant will compensate by producing new fruiting organs once the stress is relieved.
- Thirdly, the root system of cotton is taprooted, allowing deep soil exploration and access to water up to 1.5 m deep.
- Fourthly, cotton is an anisohydric plant: it does not close its stomata during moderate water deficits. This characteristic allows it to maintain photosynthesis to produce more roots, explore more soil and access deeper water resources, in relation to the previous characteristic.

It is also a so-called C3 plant,[7] less efficient in terms of CO_2 than C4 plants, many of which are found in tropical regions (maize, sorghum, sugarcane, etc.). Therefore, an increase in atmospheric carbon will be very favourable to cotton (Mauney et al. 1994).

For all these reasons, climate changes will have less negative effects on cotton than on other plants, especially if water resources are available through irrigation, abundant rainfall or deep water tables. On the other hand, under strict and limiting rainfed conditions, such as in the cotton-growing areas of semi-arid and sub-humid Africa, negative effects of high temperatures are to be feared. They will only be offset by the enrichment of atmospheric CO_2 if rainfall is sufficient.

In China, the current leading global producer, the effects of climate change will on average be negative, with regional disparities: negative in the Yangtze valley and neutral or positive in the north-west regions and the Yellow River (Chen et al. 2015).

In the United States, cotton is grown in the southern states. The forecasts of the effects of climate change there are mixed between yield losses and gains (Sharma et al. 2022), due to a negative effect of an increase in the frequencies of extreme events and a positive effect of the rise in CO_2 and precipitation.

Scientific studies of the effects of climate change in sub-Saharan Africa are scare. The major cotton production basins are the Sudanian and Sudano-Sahelian zones of West and Central Africa. The IPCC forecasts show contrasting situations between the west and the east. Rainfall deficits in Senegal, Mali and Burkina Faso will cause yield decreases. On the other hand, increases in rainfall in Nigeria, Cameroon and Chad, and uncertain intermediate situations between these two zones (Ivory Coast, Ghana, Benin, Togo) could allow for yield increases. In Benin, Amouzou et al. (2018) predicts variable effects, mostly positive, ranging between

---

[6] Base temperature (or zero vegetation): the temperature below which a crop cannot grow and develop.

[7] C3 plant: a plant whose photosynthetic mechanism is based on the production of three-carbon carbohydrates.

−7% and +41% on yields, while drawing attention to difficulties in soil nitrogen availability in case of high demand linked to the increase in biomass. Positive effects are also predicted in Cameroon (Gérardeaux et al. 2013).

In Europe, cotton is grown in Spain and in Greece with high yields thanks to irrigation and mechanisation. Low spring temperatures are a significant limiting factor for crop growth and global warming could bring favourable changes here. Adaptation of farming systems will be possible to take advantage of this (Engonopoulos et al. 2021). Moreover, water resources are in competition with the needs of the population. The crop could develop in countries further north such as France, Italy or Bulgaria.

In Australia, studies have shown contrasting effects on the yields and on fibre quality, even for rainfed cropping systems (Luo et al. 2016; Williams et al. 2015). Extreme events (droughts, floods, heatwaves) will become more frequent with more significant negative impacts.

In Central Asia, cotton is grown in plains subjected to very cold winters and very hot summers. The crops are irrigated, but water resources are already heavily exploited. Predictions of the effects of global warming include an increase in water demand and therefore a decrease in yields and cotton areas, which could disappear from certain zones (Schlubach 2021).

Finally, cotton is also produced in India: a decline in yields is expected in the North and neutral to positive effects in the central and southern parts (Hebbar et al. 2013). It is likely that yields will increase, but so will pest pressure.

In conclusion, without even considering possible adaptations, climate changes will not drastically alter the map and growing conditions of cotton. Some regions of the world will benefit, such as southern Europe, central Africa, southern India, and others will lose out, such as West Africa, Central Asia, northern India, the Yangtze valley in China.

13.5.3 The Paths of Adaptation and Mitigation of Cotton Cultivation in a Climate Change Context

Breeding associated with the choice of optimal sowing dates is an adaptive research activity that involves matching the growth phases of a crop to those of resource availability (Zimmermann et al. 2017). Moreover, a temperature increase of 2.5 °C will cause an acceleration of crop development and mechanically a shortening of the cycle of varieties by 5 days for flowering and 10 days for maturity. This can significantly disrupt the synchronism between periods when water needs are covered by rainfall. To counteract this effect, selection combined with the search for optimal sowing dates can adapt by choosing varieties with longer or shorter cycles, in order to bring flowering and maturity back to the desired dates (Wu et al. 2023).

In some cotton-growing areas, climate changes will create conditions favourable to cotton growth. But for these favourable conditions to result in yield improvement, good plant nutrition must be ensured to hope for productivity gains. The biomass gains allowed by the favourable conditions must be covered by an improvement in the soil's nutrient supply. In the family farming systems of Africa, where access to the input market is limited and soils are fragile, it is highly likely that these potential gains will be nullified. One promising way to improve soil fertility without relying solely on external inputs is to practice agroecological farming systems, involving rotations or associations with legumes, buried or left on the soil to form a mulch of cover crops. Again, the capacities of small family farms to modify their farming systems, to access seed and agricultural equipment markets remain limited. The implementation of these adaptation paths will depend on public support policies.

Like most crops, cotton cultivation emits greenhouse gases through nitrogen inputs, fuel and electricity for tractors, irrigation, harvesters and ginning factories. It is estimated that 1.6 tonnes of GHGs are emitted to produce one ton of fibres (Hedayati et al. 2019). However, if we stop our balance at the year of production, the carbon stored in the fibres and in the roots is greater than that emitted to produce it. Unfortunately, over time, the carbon stored in the fibres is released into the atmosphere through the life cycle of the fabrics (Cotton Incorporated 2009). Research in agroforestry or agroecology aiming to improve the resilience of farming systems to climate changes by improving soil fertility and reducing dependence on synthetic nitrogen inputs also contributes to improving GHG balances. Experiments are underway all over the world, we can mention those conducted in the Desira-UE Innovac project in North Cameroon, which aim to qualify the effects of agroforestry parks on crop sustainability.

13.6 Conclusion: Tropical Crops Facing Climate Change, Knowledge to Share, Steps to Master, Governance to Rethink

Alexia Prades

Cirad, DGD-RS, Montpellier, France

The solutions recommended for food or commercial crops presented in this chapter are of three types. The first includes short-term solutions, often already adopted by farms: optimisation of water resource management, valorisation of local cultivated biodiversity, change in farming practices often including a shift towards agroecological systems, such as trials of innovative crop combinations. The second type of solution includes more radical measures, achievable in the medium term, such as moving production areas. Finally, certain long-term solutions rely, on varietal improvement combined with an increasingly refined understanding of plant functioning in a complex environment. Complexity arises from the use of several companion plants or agroforestry systems for example, to which is added the constraint of climate instability.

These three types of solutions are implemented at different time scales (see Chap. 20), but they do have one thing in common. All three are still under-researched, even though they have shown promising results (Côte et al. 2018). Actually, this research requires the mobilisation of coordinated multi-scale governance and support from public policies to catalize change and mitigate exposure to risk, especially for smallholders. Some of the solutions will disrupt the trajectories of entire territories and, in particular, their populations. Multidisciplinary and cross-value chains research will play a pivotal role for example, when studying certain forms of crop associations or replacing one crop with another. In addition, partnership research practices should be encouraged by mobilizing participatory approaches. The solutions exist and research actors, along with stakeholders, are ready to support the transformation of these territories. "Although they play a central role in feeding the planet, small producers only have access to 1.7% of climate action funding," reminds the International Fund for Agricultural Development (IFAD), in an article published on its website on 4 April 2022. There is an urgent need to change our way of thinking, to modify our perspectives, our practices, our institutions and to finally trust small and medium scale producers who play a key role in the resilience of tropical territories facing climate change.

References

Adam M, MacCarthy DS, Traoré PCS, Nenkam A, Freduah BS, Ly M, Adiku SGK (2020) Which is more important to sorghum production systems in the Sudano-Sahelian zone of West Africa: climate change or improved management practices? Agric Syst 185:102920. https://doi.org/10.1016/j.agsy.2020.102920

Adom F, Maes A, Workman C, Clayton-Nierderman Z, Thoma G, Shonnard D (2012) Regional carbon footprint analysis of dairy feeds for milk production in the USA. Int J Life Cycle Assess 17:520–534. https://doi.org/10.1007/s11367-012-0386-y

Ahmadi N, Baroiller J-F, D'Cotta H, Morillon R (2015) Adaptation to salinity. In: Torquebieau E (coord) Climate change and world agriculture. Versailles, Quae editions, pp 50–62

Amadou T, Falconnier GN, Mamoutou K, Georges S, Alassane BA, François A et al (2022) Farmers' perception and adaptation strategies to climate change in Central Mali. Weather Clim Soc 14:95–112. https://doi.org/10.1175/WCAS-D-21-0003.1

Amouzou KA, Naab JB, Lamers JPA, Borgemeister C, Becker M, Vlek PLG (2018) CROPGRO-cotton model for determining climate change impacts on yield, water- and N- use efficiencies of cotton in the dry savanna of West Africa. Agric Syst 165:85–96

Aurand TC, Sunthornvarabhas J, Sriroth K (2022) Value addition through diversification of the sugar industry from farm to mill. Sugar Tech 24(4):1155–1166

Carvalho AL d, Menezes RSC, Nóbrega RS, Pinto A d S, Ometto JPHB, von Randow C, Giarolla A (2015) Impact of climate changes on potential sugarcane yield in Pernambuco, northeastern region of Brazil. Renew Energy 78:26–34

Chen C, Pang Y, Pan X, Zhang L (2015) Impacts of climate change on cotton yield in China from 1961 to 2010 based on provincial data. J Meteorol Res 29:515–524

Cherubin MR, Carvalho JLN, Cerri CEP, Nogueira LAH, Souza GM, Cantarella H (2021) Land use and management effects on sustainable sugarcane-derived bioenergy. Land 10(1):72

Chevalier L, Christina M, Février A, Jourdan C, Ramos M, Poultney D, Versini A (2023) Sugarcane responds to nitrogen fertilisation by reducing root biomass without modifying root accumulation. Presented at xxxi International Society of Sugar Cane Technologists (ISSCT) Congress, Hyderabad

Christina M, Mézino M, Le Mézo L, Todoroff P (2024) Modelled impact of climate change on sugarcane yield in Réunion, a tropical Island. Sugar Tech 8:646

Côte F-X, Poirier-Magona E, Perret S, Roudier P, Rapidel B, Thirion M-C (2018) The agroecological transition of agriculture in the south. Quæ editions., 368 p, Versailles. https://doi.org/10.35690/978-2-7592-2824-9

Cotton Incorporated (2009) Summary of life cycle inventory data for cotton (Field to Bale—version 1.1–2 July 2009). Cotton Incorporated

de Figueiredo EB, Panosso AR, Romão R, La Scala N (2010) Greenhouse gas emission associated with sugar production in southern Brazil. Carbon Balance Manag 53

Defrance D, Sultan B, Castets M, Famien AM, Baron C (2020) Impact of climate change in West Africa on per capita cereal production in 2050. Sustainability 12:7585. https://doi.org/10.3390/su12187585

Engonopoulos V, Kouneli V, Mavroeidis A, Karydogianni S, Beslemes D, Kakabouki I et al (2021) Cotton versus climate change: the case of Greek cotton production. Notulae Botanicae Horti Agrobotanici Cluj-Napoca 49:12547

FAO (2025) FAOSTAT. https://www.fao.org/faostat/en/#data/QCL

Faye B, Webber H, Naab JB, MacCarthy DS, Adam M, Ewert F et al (2018) Impacts of 1.5 versus 2.0 °C on cereal yields in the West African Sudan Savanna. Environ Res Lett 13:034014. https://doi.org/10.1088/1748-9326/aaab40

Gbenou GX, Assouma MH, Bastianelli D, Kiendrebeogo T, Bonnal L, Zampaligre N et al (2024) Supplementing zebu cattle with crop co-products helps to reduce enteric emissions in West Africa. Arch Anim Nutr:1–17. https://doi.org/10.1080/1745039X.2024.2356326

Gérardeaux E, Sultan B, Palai O, Guiziou C, Oettli P, Naudin K (2013) Positive effect of climate change on cotton in 2050 by CO_2 enrichment and conservation agriculture in Cameroon. Agron Sustain Dev 33:485–495

Goebel F-R, Sallam N (2011) New pest threats for sugarcane in the new bioeconomy and how to manage them. Curr Opin Environ Sustain 3(1–2):81–89

Gomathi R, Rao PNG, Chandran K, Selvi A (2015) Adaptive responses of sugarcane to waterlogging stress: an overview. Sugar Tech 17(4):325–338

Grandis A, Fortirer JS, Navarro BV, de Oliveira LP, Buckeridge MS (2024) Biotechnologies to improve sugarcane productivity in a climate change scenario. Bioenergy Res 17(1):1–26

Hebbar KB, Venugopalan MV, Prakash AH, Aggarwal PK (2013) Simulating the impacts of climate change on cotton production in India. Clim Chang 118:701–713

Hedayati M, Brock PM, Nachimuthu G, Schwenke G (2019) Farm-level strategies to reduce the life cycle greenhouse gas emissions of cotton production: an Australian perspective. J Clean Prod 212:974–985

Impa SM, Perumal R, Bean SR, John Sunoj VS, Jagadish SVK (2019) Water deficit and heat stress induced alterations in grain physico-chemical characteristics and micronutrient composition in field grown grain sorghum. J Cereal Sci 86:124–131. https://doi.org/10.1016/j.jcs.2019.01.013

IPCC (2021a) Climate change 2021—the physical science basis

IPCC (2021b) Contribution to the sixth assessment report of the intergovernmental panel on climate change. Working Group I. https://www.cambridge.org/core/books/climate-change-2021-the-physical-science-basis/415F29233B8BD19FB55F65E3DC67272B

Jones MR, Singels A, Ruane AC (2015) Simulated impacts of climate change on water use and yield of irrigated sugarcane in South Africa. Agric Syst 139:260–270

Kuyah S, Muoni T, Bayala J, Chopin P, Dahlin AS, Hughes K et al (2023) Grain legumes and dryland cereals contribute to carbon sequestration in the drylands of Africa and South Asia. Agric Ecosyst Environ 355:108583. https://doi.org/10.1016/j.agee.2023.108583

La Scala Junior N, De Figueiredo EB, Panosso AR (2012) A review on soil carbon accumulation due to the management change of major Brazilian agricultural activities. Braz J Biol 72(3):775–785

Lafarge T, Julia C, Baldé A, Ahmadi N, Muller B, Dingkuhn M (2015) Rice adaptation strategy in response to heat at the flowering stage. In: Torquebieau E (coord) Climate change and world agriculture. Versailles, Quae editions, pp. 37–49

Linnenluecke MK, Nucifora N, Thompson N (2018) Implications of climate change for the sugarcane industry. WIREs Clim Change 9(1):e498

Loarie SR, Lobell DB, Asner GP, Mu Q, Field CB (2011) Direct impacts on local climate of sugarcane expansion in Brazil. Nat Clim Chang 1(2):105–109

Luo Q, Bange M, Johnston D (2016) Environment and cotton fibre quality. Clim Chang 138:207–221

Luquet D, Perrier L, Clément-Vidal A, Jaffuel S, Verdeil J-L, Roques S et al (2019) Genotypic covariations of traits underlying sorghum stem biomass production and quality and their regulations by water availability: insight from studies at organ and tissue levels. GCB Bioenergy 11:444–462. https://doi.org/10.1111/gcbb.12571

Macedo IC, Seabra JE, Silva JE (2008) Greenhouse gas emissions in the production and use of ethanol from sugarcane in Brazil: the 2005/2006 averages and a prediction for 2020. Biomass Bioenergy 32(7):582–595

Mauney JR, Kimball BA, Pinter PJ, LaMorte RL, Lewin KF, Nagy J, Hendrey GR (1994) Growth and yield of cotton in response to a free-air carbon dioxide enrichment (FACE) environment. Agric For Meteorol 70:49–67

Park S, Howden M, Horan H (2007) Evaluating the impact of and capacity for adaptation to climate change on sectors in the sugar industry value chain in Australia. In: XXVI Congress, International Society of Sugar Cane Technologists, ICC, Durban, South Africa, 29 July–2 August 2007, pp 312–326

Poser C, Barau L, Mézino M, Goebel F-R, Ruget F (2020) Effect of the germination threshold temperature on the geographical distribution of the variety R583 in Reunion Island. Int Sugar J 122(1461):640–657

Rotundo JL, Salinas A, Gomara N, Borras L, Messina C (2024) Maize outyielding sorghum under drought conditions helps explain land use changes in the US. Field Crops Res 308:109298. https://doi.org/10.1016/j.fcr.2024.109298

Schlegel AJ, Lamm FR, Assefa Y, Stone LR (2018) Dryland corn and grain sorghum yield response to available soil water at planting. Agron J 110:236–245. https://doi.org/10.2134/agronj2017.07.0398

Schlubach J (2021) Downscaling model in agriculture in Western Uzbekistan climatic trends and growth potential along field crops physiological tolerance to low and high temperatures. Heliyon 7:e07028

Sharma RK, Kumar S, Vatta K, Dhillon J, Reddy KN (2022) Impact of recent climate change on cotton and soybean yields in the southeastern United States. J Agric Food Res 9:100348

Sinha R, Soni P, Perret S (2020) Environmental and economic assessment of paddy based cropping systems in middle indo-Gangetic plains, India. Environ Sustain Ind 8:100067. https://doi.org/10.1016/j.indic.2020.100067

Soulé M, Mansuy A, Chetty J, Auzoux S, Viaud P, Schwartz M et al (2024) Effect of crop management and climatic factors on weed control in sugarcane intercropping systems. Field Crop Res 306:109234

Sourisseau J-M (2014) Family farming and future worlds. Quæ editions, Versailles. 364 p

Subbarao GV, Nakahara K, Ishikawa T, Ono H, Yoshida M, Yoshihashi T et al (2013) Biological nitrification inhibition (BNI) activity in sorghum and its characterisation. Plant Soil 366:243–259. https://doi.org/10.1007/s11104-012-1419-9

Sultan B, Defrance D, Iizumi T (2019) Evidence of crop production losses in West Africa due to historical global warming in two crop models. Sci Rep 9:1–15. https://doi.org/10.1038/s41598-019-49167-0

Thomas HL, Pot D, Jaffuel S, Verdeil J-L, Baptiste C, Bonnal L et al (2021) Mobilising sorghum genetic diversity: biochemical and histological-assisted design of a stem ideotype for biomethane production. GCB Bioenergy 13:1874–1893. https://doi.org/10.1111/gcbb.12886

Tongwane M, Mdlambuzi T, Moeletsi M, Tsubo M, Mliswa V, Grootboom L (2016) Greenhouse gas emissions from different crop production and management practices in South Africa. Environ Dev 19:23–35

Traore A, Falconnier GN, Couëdel A, Sultan B, Chimonyo VGP, Adam M, Affholder F (2023) Sustainable intensification of sorghum-based cropping systems in semi-arid sub-Saharan Africa: the role of improved varieties, mineral fertiliser, and legume integration. Field Crops Res 304:109180. https://doi.org/10.1016/j.fcr.2023.109180

Trébuil G, Hossain M (2004) Rice, ecological and economic challenges. Belin, Paris. 263 p

Tuong TP, Bouman BAM (2003) Rice production in water-scarce environments. In: Kijne JW, Barker R, Molden D (eds) Water productivity in agriculture: limits and opportunities for improvement. CABI Publishing, pp 53–67. https://www.cabidigitallibrary.org/doi/epdf/10.1079/9780851996691.0053

Warren R, Price J, Forstenhäusler N, Andrews O, Brown S, Ebi K et al (2024) Risks associated with global warming of 1.5 to 4 °C above pre-industrial levels in human and natural systems in six countries. Clim Chang 177(3):48

Williams A, White N, Mushtaq S, Cockfield G, Power B, Kouadio L (2015) Quantifying the response of cotton production in eastern Australia to climate change. Clim Chang 129:183–196

Wu F, Guo S, Huang W, Han Y, Wang Z, Feng L et al (2023) Adaptation of cotton production to climate change by sowing date optimisation and precision resource management. Ind Crop Prod 203:117167

Zimmermann A, Webber H, Zhao G, Ewert F, Kros J, Wolf J et al (2017) Climate change impacts on crop yields, land use and environment in response to crop sowing dates and thermal time requirements. Agric Syst 157:81–92

Open Access This chapter is licensed under the terms of the Creative Commons Attribution-NonCommercial-NoDerivatives 4.0 International License (http://creativecommons.org/licenses/by-nc-nd/4.0/), which permits any noncommercial use, sharing, distribution and reproduction in any medium or format, as long as you give appropriate credit to the original author(s) and the source, provide a link to the Creative Commons license and indicate if you modified the licensed material. You do not have permission under this license to share adapted material derived from this chapter or parts of it.

The images or other third party material in this chapter are included in the chapter's Creative Commons license, unless indicated otherwise in a credit line to the material. If material is not included in the chapter's Creative Commons license and your intended use is not permitted by statutory regulation or exceeds the permitted use, you will need to obtain permission directly from the copyright holder.

Chapter 14
Oil Palm: Building Climate Resilience

Alain Rival and Cécile Chéron-Bessou

Abstract Oil palm cultivation, still based on intensive monoculture, raises ongoing concerns about its sustainability and impact on climate change. Despite slower expansion, the sector remains globally interconnected, with rising involvement from Southern and Asian countries in processing and biofuel production. Labor shortages and climate vulnerabilities, especially during droughts, threaten productivity and demand mechanisation. Climate change weakens palm productivity through disrupted photosynthesis and flowering patterns. Additionally, biodiversity loss from deforestation.

The oil palm (*Elaeis guineensis* and *oleifera*) still represents the archetype of intensive monoculture developed in perennial plantations. Cultivated on 24 Mha in the intertropical belt, its exploitation system has changed little over a century (Rival and Chalil 2023). The large-scale development of this crop, even though it has significantly slowed down over the last decade (Gaveau et al. 2022), still raises many recurrent research questions concerning the sustainability of the production.

The controversy that has accompanied the development of the oil palm for over 30 years also concerns the management of climate change in a highly globalised and rapidly changing sector. The colonial legacy has thus shaped a geopolitics based on the exploitation of resources in the South (plantations) and transformation in the North (refining, industrial processing and distribution). This landscape is changing

---

A. Rival (✉)
Cirad, UMR ABSys, Montpellier, France

UMR ABSys, University of Montpellier, Cirad, INRAE, Agro Institute of Montpellier, Montpellier, France

Cirad DRASEI, Jakarta, Indonesia
e-mail: alain.rival@cirad.fr

C. Chéron-Bessou
Cirad, UMR ABSys, Montpellier, France

UMR ABSys, University of Montpellier, Cirad, INRAE, Agro Institute of Montpellier, Montpellier, France

rapidly: Southern industrialists (Indonesia, Malaysia, Colombia, Thailand) are gradually investing in downstream activities of the sector, including the production of biofuels. The oil palm sector remains linked to intercontinental exchanges, with an increasing role assigned to Asian giants (China and India), both in the consumption of crude oil and in its transformation. The scarcity of rural labour (there was a 20% shortfall in the available workforce in the plantations in the years preceding Covid-19) will force the sector to undergo rapid changes, particularly in terms of mechanisation and automation.

The low climate resilience of current cultivation systems is revealed when extreme droughts occur, with measurable consequences on productivity. In Southeast Asia, the El Niño episodes of 2015 and then 2019 highlighted the low climate resilience of the current operating systems, both village and industrial. Intense droughts directly result in blocking gaseous exchanges and the photosynthetic capacity of palms. Simultaneously, the haze generated by forest fires and uncontrolled slash-and-burn farming significantly reduce the productivity of plantations. The oil palm also has the property of branching into male flowering when agroclimatic conditions become unfavourable, with an impact on productivity measurable over several years (Monzon et al. 2021).

The biodiversity losses linked to the expansion of plantations at the expense of tropical forests have a direct impact on the functioning of ecosystems within and around the plantations (Meijaard et al. 2020). Thus, oil palm plantations can contribute to climate change, particularly through deforestation, while being weakened by the risks associated with this change. These imbalances in ecosystems affect not only the frequency and intensity of rainfall to which the palm is very sensitive, but also the procession of pests (or, conversely, pollinators) making the plant more vulnerable to climate changes.

14.1 State of the Sector

Palm oil is the most consumed vegetable oil in the world; demand has accelerated with the emergence of new outlets in the biofuel sector, in addition to traditional food and oleochemical uses (Rival and Levang 2013). This strong growth has undoubtedly contributed to the economic development of the main producing countries, mainly Indonesia and Malaysia, which now supply 83% of global demand (USDA 2023). The sector is a key source of foreign exchange reserves, as well as a major instrument for poverty reduction and rural economic development (Feintrenie et al. 2010; Rist et al. 2010). Globally, palm oil provides 40% of the world's vegetable oil demand on just under 6% of the land used to produce all vegetable oils. To obtain the same amount of oil from other sources such as soy, rapeseed or sunflower, it would require cultivating between four and ten times more land. The comparative advantages of palm oil over competing vegetable oils still rely on low production costs which structurally result from the abundance of arable land, the high natural productivity of the crop and the low cost of labour (Meijaard et al. 2024).

Global palm oil production is now approaching 80 Mt., for a total mapped area of 24 Mha, of which 40% are industrial oil palm plantations and 60% are smallholders. These two types of operators have very different direct impacts on climate change. Small family farmers (less than 40 ha) follow different operating systems from those of large agro-industrial perimeters, based on the intensive use of inputs (mainly chemical fertilisers). Smallholders remain hampered by the lack of access to suitable and selected plant material, even though seed sectors have made enormous progress in their ability to disseminate genetic progress. The capacities for adaptation and mitigation, due to lack of means and resources, will be much lower in village plantations than in large agro-industrial perimeters. Some plantation companies, in Latin America as well as in Southeast Asia, have developed irrigation systems to cope with recurrent and more intense droughts.

In Africa, the proportion of palm oil production provided by small farmers is over 80% in most producing countries (Ivory Coast, Cameroon, Nigeria). These systems use very little, if any, chemical inputs and rely on the exploitation of unimproved plant material, leading to low yields and a pronounced vulnerability to extreme climatic events.

Over the past decade, Indonesia has managed to remarkably reduce deforestation for the production of palm oil. In 2018–2022, deforestation for palm oil was 32,406 ha/year, representing 18% of the peak reached 10 years earlier. It is important to note that deforestation decreased during a period of continuous expansion of palm oil production. Although deforestation remained lower than that recorded in previous years, it has continued. The forest provinces of Kalimantan were the hardest hit, representing 72% of all deforestation linked to palm oil in Indonesia for the period 2018–2022. The island of Sumatra saw the deforestation for palm oil multiplied by 3.7 in 2022 compared to 2020.

The new European legislation, the EUDR regulation (European Union Deforestation Regulation), prohibits the sale of goods resulting from deforestation or forest degradation. This measure aims to limit the environmental impact of products, particularly agricultural and forestry, imported into the EU (Gilbert 2024). As a rule, operators (including non-SMEs) must exercise due diligence with respect to all products falling within the scope of the regulation, from each of their suppliers. Operators sourcing entirely from commodities in low-risk areas will be subject to simplified due diligence obligations. While it was due to apply at the end of 2024, many voices, in Europe and in the South countries, have called for a postponement of the regulation. Indeed, the new regulation requires the establishment *de novo* of a system of supply chain traceability, regularly updated and supported by reliable mapping of plots. On the ground, the main difficulties concern the absence of land surveys and property titles, mainly for independent smallholders. The main producing countries, in particular Indonesia and Malaysia, have thus requested a postponement of the regulation, deemed necessary for the implementation of verification measures. The European Commission had proposed to postpone by 1 year its entry into force, from 30 December 2024 to 30 December 2025. The European Parliament then voted to create a new category of countries considered as "risk-free", and which would be exempt from certain obligations. The postponement of the law was finally published in the Official Journal of the EU on 23 December 2024, so that the European regulation on export controls will be applied from 30 December 2025 by medium and large companies, and from 30 June 2026 by small and microenterprises. The future of the EUDR regulation in the long term therefore remains uncertain, as it concerns one of the most important texts of the European Union in environmental matters.

The introduction of biodiversified systems likely to offer a better climate resilience than the monospecific plantations developed on the postcolonial model is being experimented in several countries (Rival et al. 2023; Zemp et al. 2023; Masure et al. 2023).

14.2 The Impacts of Climate Change in the Case of the Oil Palm

14.2.1 Impacts on Yields

The low climate resilience of palm oil production systems is observable in Fig. 14.1, which shows a drop in palm oil production particularly following the El Niño-Southern Oscillation (ENSO) episode that occurred in Southeast Asia in 2015.

In 2015, Malaysian plantations had to face the double impact caused by extreme drought. They were affected directly by responses induced by water stress, such as changes in the sex-ratio which will have delayed repercussions in time, and the

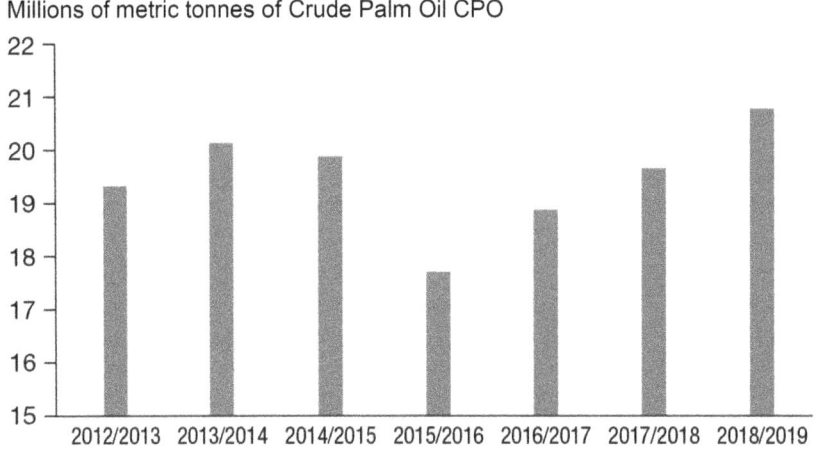

Fig. 14.1 Production of crude palm oil in Malaysia. (Source: USDA-FAS (2012–2019)). The climate sensitivity of oil palm cultivation is revealed on this graph, which shows the direct effects of the El Nino phenomenon—lingering over two campaigns—on the productivity of palm oil in Malaysia. Crude palm oil is the circulating form of this commodity, traded since colonial times, on a generally South-North pattern

decrease in fruit development and oil synthesis. In addition, they indirectly suffered the impact of bushfires linked to slash-and-burn activities in and around the plantations, which limited the photosynthetic activity of the palms for months, when thick dry fogs covered the entire region (Khor et al. 2021).

Stiegler et al. (2019) showed that the haze conditions measured during the ENSO episode led to a complete pause in the net carbon assimilation in the oil palm, which lasted nearly a month and a half, and this situation turned out to be the cause of a 35% decrease in oil palm yield. The model developed by Stiegler et al. (2019) also demonstrated that an intensification of drought could significantly reduce the net absorption of CO_2. Thick dry fogs, when exacerbated by drought, can lead to substantial losses in productivity and net absorption of CO_2 by oil palm plantations.

On the other hand, Eycott et al. (2019) hypothesised that biodiversity losses associated with the expansion of oil palm plantations at the expense of rainforests have led to a series of measurable impacts on the ecosystem functioning within the palm plantation and on the resilience of these functions to changes in rainfall patterns, with a real impact on yield.

14.2.2 Impacts on Pest Pressure

Climate change affects all living beings, including pests or crop pathogens. For decades, oil palm plantations have been protected by confining pathogens to specific, geographically limited contexts. Cultivar selection has allowed the development of resistant/tolerant genotypes in response to these constraints. Moreover, quarantine measures have limited contamination risks. Climate change is highly likely to alter these partitions. If favourable cultivation areas shift, the pathogens will follow, while dynamics within prey-predator assemblages may also be directly affected by the disruptions.

The oil palm is historically sensitive to three main pathogens, the fungi *Ganoderma* in Asia, *Fusarium* in Africa, and heart rot (*bud rot*) in Latin America, due to varying assemblages of pathogens (Mercière et al. 2017; Paterson et al. 2013). These authors modelled how climate change could influence the impact of these pathogens, simulating both the direct effects on them and the indirect effects related to the displacement of plantations to new areas. In current production areas, pressure is likely to increase due to increased vulnerability of palms in rapidly changing and deteriorating contexts (Paterson et al. 2013; Paterson 2019). Furthermore, some pathogens are transmitted by vectors such as insects, whose development is stimulated by rising temperatures, thus increasing parasitic risks for the oil palm (Paterson and Lima 2010). Finally, the movement of humans and their means of production, due to climate migrations, will lead to the displacement of pathogens, the breakdown of risks and the spread of existing pathogens or the adaptation of new vectors. The negative effects could be partially offset by the spread of palm plantations into new parasite-free areas—a phenomenon identified as *parasites lost*—but the probability and extent of this phenomenon are difficult to quantify (Paterson et al. 2013).

Due to the current lack of knowledge about the development and infestation cycles of pathogens and the many uncertainties about the indirect effects of climate change, it is difficult to predict global trends regarding pest pressure. Nevertheless, it is undeniable that uncertainties about risks will greatly increase, thereby reducing known margins for maneuvering. Current scientific and agronomic gaps are already detrimental to the solutions to be found any time soon.

14.3 Carbon Balance and Mitigation Solutions

The approach implemented in life cycle analysis (LCA) or in carbon balance calculators based on LCA allows quantifying all sources of greenhouse gases (GHGs). Various studies have highlighted the main sources of GHGs and mitigation possibilities, including the PalmGHG calculator based on LCA standards (ISO 14040 and 14044, 2006). This calculator was developed by the RSPO (Roundtable on Sustainable Palm Oil) to allow palm oil producers to estimate and monitor their net GHG emissions and thus identify and control risk areas in their production chain (Bessou et al. 2014).

The main source of GHGs occurs at the time of plantation establishment, if these are established after deforestation or drainage of peatlands or marshes (Bessou et al. 2014; Cooper et al. 2020; Schmidt 2010). Emission factors for converted peat forests range between 70 and 117 t eqCO_2 /ha/year (95% confidence interval [CI]), with CO_2 and N_2O accounting for about 60% and 40% of this value, respectively. These emissions suggest that the conversion of peatlands in Southeast Asia contribute between 16.6% and 27.9% of the total combined national GHG emissions from Malaysia and Indonesia, equating to between 0.44% and 0.74% (95% CI) of annual global emissions (Cooper et al. 2020).

Furthermore, oil palm plantations require fertiliser inputs that account for 46–85% of production costs and significantly contribute to environmental impacts such as land acidification and climate change. These fertiliser inputs are the second source of GHGs, particularly for highly intensive systems, and can even be the primary source in plantations without deforestation and on mineral soils. Depending on the types of mineral fertilisers and their origins, the production and transport stages contribute to varying degrees to the carbon balance. However, the majority of GHGs are linked to field emissions following their application. Agroecological practices can reduce inputs throughout the cultivation cycle. During the immature phase of the crop, a temporary soil cover with legumes has the advantages of recycling nutrients from the decomposition of the trunks of the previous harvest and preventing the development of weeds. Then, throughout the cultivation cycle, recycling abundant and diverse by-products in the plantations improves the nutrient content of the soil and its physico-chemical and biological properties (Bessou et al. 2017). Oil palm plantations can generate about 16 t/ha/year of by-products, in addition to the production of palm and palm kernel oil (about 5 t/ha/year). In large agro-industrial plantations, the empty fruit bunches are most often co-composted, particularly with the liquid effluents produced from the oil extraction factories, thus increasing the nutritional value and stability of the amendment while reducing transport costs and the environmental impacts of effluent treatment (Baron et al. 2019). The results of the LCA have highlighted that composting organic residues can replace 10–25% of synthetic fertilisers, while significantly reducing the impact on climate change. Despite the large quantities of by-products generated, demand within or outside the value chain may exceed supply, so that issues of competition and fertility transfer will need to be precisely studied to characterise sustainable practices at the landscape scale.

The last significant source of GHGs is the emissions of CH_4 during the anaerobic treatment of liquid effluents from oil mills. Capture and recycling techniques exist and are rapidly developing in large industries. On a smaller scale, the best option is to reuse these effluents during composting, which requires regular moistering.

The sources of GHGs and mitigation levers for palm oil production are known. The criteria of the RSPO have indeed been adjusted during updates taking into account these results with, notably, much more stringent criteria concerning plantations on peatlands and obligations for GHG reduction plans (RSPO 2018). Nevertheless, at the global scale, risks remain either in terms of expansion of cultivated lands in a non-climate-smart way for political and economic reasons, or due to a lack of knowledge or access to information concerning soil quality and climate-smart inputs, combining recycled organic inputs and maintaining a protective soil cover.

14.4 Predicted Evolution of the Production Zone

Tropical regions continue to be explored to extend oil palm cultivation in response to increasing food and energy needs (Tapia et al. 2021), while modelling studies indicate that the climate will gradually become less suitable for oil palm cultivation (Paterson et al. 2017). These authors estimate that the proportion of unsuitable areas would increase by 6%, while highly favourable areas would decrease by 22% by 2050. A strong decrease in suitability for cultivation is anticipated, with a dramatic drop by 2100, suggesting the emergence of regions totally unsuitable, although they are currently conducive to oil palm cultivation. Many producing regions in Latin America and Africa (Brazil, Colombia and Nigeria) should thus experience a dramatic regression of areas suitable for oil palm cultivation. Conversely, other subtropical regions could become exploitable for oil palm and interesting in terms of virgin parasite history (Paterson and Lima 2011). According to the FAO and IIASA models, with the GAEZ v4.0 model (2021), potential palm oil yields in 2040 would globally decrease with more areas where yields would decrease compared to those where conditions would allow new production or an increase in yield compared to 2010 (Fig. 14.2).

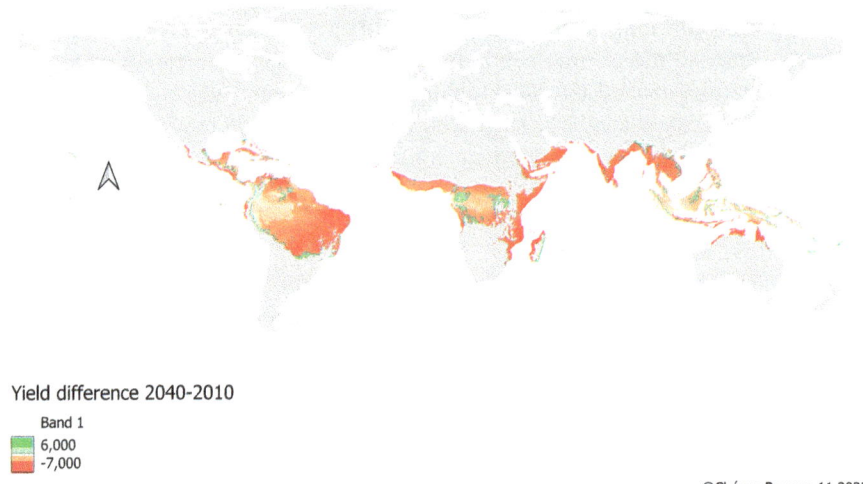

Fig. 14.2 Differences in potential palm oil yields between 2010 and 2040. (Source: Chéron-Bessou; FAO/IIASA data, GAEZ v4.0 model (2021)). In kg of CPO oil—Crude Palm Oil—per hectare, without irrigation, high level of inputs; red signifies a yield decrease, green an increase. Map showing the differences in potential palm oil yields between 2010 and 2040 (in kg of CPO oil, that is, Crude Palm Oil, per ha, without irrigation and with a high level of inputs). The areas coloured in red show a decrease in yield, those in green, an increase. The data are extracted from the GAEZ v4.0 (2021) model by FAO and IIASA (FAO and IIASA 2021)

The recent IPCC report (Shaw et al. 2022) highlights that Southeast Asian countries (where nearly 90% of the world's palm oil production is concentrated) are regularly identified as the most vulnerable to climate risks, with key sectors such as agriculture, cities, infrastructure, or terrestrial ecosystems, expected to be heavily exposed to multiple hazards. Due to their rapid development and large population, Asian countries are emitting more and more GHGs, even if per capita emissions and cumulative emissions are relatively lower than in developed economies.

The oil palm is a plant of relatively recent domestication and large-scale exploitation. Its rapid development, among smallholders as well as agro-industries, has led (as for all plant species) to a reduction in the useful genetic base, mainly driven by profitability per hectare. The development of cultivars resistant to *Fusarium* and then to *Ganoderma* has further narrowed this genetic base, further reducing the exploitable agrobiodiversity (Rival 2017). In the oil palm, cultivar improvement progresses at the pace imposed by a perennial crop with a long cycle. Creating, multiplying and evaluating a new cultivar will require nearly 50 years of research: the shift in research and development programmes towards drought resistance is still ongoing for a long time…

14.5 Conclusion

Faced with the urgency of the responses needed to counter and adapt to climate change, palm oil production sectors are now at a crossroads (Rival and Chalil 2023; Rival and Bessou 2023). In a completely globalised sector and faced with multiple challenges, the *status quo* is no longer an option: urgent and concrete responses are needed to deeply transform production systems.

The European regulation on imported deforestation[1] aims to put an end to the importation of products resulting from deforestation on the continent such as cocoa, coffee, soy, palm oil, wood, beef, but also rubber. The regulation on deforestation could allow a decrease in global deforestation by 10%, according to the impact study of the European Commission, and would make the European Union a pioneer on the subject, but the uncertainties around its application remain numerous.

Climate change acts as a revealer of the proven weaknesses of a solid sector, but which must initiate deep changes. Faced with increasing uncertainties related to the global geopolitical context, the agroecological transition of production systems must be a priority for scientists, policy makers and the oil palm industry in all producing countries.

References

Baron V, Saoud M, Jupesta J, Praptantyo IR, Admojo HT, Bessou C, Caliman JP (2019) Critical parameters in the life cycle inventory of palm oil mill residues composting. IJoLCAS 3(1):19

Bessou C, Chase LDC, Henson IE, Abdul-Manan AFN, Milà I, Canals L et al (2014) Pilot application of PalmGHG, the RSPO greenhouse gas calculator for oil palm products. J Clean Prod 73:136–145. https://doi.org/10.1016/j.jclepro.2013.12.008

Bessou C, Verwilghen A, Beaudoin-Ollivier L, Marichal R, Ollivier J et al (2017) Agroecological practices in oil palm plantations: examples from the field. OCL 24(3). https://doi.org/10.1051/ocl/2017024

Cooper H, Evers S, Aplin P, Crout N, Dahalan MPB, Sjogersten S (2020) Greenhouse gas emissions resulting from conversion of peat swamp forest to oil palm plantation. Nat Commun 11:407. https://doi.org/10.1038/s41467-020-14298-w

Eycott AE, Advento AD, Waters HS, Luke SH, Aryawan AAK, Hood AS et al (2019) Resilience of ecological functions to drought in an oil palm agroecosystem. Environ Res Commun 1(10):101004

FAO and IIASA (2021) Global Agro-Ecological Zones (GAEZ v4)—Data Portal user's guide. Rome. https://doi.org/10.4060/cb5167en

Feintrenie L, Chong WK, Levang P (2010) Why do farmers prefer oil palm? Lessons learnt from Bungo district, Indonesia. Small Scale For 9:379–396

Gaveau DL, Locatelli B, Salim MA, Husnayaen MT, Descals A et al (2022) The slowdown in deforestation in Indonesia is linked to a decrease in oil palm expansion and lower oil prices. PLoS One 17(3):e0266178

---

[1] https://www.importeddeforestation.ecologie.gouv.fr/european-regulation-against-deforestation-and-forest-degradation/article/european-regulation-against-deforestation-and-forest-degradation

Gilbert CL (2024) The EU deforestation regulation. EuroChoices 23(3):64–70. https://doi.org/10.1111/1746-692X.12436

Khor JF, Ling L, Yusop Z, Tan WL, Ling JL, Soo EZX (2021) The impact of El Niño on oil palm yield in Malaysia. Agronomy 11(11):2189

Masure A, Martin P, Lacan X, Rafflegeau S (2023) Advocating for oil palm-based agroforestry systems: a valuable asset for the sustainability of the industry. Cah Agric 32:16

Meijaard E, Brooks TM, Carlson KM, Slade EM, Garcia-Ulloa J, Gaveau DL et al (2020) The environmental impacts of palm oil in context. Nat Plants 6(12):1418–1426

Meijaard E, Virah-Sawmy M, Newing HS, Ingram V, Holle MJM, Pasmans T et al (2024) Exploring the future of vegetable oils. Implications of oil crops—fats, forests, forecasts, and futures. IUCN, and SNSB, Gland, Switzerland. https://doi.org/10.2305/KFJA1910

Mercière M, Boulord R, Carasco-Lacombe C, Klopp C, Lee Y-P, Tan J-S et al (2017) On *Ganoderma boninense* in oil palm plantations of Sumatra and peninsular Malaysia: ancient population expansion, extensive gene flow and large-scale dispersion ability. Fungal Biol 121:529–540

Monzon JP, Slingerland MA, Rahutomo S, Agus F, Oberthür T, Andrade JF et al (2021) Encouraging a climate-smart intensification for oil palm. Nat Sustain 4(7):595–601

Paterson RRM (2019) *Ganoderma boninense* disease of oil palm is expected to significantly reduce production after 2050 in Sumatra if projected climate change occurs. Microorganisms 7(1):24. https://doi.org/10.3390/microorganisms7010024

Paterson RRM, Lima N (2010) How will climate change affect mycotoxins in food? Fd Res Int 43:1902–1914

Paterson RRM, Lima N (2011) Additional mycotoxin effects from climate change. Fd Res Int 44:2555–2566

Paterson RRM, Kumar L, Shabani F, Lima N (2017) Global climate suitability projections to 2050 and 2100 for oil palm cultivation. J Agric Sci 155(5):689–702

Paterson RRM, Sariah M, Lima N (2013) How will climate change affect oil palm fungal diseases? Crop Prot 46:113–120. https://doi.org/10.1016/j.cropro.2012.12.023

Rist L, Feintrenie L, Levang P (2010) The livelihood impacts of oil palm: smallholders in Indonesia. Biodivers Conserv 19:1009–1024

Rival A (2017) Breeding the oil palm (*Elaeis guineensis* Jacq.) for climate change. OCL 24(1):D107. https://doi.org/10.1051/ocl/2017001

Rival A, Ancrenaz M, Lackman I, Burhan S, Zemp C, Firdaus M, Djama M (2023) Innovative planting designs for oil palm-based agroforestry. Agrofor Syst 99(1):27

Rival A, Bessou C (2023) Climate change is challenging oil palm (*Elaeis guineensis* Jacq.) production systems. In: Abul-Soad AA, Al-Khayri JM (eds) Cultivation for enhanced climate change resilience, Tropical fruit trees, vol 1. CRC Press, pp 109–126

Rival A, Chalil D (2023) Oil palm plantation systems at a crossroad. OCL 30:28. https://doi.org/10.1051/ocl/2023029

Rival A, Levang P (2013) The palm of controversies: oil palm and development issues. Quæ editions, Versailles

RSPO (2018) Principles and criteria for the production of sustainable palm oil, endorsed by the RSPO Board of Governors and adopted at the 15th annual general assembly by RSPO members on 15 November, 2018

Schmidt JH (2010) Comparative life cycle assessment of rapeseed oil and palm oil. Int J Life Cycle Assess 15(2):183–197. https://doi.org/10.1007/s11367-009-0142-0

Shaw R, Luo Y, Cheong TS, Abdul HS, Chaturvedi M, Hashizume GE et al (2022) Asia. In: Climate change 2022: impacts, adaptation and vulnerability. Contribution of Working Group II to the Sixth Assessment Report of the Intergovernmental Panel on Climate Change Cambridge University Press, Cambridge, UK/New York, NY, pp 1457–1579. https://doi.org/10.1017/9781009325844.012

Stiegler C, Meijide A, Fan Y, Ashween AA, June T, Knohl A (2019) El Nino Southern Oscillation (ENSO) event reduces CO_2 uptake of an Indonesian oil palm plantation. Biogeosciences 16(14):2873–2890

Tapia JFD, Doliente SS, Samsatli S (2021) How much land is available for sustainable palm oil? Land Use Policy 102:105187

USDA (2023) United States Department of Agriculture Foreign Agricultural Service. htttps://apps.fas.usda.gov/psdonline/app/index.html#/app/home

Zemp DC, Guerrero-Ramirez N, Brambach F, Darras K, Grass I, Potapov A et al (2023) Tree islands enhance biodiversity and functioning in oil palm landscapes. Nature 618:316–321. https://doi.org/10.1038/s41586-023-06086-5

Open Access This chapter is licensed under the terms of the Creative Commons Attribution-NonCommercial-NoDerivatives 4.0 International License (http://creativecommons.org/licenses/by-nc-nd/4.0/), which permits any noncommercial use, sharing, distribution and reproduction in any medium or format, as long as you give appropriate credit to the original author(s) and the source, provide a link to the Creative Commons license and indicate if you modified the licensed material. You do not have permission under this license to share adapted material derived from this chapter or parts of it.

The images or other third party material in this chapter are included in the chapter's Creative Commons license, unless indicated otherwise in a credit line to the material. If material is not included in the chapter's Creative Commons license and your intended use is not permitted by statutory regulation or exceeds the permitted use, you will need to obtain permission directly from the copyright holder.

Chapter 15
Horticultural Production in the Face of Climate Change

Éric Malézieux, Damien Beillouin, Raphaël Belmin, Isabelle Grechi, Rémi Kahane, Thibaud Martin, and Fabrice Le Bellec

Abstract Over two billion people suffer from micronutrient deficiencies, and many cannot afford a healthy diet. Horticultural products like fruit and vegetables are vital for nutrition but are under-consumed in most countries, low-income ones in particular. Climate change threatens horticulture due to the perishability and sensitivity of these crops and alters the incidence of pests. The chapter explores how crop diversity can help farmers adapt to climate risks. It also examines global adaptation strategies within horticultural systems.

15.1 Introduction: Climate Change, a Threat to Global Food and Nutrition Security

Today, more than two billion individuals worldwide suffer from micronutrient deficiencies, and nearly three billion people cannot afford a healthy and nutritious diet (FAO 2020). Horticultural food products, including fruit, vegetables, aromatic and medicinal plants, with their richness in fiber, vitamins, and essential mineral compounds such as iron, contribute significantly to a diverse and nutritious diet. However, the FAO and WHO recommendation to consume more than 400 g of fresh

É. Malézieux (✉)
Cirad, UPR HortSys, Capesterre-Belle-Eau, Guadeloupe, France

HortSys, Cirad, University of Montpellier, Montpellier, France
e-mail: eric.malezieux@cirad.fr

D. Beillouin
HortSys, Cirad, University of Montpellier, Montpellier, France

Cirad, UPR HortSys, Caribbean Agro-Environmental Campus (CAEC), Martinique, France

R. Belmin
HortSys, Cirad, University of Montpellier, Montpellier, France

Cirad, UPR HortSys, Senegalese Institute of Agricultural Research, Dakar, Senegal

I. Grechi · R. Kahane · T. Martin · F. Le Bellec
Cirad, UPR HortSys, University of Montpellier, Montpellier, France

fruit and vegetables per day per person is only achieved in certain regions of Asia and among the affluent populations of industrialised countries, unlike low-income countries, which are particularly deficient. Therefore, the regular consumption of a variety of fruit and vegetables is a central concern for global food security. Climate change directly and indirectly affects horticultural products due to their perishability and sensitivity to numerous biotic and abiotic factors. In this context, how can the great diversity of cultivated fruit and vegetables offer farmers more opportunities to adapt and become less vulnerable to climate hazards? This chapter analyses the main impacts and risks induced by climate change and attempts to identify the adaptation pathways in progress in different horticultural systems worldwide.

15.2 The Impacts of Climate Change on Horticultural Production

Horticultural production will be affected by climate change through three main pathways: by altering the physiological efficiency of plants, their phenology, and ultimately the available quantities and quality of harvested products; by changing the global geographical distribution of certain horticultural species; and by altering biotic constraints related to pests and diseases and their damage to crops, as well as populations of beneficial organisms, pollinators, and soil microorganisms.

15.2.1 Impacts on Plant Physiology and Phenology, on Yields and Product Quality

The increase in CO_2 concentrations promotes higher crop yields, particularly in C3 plants, where photosynthetic efficiency is limited by current atmospheric CO_2 concentrations (Scheelbeek et al. 2018). This is due to an increase in photosynthesis and improved water use efficiency. However, these effects can be mitigated in the presence of other environmental stress factors (such as increased OO_3 and temperatures), or by long-term photosynthetic acclimation (Kimball et al. 2007). Increased CO_2 concentrations also have negative effects on the nutritional quality of C3 plants, with a decrease in major mineral content (Loladze 2014).

In tropical regions, temperatures already reach or exceed the optimal thresholds for plants at certain times of the year. These thresholds partly depend on their carbon metabolism: C3 plants, like tomatoes and most horticultural plants, thrive at optimal temperatures of 20–32 °C, while C4 plants, like sweet corn, tolerate temperatures up to 34 °C. During heatwaves, photosynthetic efficiency decreases, and plant respiration increases. Heatwaves have detrimental effects on the number, morphology, and functioning of plant reproductive organs. A rise in temperatures can also lead to significant yield losses by affecting pollination. This can occur through an alteration of stigma receptivity or by desynchronisation between the flight periods of pollinators and the flowering periods of crops (Duchenne et al. 2019). Crop exposure to high temperatures (El Niño 2015–2016 and 2023–2024 in Central and South America) can also cause physiological disorders that alter the appearance of products and change their organoleptic and nutritional quality (Moretti et al. 2010; Christopoulos and Ouzounidou 2021; Kishor et al. 2023). Plants grown under shelter are also exposed to this risk (Fink et al. 2009). Moreover, the lack of a colder period can disrupt the dormancy break of certain fruit trees, perennial or biennial vegetables, such as onions, and floral induction in certain species.

Climate change alters the availability of water for the plants, which can lead to water stress that generally has a negative effect on flowering, fruit set, fruit retention and fruit growth. A lower water availability decreases yields. Average decreases of around 20% in the production of fruit, nuts and seeds have been observed when water availability was reduced by 50% (Alae-Carew et al. 2020). Water stresses also affect the contents of sugars, acids and carotenoids, positively or negatively depending on the stage of fruit development at which this stress is applied and the genotype (Ripoll et al. 2016).

Emissions of pollutants, by causing an increase in concentrations of tropospheric ozone, pose an additional climate challenge for horticultural crops. This gas can penetrate plants through their stomata, damage the cellular structures of leaves and lead to yield decreases, a delay in fruit ripening and a decrease in their sugar content.

Thus, climate change, through interactions between CO_2, temperature, water availability and other climate manifestations, complexly alters both the volumes and quality of horticultural products. Interactions with other factors (such as the biological cycle, genetic material, geographical location, cultivation method and agricultural practices, the stage of the crop during which the climate event occurs) make formulating a clear projection of the impact of climate change even more difficult.

15.2.2 Future Distributions of Horticultural Species: Between Threats and Opportunities

Climate change, by altering plant growth conditions, has over the decades transformed the distribution of horticultural species cultivation zones. This geographical shift ultimately affects food and nutritional security of populations. With a global warming of +2 °C, half of the current arable lands would see the potential diversity of crops decrease, while higher latitudes would, on the other hand, experience an increase in the potential diversity of crops, thus offering opportunities for adaptation to climate change (Mahaut et al. 2022). Overall, species originating from warm climates are already thriving in these new areas (Lenoir et al. 2008). In Spain, subtropical crops such as avocados and mangoes now occupy 4% of the total orchard area (Esyrce 2018). In Central Africa, climate change would have rather positive effects on the potential cultivation areas for root vegetables, tubers and bananas, but would negatively affect potato cultivation areas (Manners et al. 2021). The majority of scientific data on changes in crop distribution areas, however, focus on four main staple crops: rice, corn, wheat and soy. Our understanding of the impacts of climate change on fruit and vegetable crops is still limited.

15.2.3 Impact on Pests

Climate change directly influences the proliferation and distribution of phytophagous arthropods as well as the viral and bacterial diseases they transmit (Muilenburg and Herms 2013; Skendžić et al. 2021). Arthropods from temperate zones would see their populations increase, while those from tropical regions, living near their thermal optimum, could be negatively affected if temperatures exceed this optimum or reach their lethal threshold (Bonebrake and Deutsch 2012). Their ability to disperse would allow some species to move to higher latitudes and altitudes, where temperatures become more favourable.

Changes in precipitation patterns also influence arthropods (e.g., leaching of individuals by heavy rains or mortality induced by water stagnation in the soil during their telluric phase). For example, the mortality rate of *Bactrocera zonata* pupae significantly increases with soil water content (El-Gendy and AbdAllah 2019). According to Gely et al. (2020), most insect guilds would be negatively affected by water stress, with the exception of wood borers, bark beetles, sap –sucking insects and leaf miners.

Phytophagous arthropods will also be affected by the availability and quality of the host plant and by the disruption of the trophic network (Eigenbrode and Adhikari 2023; Muilenburg and Herms 2013; Skendžić et al. 2021).

Warming can increase or decrease the phenological synchronisation between arthropods, their host plants and their natural enemies (Forrest 2016). However, phenological asynchronies should affect tropical regions less.

Plants exposed to a CO_2-enriched environment present physiological status (C/N ratio, nitrogen concentration, secondary compounds) that can reduce their nutritional quality or act as defenses against pests. Conversely, some chewing insects may consume more tissue plants to compensate for the lower nutritional quality of these plants.

Pathogens are sensitive to abiotic conditions, particularly temperature and relative humidity, which directly influence their life cycle, or indirectly the physiology and immune response of plants. The effects of climate change on diseases of tropical fruit crops vary, with severity that could increase (in the case of the banana and the fungus *Fusarium oxysporum*) or decrease (in the case of the banana and the fungus *Mycosphaerella fijiensis*) (Ghini et al. 2011). At high temperatures, plants may be more susceptible to pathogens due to the suppression of their immunity (Singh et al. 2023).

Climate change alters the incidence of pests on crops. The extent of these changes is difficult to predict, especially for horticultural crops, which exhibit a great diversity of species and varieties. Pests and their natural enemies will not be uniformly affected, as climate effects vary by species and plant-pest systems. Effects depend also on the interaction of climatic factors and environmental context. Climatic factors will also have complex effects on pollinators, increasing or reducing their efficiency depending on the context (Eigenbrode and Adhikari 2023; IPBS 2016).

15.3 Adapting to Climate Change: Ongoing Innovations

The adaptation of horticultural systems to climate change involves three main adaptation strategies: increasing and enhancing cultivated diversity and genetic selection for less demanding species, creating a more favourable environment around crops (from mulch to above-ground greenhouses), and good resource management, including the essential management of water. These adaptation strategies also rely on public policies and territorial innovations.

15.3.1 Adapting Cropping Systems

15.3.1.1 The Benefits of Diversity

Horticulture encompasses species with short cycles (annual or biennial) and long cycles (5–10 years, or even more than 10 years), with great biological diversity (vegetative or sexual reproduction, herbaceous or woody plants). Fruit and vegetables, in addition to their diversity, are grown in a wide range of cropping and production systems, ranging from multi-species subsistence systems in permaculture without chemical inputs (such as Creole gardens, micro-food gardens) to intensive monoculture systems often mono-varietal, or even monoclonal (bananas, pineapples) using large quantities of chemical inputs, or even above-ground greenhouses with controlled atmosphere and nutrient flows, real "plant factories". The resilience and adaptation of these different systems to climate change are obviously very different. Without being able to detail all potential forms of adaptation, it is crucial to underline the importance of species diversity in a given system to provide ecosystem services and to ensure its resilience in the face of climate hazards (Kahane et al. 2013; Malézieux et al. 2009; Malézieux et al. 2022; Beillouin et al. 2021).

Intraspecific diversity and genetic improvement pathways constitute another form of adaptation. To cope with climate change, the robustness of traditional cultivars, whose heritage (biological and intangible such as peasant know-how) should be preserved, and varietal selection constitute two complementary pathways. Highly improved varieties, often hybrids, mostly require intensive techniques (irrigation, fertilisation, plant protection) and high financial investments, exposed to climate risks. So-called "farmer seeds" benefit small farms thanks to their adaptation potential: jojoba, argan tree, pigeon pea, amaranth for aridity, cowpea, mung bean for short cycles, and rustic traditional varieties of okra, roselle, African eggplants. Some horticultural crops, forgotten or neglected, vegetable or fruit, aromatic or medicinal, thus offer new opportunities. "Informal" farmer seed systems constitute a huge reservoir of genes of interest in the face of climate change and promote the adoption of beneficial agroecological practices for small-scale farmers. Farmers also have room for maneuver by adjusting planting and sowing calendars, promoting crop rotations and associations, including agroforestry. These strategies combine the preservation of biological and intangible heritage, with innovations allowing better adaptation of crops to climate challenges.

15.3.1.2 Water Resource Management

The issue of water supply is particularly acute in Maghreb and sub-Saharan Africa, where water resources are poorly distributed and increasingly unpredictable. In this context, costly drilling and pumping are sometimes implemented. In large production basins like the Senegal River valley, the expansion of irrigated areas is supported by policies funded by international aid. While this choice meets a short-term need, it outlines a fragile and vulnerable model for the future, as it relies on water captured and transported with significant use of fossil energy (Fig. 15.1). Other projects encourage the adoption of water-saving irrigation techniques, such as drip irrigation, and attempt to integrate the expertise of peasant communities into agroecological design approaches (Belmin et al. 2022, 2023a). Traditional oasis agrosystems in the Maghreb and sub-Saharan Africa, for example, rely on multilayered crops (palms, fruit trees, and vegetable crops) that conserve water and reduce evapotranspiration. In the semi-arid areas of Burkina Faso, farmers use the "zaï" technique, which involves dry sowing in holes amended with manure, which then fill with water when the first rains arrive (Belmin et al. 2023a). In some regions, water resource management benefits from new technological inputs (drip irrigation, moisture sensors, remote triggers, etc.), but to be successful it must primarily rely on human, social, and solidarity governance to make its effects sustainable, from the plot to the watersheds.

15.3.1.3 Protected Crops

Greenhouses, in which climate parameters are controlled, are characteristic of horticulture. The artificialisation of the environment can even become complete in extreme climatic situations such as deserts (California, Israel), polluted or fully urbanised areas (Cairo, Singapore). These "plant factories" cater to specific contexts and for species with high economic or cultural value (herbs, strawberries, chilies). Horticulture becomes a technological agriculture, sometimes vertical (and therefore with an extremely limited spatial footprint), using the most recent

Fig. 15.1 In major production basins like the Senegal River Valley, the expansion of irrigated areas is supported by development policies funded by international aid. But this choice, while meeting a short-term need, outlines a fragile and vulnerable agricultural model in the long term, as it relies on water captured and transported with significant use of fossil energy. (Credit: photo by R. Belmin, Cirad). Photograph of an open-air irrigation canal in the Senegal River Valley

scientific advances in the fields of digital technology, robotics, metabolomics, energy (LED lamps, photovoltaics, cogeneration, etc.). However, questions of profitability, scale change, energy consumption and social acceptability of these highly intensified systems remain. "Low-tech" solutions are also emerging in densely populated areas, based on the recycling of water, organic matter and local materials (hydroponics, aeroponics, aquaponics, vertical crops, rooftop crops) or even in rural areas, for protection against insects, particularly in sub-Saharan Africa (Nordey et al. 2017). These systems, by freeing themselves from climate hazards, contribute to nutrition security either directly through the consumption of fresh products, leafy vegetables or fruit or through the income from their sale (Orsini et al. 2013).

The last 20 years have been marked by the development of new practices that allow production with fewer chemical inputs. Traditional practices based on biological diversity (agroforestry, Creole gardens) are reviewed and adapted to fit into new economic circuits and to meet new needs. Agroecological practices can also use new information and communication technologies (ICTs), particularly to anticipate unfavourable climatic episodes, to avoid pre- and post-harvest losses, to optimise water management, and to control, through sensors and digital imaging, protected surfaces or volumes. Their applicability in economically and socially fragile production conditions is still being questioned.

15.3.2 Organisational Adaptations (Supply Chains, Food Systems, Public Policies)

While new agroecological cultural practices are necessary to cope with climate change, it is also essential to consider the downstream of the supply chains, a significant source of losses for horticultural products, particularly perishable and sensitive to climatic variations. The impact is particularly strong in low-income countries, where infrastructure (roads, storage and distribution facilities) is deficient. This is particularly true for local supply chains, sometimes derived from export chains, as is the case for mangoes in Côte d'Ivoire (Fig. 15.2). At another level, if new practices need to be implemented in horticultural supply chains, from upstream to downstream, these will only be effective on a global scale if food consumption practices also change (diversification and vegetalisation of diets), and if public policies support this movement (Belmin et al. 2023b).

Fig. 15.2 Mango harvest in Korhogo, Ivory Coast. Mangoes destined for local markets are transported under conditions that do not preserve the quality of the product. This results in significant losses in case of adverse weather conditions (phases of sun exposure, heavy rains) (Credit: Photo by É. Malézieux). Photograph of a mango harvest scene in an orchard in Ivory Coast. Mangoes destined for local markets are often transported under conditions that do not preserve the quality of the product. This results in significant losses in case of adverse weather conditions (phases of sun exposure, heavy rains). Here we see two agricultural workers filling a small motor-truck just after the fruit harvest. The fruits will then be dumped into other means of transport, most often trucks without cardboard or refrigeration, to be transported to the cities

15.4 Mitigation Strategies

Compared to large-scale farming or livestock rearing, horticulture, due to its smaller scale, is not a major contributor to climate change. The carbon sequestration capacity is modest compared to sectors like forestry. However, horticulture often consumes many chemical inputs (plastics, fertilisers, pesticides) derived from petroleum. Mitigation strategies in horticulture mainly focus on reducing GHG emissions associated with agricultural practices. Recycling effluents in soilless culture, precise application of fertilisers and organic nutrient sources can minimise pollutant emissions. Adopting reduced tillage practices can sequester carbon, improve soil structure, and reduce emissions due to soil disturbance. Transitioning to organic farming practices reduces the use of synthetic fertilisers and pesticides, thereby reducing the carbon footprint of horticultural production. Implementing renewable energy sources such as solar panels and wind turbines on horticultural farms can help reduce GHG emissions associated with energy consumption. Developing short supply chains, consuming seasonal products, reducing losses and food waste along the horticultural supply chain, from production to consumption, will also reduce emissions. Choosing eco-friendly packaging materials and reducing excessive packaging can also contribute to reducing the carbon footprint of horticultural products.

15.5 Conclusion

At a time when malnutrition problems persist globally and climate change is accelerating, new paths must be found for agriculture. Food horticultural products, essential for a healthy diet, are at the heart of these challenges. However, increasing production alone will not solve malnutrition. In many urban populations, subjected to new nutritional pathologies (obesity, type II diabetes, cardiovascular diseases), the processes of overconsumption of processed foods, energy and chemical inputs work in synergy. Micronutrient deficiencies systematically affect disadvantaged populations in all countries. In the most agriculture-dependent economies, climate change will affect all components of food security, i.e., accessibility, availability, transformation and food stability. Increasing food diversity for the poorest is a decisive step to improve nutrition (Malézieux et al. 2024). Research should aim to increase the accessibility and availability of fruit and vegetables in tropical regions, by increasing and diversifying production, and limiting losses. Faced with the climate challenge and biodiversity loss, it is necessary to broaden current research themes to include interactions between agriculture, food and health, in order to better understand the close links between food production, human health and sustainable environment.

References

Alae-Carew C, Nicoleau S, Bird FA, Hawkins P, Tuomisto HL, Haines A et al (2020) The impact of environmental changes on the yield and nutritional quality of fruits, nuts and seeds: a systematic review. Environ Res Lett 15(2):023002

Beillouin D, Ben-Ari T, Malézieux E, Seufert V, Makowski D (2021) Positive but variable effects of crop diversification on biodiversity and ecosystem services. Glob Chang Biol 27(19):4697–4710

Belmin R, Malézieux E, Basset-Mens C, Martin T, Mottes C, Della RP et al (2022) Designing agroecological systems across scales: a new analytical framework. Agron Sustain Dev 42:3

Belmin R, Sawadogo H, Ndiénor M (2023a) Cultivating with little or no water: the zaï technique in the Sahel. The Conversation

Belmin R, Paulin M, Malézieux E (2023b) Adapting agriculture to climate change: which pathways behind policy initiatives? Agron Sustain Dev 43(5):59

Bonebrake TC, Deutsch CA (2012) Climate heterogeneity modulates impact of warming on tropical insects. Ecology 93(3):449–455

Christopoulos M, Ouzounidou G (2021) Effects of climate change on the perceived and nutritional quality of fruits and vegetables. J Innov Econ Manag 34:79–99

Duchenne F, Thébault E, Michez D, Elias M, Drake M, Persson M et al (2019) Phenological shifts alter the seasonal structure of pollinator assemblages in Europe. Nat Ecol Evol 4(1):115–121

Eigenbrode SD, Adhikari S (2023) Climate change and the management of insect pests and beneficials in agricultural systems. Agron J 115:2194–2215

El-Gendy IR, AbdAllah AM (2019) Effect of soil type and soil water content levels on pupal mortality of the peach fruit fly [*Bactrocera zonata* (Saunders)] (Diptera: Tephritidae). Int J Pest Manag 65(2):154–160

ESYRCE (2018) Survey on crop areas and yields. MAPA, Fisheries and Food. https://www.mapa.gob.es/es/estadistica/temas/estadisticas-agrarias/boletin2018_tcm30-504212.pdf

FAO (2020) The state of food and agriculture 2020. Meeting the challenge of water in agriculture, Rome. https://doi.org/10.4060/cb1447fr

Fink M, Kläring HP, George E (2009) Horticulture and climate change. In: Dirksmayer W, Sourell H (eds) Water in horticulture, vol 328. Landbauforschung, Sdh, pp 1–9

Forrest JRK (2016) Complex responses of insect phenology to climate change. Curr Opin Insect Sci 17:49–54

Gely C, Laurance SG, Stork NE (2020) How do herbivorous insects respond to drought stress in trees?. Biol Rev 95(2):434–448

Ghini R, Bettiol W, Hamada E (2011) Diseases in tropical and plantation crops as affected by climate changes: current knowledge and perspectives. Plant Pathol 60:122–132

IPBES (2016) Assessment report on pollinators, pollination and food production. Summary for policymakers. https://files.ipbes.net/ipbes-web-prod-public-files/downloads/2016_spm_pollination-fr.pdf

Kahane R, Hodgkin T, Jaenicke H, Hoogendoorn C, Hermann M, Keatinge JDH et al (2013) Agrobiodiversity for food security, health and income. Agron Sustain Dev 33(4):671–693. https://doi.org/10.1007/s13593-013-0147-8

Kimball BA, Idso SB, Johnson S, Rillig MC (2007) Seventeen years of carbon dioxide enrichment of sour orange trees: final results. Global Change Biol 13:2171–2183

Kishor PBK, Guddimalli R, Kulkarni J, Singam P, Somanaboina AK, Nandimandalam T et al (2023) Impact of climate change on altered fruit quality with organoleptic, health benefit, and nutritional attributes. J Agric Food Chem 71:17510–17527

Lenoir J, Gégout JC, Marquet PA, De Ruffray P, Brisse H (2008) A significant upward shift in plant species optimum elevation during the 20th century. Science 320(5884):1768–1771. https://www.science.org/doi/abs/10.1126/science.1156831

Loladze I (2014) Hidden shift of the ionome of plants exposed to elevated CO_2 depletes minerals at the base of human nutrition. elife 3:e02245

Mahaut L, Pironon S, Barnagaud JY, Bretagnolle F, Khoury CK, Mehrabi Z et al (2022) Matches and mismatches between the global distribution of major food crops and climate suitability. Proc R Soc B 289(1983):20221542

Malézieux E, Beillouin D, Makowski D (2022) Feeding the world better: crop diversification to build sustainable food systems. Perspective 58:1–4

Malézieux E, Crozat Y, Dupraz C, Laurans M, Makowski D, Ozier-Lafontaine H et al (2009) Mixing plant species in cropping systems: concepts, tools and models. A review. Agron Sustain Dev 29:43–62

Malézieux E, Verger EO, Avallone S, Alpha A, Biu Ngigi P, Lourme-Ruiz A et al (2024) Biofortification versus diversification to combat micronutrient deficiencies: an interdisciplinary review. Food Sec 16:261–275

Manners R, Vandamme E, Adewopo J, Thornton P, Friedmann M, Carpentier S et al (2021) The suitability of root, tuber, and banana crops in Central Africa could be enhanced under future climates. Agric Syst 193:103246

Moretti CL, Mattos LM, Calbo AG, Sargent SA (2010) Climate changes and potential impacts on postharvest quality of fruit and vegetable crops: a review. Food Res Int 43:1824–1832

Muilenburg VL, Herms DA (2013) Responses of insect pests to climate change: effects and interactions of temperature, CO_2, and soil quality. In: Sivakumar MVK, Lal R, Selvaraju R, Hamdan I (eds) Climate change and food security in West Asia and North Africa. Springer, Dordrecht. 423 p

Nordey T, Basset-Mens C, De Bon H, Martin T, Deletre E, Simon S et al (2017) Protected cultivation of vegetable crops in sub-Saharan Africa: limits and prospects for smallholders. A review. Agron Sustain Dev 37(6):e53. https://doi.org/10.1007/s13593-017-0460-8

Orsini F, Kahane R, Nono-Womdim R, Gianquinto G (2013) Urban agriculture in the developing world: a review. Agron Sustain Dev 33:695–720. https://doi.org/10.1007/s13593-013-0143-z

Ripoll J, Urban L, Brunel B, Bertin N (2016) Water deficit effects on tomato quality depend on fruit developmental stage and genotype. J Plant Physiol 15(190):26–35. https://doi.org/10.1016/j.jplph.2015.10.006

Scheelbeek PF, Bird FA, Tuomisto HL, Green R, Harris FB, Joy EJ et al (2018) Effect of environmental changes on vegetable and legume yields and nutritional quality. Proc Natl Acad Sci 115(26):6804–6809

Singh BK, Delgado-Baquerizo M, Egidi E, Guirado E, Leach JE, Liu H, Trivedi P (2023) Climate change impacts on plant pathogens, food security and paths forward. Nat Rev Microbiol 21:640–656

Skendžić S, Zovko M, Živković IP, Lešić V, Lemić D (2021) The impact of climate change on agricultural insect pests. Insects 12:440

Open Access This chapter is licensed under the terms of the Creative Commons Attribution-NonCommercial-NoDerivatives 4.0 International License (http://creativecommons.org/licenses/by-nc-nd/4.0/), which permits any noncommercial use, sharing, distribution and reproduction in any medium or format, as long as you give appropriate credit to the original author(s) and the source, provide a link to the Creative Commons license and indicate if you modified the licensed material. You do not have permission under this license to share adapted material derived from this chapter or parts of it.

The images or other third party material in this chapter are included in the chapter's Creative Commons license, unless indicated otherwise in a credit line to the material. If material is not included in the chapter's Creative Commons license and your intended use is not permitted by statutory regulation or exceeds the permitted use, you will need to obtain permission directly from the copyright holder.

Chapter 16
Livestock Systems Facing the Challenges of Climate Change

Vincent Blanfort, Christian Corniaux, Véronique Alary, and Guillaume Duteurtre

Abstract Livestock production is expanding globally due to farm concentration and industry consolidation, contributing around 12% of total human-caused greenhouse gas emissions. While often viewed as a climate threat, livestock systems also provide key services like nutrient recycling, carbon storage, and organic fertilization. These multifunctional roles are increasingly at risk from climate change. The chapter argues for recognizing livestock as part of both the climate problem and the solution. It examines global trends, climate impacts, and strategies for integrating livestock into adaptation and mitigation efforts.

Animal production systems are undergoing significant restructuring in both industrialized nations and in developing and emerging economies. Over the past few decades, structural changes—particularly the concentration of livestock farms and the consolidation of processing and distribution industries—have dramatically expanded the global footprint of livestock production.

Today, the livestock sector is widely recognized as a major contributor to climate change within the broader agricultural sector. According to FAO (2023), greenhouse gas (GHG) emissions from livestock systems—including cattle, buffalo, sheep, goats, pigs, and poultry—account for approximately 12% of total anthropogenic emissions.

However, beyond quantitative estimates of emissions, these impacts must be analyzed in light of the diversity of livestock systems and the multiple functions they serve. The objectives of livestock production have evolved. While ensuring food security remains a central goal, growing attention is now paid to the broader social and environmental services provided by livestock systems (see Chap. 10). These include, for example, the recycling of crop residues through animal feeding, the provision of organic fertilizers, and the storage of carbon in soils—some of which directly contribute to mitigating the environmental impact of agriculture. This is the

---

V. Blanfort (✉) · C. Corniaux · V. Alary · G. Duteurtre
Cirad joint research unit Selmet "Tropical and Mediterranean Animal Production Systems", Montpellier Univ, Inrae, Institut Agro, Montpellier, France
e-mail: vincent.blanfort@cirad.fr

case, for example, with the recycling of crop residues by livestock, the provision of organic fertilisers, or the storage of carbon in grassland soils.

Yet, these multifunctional services are increasingly vulnerable in the face of climate change. It is therefore essential to move beyond viewing livestock solely as an environmental problem. Livestock systems must also be seen as part of the solution, given their potential contributions to both climate change mitigation and adaptation.

Following a synthetic overview of the current state and global trends in animal production", this chapter explores the major impacts of climate change across various livestock sectors (excluding aquaculture) and show how livestock systems can be integrated into multi-level governance strategies for both adaptation and mitigation.

16.1 Global Dynamics of Animal Production Chains

Livestock farming is found across the globe. Regardless of latitude or climate—whether in lowlands or mountainous regions, in rural or increasingly urban areas—men and women are engaged in the production, processing, and consumption of animal products. Behind the broad term "livestock farming" lies a vast diversity of situations and practices.

There is a wide range of domestic animal species involved: cattle (including cows, buffaloes, yaks, and others), sheep, goats, camelids (such as dromedaries, camels, and llamas), pigs, poultry (chickens, ducks, rabbits, guinea fowl, etc.), equines (horses, donkeys, etc.), reindeer, etc. This diversity of species is accompanied by a wide variety of animal products: white and red meats, offal, milk, eggs, wool, hides and skins, as well as manure and slurry. Different species are also raised for their draught power—as pack or transport animals—or for their role as a store of wealth, particularly in countries of the Global South. In addition to producing food and materials, certain species are also kept for their draught power—used for transport or agricultural labour, especially in contexts where fewer than 10% of farms globally have access to mechanisation—and for their role as a store of value, particularly in countries of the Global South. Livestock production systems themselves are highly diverse. They range from very extensive systems, such as transhumant or nomadic herding, to highly intensive operations including battery pig and poultry farms, large-scale dairy farms, and feedlots.[1] In between, there are mixed crop–livestock family farms and pasture-based systems such as extensive ranching or rotational grazing models

.Nevertheless, as with the wider agricultural sector, livestock systems are undergoing a general process of intensification at both global and continental scales. Production levels have increased markedly over recent decades. Between 1980 and 2020, while the world's human population grew by a factor of 1.76 (Fig. 16.1),

---

[1] Industrial and intensive cattle fattening parks exploited for meat production.

16 Livestock Systems Facing the Challenges of Climate Change

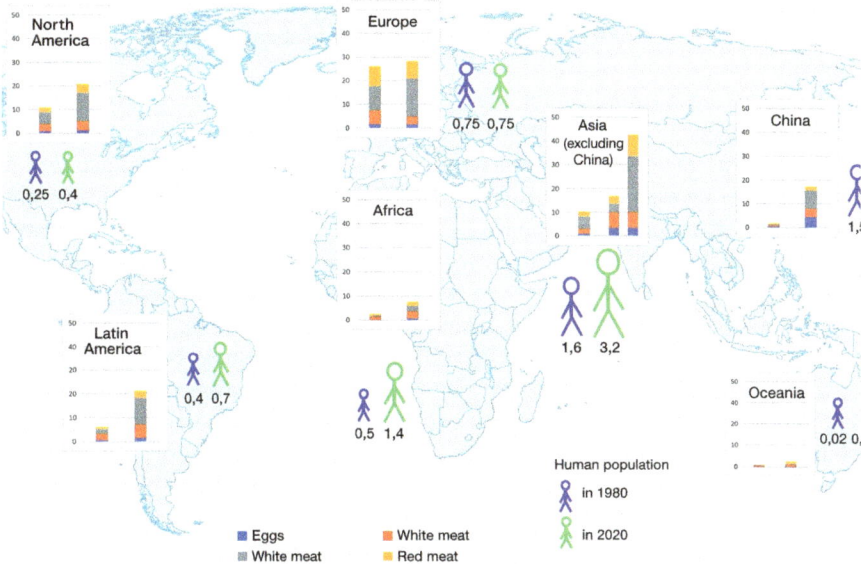

Fig. 16.1 Evolution of the world population (in billions of individuals) and animal protein production from 1980 to 2020 (in millions of tonnes, Mt). (Source: Based on FAOSTAT and our calculations)

chicken production rose by a factor of 5.3, egg production by 3.4, small ruminant production by 2.2, pig production by 2.1, milk production by 2.0, and cattle production by 1.6 (FAOSTAT 2023). In other words, livestock production has grown faster than the global population—most notably in poultry, dairy, and pig sectors—with the notable exception of beef. This trend lies at the heart of ongoing debates around livestock farming and its environmental, social, and economic impacts.

In 2023, global white meat production—primarily poultry (120 million tonnes) and pork (108 million tonnes), both derived from monogastric animals—reached 246 million tonnes, significantly surpassing red meat production from ruminants, which totalled 114 million tonnes (mainly beef at 80 million tonnes, and sheep and goat meat at 16 million tonnes). The rise of poultry is evident across all continents, particularly in Asia—and most notably in China—where production has increased tenfold since 1980. This growth has been driven by the expansion of highly intensive industrial systems, which contribute substantially to greenhouse gas (GHG) emissions in these regions (see Chap. 10). A similar trend of expansion in livestock production has been observed in Latin America and North America and has been more limited in Europe. While red meats have been increasingly replaced by white meats in European diets, Europe—once the global leader in animal production in 1980—has now been overtaken by Asia. Against this global backdrop, Africa has seen an increase in animal production, though levels remain insufficient, especially with regard to white meat. The ratio of animal production to human population is notably low compared to other continents, reflecting low levels of productivity and intensification. Although this contributes to a relatively low climate footprint, it is inadequate to meet the nutritional needs of a population expected to double within the next two decades. Pastoralism currently accounts for 70% of milk and half of the beef and small ruminant meat consumed in Sahelian countries.

Moreover, these strong evolutions in livestock farming driven by demand (see Chap. 10) challenge traditional systems of production due to economic pressures, health concerns, and standardisation requirements imposed by dominant public and private actors (Teyssier d'Orfeuil et al. 2015). These changes are taking place within national and international governance frameworks that remain inadequate and heavily influenced by commercial interests. In developing countries, particularly in Africa, national livestock policies are scarce, as public priorities tend to focus on staple crop production. In Europe, ambitions regarding livestock farming remain uncertain, shaped by various social and environmental pressures—including vegetarianism, animal welfare concerns, and animal rights movements. In contrast, countries with strong productivist and export-oriented agendas tend to implement robust national livestock policies, where environmental considerations are rarely, if ever, addressed. This is notably the case for Canada, the United States, China, Russia, Argentina, India, and Brazil.

These changes—whether in terms of herd size or production concentration—have made livestock systems (including cattle, buffalo, sheep, goats, pigs and poultry) a major contributor to global greenhouse gas (GHG) emissions from the agricultural sector (see Chap. 10). According to FAO estimates[2] for the reference year 2015 (FAO 2023), emissions from livestock reached 6.2 Gt CO_2-equivalent per year, representing 12% of total anthropogenic GHG emissions. This accounts for more than half of the emissions from the AFOLU sector (Agriculture, Forestry and Other Land Use). More specifically, 54% of these emissions are attributable to methane (CH_4), 31% to carbon dioxide (CO_2), and 15% to nitrous oxide (N_2O). With regard to the products concerned, approximately two-thirds of global emissions are attributable to meat production across all species, while one-third is linked to the production of animal feed, including the use of fertilisers and pesticides.

16.2 The Impacts of Climate Change on the Livestock Sector

Major developments in the livestock sector have taken place within a highly disrupted global context. The impacts of climate change on livestock are numerous and complex, and they have intensified over the past 30 years. They affect both feed resources and livestock systems, either directly or indirectly, and influence the spread of disease vectors (Richard et al. 2013), as well as causing disruptions in the

---

[2] Life Cycle Assessment (LCA) approach, using the FAO's GLEAM tool quantifying emissions associated with livestock farming and integrating enteric fermentation, indirect emissions from upstream activities, part of the downstream processes, including transport, processing and marketing of raw products after farming.

Table 16.1 Overview of the different effects of climate change and its impacts on livestock systems

	Phenomena	Impacts
Thermal and water constraints	Droughts or irregularities of rainfall episodes combined with an increase in water needs due to rising temperatures.	Decrease in food consumption, hence decrease in feed intake, leading to reduced productivity. Increase in mortality rates outside controlled systems. Episodic losses of millions of animals in sub-Saharan Africa and the Horn of Africa, especially in arid and semi-arid regions (Blanfort et al. 2015).
Degradation of ecosystems and forage resources	Degradation of land used for livestock: vegetation, soils, and related resources (water, nutrients). Climate "anomalies" acting as drivers or cofactors of rangeland or feed degradation	Since 1945, significant soil degradation has affected 20% of global pastures, including 70% in arid zones (UNEP 1991 in Steinfeld et al. 2006) (see Chap. 1). Decline in the productivity and nutritional quality of natural and cultivated forages. Proliferation of hardy, less palatable, or invasive plant species to the detriment of forage crops. In temperate regions: expansion of invasive plants into areas where they were previously absent.
Climate and animal health	Climate change has complex consequences for animal health, arising from multiple, interacting effects.	Direct effects on animals: physiological and pathological disturbances. Indirect effects: changes in exposure to pathogens, particularly those carried by vectors (invertebrates or wildlife), whose distribution may shift due to climate change (Richard et al. 2013). System intensification: the scale and complexity of supply and trade chains in a globalised economy amplify health risks.

organisation of markets and supply chains (see Chap. 10). The impacts identified in the literature can be categorised according to the bioclimatic phenomena from which they originate (see Table 16.1).

16.3 The Adaptive Capacities of Different Livestock Farming Systems

The diversity of situations and developments within the livestock sector gives rise to a wide range of adaptation strategies and forms (see Table 16.2). Livestock farming also emerges as an adaptation mechanism in its own right—particularly for vulnerable and low-income populations facing increasingly variable environments and associated risks (Alary et al. 2011; Vigne et al. 2015).

Table 16.2 Characteristics, levers and adaptation constraints of different livestock systems (LS)

Type of LS	Adaptation characteristics	Levers, constraints, directions
LS in "challenging" environments *(e.g. drylands, mountainous areas)*	Animals' capacity to exploit a wide range of plant resources. Adaptive mechanisms such as body reserve management and tolerance to environmental stressors.	In dry regions, mobility reduces vulnerability compared to reliance on stored cereals and forages (Alary et al. 2011).
Mixed crop–livestock systems	Diversity and complementarity between plant and animal production help to **reduce the impacts of extreme climatic events**. Flexibility and gradual system transformation over time. Use of negative impacts as inputs or resources for other parts of the system (Franzluebbers 2014). In arid areas, small ruminants increasingly substitute for cattle. Multispecies herding supports complementarities (or substitutions) in the use of resources and income management (Aboul-Naga et al. 2014).	Crop–livestock integration improves nutrient cycling and resource flows between plant and animal production. **Silvopastoralism**: the presence of trees in pastures improves productivity (by delaying dry-season stress) and offers shade. In North Africa: better management practices, protective strategies, and restoration of natural vegetation cover (Huguenin et al. 2015).
Grazing systems of temperate and humid tropical zones	Grazing calendars adjusted in response to changes in pasture productivity (Vall and Diallo 2009). Strategic selection of species and plant varieties to ensure resilient ground cover and maintain soil health. Rotational grazing with adaptable animal stocking rates.	Ecological intensification of grazing systems: rotational grazing using plot division, tree planting, legume intercropping, and simplified soil management techniques (e.g. in the Amazon; Aubron et al. 2022).
Intensive monogastric and ruminant farming systems	Direct effects of climate change, such as high temperatures, impacting livestock. High importance of maintaining sanitary conditions in densely stocked systems.	A key topic in intensive farming systems (Renaudeau et al. 2004 in Richard et al. 2013). Development of early warning systems and targeted interventions for emerging diseases.
All livestock farming systems	**Mobilisation of animal biodiversity**, drawing on the specific genetic traits of breeds and species that confer adaptive advantages: heat tolerance, disease and parasite resistance, and dietary flexibility.	Use of hardy but often less productive breeds better suited to harsh environments. Represents a trade-off between productivity and climate adaptation, while supporting biodiversity conservation in many countries.

Mixed crop–livestock systems appear to be more resilient to the impacts of climate change than specialised systems, owing to the diversity of their activities (such as mixed cropping and livestock units) and the range of resources they draw upon. These systems also represent the dominant form of agriculture worldwide. They provide half of the world's food supply, use 90% of cultivated land, produce 88.5% of global cattle outputs (meat and milk), 61% of pork, and 26% of poultry, and employ 84% of the agricultural population in Europe and Asia (Herrero et al. 2010). Grazing systems, whether extensive or intensified, in temperate and humid tropical regions are based on the use of cultivated forage resources—either grazed or harvested—and are highly sensitive to the impacts of climate change. Intensive livestock systems for both monogastrics and ruminants are heavily reliant on feed resources, whether locally produced or sourced from global markets. These markets are extremely vulnerable to climate-related fluctuations and political instability. Reducing external dependence—by improving on-farm feed autonomy—represents a key adaptation pathway for enhancing the resilience of such systems.

Mobility is a key adaptive feature of livestock systems in drylands and mountainous regions facing severe environmental constraints. However, this crucial asset is increasingly challenged by new climatic pressures, which interact with social and political constraints that marginalise nomadic populations and generate conflicts with other land uses—particularly in Africa (see Chap. 12). In West Africa, these systems are evolving towards agropastoral models, which may include off-ground systems aimed at improving animal productivity through more controlled rearing conditions—for example, peri-urban dairy systems in Mali and Burkina Faso (Vigne et al. 2014 in Blanfort et al. 2019).

Among the common adaptation strategies across livestock systems, the use of animal and plant biodiversity has always been an integral part of livestock farming. Its preservation is essential for maintaining a valuable genetic pool, as documented by the FAO.[3] The FAO's conservation efforts—through its Global Plan of Action for Animal Genetic Resources—cover 7745 local breeds, yet these efforts remain insufficient, with 26% of breeds currently at risk of extinction. Major animal breeding programmes have largely overlooked tropical regions, due in part to limited knowledge—for instance, concerning adaptive mechanisms related to feed intake and reproduction. In particular, the adaptive potential of local goat and sheep breeds remains poorly understood, despite rapid herd growth in regions such as Africa and Central Asia. Similarly, camelids are recognised as highly relevant species for climate change adaptation in arid and high-altitude areas (Faye 2020), yet the performance characteristics of their different breeds remain inadequately studied.

By contrast, the adaptive capacities of zebu breeds compared to taurine cattle—such as greater heat and parasite resistance—have been the subject of substantial research (Richard et al. 2013). Monogastric species have also been the focus of intensive genetic research, particularly in poultry, leading to the development of solutions better suited to hot climates.

---

[3] FAO (2019): http://www.fao.org/3/CA3129EN/CA3129EN.pdf

16.4 Transforming Livestock Systems for Climate Sustainability

The assessment presented in this chapter highlights that animal production chains face major and complex challenges, particularly in relation to climate change. The sector is responsible for a significant share of greenhouse gas emissions, and it is difficult to envision its future development without undergoing a fundamental transition towards low-carbon livestock systems. According to the FAO (2023), greenhouse gas emissions from livestock could be reduced globally by up to 30% through the adoption of improved practices. However, this mitigation potential varies significantly between the Global North and South, and across regions (see Fig. 16.2). At the same time, livestock activities must address the specific economic, social, environmental and political challenges faced in different parts of the world.

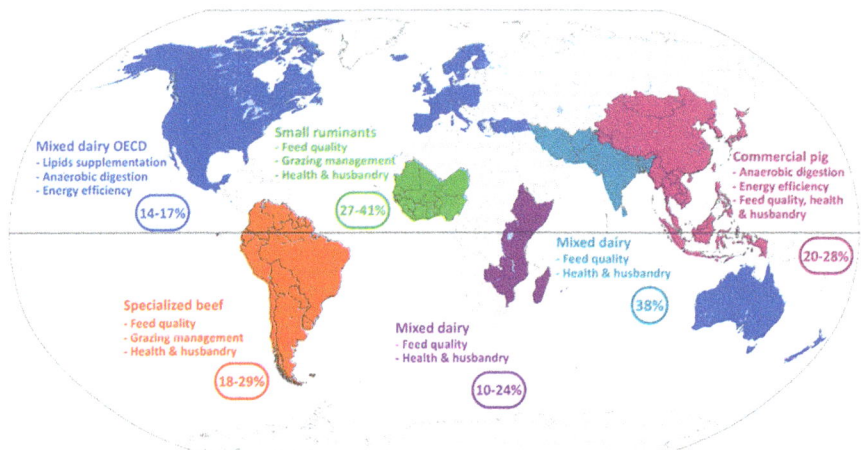

Fig. 16.2 Opportunities for action to reduce GHG emissions in different regions of the world. (Source: Gerber et al. 2013)This global map illustrates an assessment of potential strategies to mitigate greenhouse gas (GHG) emissions across various livestock systems worldwide. It presents the estimated percentage reduction in GHG emissions for each system by major region and outlines potential measures for action. The following provides a detailed interpretation of the map:North America: In mixed dairy farming systems (OECD countries), the potential for reducing GHG emissions is estimated at 14–17%, primarily through lipid supplementation, anaerobic digestion, and improvements in energy efficiencySouth America: In specialised beef cattle systems, the mitigation potential ranges from 18% to 29%, with key strategies including enhanced feed quality, improved pasture management, and better animal health and welfare practicesWest Africa: For small ruminant systems, the potential reduction is estimated at 27–41%, mainly through improvements in feed quality, grazing practices, and animal health careEast Africa: In mixed dairy systems, GHG emissions could be reduced by 10–24% through better feed quality and strengthened animal health and careIndia and the Middle East: Mixed dairy systems show a mitigation potential of around 38%, largely dependent on feed quality improvements and enhanced animal health and welfareChina and Southeast Asia: In commercial pig production, reductions of 20–28% are possible through a combination of anaerobic digestion, energy efficiency measures, improved feed quality, and animal health interventions

Research plays a key role in identifying, characterising, and evaluating these solutions. It helps to demonstrate and clarify the capacity of certain types of livestock systems to offer a wide range of mitigation options across multiple scales—from the individual animal and the farm, to the landscape, value chains, national, regional, and global levels. Several categories of action levers are involved: reducing greenhouse gas (GHG) emissions; transferring and storing carbon from the atmosphere into stable terrestrial compartments (see Chap. 17); improving efficiency through integrated approaches that promote synergies between mitigation and adaptation; and recycling livestock by-products, such as manure and slurry. These actions are governed by regulations and supported by public policies (see Chap. 25), which require further reinforcement and improvement. The knowledge generated within this broad and complex framework encompasses: (1) biotechnical mechanisms; (2) organisational processes; and (3) the socio-technical environment, including the household level, stakeholders across the entire value chain, and public policy frameworks.

While not claiming to provide an exhaustive overview of all possible livestock development pathways compatible with climate change mitigation, this paragraph highlights research that positions livestock as a source of solutions along two main axes: the reduction of emissions and carbon sequestration.

16.5 Improving Livestock Practices to Reduce GHG Emissions

Methane (CH_4) emissions are considered one of the key drivers of the climate crisis (see Chaps. 1 and 24), prompting strong expectations for mitigation in major sectors such as livestock. In Europe, recent radical proposals—such as reducing cattle herds—have sparked controversy within the farming community. Such measures are viewed as potentially undermining food sovereignty, due to the increased reliance they could create on imports from countries lacking self-sufficiency in animal products—many of which engage in agricultural practices that carry significant environmental risks, such as imported deforestation. Reducing beef consumption has also been promoted in countries with high levels of animal protein intake. Yet beyond these relatively narrow solutions, research and extension institutions in the Global North have investigated feed strategies for ruminants that reduce enteric methane emissions[4]—for example, the use of concentrate feeds and crop by-products (Gerber et al. 2013; Doreau et al. 2013). However, the widespread use of imported concentrates is not applicable to all livestock systems (Corniaux et al. 2012). To address the knowledge gaps in the Global South, ongoing research is being conducted in agropastoral systems in sub-Saharan Africa (see Chap. 24).

Carbon dioxide (CO_2) and nitrous oxide (N_2O) emissions together account for 45% of emissions from the livestock sector, with one third of these linked to fossil fuel combustion—either directly for herd management or indirectly through the production of inputs (Gerber et al. 2013). Promoting systems with improved energy

---

[4] This refers to the methane emitted by the digestive system of ruminants (see Chap. 10).

efficiency offers a promising pathway for reducing emissions, particularly through the recycling and valorisation of livestock effluents. Stronger crop–livestock integration can also help reduce dependence on industrially produced mineral fertilisers, which are costly, highly reliant on non-renewable energy resources, and significant contributors to GHG emissions (Vigne et al. 2015; Bénagabou et al. 2017).

16.6 Grazing Systems and Soil Carbon Storage: A Critical Mitigation Pathway

In the context of climate change mitigation, the issue of carbon (C) was initially addressed through the forestry sector, which was identified as the largest reservoir of above-ground biomass carbon. Subsequently, as soil carbon storage emerged as a significant additional option (see Chap. 17), the cropping sector also became involved, given its long-standing focus on soil organic matter management—of which carbon accounts for around 50%. Today, soil management is recognised as a key factor in controlling carbon fluxes in the fight against climate change.

According to Gerber et al. (2013), soil carbon management offers the greatest mitigation potential in agriculture, particularly within livestock systems that rely on grazed and harvested forage resources. As such, this process is increasingly being incorporated into the sustainable development strategies of livestock value chains.

However, the standard metrics and methodologies currently used may not be suitable for accurately assessing grassland ecosystems—especially in tropical regions, where the overall sequestration potential is high, given the extent of the land involved (see Chap. 17). Research has already contributed to characterising carbon sequestration processes in tropical grasslands (Assouma 2016; Blanfort et al. 2015, 2022; Idrissou et al. 2024).

While soil carbon sequestration clearly represents a proven mitigation opportunity for grazing-based livestock systems, it also has limitations. Soil carbon stocks are fragile and can be significantly affected by land-use change, rising temperatures, fertilisation practices, and soil disturbance (Edouard-Rambaut et al. 2022).

> **Box 16.1 The Issue of Metrics and the Assessment of Livestock Systems**
>
> A regional analysis conducted by Gerber et al. (2013) shows that sub-Saharan Africa—due to large herd sizes combined with low productivity—and Latin America and the Caribbean—primarily because of the conversion of primary forests into pastureland and feed crops—are the highest emitters in terms of greenhouse gases per kilogram of carcass produced (70 kg CO_2-eq/kg). However, reliable data on the specific contribution of livestock systems (LS) in low- and middle-income countries to global GHG emissions remain limited. Such aggregated figures often mask the diversity of systems, which has been increasingly revealed through more detailed assessments led by Cirad and its partners.

16.6.1 Carbon (C) Neutral Balances in Sahelian Pastoral Livestock Systems

Livestock farming in the Sahel is often perceived as having some of the highest GHG emission intensities, based on calculations of emissions per unit of product (meat, milk, etc.). This method tends to penalise extensive systems—such as pastoral and agropastoral systems—which are characterised by low productivity. However, an ecosystem-based approach (Assouma 2016; Assouma et al. 2019; Blanfort et al. 2022) that integrates both GHG emissions from livestock activities and carbon (C) storage in pastoral ecosystems shows that Sahelian pastoral systems follow a trajectory of carbon neutrality, with a low but measurable potential for carbon sequestration of 40 ± 6 kg C/ha/year (Assouma et al. 2019). This research forms part of the CASSECS* research and development project (supporting innovation for the resilience of pastoral and agropastoral livestock systems in Sahelian CILSS countries; www.cassecs.org). Since 2020, it has supported this shift in perspective by: (i) producing detailed assessments of emission factors and carbon storage potentials in Sahelian ecosystems; (ii) supporting public policy development in CILSS countries and contributing to the strengthening of their Nationally Determined Contributions (NDCs).

16.6.2 Landscape Restoration and Ecosystem Services in Livestock Areas of the Amazon

In the Brazilian Amazon, most livestock farmers ceased deforestation between 2008 and 2011. Since then, deforestation has remained largely confined to new frontier zones. In other Amazonian regions where deforestation has stopped entirely, research is supporting a transition towards low-carbon development pathways (see Box 22.4). The renewable resources in these regions—solar radiation, rainfall, and soils—can be efficiently harnessed for productive grassland systems that also provide carbon sequestration, as demonstrated by studies conducted in French Guiana under the CARPAGG project. Drawing on this knowledge, the ACCT-DOM Guyane tool was developed as a local adaptation of the European ACCTools (AgriClimateChange) instrument (Dallaporta et al. 2016). It calculates the annual energy balance and greenhouse gas (GHG) emissions of a farm, while also accounting for carbon storage**. This life cycle–based approach has also been implemented in Brazil (ACCT-PARÁ Da Cruz Corrêa et al. 2025) and in New Caledonia.

* CaSSECS: Carbon sequestration and greenhouse gas emissions in the (agro) silvopastoral ecosystems of Sahelian states. https://www.cassecs.org/content/download/4790/37072/version/1/file/Policy+brief+COP28+CaSSECS.FR.pdf.

** https://www.terramaz.org/; ** https://solagro.org/travaux-et-productions/outils/acctool-acct-simplified-version-acct-dom.

16.7 Conclusion

In the face of climate change, transforming livestock systems into more resilient models with minimal negative impacts has become a priority. Numerous short-term adaptations have already taken place. These include adjustments in feeding systems, animal genetics, and husbandry practices. For example, some systems are seeking to reduce their reliance on imported feed and fodder, or to improve the recycling of products and by-products through stronger integration between livestock and crop production. In contrast to the trend towards increasingly standardised industrial systems, these adaptations rely on preserving the rich diversity of territories. Owing to the wide range of functions it fulfils, livestock plays a central role in the development of sustainable food systems. This diversity in livestock systems also represents a social, cultural, and ecological asset that must be preserved (Alary et al. 2024).

To date, international reports have tended to highlight the urgent challenges that intensive livestock systems pose for global warming, despite the fact that only 3.7% of cattle are raised in feedlots, contributing just 5% of the total protein derived from cattle (Ickowicz and Moulin 2022). Some indicators and metrics have persisted for decades without critical reassessment, continuing to shape sectoral discourse in a reductive way. Yet more recent research that takes into account environmental and local contexts has provided greater insight into the differentiated—and at times positive—contributions of livestock depending on socio-economic and ecological conditions (Alary and Gautier 2023; Alary et al. 2024).

Revisiting the metrics used to assess livestock systems—taking into account contextual diversity and the sector's multiple contributions—could help provide a more balanced and nuanced understanding of livestock's role in agricultural transitions. The changes required will depend on improving how carbon impacts are evaluated and on identifying effective solutions to support mitigation. Among these solutions, public policies (see Chap. 25) and targeted investments (see Chap. 26) are needed to steer livestock systems towards greater forage autonomy, reduced dependence on chemical inputs, increased local feed production, and management practices with lower carbon footprints. It is also essential to strengthen the position of livestock within the international climate agenda. While agriculture was included in international climate negotiations as early as the 2000s, livestock only began to receive attention in 2011. At the national level as well, Nationally Determined Contributions (NDCs) should more effectively reflect the mitigation potential of the livestock sector. As of 2022, only 54% of countries—88 out of 164—explicitly mentioned livestock in their NDCs (Rose et al. 2021).

References

Aboul-Naga A, Osman MA, Alary V, Hassan F, Daoud I, Tourrand JF (2014) Raising goats as adaptation process to long drought incidence at the Coastal Zone of Western Desert in Egypt. Small Rumin Res 121(1):106–110

Alary V, Gautier D (2023) Evaluating the contribution of livestock to the development of dry regions: indicators for adapted public policies. Perspective 60:1–4. https://doi.org/10.19182/perspective/37106

Alary V, Duteurtre G, Faye B (2011) Livestock and societies: the multiple roles of livestock in tropical countries. INRAE Anim Prod 24(1):145–156. https://doi.org/10.20870/productions-animales.2011.24.1.3246

Alary V, Frija A, Rueda GA (2024) Importance of livestock for rural territories social development. In: Contribution of the livestock sector to food security and sustainable agrifood systems—benefits, constraints, synergies and trade-offs. FAO, Rome

Assouma MH (2016) Ecosystem approach to the greenhouse gas balance of a Sahelian silvopastoral territory: contribution of livestock. AgroParisTech, Paris, Montpellier, 230 p. PhD thesis. http://agritrop.cirad.fr/593394/

Assouma MH, Hiernaux P, Lecomte P, Ickowicz A, Bernoux M, Vayssières J (2019) Contrasted seasonal balances in a Sahelian pastoral ecosystem result in a neutral annual carbon balance. J Arid Environ 162:62–73. https://doi.org/10.1016/j.jaridenv.2018.11.013

Aubron C, Huguenin J, Nozières-Petit M-O, Poccard-Chapuis R (2022) Adaptation trajectories of livestock farming in territories: what role for grazing? What determinants? In: Ickowicz A, Moulin C-H (eds) Grazing livestock and sustainable development of Mediterranean and tropical territories. Recent knowledge about their strengths and weaknesses. Quæ Editions, Versailles, pp 73–80

Bénagabou OI, Blanchard M, Bougouma Yaméogo VMC, Vayssières J, Vigne M, Vall E et al (2017) Does crop-livestock integration improve energy-use efficiency, recycling and self-sufficiency of smallholder farming systems in Burkina Faso? Rev Elev Med Vet Pays Trop 70(2):31–41. https://doi.org/10.19182/remvt.31479

Blanfort V, Vigne M, Vayssières J, Lasseur J, Ickowicz A, Lecomte P (2015) The agronomic roles of livestock in the contribution to adaptation and mitigation of climate change in the North and South. Agron Environ and Soc 5(1):107–115

Blanfort V, Assouma MH, Bois B, Edouard-Rambaut LA, Vayssières J, Vigne M (2022) Efficiency to account for the complexity of the contributions of grazing livestock systems to climate change. In: Ickowicz A, Moulin C-H (eds) Grazing livestock and sustainable development of Mediterranean and tropical territories. Recent knowledge about their strengths and weaknesses. Quæ Editions, Versailles, pp 86–104

Corniaux C, Alary V, Gautier D, Duteurtre G (2012) Dairy producer in West Africa: a modernity dreamed of by technicians tested in the field. Elsewhere 62:17–36

Da Cruz Corrêa C, Poccard-Chapuis R, Blanfort V, Bochu JL, Lescoat P (2025) Impacts of cattle farming practices and associated livestock systems on energy balances and greenhouse gas emissions in the municipality of Paragominas – State of Pará – Amazonia. In: The role of pastoral livestock and products in climate change. Pastoralism: research, policy and practice

Dallaporta D, Bochu JL, Vigne M, Ouliac B, Zoogones LJ, Lecomte P, et al (2016) Taking into account carbon sequestration of pasture in carbon balance of cattle ranching systems established after deforestation in Amazonia. Proceedings of the 10th International Rangeland Congress, Saskatoon, Canada, 6–22 July 2016, pp 399–400

Doreau M, Makkar HPS, Lecomte P (2013) The contribution of animal production to agricultural sustainability. In: Energy and protein metabolism and nutrition in sustainable animal production, vol 134. Springer, pp 475–485

Edouard-Rambaut L-A, Vayssières J, Versini A, Salgado P, Lecomte P, Tillard E (2022) 15-year fertilisation increased soil organic carbon stock even in systems reputed to be saturated like permanent grassland on andosols. Geoderma 425:116025. https://doi.org/10.1016/j.geoderma.2022.116025

FAO (2023) Pathways towards lower emissions—a global assessment of the greenhouse gas emissions and mitigation options from livestock agrifood systems. FAO, Rome. https://doi.org/10.4060/cc9029en

Faye B (2020) How many large camelids in the world? A synthetic analysis of the world camel demographic changes. Pastoralism 10:25

Franzluebbers A (2014) Climate change and integrated crop-livestock systems in temperate-humid regions of North and South America: mitigation and adaptation. In: Fuhrer J, Gregory P (eds) Climate change impact and adaptation in agricultural systems. CAB International, pp 124–139

Gerber PJ, Steinfeld H, Henderson B, Mottet A, Opio C, Dijkman J et al (2013) Tackling climate change through livestock—a global assessment of emissions and mitigation opportunities. FAO, Rome, 115 p

Herrero M, Thornton PK, Notenbaert AM, Wood S, Msangi S, Freeman HA et al (2010) Smart investments in sustainable food production: revisiting mixed crop-livestock systems. Science 327(5967):822–825. https://doi.org/10.1126/science.1183725

Huguenin J, Hammouda RF, Jemaa T, Capron JM, Julien L (2015) Evolution of steppe livestock systems in the Maghreb: adaptation or metamorphosis? In: Pastoral spaces, specific socioeconomic spaces. Pastoralisms of the world. Ed. de la Cardère, Avignon, pp 28–31

Ickowicz A, Moulin C-H (2022) Family ruminant farming in Mediterranean and tropical grazing areas facing the challenges of sustainable development. In: Ickowicz A, Moulin C-H (eds) Grazing livestock and sustainable development of Mediterranean and tropical territories. Recent knowledge about their strengths and weaknesses. Quæ Editions, Versailles, pp 12–31

Idrissou Y, Vall E, Blanfort V, Blanchard M, Alkoiret TI, Lecomte P (2024) Integrated crop-livestock effects on soil carbon sequestration in Benin, West Africa. Heliyon 10(7):e28748

Richard D, Dourmad JY, Coulon JB, Picon-Cochard C (2013) Livestock and animal sectors. In: Soussana J-F (ed) Adapting to climate change: agriculture, ecosystems and territories. Quæ Editions, Versailles

Rose S, Khatri-Chhetri A, Dittmer K, Stier M, Wilkes A, Shelton S et al (2021) Livestock management ambition in the new and updated nationally determined contributions: 2020–2021: analysis of agricultural sub-sectors in national climate change strategies. Updated October 2022. CCAFS Info Note. CGIAR Research Program on Climate Change, Agriculture & Food Security (CCAFS), Wageningen

Steinfeld H, Gerber P, Wassenaar T, Castel V, Rosales M, De Haan C (2006) Livestock's long shadow: environmental issues and options. FAO, Rome, 414 p

Teyssier d'Orfeuil J, Berger Y, Lejeune H (2015) Mapping of international livestock influence initiatives (Mission° 14098). General Council of Food, Agriculture and Rural Areas (CGAAER), Ministry of Agriculture of Food Sovereignty and the Forest. https://agriculture.gouv.fr/cartographie-des-initiatives-dinfluence-en-matiere-delevage-au-niveau-international

Vall E, Diallo M (2009) Local technical knowledge and practices: the management of herds at pasture (BF). NSS 17(2):122–135

Vigne M, Blanfort V, Vayssières J, Lecomte P, Steinmetz P (2015) Constraints on livestock in the South: ruminants between adaptation and mitigation. In: Torquebiau E (ed) Climate change and world agriculture. Quæ Editions, Versailles, pp 123–135

Open Access This chapter is licensed under the terms of the Creative Commons Attribution-NonCommercial-NoDerivatives 4.0 International License (http://creativecommons.org/licenses/by-nc-nd/4.0/), which permits any noncommercial use, sharing, distribution and reproduction in any medium or format, as long as you give appropriate credit to the original author(s) and the source, provide a link to the Creative Commons license and indicate if you modified the licensed material. You do not have permission under this license to share adapted material derived from this chapter or parts of it.

The images or other third party material in this chapter are included in the chapter's Creative Commons license, unless indicated otherwise in a credit line to the material. If material is not included in the chapter's Creative Commons license and your intended use is not permitted by statutory regulation or exceeds the permitted use, you will need to obtain permission directly from the copyright holder.

Part III
Mitigating and Adapting Agricultural and Food Systems: What Solutions, What Synergies?

Chapter 17
Soil Carbon Sequestration: A Solution to Mitigate and Adapt to Climate Change?

Julien Demenois, Damien Beillouin, David Berre, Vincent Blanfort, Rémi Cardinael, Abigail Fallot, Frédéric Feder, Christophe Jourdan, Dominique Masse, and Tom Wassenaar

Abstract Soil, a non-renewable resource crucial for food production, is severely threatened by degradation and climate change, particularly through the loss of soil organic carbon (SOC). As the largest terrestrial carbon reservoir, SOC plays a vital

---

J. Demenois (✉) · D. Berre
Cirad, UPR Agroecology and Sustainable Intensification of Annual Crops (Aïda), Montpellier, France

Aïda, Univ. Montpellier, Cirad, Montpellier, France
e-mail: julien.demenois@cirad.fr

D. Beillouin
Cirad, UPR HortSys, Caribbean Agro-Environmental Campus (CAEC), Martinique, France

HortSys, Cirad, University of Montpellier, Montpellier, France

V. Blanfort
Cirad joint research unit Selmet "Tropical and Mediterranean Animal Production Systems", Montpellier Univ, Inrae, Institut Agro, Montpellier, France

R. Cardinael
Cirad, UPR Aïda, Univ Montpellier, Montpellier, France

Department of Plant Production Sciences and Technologies, University of Zimbabwe, Harare, Zimbabwe

A. Fallot
Cirad, UMR SENS, Montpellier, France

SENS, Univ Montpellier, Cirad, Montpellier, France

F. Feder · T. Wassenaar
Cirad, UPR Recycling and Risk, Univ. Montpellier, Montpellier, France

C. Jourdan
Cirad, UMR Eco&Sols, Montpellier, France

Eco&Sols, Univ Montpellier, Cirad, INRAE, IRD, InstitutAgro, Montpellier, France

D. Masse
Eco&Sols, Univ Montpellier, Cirad, INRAE, IRD, InstitutAgro, Montpellier, France

IRD, UMR Eco&Sols, Montpellier, France

Research Pole Ecological Intensification of Cultivated Soils in West Africa (IESOL), IRD, ISRA, Dakar, Senegal

© The Author(s) 2026
V. Blanfort et al. (eds.), *Climate Impacts and Challenges in Agriculture, Forests and Food Systems*, https://doi.org/10.1007/978-3-032-04331-3_17

role in climate regulation, yet human activities and land use changes have caused massive carbon losses. Soil carbon sequestration offers a promising solution to mitigate climate change while enhancing food security and biodiversity. Effective SOC management depends not only on technical practices but also on socio-economic and political factors. This chapter explores both technical solutions and territorial approaches, especially in West Africa, while assessing potential maladaptation risks

Soil is a non-renewable resource on a human scale, which has been feeding humanity since the beginning. Today, 95% of the nutrients in our food come from it (FAO 2022). However, the ability of the soil to continue to feed humanity is threatened by climate change and is already altered by various types of degradation: urbanisation, erosion, pollution, salinisation. Soil degradation and vegetation loss already affect 75% of the world's land and threaten the livelihoods of more than a billion people (Cherlet et al. 2018). This degradation includes the loss of soil organic carbon (SOC), the main constituent of soil organic matter. The SOC ensures many environmental and agricultural functions (fertility, biodiversity, water retention, etc.). Indeed, the soil plays a major role in the carbon cycle and in climate regulation.

SOC stock, excluding permafrost, is three times higher than that of the atmosphere. Thus, the soil is the largest terrestrial carbon reservoir: 2400 GtC between 0 and 2 m deep globally. This stock is predominantly found in forest soils (30%) and grasslands (30–35%), and to a lesser extent in the soil of croplands (15%) (Lal et al. 2012), where it is nevertheless more subject to changes. To this is added the carbon of the permafrost (more than 1600 Gt between 0 and 3 m deep). The melting of permafrost linked to the average increase in global surface temperature (+ 1.48 °C compared to the period 1850–1900) releases large quantities of greenhouse gases (GHGs), far exceeding the quantities of carbon that can be absorbed by photosynthesis and by vegetation growth, thereby exacerbating climate change.

Land use changes are another major source of net GHG emissions. Over the past decades (1960–2020), nearly a third of the global land surface has undergone a change in land use (Winkler et al. 2021). The conversion of natural ecosystems into agricultural land has thus led to a cumulative loss of 116 GtC in the top two metres of soil (Sanderman et al. 2017). In addition to ecosystem conversion, human activities have a profound impact on SOC, reducing its contribution to ecosystem services (Ma et al. 2023).

SOC content can nevertheless be modulated, allowing the soil to sequester carbon, thus contributing to the mitigation of climate change by reversing these trends of degradation and loss of SOC. Soil carbon sequestration is defined as the "process of transferring carbon from the atmosphere to the soil through plants or other organisms, which is retained as organic carbon in the soil, resulting in an increase in the global soil carbon stock" (Olson et al. 2014).

In terms of climate regulation, SOC would represent 9% of the mitigation potential of forests, and 47% of that of agriculture and grasslands according to Bossio et al. (2020). The maintenance or increase in SOC stock is one of the few options identified by the IPCC (2019) that can contribute simultaneously to climate change mitigation, combating land degradation and biodiversity loss, while improving food security. These synergies led to the launch of the international initiative "4 per 1000: Soils for Food Security and Climate" at the COP21 climate conference in 2015.

SOC stock depends on many factors at different scales, from climate to soil physico-chemical properties, through land use and management practices (Wiesmeier et al. 2019). Land use and management practices shape the inputs and outputs of carbon at the plot scale as well as the quality of carbon inputs, and can modify the renewal of the SOC stock (Fujisaki et al. 2018). These management practices and land use choices are dependent on socio-economic and political factors, whether for example their implementation costs or the level of skills they require, more so than technical factors (Amundson and Biardeau 2018). Indeed, the territory, defined as a socio-ecosystem, is the focal scale where land and soil management issues can be discussed and their consequences evaluated on soil carbon sequestration (Demenois et al. 2020).

In the first part, this chapter presents the impacts of climate change, land use change and agricultural practices on SOC and identifies technical solutions for preserving or storing more carbon in the soil. In the second part, territorial approaches are discussed through the example of semi-arid and sub-humid West Africa, illustrating the importance of socio-economic and political dimensions. Finally, a critical look is taken at soil carbon sequestration through the prism of the risks of maladaptation to climate change.

17.1 The Impact of Human Activities on Soil Organic Carbon and its Implications for Climate Regulation

Land use and management practices are major levers for SOC sequestration. The preservation and restoration of SOC alone represent up to 25% of the potential of all nature-based solutions (NBS) to combat climate change (Bossio et al. 2020). The first lever therefore lies in the preservation of carbon-rich ecosystems such as peatlands, ancient forests, wetlands and mangroves, which together hold at least 10% of the global soil carbon stock (260 Gt) (Noon et al. 2022). As this stock can not be regenerated within a generation, their preservation is essential.

Recent studies (Beillouin et al. 2023) have shown that the effects of land use change on SOC are seven to ten times greater per unit area than the direct effects of climate change (such as temperature increase, CO_2 increase). For instance, the conversion of forests, grasslands, or wetlands into croplands results in an average SOC loss of around 25%, 16%, and 25%, respectively. These losses, combined with the conversion of vast areas into croplands over the past decades (about one million square kilometres over the last sixty years), have substantially contributed to the increase in atmospheric CO_2. The conversion of natural habitats into cultivated land has long-term consequences, as the lost SOC will take decades, or even longer, to return to its initial level, if practices that allow for the restoration of the stock are adopted again and sustained (Dignac et al. 2017). For example, in French Amazonia (French Guiana), in grasslands resulting from deforestation without the use of fire, Stahl et al. (2017) demonstrated that the SOC stock under grassland managed with rotational grazing and legumes returns to its pre-deforestation level after 24 years and continues to increase, particularly in depth (less than 1 m) (Stahl et al. 2016). The increase in SOC in these grassland systems potentially reaches a high level (1.27 ± 0.37 tC/ha/year) (Blanfort et al. 2022). Yet the loss of carbon from forest biomass due to deforestation is permanent. However, the effect of land use change on SOC has rarely been the subject of systematic studies, whether in North Africa, Central Africa, the Middle East, or Central Asia.

The restoration of the SOC stock in cultivated ecosystems is a second important lever for sequestering carbon in the soil (Cook-Patton et al. 2021). Numerous agricultural practices can locally increase SOC, such as the addition of exogenous organic matter or the *in situ* production of organic matter. The origin of the organic matter, the effects per unit area of these practices on carbon, and the available areas for their implementation will determine the global carbon sequestration. Recent studies are trying to quantify these effects more precisely and identify the most promising practices.

Agroforestry is identified as one of the most effective options for *in situ* increase of organic matter: trees increase the SOC in arable land by an average of 20% (Beillouin et al. 2023) to 25% (Cardinael et al. 2018a) compared to systems without trees (Box 17.1). The global agricultural area under agroforestry is estimated at 1.6 billion hectares, with about 78% in tropical regions and 22% in temperate regions (Nair et al. 2021), with a strong potential to further increase these areas (Zomer et al. 2022). Thus, global estimates of the mitigation potential of agroforestry range from 0.12 GtC/year to 0.31 GtC/year, making it one of the main nature-based solutions to combat climate change (Terasaki et al. 2023).

Other crop diversification practices can also increase SOC. Thus, the integration of cover crops, the retention of crop residues in the fields, or the use of crop rotations can increase SOC by an average of 11.6%, 13%, and 6.5% per hectare (Beillouin et al. 2023). Some authors estimate that up to 50% of the 800 Mha of cultivated land are suitable for the use of cover crops (Bossio et al. 2020), resulting in a potential carbon sequestration of approximately 0.11 GtC/year.

The addition of exogenous carbon also allows for significant local increases in SOC stocks in croplands and grasslands (Beillouin et al. 2023). However, the interest of exogenous carbon inputs in mitigating climate change may be limited depending on the origin of these organic materials (Don et al. 2023) and the technical and economic feasibility of their production. Beillouin et al. (2023) note that biochar[1] leads to the highest observed increase per unit area, with an average SOC gain of 67% in croplands and by 32% in grasslands. Therefore, the application of biochar

---

[1] Biochar: a carbon material produced from biomass and whose use allows the carbon it contains to constitute a long-term "carbon sink" according to the European Biochar Council. Biochar is made by pyrolysis of biomass at high temperatures (between 350 °C and 1000 °C) in the absence of oxygen.

Box 17.1 Two examples of agroforestry systems

Many types of agroforestry are effective in increasing SOC stocks: the integration of trees into agricultural lands leads to an average increase in SOC of 33% for multilayered systems, 32% for wooded parks, 21% for intra-plot agroforestry, 19% for improved fallows (implemented in tropical regions) and 17% for hedges (Beillouin et al. 2023). The planting of trees in grasslands (i.e., silvopastoralism) also leads to a significant increase in SOC of 26%. It should be noted that agroforestry can also store significant amounts of carbon in the aboveground and belowground biomass of trees (Cardinael et al. 2018a). Agroforestry exists and can be implemented in both Northern and Southern countries (see below an example in southern France and an example in Senegal). Countries regularly mention agroforestry for climate change mitigation. For example, 40% of the 147 non-Annex I countries of the Kyoto Protocol propose agroforestry as a GHG emission mitigation option in their nationally determined contributions (Rosenstock et al. 2019). The agroforestry system of Restinclières in southern France The Restinclières estate has been home to one of the oldest intra-plot agroforestry experimental sites in France since 1995, combining hybrid walnut with durum wheat (Fig. 17.1). An adjacent agricultural plot without trees serves as a reference. On this site, a hundred soil cores were taken up to 2 m deep in each plot to quantify the SOC stocks. The inputs of organic carbon (OC) to the soil, both aboveground (leaf fall) and belowground (fine roots of trees and crops), were also quantified. An increase in the SOC stock of 6.3 tC/ha was observed in agroforestry after eighteen years. This corresponds to a storage rate of 248 ± 31 kgC/ha/year in the first thirty centimetres of soil, and 350 ± 41 kgC/ha/year in the first metre (Cardinael et al. 2015), an increase of 7‰/year. This is in addition to an accumulation of carbon in the tree biomass of 770 ± 110 kgC/ha/year (Cardinael et al. 2017). This additional SOC storage is explained by an increase in OC inputs to the soil of about 40% compared to the agricultural plot (Cardinael et al. 2018b). The herbaceous vegetation developing on the tree line plays an important role. There is a high spatial heterogeneity, with the SOC being mainly stored on the tree lines. The majority of the additional carbon is in the form of labile particulate organic matter, and therefore sensitive to any future land use change (Cardinael et al. 2015, 2018b).

The Faidherbia agroforestry park in Niakhar (Senegal): a case study of the impact of the root systems of *Faidherbia albida* on the carbon cycle.

The impact of Sahelian agro-silvopastoral systems on the carbon cycle and GHG fluxes is poorly documented. Subterranean litter inputs are crucial, but little is known about the distribution, productivity, and contribution of the root system to carbon inputs in the soil. To increase carbon storage in cultivated soils, it is necessary to better understand carbon dynamics, particularly in depth, where the decomposition of root litter appears slower than in the arable layer. In an agro-silvopastoral park in the peanut basin in Senegal, dominated by *Faidherbia albida* in intercropping with millet and peanuts in interannual

(continued)

Box 17.1 (continued)

rotation, the temporal and spatial variability of GHG fluxes and the dynamics of carbon input from roots into the soil have been monitored over several years.

Unexpectedly, the root biomass of the trees was higher more than thirty meters from the trunk than under the tree in the 30 to 100 cm soil layer, suggesting a long-distance influence of *Faidherbia* (Siegwart et al. 2023). Up to 150 cm deep, the contribution of tree root litter to soil carbon stocks represents 6.5% of total soil carbon inputs. In this type of park in a semi-arid area with low soil carbon stocks, increasing tree density or selecting deep-rooting crops should be encouraged to mitigate CO_2 emissions.

Fig. 17.1 Experimental site at Restinclières in intra-plot agroforestry combining hybrid walnut with durum wheat. (Credit: photo by R. Cardinael)

has a strong potential for mitigating climate change (Lehmann et al. 2021), but this varies depending on the application rate and the physicochemical properties of the biochars. The application of biochar allows for longer-term carbon storage than the direct application of the biomass from which it is derived. This is true even when considering the GHG emissions that occur during its production and handling. However, the scarcity of biomass in some regions or its competition with livestock feed, particularly in sub-Saharan African countries, does not necessarily make large-scale biochar production possible or desirable. It is also necessary to consider the potential harmful effects of biochar on soil properties (such as the toxicity of substances present in the biochar that can be generated during the pyrolysis process) and on biodiversity (Brtnicky et al. 2021). Organic amendments applied in croplands, other than biochar, allow for an average increase in SOC of 29%, and 34% in grasslands. In Reunion Island, Edouard-Rambaut et al. (2022) demonstrated that the application of organic amendments for 15 years in mowed grasslands continues to induce an increase in SOC, particularly in the deeper soil horizons. Numerous practices such as soil tillage, different grazing methods, and mineral fertilisation can also have significant impacts on SOC. The effects of different practices vary depending on pedoclimatic conditions and agricultural practices. They can be combined, and often bring other co-benefits, such as biodiversity preservation or yield improvement. Farmers' practices should not be considered individually, but also taking into account the different componentsof the ecosystem (animals, soil, vegetation) and different spatial scales (plot, farm, territory). Thus, landscape-scale studies show that pastoral landscapes in the Sahel can have an unexpected potential for mitigating climate change. For example, the carbon balance of a silvopastoral territory in North Ferlo (Senegal) appears to be in equilibrium when SOC is taken into account in this balance (Assouma et al. 2019), even though these grazing lands are traditionally considered as sources of GHGs.

17.2 In Semi-arid and Sub-humid West Africa, Territorial Approaches to Consider the Use of Biomass and Nutrient Fluxes

The agro-silvopastoral systems of West Africa are defined by the integration of agriculture with livestock activities and by a central role of trees whether through wooded savannas, fallow lands or agroforestry parks. The spatial organisation of territories, closely associating the cultivated area and the non cultivated area, generates significant transfers of soil fertility and strongly influences the organic status of soils and soil carbon stocks as well as the sustainability of agricultural production systems.

Cirad and its partners are conducting numerous projects in the semi-arid and sub-humid zone of West Africa,[2] where the issue of soil carbon sequestration is linked to maintaining soil fertility, before being a concern for climate change mitigation. In this region where agricultural yields are strongly limited by access to inputs, limited mechanisation, low soil fertility and rainfall (Giller et al. 2021), carbon sequestration is constrained by the small amount of carbon stored by photosynthesis. In this

---

[2] https://www.fair-sahel.org/, https://www.cassecs.org/, https://ur-aida.cirad.fr/our-research/projects-and-expertise/3f2

context, the valorisation of crop residues appears as a necessity to recycle carbon, and two historically opposing paths exist: returning crop residues to the soil or using them to feed livestock, the latter being able to provide organic manure useful for maintaining soil fertility (Naudin et al. 2015).

In these agro-silvopastoral systems, it is difficult to initiate a virtuous cycle of biomass management through farm-scale agriculture-livestock integration. The lack of biomass leads animals to leave the farm to seek residues in neighbouring fields or even in more humid regions during large transhumances. In this context of expanding cultivated areas and the presence of wastelands in open landscapes, residues are collected by producers to meet the forage needs of their herds. However, Zoungrana et al. (2023) showed that up to 60% of these residues were left in the field or burned due to lack of resources (labour, equipment). These production areas are marked by a significant presence of cotton cultivation (synonymous with access to inputs for producers), whose residues are poorly digestible and therefore difficult to valorise by ruminants. In a more arid context in northern Burkina, Assogba et al. (2023) also demonstrated that stored crop residues do not cover the needs of animals (energy and protein) during the dry season.

It is then at the regional scale that it is appropriate to reflect on the optimisation of nutrient fluxes between cropping systems and animal production systems, in interaction with other landscape units. Common pastoral areas constitute a crucial source of biomass for the balance of ecosystems, in areas sometimes difficult to cultivate, which herds utilise by grazing and fertilise by depositing faeces. However, it should be noted that in some cases, cow dung deposited in these grazing areas is also collected and brought back to the cultivated areas, jeopardising the fertility of these areas in the long term (Sattari et al. 2016). This transfer of fertility between grazing and croplands must therefore necessarily be done respecting the nutrient balances of each part, otherwise it will not be able to sustainably compensate for the low productivity of the region's farming systems.

Conflicts over these pastoral areas, or their gradual disappearance in favour of crops, have endangered other spaces such as protected forests in southern Burkina Faso (Koumbia). In response to constraints on access to cultivated forage resources or grazing areas, farmers in this region come to the forest to graze their animals illegally. Orounladji et al. (2024) also demonstrated that common resources (water points, crop residues, grazing areas) are essential for the viability of agro-silvopastoral systems. The analysis must therefore include political and territorial dimensions.

To identify the conditions for carbon sequestration and specify the corresponding practices, it is therefore necessary to question the possibilities at the scale of a territory to optimise nutrient flows. To quantify these flows, modelling is a relevant tool that allows for the spatialisation of localised results and the characterisation of temporal dynamics of fertility related to these flows. Berre et al. (2021) used multi-agent modelling to demonstrate that mulching based on crop residues led to a decrease in fertility at the territorial scale in southern Burkina Faso. Assogba et al. (2023) showed using this type of model that the introduction of legumes indeed

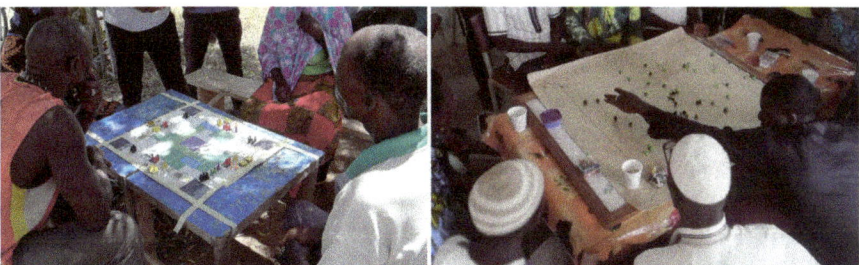

Fig. 17.2 Implementation of a serious game with different types of stakeholders to discuss biomass flows (left) or constraints of farmers (water points, grazing areas) (right). (Credit: photos by G. Assogba (left photo) and D. Dabire (right photo)). The illustration shows two situations where local stakeholders are discussing around a "serious game" board. On the left, the researcher is standing and holding cards that he will assign to the players based on the choices they will make in the game. Two male players and two female players can be seen. Sheets on the upper part of the image suggest that the session is taking place under a tree. The game board is taped to a wooden table and different types of tokens can be seen on it. On the right illustration, many stakeholders are having lively exchanges around a game board, two of them pointing to parts of the game board, while others seem to be discussing. The game board is large and tokens are arranged all over its surface. The session is taking place inside a building

improves soil fertility, but at the expense of household food security. This largely depends on cereals and the coverage of animal forage needs. The same authors demonstrate that it is the grazing areas outside the territory that ensure the coverage of forage needs and that the input of exogenous nitrogen is essential to maintain the chemical fertility of the soils (Falconnier et al. 2023).

In order to grasp the processes at play and their consequences at the territorial scale, these reflections must integrate the perceptions of a diversity of stakeholders. Among the innovative participatory methods, the development of serious games has allowed actors to come together around issues at the territorial scale (Fig. 17.2).

This description highlights the importance of a territorial understanding of issues related to the local dynamics of soil organic matter. Some soils can contribute to biomass production and thus capture carbon while other soils will be receptacles for organic matter and carbon. Just like reasoning the sustainability of systems and the food security of populations, soil carbon sequestration is analysed by taking into account the complexity of agricultural production systems at the scale of a territorial unit. Similarly, the history of these agro-silvopastoral systems in arid and semi-arid sub-Saharan regions shows that environmental, social and economic constraints are not linear or immutable. This requires being able to estimate the long-term risks of changes that can affect this complexity and *ultimately* the proces of soil carbon sequestration, as well as of course the sustainability of these agroecosystems.

The need to apprehend different spatial scales to contribute to climate change mitigation, particularly *via* the soil, and to innovate in terms of space is also illustrated through the example of the TerrAmaz project in Amazonian territories of Brazil, Colombia, Ecuador and Peru (see Chap. 22).

17.3 Soil Carbon Sequestration and the Risk of Maladaptation

In addition to the contribution to mitigation, described previously, soil carbon sequestration contributes to climate change adaptation, as it enhances the soil's capacity to help a region cope with climatic hazards. For instance, if SOC increases, we can expect healthier soil, improved productivity, a more efficient nutrient cycle that may involve less use of mineral fertilisers, plants less susceptible to pathogen attacks thus reducing pesticide use, better preserved biological activities and biodiversity, etc. All of this contributes to reducing the vulnerability of the agroecosystem in question to current or future climate changes.

Thinking about adaptation leads us to question more broadly the conditions and scope of actions taken to sequester carbon in the soil. What is the vulnerability of the area in question to the consequences of climate change? To what extent does soil carbon sequestration contribute to reducing this vulnerability? What are the other determinants that could change the vulnerability of a region? Indeed, like any climate action measure, carbon sequestration is one objective among many, including socio-economic ones, for the same region. It is necessary to take into account these different processes that operate and evolve the socio-ecosystem in which carbon sequestration is considered.

If, due to these processes or an unexpected event, an action aimed at climate change adaptation has harmful effects contrary to the expected positive effects, adaptation becomes maladaptation (Reckien et al. 2023; *IPCC* 2023). This could be the case for a soil restoration measure that does not take into account the basic needs of local populations for food crops and firewood, a measure that increases agricultural yields but reduces biodiversity or indirectly increases GHG emissions.

Beyond the idea that interventions are "good" or "bad", the identification of a continuum between adaptation and maladaptation highlights how interventions can, in a complex and changing way, have mixed results based on a number of evaluation criteria. These criteria include the consequences for people (who benefits, who is negatively affected?) and inequalities (are they reduced, exacerbated?), for ecosystem services (which are improved or preserved, which are degraded or endangered?), for the overall balance of GHGs. It is necessary to propose an evaluation framework that takes into account these different criteria. This framework will need to be constructed with all stakeholders. Indeed, an evaluation depends on the local context in which a transformation planning is implemented, the people who implement these changes, and those who evaluate and decide on the relevance of an adaptation.

In this evaluation framework, the question of time is major. Adaptation processes sometimes have very long-term impacts. This is the case for the constitution of a significant organic matter stock which takes several years, or even several decades. Difficulties then arise in being able to evaluate the storage potential in the soil as well as the balance of GHG fluxes that will allow the objective to be achieved and the possibility of measuring it. Similarly, throughout this process, environmental constraints may arise, modifying the biophysical processes involved. For example, climatic changes can retroactively affect the biological dynamics that determine the constitution of the soil's organic compartment. But more significantly, there are significant risks of changes in the socio-economic conditions of the stakeholders involved, forcing them to modify practices, such as removing trees in an agroforestry system or having to adapt to a lack of workforce, etc. All these factors are difficult to determine over the long term. However, it is possible to establish, from the criteria for differentiating between adaptation and maladaptation, points of vigilance that will need to be observed and that will allow corrective actions to be taken or a new adaptation process to be launched.

17.4 Conclusion

This chapter has shown that soil carbon sequestration offers solutions to mitigate the effects of climate change, in the North as well as in the South, provided that it is implemented in a contextualised manner. It is essential to consider the territorial dimension, with soil being seen as a land asset, as it ensures the consideration of socio-economic and political dimensions and spatial interactions. Finally, analysing the risks of maladaptation related to soil carbon sequestration is essential for managing short and long-term issues and for seeking social justice, the indispensable condition for climate action. It is under these conditions that soil carbon sequestration will be a solution to mitigate climate change and adapt to it.

References

Amundson R, Biardeau L (2018) Soil carbon sequestration is an elusive climate mitigation tool. PNAS 115:11652–11656. www.pnas.org/cgi/doi/10.1073/pnas.1815901115

Assogba GGC, Berre D, Adam M, Descheemaeker K (2023) Can low-input agriculture in semi-arid Burkina Faso feed its soil, livestock and people? Eur J Agron 151:126983

Assouma MH, Hiernaux P, Lecomte P, Ickowicz A, Bernoux M, Vayssières J (2019) Contrasted seasonal balances in a Sahelian pastoral ecosystem result in a neutral annual carbon balance. J Arid Environ 162:62–73. https://doi.org/10.1016/j.jaridenv.2018.11.013

Beillouin D, Corbeels M, Demenois J, Berre D, Boyer A, Fallot A et al (2023) A global meta-analysis of soil organic carbon in the Anthropocene. Nat Commun 14(1):3700

Berre D, Diarisso T, Andrieu N, Le Page C, Corbeels M (2021) Biomass flows in an agro-pastoral village in West-Africa: who benefits from crop residue mulching? Agric Syst 187:102981. https://doi.org/10.1016/j.agsy.2020.102981

Blanfort V, Assouma MH, Bois B, Edouard-Rambaut LA, Vayssières J, Vigne M (2022) Efficiency to account for the complexity of contributions of grazing livestock systems to climate change. In: Ickowicz A, Moulin C-H (eds) Grazing livestock and sustainable development of Mediterranean and tropical territories. Recent knowledge about their strengths and weaknesses. Quæ editions, Versailles, pp 86–104

Bossio DA, Cook-Patton SC, Ellis PW, Fargione J, Sanderman J, Smith P et al (2020) The role of soil carbon in natural climate solutions. Nat Sustain 3(5):391–398. https://doi.org/10.1038/s41893-020-0491-z

Brtnicky M, Datta R, Holatko J, Bielska L, Gusiatin ZM, Kucerik J et al (2021) A critical review of the possible adverse effects of biochar in the soil environment. Sci Total Environ 796:148756

Cardinael R, Chevallier T, Barthès BG, Saby NPA, Parent T, Dupraz C et al (2015) Impact of alley cropping agroforestry on stocks, forms and spatial distribution of soil organic carbon—a case study in a Mediterranean context. Geoderma 259–260:288–299. https://doi.org/10.1016/j.geoderma.2015.06.015

Cardinael R, Chevallier T, Cambou A, Béral C, Barthès BG, Dupraz C et al (2017) Increased soil organic carbon stocks under agroforestry: a survey of six different sites in France. Agric Ecosyst Environ 236:243–255. https://doi.org/10.1016/j.agee.2016.12.011

Cardinael R, Guenet B, Chevallier T, Dupraz C, Cozzi T, Chenu C (2018b) High organic inputs explain shallow and deep SOC storage in a long-term agroforestry system—combining experimental and modeling approaches. Biogeosciences 15:297–317. https://doi.org/10.5194/bg-2017-125

Cardinael R, Umulisa V, Toudert A, Olivier A, Bockel L, Bernoux M (2018a) Revisiting IPCC tier 1 coefficients for soil organic and biomass carbon storage in agroforestry systems. Environ Res Lett 13:1–20. https://doi.org/10.1088/1748-9326/aaeb5f

Cherlet M, Hutchinson C, Reynolds J, Hill J, Sommer S, von Maltitz G (eds) (2018) World atlas of desertification. Publication Office of the European Union, Luxembourg

Cook-Patton SC, Drever CR, Griscom BW, Hamrick K, Hardman H, Kroeger T et al (2021) Protect, manage and then restore lands for climate mitigation. Nat Clim Chang 11:1027–1034. https://doi.org/10.1038/s41558-021-01198-0

Demenois J, Torquebiau E, Arnoult MH, Eglin T, Masse D, Assouma MH et al (2020) Barriers and strategies to boost soil carbon sequestration in agriculture. Front Sustain Food Syst 4(37). https://doi.org/10.3389/fsufs.2020.00037

Dignac M-F, Derrien D, Barré P, Barot S, Cécillon L, Chenu C et al (2017) Increasing soil carbon storage: mechanisms, effects of agricultural practices and proxies. A review. Agron Sustain Dev 37:14. https://doi.org/10.1007/s13593-017-0421-2

Don A, Seidel F, Leifeld J, Kätterer T, Martin M, Pellerin S et al (2023) Carbon sequestration in soils and climate change mitigation—definitions and pitfalls. Glob Chang Biol e16983. https://doi.org/10.1111/gcb.16983

Edouard-Rambaut L-A, Vayssières J, Versini A, Salgado P, Lecomte P, Tillard E (2022) 15-year fertilisation increased soil organic carbon stock even in systems reputed to be saturated like permanent grassland on andosols. Geoderma 425:116025. https://doi.org/10.1016/j.geoderma.2022.116025

Falconnier GN, Cardinael R, Corbeels M, Baudron F, Chivenge P, Couëdel A et al (2023) The principle of input reduction in agroecology is incorrect in relation to mineral fertiliser use in sub-Saharan Africa. Outlook Agric 52(3):311–326. https://doi.org/10.1177/00307270231199795

FAO (2022) Soils for nutrition: current understanding. Rome. https://doi.org/10.4060/cc0900en

Fujisaki K, Chevallier T, Chapuis-Lardy L, Albrecht A, Razafimbelo T, Masse D et al (2018) Changes in soil carbon stocks in tropical croplands are primarily driven by carbon inputs: a synthesis. Agric Ecosyst Environ 259:147–158. https://doi.org/10.1016/j.agee.2017.12.008

Giller KE, Delaune T, Silva JV, van Wijk M, Hammond J, Descheemaeker K et al (2021) Small farms and development in sub-Saharan Africa: farming for food, for income or due to lack of better options? Food Secur:1–24. https://doi.org/10.1007/s12571-021-01209-0

IPCC (2019) Summary for policymakers. In: Climate Change and Land: An IPCC special report on climate change, desertification, land degradation, sustainable land management, food security, and greenhouse gas fluxes in terrestrial ecosystems. Cambridge University Press, pp 1–41

IPCC (2023) Climate change 2023: synthesis report. Contribution of working groups I, II and III to the sixth assessment report of the intergovernmental panel on climate change. IPCC, Geneva, Switzerland., 184 p. https://doi.org/10.59327/IPCC/AR6-9789291691647

Lal R, Lorenz K, Hüttl RF, Schneider BU, von Braun J (2012) Terrestrial biosphere as a source and sink of atmospheric carbon dioxide. In: Lal R, Lorenz K, Hüttl RF, Schneider BU, von Braun J (eds) Recarbonization of the biosphere: ecosystems and the global carbon cycle. Springer, Dordrecht, pp 1–15

Lehmann J, Cowieg A, Masiello CA, Kammann C, Woolf D, Amonetteg JE et al (2021) Biochar in climate change mitigation. Nat Geosci 14:883–892

Ma Y, Woolf D, Fan M, Qiao L, Li R, Lehmann J (2023) Global crop production increase by soil organic carbon. Nat Geosci 16:1159–1165. https://doi.org/10.1038/s41561-023-01302-3

Nair PKR, Kumar BM, Nair VD (2021) Global distribution of agroforestry systems. In: An introduction to agroforestry. Springer, Cham. https://doi.org/10.1007/978-3-030-75358-0_4

Naudin K, Bruelle G, Salgado P, Penot E, Scopel E, Lubbers M et al (2015) Trade-offs around the use of biomass for livestock feed and soil cover in dairy farms in the Alaotra lake region of Madagascar. *Agric Syst* 134:36–47. https://doi.org/10.1016/j.agsy.2014.03.003

Noon ML, Goldstein A, Ledezma JC, Roehrdanz PR, Cook-Patton SC, Spawn-Lee SA et al (2022) Mapping the irrecoverable carbon in earth's ecosystems. *Nat Sustain* 5:37–46. https://doi.org/10.1038/s41893-021-00803-6

Olson KR, Al-Kaisi MM, Lal R, Lowery B (2014) Experimental considerations, treatments, and methods in determining soil organic carbon sequestration rates. Soil Sci Soc Am J 78(2):348–360. https://doi.org/10.2136/sssaj2013.09.0412

Orounladji BM, Sib O, Berre D, Assouma MH, Dabire D, Sanogo S, Vall E (2024) Cross-examination of agroecology and viability in agro-sylvo-pastoral systems in Western Burkina Faso. Agroecol Sustain Food Syst 48(4):581–609. https://doi.org/10.1080/21683565.2024.2307902

Reckien D, Magnan AK, Singh C, Lukas-Sithole M, Orlove B, Schipper ELF et al (2023) Navigating the continuum between adaptation and maladaptation. Nat Clim Chang 13(9):907–918. https://doi.org/10.1038/s41558-023-01774-6

Rosenstock TS, Wilkes A, Jallo C, Namoi N, Bulusu M, Suber M et al (2019) Making trees count: measurement and reporting of agroforestry in UNFCCC national communications of non-annex I countries. Agric Ecosyst Environ 284:106569

Sanderman J, Hengl T, Fiske GJ (2017) Soil carbon debt of 12,000 years of human land use. *Proc Natl Acad Sci* 114:9575–9580. https://doi.org/10.1073/pnas.1706103114

Sattari SZ, Bouwman AF, Martinez RR, Beusen AHW, van Ittersum MK (2016) Negative global phosphorus budgets challenge sustainable intensification of grasslands. Nat Commun 7:10696. https://doi.org/10.1038/ncomms10696

Siegwart L, Bertrand I, Roupsard O, Jourdan C (2023) Contribution of tree and crop roots to soil carbon stocks in a sub-Saharan agroforestry parkland in Senegal. Agric Ecosyst Environ 352:108524. https://doi.org/10.1016/j.agee.2023.108524

Stahl C, Fontaine V, Dézécache C, Ponchant L, Freycon V, Picon-Cochard C et al (2016) Soil carbon stocks after conversion of Amazonian tropical forest to grazed pasture: importance of deep soil layers. Reg Environ Chang 16:2059–2069. https://doi.org/10.1007/s10113-016-0936-0

Stahl C, Fontaine V, Klumpp K, Picon-Cochard C, Freycon V, Grise MM et al (2017) Continuous soil carbon storage of old permanent pastures in Amazonia. Glob Chang Biol 23(8):3382–3392. http://onlinelibrary.wiley.com/doi/10.1111/gcb.13573/full

Terasaki HD, Yeo S, Almaraz M, Beillouin D, Cardinael R, Garcia E et al (2023) Priority science can accelerate agroforestry as a natural climate solution. Nat Clim Chang 13:1179–1190. https://doi.org/10.1038/s41558-023-01810-5

Wiesmeier M, Urbanski L, Hobley E, Lang B, von Lützow M, Marin-Spiotta E et al (2019) Soil organic carbon storage as a key function of soils—a review of drivers and indicators at various scales. Geoderma 333:149–162. https://doi.org/10.1016/j.geoderma.2018.07.026

Winkler K, Fuchs R, Rounsevell M, Herold M (2021) Global land use changes are four times greater than previously estimated. Nat Commun 12:2501

Zomer RJ, Bossio DA, Trabucco A, Noordwijk M, Xu J (2022) Global carbon sequestration potential of agroforestry and increased tree cover on agricultural land. Circular Agric Syst 2:1–10. https://doi.org/10.48130/CAS-2022-0003

Zoungrana SR, Ouédraogo S, Sib O, Bougouma-Yameogo VMC, Fayama T, Coulibaly K et al (2023) Recycling crop and livestock co-products on agro-pastoral farms for the agroecological transition: more than 60% potentially recoverable in western Burkina Faso. Biotechnol Agron Soc Environ 27(4):270–283. https://doi.org/10.25518/1780-4507.20537

Open Access This chapter is licensed under the terms of the Creative Commons Attribution-NonCommercial-NoDerivatives 4.0 International License (http://creativecommons.org/licenses/by-nc-nd/4.0/), which permits any noncommercial use, sharing, distribution and reproduction in any medium or format, as long as you give appropriate credit to the original author(s) and the source, provide a link to the Creative Commons license and indicate if you modified the licensed material. You do not have permission under this license to share adapted material derived from this chapter or parts of it.

The images or other third party material in this chapter are included in the chapter's Creative Commons license, unless indicated otherwise in a credit line to the material. If material is not included in the chapter's Creative Commons license and your intended use is not permitted by statutory regulation or exceeds the permitted use, you will need to obtain permission directly from the copyright holder.

Chapter 18
What Solutions for Agricultural Water Management in the Face of Climate Change?

Caroline Lejars, Koladé Akakpo, Magalie Bourblanc, Emeline Hassenforder, and Pierre-Louis Mayaux

Abstract Growing food demand and competing land and water uses exert intense pressure on agricultural systems, which are further stressed by climate change. Water availability is declining and unevenly affected, with surface waters impacted faster than groundwater, creating local adaptation challenges. Agriculture competes with other sectors like energy and drinking water during shortages, especially in dry seasons. Adaptation strategies must be locally tailored, combining efficient water use, new resources, and flexible management. This chapter highlights various technical and governance-based solutions, including participatory approaches to improve coordination.

The increasing food needs, coupled with the demand for other competitive uses, exert unprecedented pressure on many agricultural production systems worldwide. Current production systems, defined as "systems at breaking points" by the FAO (2021), thus face increased competition for land and water resources, which is exacerbated by climate change (see Chap. 7).

In this context, adapting to climate change, whether it manifests as droughts or other extreme events, proves extremely complex. The impacts of climate change on water resources are poorly known at local scales, as the number of empirical studies remains limited (Moore et al. 2017). Climate change affects available resources differently: surface waters, for example, are affected more rapidly than groundwater, which often serves as "buffer" to meet the needs of irrigated crops when rain is lacking. The impacts of declining water availability also vary depending on local uses, types and diversity of available resources, and the period of reduced rainfalls. They depend on local needs which are variable: the demand for drinking water can vary seasonally in tourist cities, the needs for hydroelectricity also vary according

---

C. Lejars (✉) · K. Akakpo · M. Bourblanc · P.-L. Mayaux
Cirad, UMR Gestion de l'Eau, Acteurs, Usages (G-EAU), Montpellier, France
e-mail: caroline.lejars@cirad.fr

E. Hassenforder
Cirad, UMR G-EAU, Institut national agronomique de Tunisie (INAT), Tunis, Tunisie

to the seasons, crop needs are generally greater in summer when water is lacking. More generally, water is at the intersection of several issues: energy, food, drinking water, industrial needs and ecosystem health. When water is lacking, agriculture finds itself in competition with other sectors and the difficulties vary depending on the type of agriculture and irrigation practiced.

The mechanisms for adapting to climate change in agriculture are therefore highly dependent on contexts and local situations. Managing seasonal variations, consuming less water or more efficiently, finding new resources require the implementation of panels of locally adapted solutions that can rely on technical solutions, as well as management and governance solutions. We describe here these panels of solutions and the promising experiences of participatory approaches that would allow them to be coordinated.

18.1 Diverse Technical and Agroecological Solutions, but Poorly Evaluated and Sometimes with Unintended Impacts

In recent years, the observation of the increasing scarcity of water resources has prompted governments to turn to technical solutions and to massively reinvest in large infrastructures, particularly for irrigation (see Chap. 7). The infrastructures put in place to cope with water shortages cover both the classic works of "large-scale hydraulic systems" (large dams, inter-basin transfers, supply canals, distribution networks), but also solutions presented as innovative around wastewater reuse or desalination. These latter, which locally increase water supply, are however in many cases (such as reuse) simple mechanisms for reallocating water and in others (such as desalination) very energy-intensive solutions with potentially damaging effects on ecosystems and water quality (Williams et al. 2023). Massive investments in irrigation, while they have increased food production, have also come at the cost of growing inequalities (Mayaux et al. 2022). Beyond investments in large infrastructures, technical solutions intended to be developed at the farm and plot level have also been promoted and subsidised in many countries to optimise irrigation and best meet crop needs. Drip irrigation, a localised and water-saving irrigation system, is an emblematic example. Developed and funded in many countries, it theoretically reduces the amount of water used at the plot level. However, in the absence of control over the expansionof irrigated areas, it has led to a global increase in resource exploitation at territorial levels (Venot et al. 2017).

In parallel, the agroecological transition is gradually promoted as a solution for developing sustainable agricultural systems. The general principle is that such a transition would facilitate water retention and infiltration on plots. However, the place of water and irrigation in the agroecological transition is still little studied. To ensure sustainable food and nutritional security, many authors highlight the interest of judicious use of water available for agriculture through micro-irrigation technologies and water-saving, through conservation and climate-smart agriculture (see Chap. 3), through the development of multiple cultivars and stress-tolerant crop biotypes using biotechnological tools (see Chap. 21), and through the restoration of degraded soils (Jat et al. 2016). Agroforestry, conservation agriculture, crop diversification, climate information services are also highlighted as new climate-smart agricultural options to improve agricultural productivity, rural livelihoods and the adaptive capacity of farmers and production systems (Zougmoré et al. 2018; Partey et al. 2018). Even though these practices theoretically allow for an increase in water use efficiency (the ratio between the amount of water supplied and that actually used by the crops), it is challenging to quantify water savings, whether at the plot or territorial scale. For instance, the water savings achievable through drip irrigation or practices such as soil conservation are intrinsically linked to farmers' irrigation strategies (amounts supplied, irrigation frequency). However, these practices are not always effectively adapted or optimised on farms. Similarly, the complex interactions between trees and crops in agroforestry systems, for example, make it difficult to accurately assess their impact on available water.

In some regions, particularly in France and Europe, access to new water resources is increasingly conditioned by the adoption of agroecological practices such as soil conservation agriculture, legume-based cover crops, or organic farming, without a precise understanding of the impact of these practices on water use. The development of organic or low-input certified productions, currently promoted by several countries, could nevertheless lead to an increase in irrigation demands, as the management of irrigation schedules is a means of ensuring yields and in the absence of herbicides, can become a means of controlling weed growth. Therefore, there is a significant need for research in this area to provide solutions for farmers, while assessing their final impact on water at the plot or territorial scale (Torquebiau et al. 2018).

Regardless, the close links between agroecology and climate change are promising and provide some successful technical examples, both in the North and the South. In India, for example, the Andhra Pradesh Community Natural Farming (APCNF)—which promotes a set of agricultural practices designed to significantly reduce farmers' direct costs while increasing yields and farm health—is one of the major examples of large-scale agroecological transition in the world in terms of the number of participating farmers and involved institutions, and with interesting impacts in terms of plot-level water savings (Bharucha et al. 2020; Landy and Dorin 2022). Unfortunately, agroecology is still rarely explicitly mentioned in the nationally determined contributions (NDCs) presented by countries at the Paris Agreement in 2015 (COP21) (see Chap. 2), and some of its components are not explicitly mentioned, such as water conservation management (Torquebiau et al. 2019).

18.2 Governance Solutions Around Water Sharing Are Proving More and More Ineffective

18.2.1 Integrated Water Resources Management: A Global Diffusion with Mixed Results

During the 1990s, integrated water resources management (IWRM) gradually became a panacea in the world of water resources (Mayaux et al. 2022). It accompanied the implementation of a range of programmes and institutional tools worldwide, such as the creation of basin agencies, local management plans, and aquifer contracts. Participation is one of its key concepts, with the aim of including relevant stakeholders in the formal decision-making process. The concept of integration is multidimensional. Besides the very general integration between resource preservation, economic efficiency, and social well-being, it also needs to be operated (1) among all waters, surface and underground, and among all environments; (2) among all actors and sectors of public action related to water, namely the environment, energy, agriculture, planning, urban planning, etc.; (3) from a spatial point of view, at the watershed scale.

Ultimately, its global diffusion has, however, had mixed results (Barone and Mayaux 2019). It has undoubtedly promoted knowledge production, problem identification, and institutional capacity building. On the other hand, many basin agencies still struggle to access the human, financial, and logistical resources necessary to function properly. In South Africa, and in Africa more generally, numerous obstacles and limitations in terms of financial capacities, lack of skills, institutional conflicts, and rigidity have hindered the successful establishment of water management agencies following the decentralisation principles of IWRM. In France, limitations related to the institutional layering and financial capacities led the Court of Auditors to describe the country's water governance as "ineffective" in 2022 (Court of Auditors 2022).

More broadly, there is a glaring gap in the water sector between the complexity of the issues and the weakness of the resources allocated to the sector's management, particularly regarding the staffing of the water police (as in the case of countries as different as Morocco, Mexico, Brazil, or France). Besides the lack of technical and financial resources, water policy projects often relegate environmental issues to the background, behind economic and social issues. Water institutions are often marginalised in the system. Decision-making. Finally, the effectiveness of water policies (are they really implemented?) and their efficiency (do they achieve their objectives?) are rarely evaluated. Such evaluations of policy impact could strengthen trust between authorities and citizens.

18.2.2 Regulatory Restrictions and Tools in Response to Droughts

In the context of climate change, water restrictions, particularly in the agricultural sector, are increasingly being used as emergency regulatory measures to prevent water supply disruptions in stressed areas, whether in the Northern or the Southern countries. While these scarcity management issues are well known in the South, they are now becoming increasingly acute in Northern countries, as evidenced by the drought episode in the summer of 2022 when nearly 350 municipalities in France had to be supplied with water by tanker trucks. It is important to note here, the significance of not using this regulatory tool in a reactive manner only once the crisis is well established, but to anticipate the shortage and implement usage restrictions early enough. Indeed, in France (as in other countries), the systematic triggering of crisis tools in many departments during the summer, when they should in principle be used only two years out of ten at most, indicates deficiencies in the ordinary management of water (Barone and Mayaux 2019). The case of South Africa also illustrates how delays in making decisions about politically unpopular restriction measures inevitably lead to more drastic measures *in the end* when the crisis is established and to the promotion of new hydraulic structures to remedy it (Bourblanc 2022).

18.2.3 Incorporating Climate Change into Water Governance Instruments

From the mid-2000s, many countries or provinces began to adopt climate change adaptation (CCA) strategies including agriculture (for example France, Spain and Senegal in 2006, California and the Western Cape province in South Africa in 2008). Some innovative instruments for climate change adaptation have been implemented in several countries. Many are information instruments, which support research and innovation, through information systems, promoting the regular publication of reports at different levels (national and subnational, for example by the countries of the European Union) (see Chap. 7). Regulatory instruments are also gaining significantly. For example, a policy of quotas and restrictions on irrigation water has been implemented in France for several years (Loubier et al. 2021). These policies are sometimes accompanied by public interventions to encourage new agricultural models. Incentives have been designed to restructure certain production areas, such as subsidies for the restructuring of vineyards in Spain and southern France, but dedicated funding remains limited so far. In Spain for example, some measures, included in rural development plans and adopted by the different provinces and largely funded by the European Union, are now striving to limit water consumption in a long-term climate change adaptation perspective. In Andalusia, irrigation communities can only apply for certain grants if they demonstrate that the project they wish to fund will reduce total water consumption by at least 5% compared to the current situation. This commitment must then be regularly verified by the Andalusian agricultural administration, for a period of at least 5 years. While many controls have been developed, none of them has so far led to any penalty, despite numerous reports of unfullfilled commitments. Moreover, the farmers of this province have successfully mobilised against a much more ambitious project, that of a permanent reduction in their water rights (Boutroue et al. 2022).

Thus, political innovation towards effective climate change adaptations has so far remained limited. For example, many planning documents (for water or agriculture) do not yet consider the expected effects of climate change. Moreover, in many cases, traditional policy instruments have simply been re-labelled as instruments for climate change adaptation (Barone and Mayaux 2019). This is the case for crop yield insurance against natural disasters, subsidies for agroecological practices, drip irrigation, or inter-basin transfers in Spain, Brazil and North Africa.

18.3 The Need to Develop Territorial Approaches Towards an Agroecological Transition Incorporating the Issue of Water

18.3.1 Developing More Coherent Intersectoral and Territorialised Policies

The challenges of coordination between sectoral policies are perceived as more complex than ever with the emergence of the WELF (water, energy, land and food) nexus paradigm. Policies in the fields of water, food, land and energy are mostly developed in silos, even though the interdependencies between them are numerous and substantial. These interdependencies prove (1) complex to understand, (2) highly variable from one territory to another and (3) often not recognised in policy development. Some recent conceptual frameworks put these interdependencies at the forefront (for example, virtual water[1], ecosystem services, etc.), but they are still not very operational.

More than a decade after the introduction of Integrated Water Resources Management (IWRM), the need for territorial approaches has given rise to the new rural paradigm (OECD 2006). The OECD has shown that governments are increasingly aware of the need to improve, and sometimes even abandon, traditional sectoral policies and replace them with more appropriate instruments. OECD governments (but also some countries outside OECD) have shown increasing interest in territorial approaches to rural policy, allowing for the integration of different sectoral policies and improving the coherence and efficiency of public spending in

---

[1] This terminology refers to the water used to produce exportable goods in one location and consumed "virtually" in another space.

rural areas. . For example, in France or Tunisia, such policies have created spaces for involving local communities in decision-making and policy development (such as local water commissions, territory committees, etc.). Local planning authorities are becoming more autonomous and can promote a territorial and participatory approach. These experiences have had positive results, as shown by several experiments conducted over the past decade (Hassenforder and Ferrand 2024). However, important questions remain regarding the appropriate level of public intervention, its legitimacy, and the ability of local planning authorities to implement adequate planning processes.

18.3.2 Improving Participation and Territorial Dialogue Around Water

Participatory approaches to planning and integrated use of natural resources have been widely encouraged over the past few decades, whether in the context of IWRM, but also Sustainable Development Goals (SDGs) (public participation is target 16.7 among others). Public participation is advocated by international donors, governments, and various stakeholders themselves. Public participation is increasingly necessary to obtain funding for development projects . The last decade has also witnessed the rise of civic technologies and citizen science. In Kenya, for example, citizens can lodge complaints or provide feedback on water services through a website, MajiVoice. Citizen science initiatives are research programmes involving individuals—many of whom are not qualified scientists—in the collection, categorisation, transcription or analysis of scientific data. These trends present both opportunities and risks. On the one hand, they allow for more participatory approaches with more numerous and diversified participants, as well as greater transparency of public policies and greater accountability of decision-makers. But these trends also present risks of exclusion, marginalisation and corruption if participation is not implemented efficiently.

Even though participatory approaches have delivered on their promises in many cases, obstacles remain that can limit their implementation and impact.

- The results from participatory processes are not always taken into account in political decisions: many decision-makers still pay little attention to farmers' knowledge and do not understand the added value of public participation.
- Governments rarely have the means to involve the most marginal and disadvantaged people, especially in rural areas, as this requires individual outreach and the implementation of adapted participatory methods. Governments therefore often rely on associations or representatives of different categories of actors (whose interests are not always aligned with those of the poorest). There is also often a lack of political will to implement participatory approaches.
- The evaluation of participatory approaches is fundamentally deficient in most countries. While it sometimes includes a procedural evaluation (number of

participants, representativeness, etc.), it rarely includes the evaluation of social impacts (learning, change of practices, etc.) or environmental performance. Moreover, evaluations 5 or 10 years after the end of processes are anecdotal, and we therefore lack perspective on the impact of these approaches.
- Large-scale participation, particularly in a rural context, raises questions of data storage, aggregation and sharing (Bright and Margetts 2016; Karim et al. 2020; Tang and Liao 2019)
- There is often a gap between the participatory process and investment timelines. A participatory diagnosis must be able to proceed at its own pace, while there is often pressure to quickly spend an annual budget or a project budget.
- It remains difficult to integrate plurality, conflicts and hybridity into participatory planning approaches (Brown et al. 2017; Murray et al. 2004).

To overcome these obstacles, a set of recommendations is proposed in the literature and would allow for the implementation of coherent combinations of technical and institutional solutions. Here are a few of these recommendations.

- Implementing multi-level participatory approaches, that is, approaches that do not only transfer responsibilities at the local level, but also strengthen and support public participation in regional and national decision-making, and create links between different levels (Daniell and Barreteau 2014; Newig et al. 2017). Some participatory methods — such as role-playing games or strategic planning — can enhance interactions between several levels of governance (Hassenforder et al. 2020), particularly by involving stakeholders early in the process. In Tunisia, the Climate Change Adaptation Programme for Vulnerable Territories (2018–2022) has strengthened multi-level participatory planning, through the creation of multi-party platforms and the enhancement of the capacities of territorial facilitators (Hassenforder and Ferrand 2024).
- Developing a culture of participation, within governments and administrations, among elected officials, territorial managers and technicians. The aim is for the various actors to see participation as an opportunity to build more transparent public policies, based on more diverse and closer opinions of citizens, farmers and users. Participation must adapt to the cultural context, but the context must also adapt to the requirements of participation.
- Implementing participation charters that contribute to the transparency and accountability of public policies. Increasingly, participatory processes are designed jointly with the participants themselves. This means that the participants decide who will participate, when, with which methods and with which roles (Daniell et al. 2010; Hassenforder et al. 2020).
- Adopting more systemic approaches to participation, taking into account the complexity of the socio-environmental system within which participation takes place (Chilvers et al. 2018; Jager et al. 2020), particularly temporal inconsistencies, as very distinct temporalities (those of the participatory devices themselves, those of political decision-makers, farmers, researchers and funders) are often poorly articulated.

- Establishing meaningful systems of monitoring, evaluation and decision support, which include qualitative and quantitative data, which assess the multiple impacts of participation (learning, institutional and relational changes, justice and equity, behaviour change, etc.) and which are used to support the steering of processes and decision-making. Various approaches have been developed for such systems (Blundo Canto et al. 2018; Hassenforder and Ferrand 2024). These approaches promote the involvement of different stakeholders in the design of the evaluation process, in the collection and analysis of data and in the sharing of results.

The aim would be to move from a vision of evaluation as a control tool to an evaluation contributing to the impact of the participatory approach.

18.4 Conclusion

To cope with the pressures and decreases in water availability affecting our agri-food systems, many technical or governance solutions exist and have been partially implemented in several countries, both in the Northern and the Southern countries. But if these solutions are not made coherent with each other, they often lead to perverse effects or a form of inefficiency. Governments and the private sector, including farmers, could embrace a more proactive and participatory approach to advance the widespread adoption of sustainable and coherent land and water management practices. These solutions must be adapted and discussed at the territorial level, and their impacts must be evaluated. The implementation of sustainable, equitable and poverty-reducing land and water policies is possible, but it is a long-term effort that requires strong state capacities and citizen engagement, particularly at the territorial level. Adapting to climate change, like any public action, cannot be perfectly consensual: it will necessarily involve that some will loose more than others, at least in the short term, hence growing tensions. These tensions are not to be feared: they can be fruitful and productive, provided they can be expressed in inclusive, relatively transparent and accountable debate and decision-making spaces.

References

Barone S, Mayaux P-L (2019). Water policies. LGDJ – Clefs/Politique, 153 p

Bharucha ZP, Mitjans SB, Pretty J (2020) Towards redesign at scale through zero budget natural farming in Andhra Pradesh, India. Int J Agric Sustain 18(1):1–20. https://doi.org/10.1080/14735903.2019.1694465

Blundo CG, Barret D, Faure G, Hainzelain E, Monier C, Triomphe B, Vall E (2018) ImpresS ex ante. An approach for building ex ante impact pathways. https://doi.org/10.19182/agritrop/00013

Bourblanc M (2022) 'Water vs Wine': irrigation at the urban-rural interface in the Western Cape Province (South Africa), Work Package 3 report on "Water Allocation and Restriction" policy instrument, TYPOCLIM research project (MUSE funding), 30 p

Boutroue B, Bourblanc M, Mayaux P-L, Ghiotti S, Hrabanski M (2022) The politics of defining maladaptation: enduring contestations over three (mal)adaptive water projects in France, Spain and South Africa. Int J Agric Sustain 20(5):892–910. https://doi.org/10.1080/14735903.2021.2015085

Bright J, Margetts H (2016) Big data and public policy: can it succeed where E-participation has failed? Policy Internet 8(3):218–224

Brown G, Kangas K, Juutinen A, Tolvanen A (2017) Identifying environmental and natural resource management conflict potential using participatory mapping. Soc Nat Res 30(12):1458–1475. https://doi.org/10.1080/08941920.2017.1347977

Chilvers J, Pallett H, Hargreaves T (2018) Ecologies of participation in socio-technical change: The case of energy system transitions. Energy Res Soc Sci 42:199–210. https://doi.org/10.1016/j.erss.2018.03.020

Court of Auditors (2022) Quantitative water management in the period of climate change, Exercises 2016–2022, synthesis report. https://www.ccomptes.fr/sites/default/files/2023-10/20230717-gestion-quantitative-de-l-eau.pdf

Daniell KA, Barreteau O (2014) Water governance across competing scales: coupling land and water management. J Hydrol 519(C):2367–2380. https://doi.org/10.1016/j.jhydrol.2014.10.055

Daniell KA, White IM, Ferrand N, Ribarova I, Coad P, Rougier JE, Burn S (2010) Co-engineering participatory water management processes: theory and insights from Australian and Bulgarian interventions. Ecol Soc 15(4)

FAO (2021) The state of the world's land and water resources for food and agriculture: systems on the brink of breakdown, Synthesis report. https://www.fao.org/land-water/solaw2021/fr/

Hassenforder E, Ferrand N (coord.) (2024) Transformative participation for socio-ecological sustainability. Around the CoOPLAGE pathways. Quæ Editions, Versailles, 270 p

Hassenforder E, Barreteau O, Daniell KA, Ferrand N, Kabaseke C, Muhumusa M, Tibasiima T (2020) The effects of public participation on multi-level water governance, lessons from Uganda. Environ Manage 66:770–784. https://doi.org/10.1007/s00267-020-01348-8

Jager NW, Newig J, Challies E, Kochskämper E (2020) Pathways to implementation: evidence on how participation in environmental governance impacts on environmental outcomes. J Public Admin Res Theory 30(3):383–399. https://doi.org/10.1093/jopart/muz034

Jat ML, Dagar JC, Sapkota TB, Yadvinder-Singh B, Govaerts SL, Ridaura YS et al (2016) Chapter Three – Climate change and agriculture: adaptation strategies and mitigation opportunities for food security in South Asia and Latin America. In: Sparks DL (ed) Advances in agronomy, vol 137. Academic, pp 127–235

Karim A, Siddiqa A, Safdar Z, Razzaq M, Gillani SA, Tahir H et al (2020) Big data management in participatory sensing: issues, trends and future directions. Fut Gen Comput Syst 107:942–955. https://doi.org/10.1016/j.future.2017.10.007

Landy F, Dorin B (2022) The state to the rescue of the agroecological transition? The case of India. Movements 109:94–106. https://doi.org/10.3917/mouv.109.0094

Loubier S, Garin P, Hassenforder E, Aucante M, Lejars C (2021) Analyse économique et financière des projets de territoire de gestion de l'eau (PTGE) à composante agricole, Report 150p. https://www.inrae.fr/sites/default/

Mayaux PL, Lejars C, Farolfi S, Adamczewski-Hertzog A, Hassenforder E, Faysse N, Jamin JY (2022) Enabling institutional environments conducive to livelihood improvement and adapted investments in sustainable land and water uses. FAO, Rome, 58 p. https://doi.org/10.4060/cc0950en

Moore FC, Baldos ULC, Hertel T (2017) Economic impacts of climate change on agriculture: a comparison of process-based and statistical yield models. Environ Res Lett 12(6):065008. https://doi.org/10.1088/1748-9326/aa6eb2

Murray M, Shaffer R, Lovan WR (eds) (2004) Participatory governance – planning conflict mediation and public decision-making in civil society. Ashgate Publishing, Belfast

Newig J, Challies E, Jager NW, Kochskaemper E, Adzersen A (2017) The environmental performance of participatory and collaborative governance: a framework of causal mechanisms. Policy Stud J 46(2):269–297. https://doi.org/10.1111/psj.12209

OECD (2006) The new rural paradigm: policies and governance. OECD rural policy reviews. OECD Publishing, Paris. https://doi.org/10.1787/9789264023918-en

Partey ST, Zougmoré RB, Ouédraogo M, Campbell BM (2018) Developing climate-smart agriculture to face climate variability in West Africa: challenges and lessons learnt. J Clean Prod 187:285–295. https://doi.org/10.1016/j.jclepro.2018.03.199

Tang M, Liao H (2019) From conventional group decision making to large-scale group decision making: what are the challenges and how to meet them in big data era? A state-of-the-art survey. Omega 100:102141. https://doi.org/10.1016/j.omega.2019.102141

Torquebiau E, Rosenzweig C, Chatrchyan A, Andrieu N, Khosla R (2018) Agriculture in the face of climate change. Cah Agric 27(26):001. https://doi.org/10.1051/cagri/2018010

Torquebiau E, Roudier P, Demenois J, Saj S, Hainzelin É, Maraux F (2019) Agroecology and climate change: close links which give cause for hope. In: Côte F-X, Poirier-Magona E, Perret S, Roudier P, Rapidel B, Thirion M-C (eds) The agroecological transition of agricultural systems in the Global South. éditions Quæ, Versailles, pp 239–249

Venot J-P, Kuper M, Zwarteveen M (2017) Drip irrigation for agriculture: untold stories of efficiency, innovation and development. Routledge, 358 p. https://doi.org/10.4324/9781315537146

Williams J, Beveridge R, Mayaux PL (2023) Unconventional waters: a critical understanding of desalination and wastewater reuse. Water Altern 17. https://www.water-alternatives.org/index.php/tp1-2/1914-vol16/369-issue16-2

Zougmoré RB, Partey ST, Ouédraogo M, Torquebiau E, Campbell BM (2018) Facing climate variability in sub-Saharan Africa: analysis of climate-smart agriculture opportunities to manage climate-related risks. Cah Agric 27(3):34001. https://doi.org/10.1051/cagri/2018019

Open Access This chapter is licensed under the terms of the Creative Commons Attribution-NonCommercial-NoDerivatives 4.0 International License (http://creativecommons.org/licenses/by-nc-nd/4.0/), which permits any noncommercial use, sharing, distribution and reproduction in any medium or format, as long as you give appropriate credit to the original author(s) and the source, provide a link to the Creative Commons license and indicate if you modified the licensed material. You do not have permission under this license to share adapted material derived from this chapter or parts of it.

The images or other third party material in this chapter are included in the chapter's Creative Commons license, unless indicated otherwise in a credit line to the material. If material is not included in the chapter's Creative Commons license and your intended use is not permitted by statutory regulation or exceeds the permitted use, you will need to obtain permission directly from the copyright holder.

Chapter 19
Energy Production in Agriculture to Tackle Climate Change

François Pinta, Antoine Ducastel, Patrice Dumas, Marie Hrabanski, and Grâce Chidikofan

Abstract Agriculture heavily relies on energy, especially fossil fuels, which played a key role in the Green Revolution but also contribute significantly to greenhouse gas emissions. Agri-food systems account for about 30% of global energy use, with most consumed after harvest in transport, processing, and distribution. However, agriculture also has strong potential to produce renewable energy. This chapter examines energy generation from biomass and the integration of electricity production into farming. Special attention is given to agrivoltaics as a promising solution.

Energy is a major input in agricultural production. Historically, the green revolution which began in the 1950s was the result of the widespread use of fossil fuels and electricity in agriculture, through extensive mechanization, and the intensive use of fertilizers, and pesticides, to address global food security. As a result, agricultural production levels and the fulfilment of human food needs are closely linked to energy availability and use. However, when this energy is of fossil origin, it leads to significant greenhouse gas emissions. Agri-food systems currently account for

---

F. Pinta (✉)
Cirad, UPR BioWooEB (Biomass, Wood, Energy, Bioproducts) University of Montpellier, Montpellier, France
e-mail: francois.pinta@cirad.fr

A. Ducastel
Cirad, UMR Actors Resources Territories & Development (ART-Dev), FAO, Rome, Italy

P. Dumas
Cirad, UMR Cired, Montpellier, France

M. Hrabanski
Cirad, UMR Actors Resources Territories & Development (ART-Dev), Montpellier, France

G. Chidikofan
National University of Sciences, Technologies, Engineering and Mathematics of Abomey (ENSGEP/UNSTIM), Abomey, Benin

© The Author(s) 2026
V. Blanfort et al. (eds.), *Climate Impacts and Challenges in Agriculture, Forests and Food Systems*, https://doi.org/10.1007/978-3-032-04331-3_19

about 30% of global energy consumption[1] with about 70% of that energy used during post-harvest stages such as transport, processing, packaging, storage, marketing, and distribution.

At the same time, agriculture is also a source of renewable energy and has the potential to contribute significantly to energy production.

This chapter analyses the challenges and opportunities related to the increasing role of agriculture in green energy production. The first part focuses on energy generation from biomass (in the form of solid fuels, biogas, and liquid biofuels), while the second part explores the integration of electricity production units into agricultural systems, with particular emphasis on agrivoltaics.

19.1 Energy Production from Agricultural Biomass

Agriculture today plays a role in energy production, through various categories of biomass that can be used to generate different forms of energy. This energy can be utilized within the agricultural sector itself or externally to meet other societal needs. The three forms of energy that can be derived from agricultural biomass are: solid biomass fuels (from lignocellulosic waste), biogas (from wet agricultural waste via anaerobic digestion), and liquid biofuels (from energy crops).

The following Table 19.1 presents the main forms of energy that can be generated from agricultural biomass, whether on farms or in industrial facilities within the energy sectors.

Table 19.1 Main forms of energy produced from agricultural biomass

Biomass resource	Form of energy produced			
	Heat	Electricity	Biogas	Biofuel
Dry residues from agri-food processing	+++	+		
Lignocellulosic crop residues (e.g. straw)	+++	+		
Oilseed or carbohydrate-rich energy crops (palm, maize, rapeseed, etc.)				+++
Wood residues (pruning, hedgerows, roadside maintenance, etc.)	+++	+		
Wet biomass: livestock manure and intermediate energy crop (Cive)	+	+	++++	

---

[1] FAO, 2015: https://openknowledge.fao.org/server/api/core/bitstreams/1b142e0d-de8b-4043-9900-6d8b7c088726/content

19.1.1 Energy and Fuel Generation from Lignocellulosic Waste

This section adresses agricultural biomass, including tree pruning from agroforestry systems and arboriculture (e.g. fruit orchards), as well as wood from wooded plots of agricultural operations (Box 19.1). These resources represent significant supplies of woody biomass, also known as "dry biomass". Please note that forestry activities based on forest exploitation are not considered in this paper.

> **Box 19.1 Availability of agricultural residues for energy production: examples in West Africa (according to BOAD 2023)**
>
> In Benin, Ivory Coast and Senegal, significant deposits of agricultural and Agri-food waste have been identified, which were estimated in 2023 to be 5 Mt./year per country.
>
> Large Agri-food processing units generate cotton husks, rice husks, sugarcane bagasse, cashew nut shells, etc., of which only a small portion is currently used. For example, cashew processing plants use around 15% of the shells to fuel their boilers, leaving 85% as lignocellulosic residue available for other purposes, such as energy production.

Lignocellulosic biomass from the agricultural sector is typically too dispersed geographically, and of insufficient technological quality to be used as construction materials. The same applies to straws, husks (the outer layers of cereal grains, such as rice for example), and nut shells (e.g. hazelnuts, cashew, etc.). These materials are most effectively used as fuels for heat production via controlled combustion appropriate boilers.

When biomass must be transported, it is advantageous to convert it into pellets or compressed briquettes to ease logistics and reduce storage and transport costs. When territorial conditions are favorable, thanks to abundant resources and appropriate infrastructure, these fuels can be used in power plants (possibly with cogeneration of heat) that supply electricity to local or national grids, thereby serving both the population and local industries. Finally, in cases requiring thermal energy (e.g. drying agricultural products, or heating greenhouses), it's possible to replace fossil-derived gas (propane, butane, natural gas) with syngas. Syngas, or synthetic gas, is produced by gasifying dry biomass at high temperatures (900–1000 °C), in a gasification reactor.

The main challenges lie in mobilizing available biomass residues and overcoming economic, technical or organizational barriers. These include:

- High dispersion of biomass sources,
- High mobilization costs,
- Storage needs (due to the mismatch between residue availability and energy demand),
- Intense logistical demands and costs, etc.

Regarding the GHG emissions, the combustion or gasification of lignocellulosic residues reduces both CO_2 and CH_4 emissions that would otherwise result from natural decomposition. Even if natural decomposition can take years (depending on environmental conditions). The Life Cycle Analysis (LCA) method shows that when biomass transport distances are limited, the overall emissions balance is clearly positive. This is due to avoided emissions from natural decomposition and the substitution of fossil carbon with renewable sources.

19.1.2 Production of Biogas Through Anaerobic Digestion of Wet Biomass

Biogas is a renewable gas composed primarily of methane (CH_4) and carbon dioxide (CO_2) produced by the anaerobic digestion of organic matter in a methanisation reactor. Biogas can be used directly, or upgraded to biomethane (by removing CO_2), which can be injected into natural gas networks. Like syngas, biogas can be used to produce heat and/or electricity depending on the configuration of the facilities and energy needs. The co-product of methanisation, known as digestate, is typically used as an organic soil amendment returning organic matter and nutrients to the soil.

The raw material for biogas can include various types of wastes or dedicated agricultural crops. These may consist of energy-oriented intermediate crops (EIC) or other biomass specifically cultivated for this purpose. The waste used is diverse and includes: animal excreta, manure, slaughterhouse waste, cereal sorting waste, green waste from recycling centres, and more. Each batch of biomass has a specific methanogenic power depending on its intrinsic characteristics and the process implemented. Biogas can be used on site as seen in family farming systems in India, China, and to a lesser extent in some African countries. In these cases, biogas is used as a domestic fuel for cooking, replacing firewood or fossil fuels.

In several European Union (EU) countries, including France, methane is separated from CO_2 and injected into the natural gas grid, enabling transport to distant locations and easier storage. The EU has actively supported the development of the biogas sector through its "biogas roadmap," which helps member states implement national strategies. In Latin America, Brazil has seen the most significant growth in the methanisation sector in recent years. In Africa, several countries have launched biogas initiatives—such as the Biogas Africa program started in 2005—with South Africa, Kenya, and Nigeria leading in development.

From a GHG emissions perspective, the methanisation of agricultural waste contributes to a reduction of CO_2, methane, and nitrous oxide emissions that would otherwise result from the natural degradation of residual wet biomass. Therefore, despite the infrastructure and logistical demands, the overall balance of methanisation activities is generally very positive thanks to the significantly avoided emissions (Couturier et al. 2019). However, some conflicts and controversies persist around the use of methanisation. Experts have raised concerns about the long-term sustainability of agricultural systems integrating EICs. Additionally, some argue that biogas production can contribute to land use conflicts, with land being continuously cultivated solely for energy production. A territorialized assessment may therefore be useful to evaluate the local relevance and sustainability of biogas projects.

19.1.3 Biofuels Production

There are currently different modes of biofuel production, classified by generation, from the first to the third generation. Current biofuels are predominantly produced through the transformation of agricultural products such as sugar beet, wheat, corn, rapeseed, sunflower, these are knowned as first-generation biofuels.

The main advantage of first-generation biofuels lies in their technological maturity: production methods and costs are well established on a large scale in various regions of the world, including the European Union, Brazil and the United States. Bioethanol is primarily produced from corn (two-thirds) and sugarcane (over one quarter). It is blended with gasoline and used in the internal combustion engines of motor vehicles. Global production is close to 100 million cubic meters of ethanol in 2022, representing approximately 5–10% of total gasoline consumption worldwide. The United States is the leading producer, followed by Brazil: together, they account for 82% of the global market. However, the Asia-Pacific has shown the highest growth in ethanol production with an annual increase of over 10% in 2022. India, China and Thailand are major producers in this region.

Biodiesel, on the other hand, is derived from various sources of fatty acids, particularly soybean, rapeseed, palm and other vegetable oils. Three-quarters of biodiesel production comes from the processing of vegetable oils, especially rapeseed oil.

However, the agricultural origins of first-generation biofuels raise significant concerns and controversies, due to their competition with food crops. This mode of production can directly compete with food production, as they are produced from cereal or oilseed grains, or from the sucrose of sugar beets or sugar canes. Their large-scale cultivation may contribute to rising agricultural commodities prices and to an increased use of synthetic pesticides to maintain crop yields. For these reasons, the EU has limited the share of biofuels to an average incorporation rate of 7% at most. According to the World Bank, between 2002 and 2008, increased demand for biofuels accounted for approximately 75% of the rise in food prices and was partially blamed for the global food crisis affecting vulnerable populations, particularly in countries of warm regions.

Second-generation biofuels are produced from the nonedible parts of plants. These include lignocellulose materials from agricultural residues (e.g. straw) and forestry residues (e.g. wood), purpose-grown energy crops (e.g. fast-growing coppice), and industrial waste. These biofuels avoid direct competition with food production. Furthermore, they rely on more diverse and often lower-cost biomass resources, and they offer a better environmental production balance. However, industrial-scale production technologies for second-generation biofuels are still in the early stages of deployment and remain costly. Nevertheless, their development is supported by the EU, whose renewable energy Directive (directive 2018/2001 known as EnR2, published on 21 December 2018) sets a target of 14% in transport fuels by 2030, half of which should come from second-generation sources.

Third-generation biofuels refer to fuels produced from algae cultivated in off-ground basins. However, this technology is still at the research and development stage and lacks commercial maturity. In 2020, biofuels production remains largely dominated by first-generation sources.

Finally, it is important to note that some experts have raised concerns about land-use conflicts related to both the biogas and biofuels sectors specifically when dedicated energy crops are involved. When considering the replacement of the pre-existing vegetation on the lands used to grow energy crops, the greenhouse gas emission balances often appear negative in the short to medium term. The benefits generally emerge only in the long term (Searchinger et al. 2018).

19.2 Energy Production and Land-Use Conflicts Associated to Agrivoltaics

The second part of this chapter discusses the use of agricultural land for the installation of photovoltaic panels. Today, the decarbonisation of energy in the agricultural sector is reflected in the figures. In 2019, only 3% of global electricity demand was met by photovoltaic energy, but by 2023 this figure had risen to 6.2%. This trend is expected to continue, highlighting the global surge in photovoltaic installations, particularly in China. A difficult-to-quantify portion of these panels is installed on agricultural land, but the development of these renewable energies (RE) since the 2010s has highlighted a growing issue: competition over land use, suitable for both agricultural production and photovoltaic electricity generation. To reconcile these two objectives, the concept of 'agrivoltaics' has gradually emerged, particularly following the seminal scientific publication in 2011 (Dupraz et al. 2011). One of the stated objectives of agrivoltaic systems is to preserve agricultural land. Agricultural production is generally considered the primary objective of these systems, even if energy production proves more profitable. Solar panels, sometimes adjustable or installed at varying heights, can be integrated into a variety of agricultural practices, including horticulture, market gardening, viticulture, pasture management, and large-scale crop farming. Constraints on agricultural production vary from one country to another depending on climate, current legislation, crop type and the specific goals of the agrivoltaic system (e.g. maximizing crop yield, improving product quality, optimizing electricity generation). By extension, greenhouses equipped with solar panels may also be considered agrivoltaic systems, especially if they enhance crop production through improved climate control and water management.

Agrivoltaics is therefore presented both as a technique to prevent soil sealing due to photovoltaics and as a solution for adapting to climate change. For example, the panels can provide protection during periods of extreme heat and/or hail (Dinesh and Pearce 2016). In terms of productivity and energy efficiency, proponents of agrivoltaics emphasise that one hectare of agrivoltaic land can enable an electric car to travel a distance roughly hundred times greater than a biofuel crop (e.g. wheat or rapeseed converted into biofuel) can achieve for a combustion engine vehicle (Dupraz et al. 2011).

There are also economic benefits (Al Mamun et al. 2022), as often landowners or farmers can receive an annual lease payment for hosting panels, albeit at the risk that they may no longer be incentivised to continue agricultural activity. The development of agrivoltaics therefore also gives rise to numerous controversies or reinforces those that already existed on photovoltaics (Hrabanski et al. 2024).

This new international market is expending rapidly in Asia, and its growth is also significant in Europe and North and South America. The African continent is for the time being still little affected by this enthusiasm (Fig. 19.1).

The development potential of the sector is considered immense by some experts, who estimate that just 1% of global agricultural surface would be enough to meet the world's entire electricity production (Barron-Gafford et al. 2019). However, this potential raises questions about the future use of agricultural land. Given the particularly high economic returns from photovoltaic installations, many stakeholders in the agricultural sector—alongside citizen collectives and some environmental

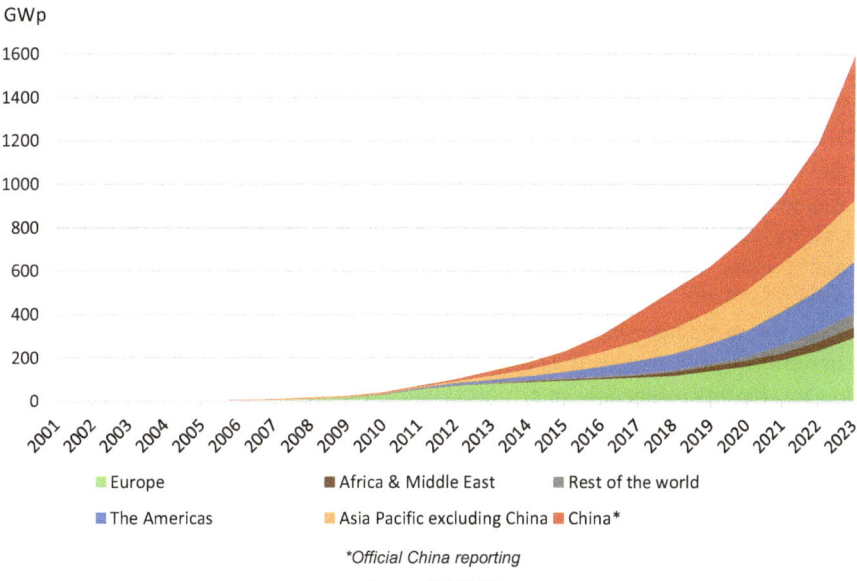

Fig. 19.1 Regional growth of photovoltaic installation capacities between 2001 and 2023 (in gigawatts) according to IEA (Source: IEA Photovoltaic Power Systems Programme 2023—Snapshot of Global PV Markets. https://iea-pvps.org/wp-content/uploads/2023/04/IEA_PVPS_Snapshot_2023.pdf)

organizations—fear that the balance may increasingly shift away from food production toward energy generation. In China and India, some authors estimate that this form of "Solar extractivism" (Hu 2023) or solar capitalism primarily benefits large industrial energy corporations, which are acquiring agricultural land on a large scale to the detriment of local socio-ecological systems. Climate change and the greening of political agendas could legitimise a particularly alarming land-grabbing process threatening global food security and the rights of local farmers (Singh 2022; Hu 2023).

19.3 Conclusion: Growing Competition Between Food and Energy Production

Climate change has brought to the forefront the need for agricultural energy consumption, particularly in the Global North. In the Global South, however, access to energy and electricity remains a major challenge for improving living conditions.

According to the FAO, every energy-consuming activity in agriculture holds potential for savings, provided stakeholders adopt a rigorous approach to identifying and implementing effective reduction strategies. The aim is to reduce energy consumption without compromising agricultural productivity. Based on an energy efficiency criterion, measured in kilowatt-hours consumed per unit of food delivered "to the plate", experts consistently highlight that reducing post-harvest losses in agri-food systems is one of the most effective strategies (see Chap. 23), while also improving global food availability.

The use of renewable energy sources such as the valorisation of residual biomass which remains abundant especially in some Southern countries, should help reduce GHG emissions in the sector, improve productivity (through enhanced local energy access) and manage feeding a rapidly growing population, particularly in sub-Saharan Africa.

However, energy production in and by agriculture, whether through lignocellulosic biomass, methanisation, agrivoltaics or wind power, can create competition between food and non-food (energy) uses.

The use of biomass or agricultural land for energy purposes also competes with other critical priorities such as preserving carbon sinks, biodiversity, and sustainable soil fertility, especially in a context where biomass availability is increasingly constrained. While renewable, biomass is not inexhaustible, and its main sources (forests and agriculture) are weakened by global warming.

Agrivoltaic installations, while capitalising on solar energy are drastically changing access to farmland in many regions. The development of renewable energy sources in agriculture also has the potential to significantly redistribute economic value within the sector, sometimes increasing existing socio-economic disparities among farmers.

Finally, these renewable energy sources may also lead to numerous conflicts over land use and impact soil quality. Their development should not, therefore, be solely driven by large industrial groups seeking short-term profitability but rather be guided by principles of social and environmental justice (Demenois et al. 2022; Demenois 2023).

At the geopolitical level, the development of agricultural energy production reinforces the issues surrounding land control. The emergence of China, as the main producer of renewable energy, associated with its growing demand for agricultural land, is likely to shift global balances of influence.

At the territorial scale, agricultural-energy projects must be integrated into broader territorial planning. These projects mobilise local resources—agricultural land, biomass, co-products, waste—and can become levers for local economic and social development.

It is therefore essential that in both the global North and South, these issues be carefully framed and regulated to prevent the destabilization of three key pillars: (1) food security, (2) farmers' rights, and (3) integrity of local ecosystems.

References

Al Mamun MA, Dargusch P, Wadley D, Zulkarnain NA, Aziz AA (2022) A review of research on agrivoltaic systems. Renew Sust Energ Rev 161:112351

Barron-Gafford GA, Pavao-Zuckerman MA, Minor RL, Sutter LF, Barnett-Moreno I, Blackett DT et al (2019) Agrivoltaics provide mutual benefits across the food–energy–water nexus in drylands. Nat Sustain 2:848–855. https://doi.org/10.1038/s41893-019-0364-5

BOAD (2023) Workshop to share the conclusions/recommendations of the study on the sustainable management of industrial household waste in UEMOA member states for energy production. Lomé, 30 June 2023. Study summary, 6p

Couturier C, Jack A, Laboubee C, Meiffren I (2019) The rural methanisation, a tool for energy and agroecological transitions. Positioning Note, Solagro

Demenois J (2023) Soils under the effects of climate change. Geopolitical perspectives, mitigation and adaptation options. Diplomacy. The Major Issues 76:20–23. https://www.areion24.news/categorie-produit/numero-grands-dossiers-de-diplomatie/

Demenois J, Dayet A, Karsenty A (2022) Surviving the jungle of soil organic carbon certification standards: an analytic and critical review. Mitig Adapt Strateg Glob Chang 27(1) 17 p. https://doi.org/10.1007/s11027-021-09980-3

Dinesh H, Pearce JM (2016) The potential of agrivoltaic systems. Renew Sust Energ Rev 54:299–308

Dupraz C, Marrou H, Talbot G, Dufour L, Nogier A, Ferard Y (2011) Combining solar photovoltaic panels and food crops for optimising land use: towards new agrivoltaic schemes. Renew Energy 36(10):2725–2732

Hrabanski M, Verdeil S, Ducastel A (2024) Agrivoltaics in France: the multi-level and uncertain regulation of an energy decarbonisation policy. In: Review of agricultural, food and environmental studies, vol 105, pp 1–27

Hu Z (2023) Towards solar extractivism? A political ecology understanding of the solar energy and agriculture boom in rural China. Energy Res Soc Sci 98:102988

Searchinger TD, Wirsenius S, Beringer T, Dumas P (2018) Assessing the efficiency of changes in land use for mitigating climate change. Nature 564:249–253. https://doi.org/10.1038/s41586-018-0757-z

Singh D (2022) 'This is all waste': emptying, cleaning and clearing land for renewable energy dispossession in borderland India. Contemp South Asia 30(3):402–419

Open Access This chapter is licensed under the terms of the Creative Commons Attribution-NonCommercial-NoDerivatives 4.0 International License (http://creativecommons.org/licenses/by-nc-nd/4.0/), which permits any noncommercial use, sharing, distribution and reproduction in any medium or format, as long as you give appropriate credit to the original author(s) and the source, provide a link to the Creative Commons license and indicate if you modified the licensed material. You do not have permission under this license to share adapted material derived from this chapter or parts of it.

The images or other third party material in this chapter are included in the chapter's Creative Commons license, unless indicated otherwise in a credit line to the material. If material is not included in the chapter's Creative Commons license and your intended use is not permitted by statutory regulation or exceeds the permitted use, you will need to obtain permission directly from the copyright holder.

Chapter 20
Adapting to Climate Change: What Innovative Practices in Tropical Production Systems?

Éric Justes, Benoit Bertrand, Hervé Etienne, Frédéric Gay, Bruno Rapidel, Philippe Thaler, and François-Xavier Côte

Abstract Agricultural and forestry systems must now balance climate change adaptation, greenhouse gas reduction, and food security. Adaptation strategies differ for annual crops, which allow incremental changes, and perennial crops, which face long-term climate stresses. Tropical crops, often grown by vulnerable family farmers, are already impacted by climate shifts, requiring changes throughout the value chain. This chapter proposes a classification of adaptation types and illustrates them with case studies on coffee and rubber. It emphasizes the importance of agroecology and climate-smart agriculture for sustainable, long-term adaptation.

From today, agricultural and forestry production systems must be designed with the aim of combining: adaptation to climate change and variability, reduction of greenhouse gas (GHG) emissions, and securing food production. Climate change requires producers to implement adaptation strategies that go beyond current practices adapted to seasonal and interannual climate variability. Some crops may even need to migrate to more suitable cultivation areas, which will likely profoundly affect

---

É. Justes (✉)
CIRAD, PERSYST Department, Montpellier, France
e-mail: eric.justes@cirad.fr

B. Bertrand · H. Etienne
CIRAD, UMR DIADE, Montpellier, France

DIADE, Université de Montpellier, IRD, CIRAD, Montpellier, France

F. Gay · B. Rapidel
CIRAD, UMR ABSys, Montpellier, France

ABSys, Université de Montpellier, CIRAD, INRAE, Institut Agro, Montpellier, France

P. Thaler
CIRAD, UMR Eco&Sols, Montpellier, France

Eco&Sols, Univ Montpellier, CIRAD, INRAE, IRD, Institut Agro, Montpellier, France

F.-X. Côte
CIRAD, DGD-RS, Montpellier, France

© The Author(s) 2026
V. Blanfort et al. (eds.), *Climate Impacts and Challenges in Agriculture, Forests and Food Systems*, https://doi.org/10.1007/978-3-032-04331-3_20

production structures and rural societies, sectors and markets. It is generally accepted that most adaptation and mitigation options will need to be compatible with each other. Therefore, trade-offs will need to be made, as it is crucial that compromises allow for the maintenance of agricultural production in a context of changing human demographics and food security requirements.

The issue of crop adaptation is a generic problem for various agricultural and forestry productions in the tropical zone, but it does not imply the same levels of risk in decisions whether one cultivates annual or perennial plants. Indeed, for annual plants, sown each year, it will probably be easier to adapt step by step (incremental adaptation) without questioning the resilience of the farm, whereas for perennial plants the question of species and variety choice is a particularly crucial point, as they are planted for long periods (including, for some, over several decades) and they will suffer the stresses that accumulate year after year.

Tropical crops have already been experiencing the impacts of climate change for several years due to rising temperatures, changes in rainfall patterns, increasingly frequent extreme weather events, and the emergence of new diseases. They are diverse in their nature and type of organisation, but they often rely on family peasant productions (see Chap. 5), which are particularly economically vulnerable and depend on the robustness and sustainability of the sectors that provide them with a vital income. If the producer is the first link in the value chain that must adapt its practices and production system to climate change, other actors in the sectors will also have to change their practices and organisations to cope with supply irregularities related to climate change, the additional costs incurred, the displacement of certain production areas and societal questions about the environmental footprint of productions (particularly productions imported from one continent or region to another).

In this chapter, we propose a reflection on a typology of adaptations of tropical agricultures to climate change. The first part describes the possible classification elements of adaptations. The second part illustrates through two case studies on perennial plants (coffee and rubber) the ongoing adaptations in these two sectors as well as the knowledge challenges to be met to accelerate these adaptation processes. The classification we describe is based on a timeline of actions to be implemented to cope with climate changes (annual adaptation, multi-year and long-term adaptation that will require more or less redesign of production systems). The solutions considered mainly refer to the concepts of agroecology and climate-smart agriculture described in Chap. 3 of this book.

20.1 Adapting Tropical Crops to Climate Change

There is an extensive literature on the concepts and potential solutions regarding the issue of adapting human societies to climate change in all its dimensions (Howden et al. 2007). Moser and Ekstrom (2010) suggested focusing primarily on the process of intentional and planned adaptation. However, this process is difficult to envisage in the long term given the uncertainties about the intensity of climate changes and their social, economic, and political consequences. Complete redesigns of production systems and supply chain organisation are foreseeable.

Debaeke et al. (2017) propose to consider measures and solutions for adapting agriculture to climate change and for adopting mitigation measures, explicitly taking into account the nature of the changes, whether they are: (1) gradual changes and autonomous responses from farmers (for example, shifting the planting date, changing crop varieties), and (2) planned and transformational changes, which require substantial investments (for example, the development of new crop varieties, the expansion of irrigation infrastructure, the introduction of agroforestry, etc.). Autonomous adjustments include farmers' efforts to optimise production without major system changes and without the involvement of other sectors (such as public policy, research). Transformational adaptations are those adopted on a larger scale (territory and region) and which alter resource use, transform or relocate production sites (Kates et al. 2012). These are major structural changes aimed at overcoming adversity caused by climate change.

A complementary categorisation to the previous one for analysing necessary adaptations is to consider the timeframes required to implement the changes. Short-term adaptations are generally implemented independently by a value chain actor, and medium and long-term adaptations are more organised and planned by various industry actors and policy makers at the territorial or global level.

20.1.1 The Different Types of Adaptations of Crop Management Systems and Production Systems

20.1.1.1 Short-Term Adaptation

Regarding short-term adaptation strategies of current practices, producers can use various technical options they are familiar with (*coping measures*). These technical solutions can be combined to reduce vulnerability and adapt cropping systems to climate change. Debaeke et al. (2017) summarise the strategies that can be implemented to cope with crop water and heat stress by their objective: evasion, avoidance (through vegetative rationing), tolerance, mitigation, resource conservation, resilience (recovery). Several types of adaptation levers can be used with a variety of technical options, such as the choice of crops (species) and varieties (cultivars, populations), the method of crop management (crop management system), the modification of the cropping system itself.

The chosen adaptations can rely on several options: (1) stress escape by shifting the sowing date to avoid water and heat stress, (2) avoidance through applications of nutrients (mineral or organic fertilisers), planting densities and spatial arrangements (for example, skipping rows) adapted to rainfall patterns and yield objectives, (3) mitigation of the effect through the use of irrigation, and (4) resource conservation through soil work and residue management to maximise soil water storage and to reduce evaporation, runoff and erosion.

Adapting the cropping system is sometimes necessary in the face of strong impacts of climate change. In this case, the objective will be to diversify crops and increase cultivated agrobiodiversity to increase resilience (rotation, spatial organisation of crops) (see Chap. 21). Various strategies are then possible, such as variety mixtures and intercropping (species mixtures), agroforestry (crop-tree or tree-tree, with or without cover crop management), flexible cropping systems with crop adaptation according to climatic conditions, or even with the implementation of double and relay intercropping. Furthermore, the redevelopment of coupled farming systems with also agro-silvopastoral systems could constitute a form of adaptation, allowing more resilience in the face of the impacts of climate change. These options based on new systems, or old systems revisited and optimised, lead to increased management complexity and access to new knowledge, and often require capacity building for producers, advisory services and research.

20.1.1.2 Long-Term Adaptation

Given the gradual change in climate in the past, producers have autonomously adapted through successive technical adjustments to climate changes and variability. With the acceleration of climate change and its major effects, it Indeed, farmers will play a significant role in the selection and implementation of adaptations, but changes concerning other long-term organisational levels, such as land use and the resulting crop productivity, will need to be considered and planned (Bindi and Olesen 2011; Kates et al. 2012; Anwar et al. 2013).

The use of new varieties, especially in perennial species, can be categorised as medium and long-term changes. The development of new varieties (see Chap. 21) indeed takes a long time (even though new genetic improvement techniques could shorten improvement cycles). These long periods will also be linked to the adaptation of the value chain to these new varieties, as well as to the alignment of the supply with the demands of consumers or processors. It is likely that agronomic solutions will also partly rely on new varieties. Current cultivars have indeed been selected for a long time with objectives related to productivity or product standardisation, different from those of adaptation to climate change (Boote et al. 2011; Ziska et al. 2012).

A transformative level of adaptation of systems may be necessary in case of a mismatch between production and the resources available determined by the "new climate". This situation will require in some cases a paradigm shift in agricultural production, for example the transformation of an irrigated system into a rainfed system, or the substitution of full sun plantations by crops and trees under natural or artificial shades.

This paradigm shift is heuristic, subjective and relative, as highlighted by Rickards and Howden (2012), who also indicate that changes for transformative adaptation vary in intensity, generality, affected spatial scale, induced or not effects on the system as a whole, level of permanence of the change or possible reversion.

This profound transformation will induce numerous cascading impacts and it will therefore be necessary to develop *ex ante* prediction tools (modelling) for decision-making and, *in itinere*, to set up monitoring-evaluation indicators of changes for the steering of transitions and the monitoring of the sustainability of new systems. A central question, beyond the technical adaptations discussed in this chapter, is of course the financing of these transitions, particularly for Southern agriculture, which is the most socially and economically vulnerable.

20.1.1.3 Typology for Characterising the Continuum of Adaptation Solutions and the Challenges to Be Met

In order to list adaptation solutions, it is possible to summarise the nature and forms of adaptation in agriculture according to an adaptation gradient integrating various concepts pragmatically synthesising the diversity of research work and action proposals in the extensive bibliography on the subject. We have adapted various representations proposed by different authors (Debaeke et al. 2017; Kates et al. 2012) into a diagram organised into a continuum of three levels of adaptation (Fig. 20.1):

- incremental "to cope", with a rather annual dimension (adaptation of sowing dates, use of new varieties, new management practices, adapted crop management systems, etc.);
- systemic "to adjust", with a multi-year dimension (agroecology solutions, agroforestry in substitution of cropping systems, etc.);

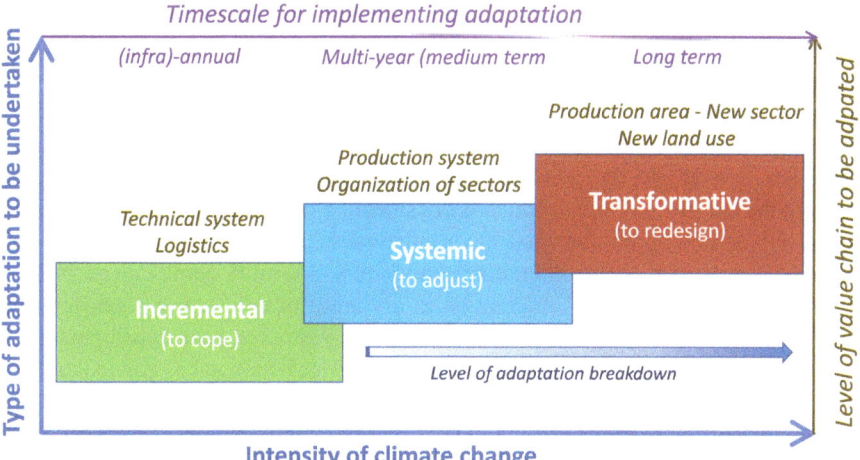

Fig. 20.1 Illustration of the continuum of forms of adaptation according to the intensity of climate change. Link with the temporal scale of the change to be implemented and with the level of the value chain that will have to transform Diagram representing the positioning of the 3 types of adaptation to climate change (incremental (cope), systemic (adjust) and transformative (redesign); cf. text) [y-axis] according to a double gradient of intensity of climate change [x-axis—bottom] and the temporal scale of its implementation [x-axis—top]

Table 20.1 Forms of adaptation and nature of the processes involved (the crosses in the table underline the intensity of these processes for adaptation)

Forms and nature of climate change adaptation options		Forms of adaptation		
		"Incremental" adaptation	"Systemic" adaptation	"Transformative" adaptation
Involved processes	Technical nature adaptation	+++	++	++
	Technical and organisational adaptation of the existing value chain		+++	+++
	Change of value chain and relocation			+++
Temporal scale of change		Annual and sub-annual	Multiannual	Decade and beyond

- transformative "to redesign" (new productions and sectors, other land uses, migration of cultivation zones).

The processes involved in this adaptation gradient are of different nature (Table 20.1), both at the technical and organisational level at the scale of the farm and the value chain. The temporal scale of the change also varies according to the form of adaptation.

20.1.1.4 Trade-Off Between Adaptation and Mitigation

Most climate change adaptation options should also have positive impacts on mitigation by improving nitrogen use efficiency (reducing greenhouse gas emissions) and carbon storage in the soil. However, some mitigation measures may require adaptations to truly contribute to the adaptive capacity of agricultural systems. Consider, for example, the contribution of legumes to reducing GHG emissions and limiting the use of fossil fuels, which is considered a relevant solution (Jensen et al. 2012). Numerous agronomic and environmental benefits of legumes have been reported (Voisin et al. 2014), their contribution to soil fertility being one of the key factors in maintaining cereal production in dry areas of developing countries and more generally in low-input areas. However, the restricted root development of most grain legume species can limit water and nutrient uptake, particularly at advanced growth stages when environmental stresses are frequent. Symbiotic N_2 fixation, which occurs in the upper soil layers, is extremely sensitive to drought (Liu et al. 2011). Therefore, the yield of food legumes grown in arid to semi-arid environments is generally variable or low due to the intense heat and end-of-cycle droughts that characterise these areas. Even in wetter environments, a water deficit can still occur over a period of a few weeks, resulting in significant yield loss. Therefore, legumes are likely to be little grown in some regions under climate change, and breeding efforts are needed to improve the adaptation potential of legume crops to thermal and water stresses concurrently with earlier sowing dates, the development of winter types, and intercropping with cereals (Cutforth et al. 2007; Vadez et al. 2012).

20.1.2 The Components of Adaptation at the Sector Level

There is little data in the scientific literature proposing a typology of climate change adaptation at the scale of the entire value chain (sector). Beyond technical adaptations, two other key adaptation processes could be considered to classify possible adaptations at the sector level. These processes concern, on the one hand, organisational adaptations and those related to the financing capacity of investments and the coverage of financial risks taken by farmers for the implementation of the transition (the "solvency" process). Developing an adaptation strategy must combine, in a concerted approach, these three components (technical, organisational, economic). The organisational innovation process aims to build multi-actor approaches, capable of bringing together a set of value chain actors, who in isolation could only partially implement the solutions necessary for adaptation to climate change. The "solvency" process, in addition to taking into account the costs and financial risks taken by farmers and other actors who engage in the transition, also concerns the recognition by markets (value creation and creation of new markets) of actors who adapt to climate change and contribute to its mitigation.

In the following part of this chapter, two examples of climate change adaptation classification have been chosen to illustrate the typologies presented here. These examples concern two perennial crops, coffee and rubber, emblematic of tropical regions and particularly affected by climate change.

20.2 Examples of Adaptation to Climate Change in Tropical Perennial Crops

20.2.1 Coffee Cultivation and Adaptation to Climate Change

Adaptation solutions for systems at different organisational levels (plant, plot, farm, landscape, sector) in coffee cultivation are presented in Table 20.2.

20.2.2 Rubber Cultivation and Adaptation to Climate Change

Adaptation solutions for systems at different organisational levels (plant, plot, farm, landscape, supply chain) in rubber are presented in Table 20.3.

Table 20.2 Adaptation solutions for coffee systems at different organisational levels

	Incremental adaptation	Systemic adaptation	Transformative adaptation
Variety-level adaptation (Côte et al. 2019)	Use of arabica and robusta varieties tolerant to drought (Sarzynski et al. 2024) New varieties adapted to emerging diseases (coffee rust, anthracnose of branches, borer beetle) New varieties (1) adapted to drought (deeper and more branched root system; bolero project), (2) with better use of mineral resources (PEPR SVA project, *flagship 4*), and (3) adapted to multispecies systems (PRCC project in Laos) Vegetative and/or sexual propagation techniques for the deployment of more resilient elite varieties (Etienne et al. 2018, 2024; Georget et al. 2017, 2019)	Varieties adapted to agroforestry, high altitudes, full sun with irrigation, high latitudes (Sarzynski et al. 2023), higher temperatures, shading (Bertrand et al. 2021, 2025) Selection of tree species adapted to agroforestry (Meylan et al. 2017; Rigal et al. 2023; Breitler et al. 2022) (European projects Breedcafs and Bolero) Varieties created using genome editing Search for genes adapted to high temperatures and drought (Alves et al. 2017; Ferreira Torres et al. 2019) Development of the CRISPR-Cas9 genome editing technique (Breitler et al. 2018; Casarin et al. 2022)	Migration of production areas with new varieties, replacement of the arabica variety with robusta in "low" arabica areas Replacement of the robusta variety with hybrid "robusta x racemosa" varieties resistant to drought Modification of the value chain (sector) related to the displacement of production areas and varietal changes (other markets)
Adaptation at the level of crop management systems, cropping systems, and organisation at the landscape scale	Synchronisation and securing of flowering by irrigation (Carr 2001) Prophylaxis for adaptation to emerging diseases Combating erosion with cover plants and live hedges (Labrière et al. 2015) Optimised irrigation (Rigal et al. 2023) Participatory design assisted by model for split fertilisation	Traditional agroecology techniques for overall soil and plant health: Use of composts, biochar, nitrogen-fixing bacteria Conservation of mosaics and landscape continuities (diversified landscapes) to promote biological regulation of pests	Digital agroecology farming with intensive use of resources, sensors and information to optimise the use of all resources without inducing impacts on human health and the environment (irrigation with effluent recovery, precisely optimised mineral fertilisation and robotisation)
And contribution to mitigation	Reduction and splitting of nitrogen fertilisation (Capa et al. 2015) Wind protection	Use of trees (agroforestry) to increase carbon sequestration, maintain moisture and reduce temperature (Koutouleas et al. 2022; Notaro et al. 2022)	Cultivation under physical shelter (nets) at very high intensity and high density (reduction of tree size by dwarfing grafting and/or selection of dwarfing genes) Effluent recycling

Adaptation at the supply chain level	Support for producers to adopt adaptation techniques (technical advice, access to suitable varieties, etc.)	Reorganisation of stakeholders to support producers and promote the adoption of crop management systems and the multiplication of new varieties to provide access to producers Organisational adaptation of coffee collectors Development of specifications and certifications generating added value (Hrabanski et al. 2013) Adaptation to new markets linked to price fluctuations, themselves linked to changes in cultivation areas	Migration of production areas to higher regions or latitudes (with cultivation under shelter) Territorial coordination between producers and stakeholders to create landscapes adapted to climate change Use of new varieties generating new markets "Agroforestry clusters" approach associating family producers, credit financiers, processors, marketers and certifiers (Meter et al. 2023)

Table 20.3 Adaptation solutions for rubber systems at different organisational levels

	Incremental adaptation	Systemic adaptation	Transformative adaptation
Adaptation at the varietal level	New clones adapted to pedoclimatic conditions and parasitic pressure (selection of clones based on the prevalence of SALB disease in South America; Rivano et al. 2015)	Clonal rootstocks (Carron et al. 2008; Carron et al. 2009)	Development of other latex-producing species adapted to hot and dry conditions (guayule)
Adaptation at the level of crop management systems, crop systems, and landscape-scale organisation	Adaptation of tapping systems to climatic constraints: (1) excess rain (Zaw et al. 2017); (2) night-time tapping for temperature increase (Seneviratne et al. 2021); (3) cessation of tapping in dry season (Chantuma et al. 2017) Ethylene stimulation practices adapted to tapping frequencies (Chambon et al. 2014) Irrigation if possible Planting techniques adapted to dry conditions Soil cover management practices to optimise water management (Clermont-Dauphin et al. 2018)	Multispecies systems more resilient to climate change Associating rubber with species that modify the microclimate for a cooling effect at the canopy level (association with other tall woody species) or under the canopy (association with low woody or herbaceous species) Adaptation of plantation management practices (management of organic matter) Replanting practices, renewal of rubber trees preserve soil health (Perron et al. 2021)	Diversified agricultural landscapes more resilient to climate change and promoting biological regulations

20.3 Conclusion

In the face of climate change, adaptation solutions are already underway, but even faster and more intense transformations of production systems must now be implemented to cope with the intensity of the negative impacts of climate change on these production systems. These modifications will need to be systemic and transformative, and the result of compromises that strengthen the multifunctional role of agriculture and take into account the vulnerability of the most fragile agricultures in the South. Moreover, it is now well documented and widely shared that the implementation of climate-smart and resilient cropping and production systems requires the implementation of integrated strategies, including relevant sets of management practices rather than the implementation of specific practices, one by one. Under certain conditions, a paradigm shift will be necessary to redesign sustainable systems (transformative adaptation). The adaptations to be implemented are part of a transition process involving technical, organisational, partnership and financial changes. In this context, two major challenges appear: (1) the challenge of coordinating the necessary adaptations between the territorial, national, global, and sectoral levels; (2) the challenge of mobilising the necessary resources for transitions (monetary, material and training for capacity building of actors) at the level of public policies and private actors, and the equitable distribution of the costs associated with mobilising these resources among the different actors.

References

Alves GSC, Ferreira TL, Déchamp E, Breitler J-C, Joët T, Gatineau F et al (2017) Differential fine-tuning of gene expression regulation in coffee leaves by CcDREB1D promoter haplotypes under water deficit. J Exp Bot 68(11):3017–3031. https://doi.org/10.1093/jxb/erx166

Anwar MR, Liu DE, Macadam I, Kelly G (2013) Adapting agriculture to climate change: a review. Theor Appl Climatol 113:225–245

Bertrand B, Villegas Hincapie AM, Marie L, Breitler JC (2021) Breeding for the main agricultural farming of Arabica coffee. Front Sustain Food Syst 5(709):901

Bertrand B, Mieulet D, Breitler JC, Leroy T, Montagnon C (2025) Breeding of new coffee varieties as a key strategy to improve coffee sustainability in response to climate change. Adv Bot Res 114:247–281. https://doi.org/10.1016/bs.abr.2024.06.001

Bindi M, Olesen JE (2011) The responses of agriculture in Europe to climate change. Reg Environ Change 11(Suppl. 1):S151–S158

Boote KJ, Ibrahim AMH, Lafitte R, McCulley R, Messina C, Murray SC et al (2011) Position statement on crop adaptation to climate change. Crop Sci 51:2337–2343

Breitler JC, Dechamp E, Campa C, Zebral RL, Guyot R, Marraccini P, Etienne H (2018) CRISPR/Cas9-mediated efficient targeted mutagenesis has the potential to accelerate the domestication of Coffea canephora. Plant Cell Tiss Org Cult (PCTOC). https://doi.org/10.1007/s11240-018-1429-2

Breitler JC, Etienne H, Léran S, Marie L, Bertrand B (2022) Description of an Arabica coffee ideotype for agroforestry cropping systems: a guideline for breeding more resilient new varieties. Plants 11(16):2133

Capa D, Pérez-Esteban J, Masaguer A (2015) Unsustainability of recommended fertilisation rates for coffee monoculture due to high N2O emissions. Agron Sustain Dev:1–9

Carr MKV (2001) The water relations and irrigation requirements of coffee. Exp Agr 37:1–36

Carron MP, Lardet L, Leconte A, Dea BG, Keli J, Granet F, et al (2009) Field trials network emphasizes the improvement of growth and yield through micropropagation in rubber tree (*Hevea Brasiliensis*, Muell. Arg.). 3rd international symposium on acclimatization and establishment of micropropagated plants, Faro, Portugal, pp 485–492

Carron MP, Nurhaimi-Haris, Sumaryono, Sumarmadji S, Granet F, Kéli J, Montoro P (2008) The rootstock clones in rubber tree: a new varietal type toward the rejuvenated bi-clone. In: Supriadi M, Suryaningtyas H, Siswanto, Haris N, Sumaryono (eds) Proceedings international workshop on rubber planting materials: Bogor, Indonesia, 28–29 October 2008, pp 89–95

Casarin T, Freitas NC, Pinto RT et al (2022) Multiplex CRISPR/Cas9-mediated knockout of the phytoene desaturase gene in Coffea canephora. Sci Rep 12:17270. https://doi.org/10.1038/s41598-022-21566-w

Chambon B, Angthong S, Kongmanee C, Somboonsuke B, Mazon S, Puengcharoen A et al (2014) A comparative analysis of smallholders' tapping practices in four rubber producing regions of Thailand. Adv Mater Res 844:34–37. https://doi.org/10.4028/www.scientific.net/AMR.844.34

Chantuma P, Lacote R, Sonnarth S, Gohet E (2017) Effects of different tapping rest periods during wintering and summer months on dry rubber yield of Hevea Brasiliensis in Thailand. J Rubber Res 20:261–272

Clermont-Dauphin C, Dissataporn C, Suvannang N, Pongwichian P, Maeght JL, Hammecker C, Jourdan C (2018) Intercrops improve the drought resistance of young rubber trees. Agron Sustain Dev 38

Côte F-X, Poirier-Magona E, Perret S, Rapidel B, Roudier P, Thirion M-C (eds) (2019) The agroecological transition of agricultural systems in the global south, agricultures and challenges of the world collection, AFD. Quæ editions, Cirad, p 360p

Cutforth HW, McGinn SM, McPhee KE, Miller PR (2007) Adaptation of pulse crops to the changing climate of the Northern Great Plains. Agron J 99:1684–1699

Debaeke P, Pellerin S, Scopel E (2017) Climate-smart cropping systems for temperate and tropical agriculture: mitigation, adaptation and trade-offs. Cah Agric 26:34002

Etienne H, Breitler J-C, Brossier J-R, Awada R, Laflaquière L, Amara I, Georget F (2024) Coffee somatic embryogenesis: advances, limitations, and outlook for clonal mass propagation and genetic transformation. Adv Bot Res 114:349–388. https://doi.org/10.1016/bs.abr.2024.04.008

Etienne H, Breton D, Breitler J-C, Bertrand B, Déchamp E, Awada R et al (2018) Coffee somatic embryogenesis: how has research, the experience gained and innovations benefited commercial propagation of elite clones in the two cultivated species? Front Plant Sci 9:1630. https://doi.org/10.3389/fpls.2018.01630

Ferreira TL, Reichel T, Déchamp E, de Aquino S, Duarte KE, Alves G et al (2019) Expression of *DREB* -like genes in *Coffea canephora* and *C. arabica* subjected to various types of abiotic stress. Trop Plant Biol 12:98–116. https://doi.org/10.1007/s12042-019-09223-5

Georget F, Courtel P, Malo GE, Hidalgo JM, Alpizar E, Breitler JC et al (2017) Somatic embryogenesis-derived coffee plantlets can be efficiently propagated by horticultural rooted mini-cuttings: a boost for somatic embryogenesis. Sci Hortic 216:177–185. https://doi.org/10.1016/j.scienta.2016.12.017

Georget F, Marie L, Alpizar E, Courtel P, Bordeaux M, Hidalgo JM et al (2019) Starmaya: the first Arabica F1 coffee hybrid produced using genetic male sterility. Front Plant Sci 10:1344. https://doi.org/10.3389/fpls.2019.01344

Howden SM, Soussana JF, Tubiello FN, Chhetri N, Dunlop M, Meinke H (2007) Adapting agriculture to climate change. PNAS 104(50):19691–19696

Hrabanski M, Bidaud C, Le Coq JF, Meral P (2013) Environmental NGOs, policy entrepreneurs of market-based instruments for ecosystem services? A comparison of Costa Rica, Madagascar and France. For Policy Econ 37:124–132

Jensen ES, Peoples MB, Boddey RM, Gresshoff PM, Hauggaard-Nielsen H, Alves BJR et al (2012) Legumes for mitigation of climate change and the provision of feedstock for biofuels and biorefineries. A review. Agron Sustain Dev 32:329–364

Kates RW, Travis WR, Wilbanks TJ (2012) Transformational adaptation when incremental adaptations to climate change are insufficient. PNAS 109(19):7156–7161

Koutouleas A, Sarzynski T, Bertrand B, Bordeaux M, Bosselmann AS, Campa C et al (2022) Shade effects on yield across different Coffea arabica cultivars—how much is too much? A meta-analysis. Agron Sustain Dev 42:55. https://doi.org/10.1007/s13593-022-00788-2

Labrière N, Locatelli B, Laumonier Y, Freycon V, Bernoux M (2015) Soil erosion in the humid tropics: a systematic quantitative review. Agr Ecosyst Environ 203:127–139

Liu Y, Wu L, Baddeley JA, Watson CA (2011) Models of biological nitrogen fixation of legumes. A review. Agron Sustain Dev 31:155–172

Meter A, Penot E, Vaast P, Etienne H, Ponçon E, Bertrand B (2023) "Coffee agroforestry business-driven clusters": an innovative social and environmental organisational model for coffee farm renovation. Open Res Eur 2:61. https://doi.org/10.12688/openreseurope.14570.2

Meylan L, Gary C, Allinne C, Ortiz J, Jackson L, Rapidel B (2017) Evaluating the effect of shade trees on provision of ecosystem services in intensively managed coffee plantations. Agric Ecosyst Environ 245:32–42

Moser SC, Ekstrom JA (2010) A framework to diagnose barriers to climate change adaptation. PNAS 107(51):22026–22031

Notaro M, Gary C, Le Coq J-F, Metay A, Rapidel B (2022) How to increase the joint provision of ecosystem services by agricultural systems. Evidence from coffee-based agroforestry systems. Agr Syst 196:103332

Perron T, Kouakou A, Simon C, Mareschal L, Frédéric G, Soumahoro M et al (2021) Logging residues promote rapid restoration of soil health after clear-cutting of rubber plantations at two sites with contrasting soils in Africa. Sci Total Environ 151:526

Rickards L, Howden SM (2012) Transformational adaptation: agriculture and climate change. Crop Pasture Sci 63:240–250. https://doi.org/10.1071/CP11172

Rigal C, Duong T, Vo C, Bon LV, Hoang QT, Chau TML (2023) transitioning from monoculture to mixed cropping systems: the case of coffee, pepper, and fruit trees in Vietnam. Ecol Econ 214(107):980

Rivano F, Maldonado L, Simbaña B, Lucero R, Gohet E, Cevallos V, Yugcha T (2015) Suitable rubber growing in Ecuador: an approach to South American leaf blight. Ind Crops Prod 66.262–270

Sarzynski T, Bertrand B, Rigal C, Marraccini P, Vaast P, Georget F et al (2023) Genetic-environment interactions and climatic variables effect on bean physical characteristics and chemical composition of *Coffea arabica*. J Sci Food Agric 103:4692–4703. https://doi.org/10.1002/jsfa.12544

Sarzynski T, Vaast P, Rigal C, Marraccini P, Delahaie B, Georget F et al (2024) Contrasted agronomical and physiological responses of five *Coffea arabica* genotypes under soil water deficit in field conditions. Front Plant Sci 15. https://doi.org/10.3389/fpls.2024.1443900

Seneviratne P, Nakandala SA, Samarasekera RK, Karunathilake PKW (2021) A study of different tapping times on latex production in smallholder rubber fields in Moneragala District in Sri Lanka. J Rubber Res Inst Sri Lanka 101:65–75

Vadez V, Berger JD, Warkentin T, Asseng S, Ratnakumar P, Rao KPC et al (2012) Adaptation of grain legumes to climate change: a review. Agron Sustain Dev 32:31–44

Voisin AS, Gueguen J, Huyghe C, Jeuffroy MH, Magrini MB, Meynard JM et al (2014) Legumes for feed, food, biomaterials and bioenergy in Europe: a review. Agron Sustain Dev 34:361–380

Zaw ZN, Sdoodee S, Lacote R (2017) Performances of low frequency rubber tapping system with rainguard in high rainfall area in Myanmar. Aust J Crop Sci:1444–1450. https://doi.org/10.21475/ajcs.17.11.11.pne593

Ziska LH, Bunce JA, Shimono H, Gealy DR, Baker JT, Newton PCD et al (2012) Food security and climate change: on the potential to adapt global crop production by active selection to rising atmospheric carbon dioxide. Proc R Soc B Biol Sci 279(1745):4097–4105

Open Access This chapter is licensed under the terms of the Creative Commons Attribution-NonCommercial-NoDerivatives 4.0 International License (http://creativecommons.org/licenses/by-nc-nd/4.0/), which permits any noncommercial use, sharing, distribution and reproduction in any medium or format, as long as you give appropriate credit to the original author(s) and the source, provide a link to the Creative Commons license and indicate if you modified the licensed material. You do not have permission under this license to share adapted material derived from this chapter or parts of it.

The images or other third party material in this chapter are included in the chapter's Creative Commons license, unless indicated otherwise in a credit line to the material. If material is not included in the chapter's Creative Commons license and your intended use is not permitted by statutory regulation or exceeds the permitted use, you will need to obtain permission directly from the copyright holder.

Chapter 21
Adapting and Innovating in Terms of Cultivated Species and Varieties: A Key Role for Cultivated and Natural Diversity? AFS

Sophie Léran, Myriam Adam, Mathieu Gonin, Cécile Grenier, Pierre Marraccini, Fabienne Micheli, Maria Camila Rebolledo, Clément Rigal, Mohamed Lamine Tékété, Michel Vaksmann, Hervé Étienne, and Delphine Luquet

Abstract Valorizing agrobiodiversity offers a sustainable way to adapt cropping systems to climate change by using genetic, species, and agroecosystem diversity. This approach optimizes resource use and increases crop resilience through techniques like intercropping and agroforestry, especially in tropical family farming. Agrobiodiversity supports agroecology by enhancing ecological and social sustainability while reducing pesticides and conserving soil. Successful adaptation requires understanding biological interactions and integrating interdisciplinary research and

---

S. Léran (✉)
Cirad, UMR DIADE, Montpellier, France

DIADE, University of Montpellier, IRD, Cirad, Montpellier, France

Nicafrance Foundation, Finca La Cumplida, Matagalpa, Nicaragua
e-mail: sophie.leran@cirad.fr

M. Adam
Cirad, UMR AGAP Institute, Montpellier, France

UMR AGAP Institute, Univ Montpellier, Cirad, INRAE, Institut Agro, Montpellier, France

Faculty of Agriculture and Food Processing, National University of Battambang, Battambang, Cambodia

M. Gonin
Cirad, UMR DIADE, Montpellier, France

DIADE, University of Montpellier, IRD, Cirad, Montpellier, France

National Coffee Research Institute, National Agricultural Research Organization, Mukono, Uganda

C. Grenier · F. Micheli · M. Vaksmann
Cirad, UMR AGAP Institute, Montpellier, France

UMR AGAP Institute, Univ Montpellier, Cirad, INRAE, Institut Agro, Montpellier, France

participatory methods. Despite its benefits, agrobiodiversity faces challenges competing with intensive monocultures.

The valorisation of agrobiodiversity appears as a particularly interesting solution in the medium and long term to adapt cropping systems to climate change in a sustainable and agroecological way. By *agrobiodiversity*, we mean the diversity of living organisms recognised as resources by farmers, and consciously managed by them. It is divided into three levels of organisation that interact with each other: genetic diversity, specific diversity and agroecosystemic diversity. The valorisation of agrobiodiversity therefore includes the establishment or use of intra or interspecific complementarities, which allow, among other things, to optimise the use of resources linked to climate change (water, CO_2), to mitigate certain climatic constraints such as evapotranspiration or heat peaks within the cultivated plot, or to increase tolerance to these constraints by creating resistant varieties (Levard 2023). These complementarities, whether physiological, biophysical or genetic, can occur at the individual level (varietal mixtures, hybrids or grafting between varieties or species, associations with the soil microbiota), or at the plot level (agroforestry or orchard-vegetable systems).

Agrobiodiversity has been the basis, for centuries, of numerous agricultural practices worldwide, such as agroforestry, interspecific associations or varietal mixtures, agro-silvopastoralism, mainly in the context of tropical family farming (Bravo-Peña and Yoder 2024). Its large-scale mobilisation to adapt cropping systems to climate change is part of a logic of food sovereignty and territorial sustainability: reduction of pesticide and water use, maintenance of soil fertility, carbon sequestration. In this respect, the use of agrobiodiversity is part of agroecology as defined by the FAO: an integrated approach that concurrently applies ecological and social notions and

---

P. Marraccini · H. Étienne
Cirad, UMR DIADE, Montpellier, France

DIADE, University of Montpellier, IRD, Cirad, Montpellier, France

M. C. Rebolledo
Cirad, UMR AGAP Institute, Montpellier, France

UMR AGAP Institute, Univ Montpellier, Cirad, INRAE, Institut Agro, Montpellier, France

Alliance Bioversity Ciat, Palmira, Colombia

C. Rigal
Cirad, UMR ABSys, Montpellier, France

UMR ABSys, University of Montpellier, Cirad, INRAE, Institut Agro, Montpellier, France

WorldAgroforestry, Vietnam Office, Hanoi, Vietnam

M. L. Tékété
Rural Economy Institute, Bamako, Mali

D. Luquet
Cirad, Bios Department, Montpellier, France

principles to the design and management of agricultural and food systems. Agroecology seeks to optimise the interactions between plants, animals, humans and their environment, while taking into account the social dimensions necessary for a food system to be sustainable and equitable.[1] However, in the face of the variability of climatic constraints, the mobilisation of agrobiodiversity as a lever for sustainable adaptation is only possible if we understand and master the biological and biophysical mechanisms at the basis of the interactions between genotype and environment (G × E) and the agronomic and ecosystemic performances of agrosystems (Nerva et al. 2022).

This chapter aims to illustrate the scientific frontiers, but also the long-standing research activities that underpin the valorisation of agrobiodiversity for the adaptation of cropping systems to climate change. In addition, it highlights the adaptive potential contained in the cultivated or related (wild) diversity, intra or interspecific, as well as the challenge of mobilising innovative interdisciplinary approaches (agronomy, physiology, genetics, selection, AI, modelling) and multi-stakeholder (science-agriculture, participatory approaches). Through examples of annual and perennial tropical crops, the results are discussed in light of the climate emergency, the challenges related to the conservation, characterisation and valorisation of cultivated diversity, the resilience of the most vulnerable agrosystems, within the framework of rapidly evolving international and European regulations. The various examples of cropping systems presented in this chapter suggest that, despite the undeniable advantages that intra and interspecific mixtures provide in terms of resilience to climate change and sustainability, they unfortunately do not dethrone the more intensive and productive monoculture systems.

21.1 The Diversification of Cropping Systems as a Sustainable Lever for Adaptation to Climate Change

The potential offered by interspecific cropping systems for sustainable adaptation to climate change is now well recognised. However, the biological processes involved in the resilience and productivity of these systems are complex and still poorly understood, which limits their generalisation and large-scale deployment in the context of the agroecological transition (Benitez-Alfonso et al. 2023).

Agroforestry systems (AFS) are known for their adaptive capacity to climatic constraints (extreme temperatures, water deficit), for their ability to promote the health and fertility of ecosystems, and for their contribution to the maintenance of biodiversity and carbon sequestration (Rigal et al. 2022). AFS based on coffee, for example, vary from simple configurations with a single shade tree species, to the most rustic, where coffee trees are planted in forest environments. Shade trees play a crucial role in tempering the microclimate, allowing photosynthesis to continue

---

[1] www.fao.org/agroecology/fr; agroecology@fao.org

during the hottest hours. Climate models show that the expansion of AFS could mitigate the effects of climate change in the warmest low-altitude areas (500–800 m), thus preserving conditions favourable for coffee cultivation (Rahn et al. 2018). For cocoa trees, plantation systems can exhibit great heterogeneity, both in terms of associated plant species and shading (Saj et al. 2023). While globally cocoa-based AFS are gradually being replaced by unshaded monocultures (Heming et al. 2022), several studies show that traditional AFS (like the cabruca system in Brazil) can reduce the negative impacts of climate change, and their conservation should be an important objective of regional agricultural policies. Work is underway to understand and optimise the cocoa tree's response to drought and shading and to assist in the co-design of more efficient and sustainable agroforestry systems through modelling (Cacao4Future[2]). Adapting food crops to climate change has become a daily reality in regions where food security is a major concern, such as in the Sudan-Sahel region of Africa. Rainfed crops (cereals and legumes) as well as some traditional agroecological practices are increasingly being used, guaranteeing a harvest per year, while offering ecosystem services: pest regulation, soil fertility. Despite their advantages, these agroecological practices are not attractive, as they compete with more productive intensive methods. Crop association is practised, among others, with the sorghum-cowpea model, which uses a wide diversity of local or selected varieties for their adaptation to local pedoclimatic contexts (Ganeme et al. 2022) (Oracle project[3]). It has been shown that yields are more stable in association than in pure culture and that a sorghum-cowpea association in the same seed hole improves productivity. While this sorghum-cowpea association remains the most practised, the diversification of sorghum-based systems could involve the introduction of other species, such as mung beans, which are very well adapted to the semi-arid conditions of Burkina Faso (Raboin et al. 2023).

21.2 Agrobiodiversity as a Resource for Plant Breeding Programmes in the Face of Climate Change

21.2.1 Using Intraspecific Diversity to Adapt Breeding Programmes to Agroforestry Contexts and Drought: The Case of Coffee

Most of the varieties used today have been selected for cultivation in full sun and suffer severe yield losses (about 40%) when grown under shade (Haggar et al. 2011). It is therefore important to take into account the environmental components of agroforestry systems in current breeding programmes (low light, competition for nutrients and soil water). In the case of Arabica coffee (*Coffea arabica*), breeders

---
[2] https://www.cocoa4future.org/
[3] https://www.fondationavril.org/projects/cirad-oracle/

face a major obstacle, namely the very low genetic diversity available within cultivated varieties (Steiger et al. 2002). To overcome this problem, one improvement route is the creation of intraspecific hybrids, resulting from crosses between cultivated varieties (so-called American lines) with wild individuals from the center of origin (Ethiopia). These shade-adapted Arabica hybrids have better organoleptic quality, a productivity 30% higher than traditional homozygous varieties and sometimes increased drought tolerance (Breedcafs project[4]) (Sarzynski et al. 2024); they are currently being disseminated in many countries (Turreira-García 2022).

To guide breeding programmes in Arabica and Robusta (*Coffea canephora*) cultivated species in the choice of drought-tolerant varieties, different approaches have been implemented, such as the development of a rapid phenotyping method to assess the drought resistance of a large number of plants (Nestlé-Cirad-Nicafrance Foundation collaboration[5] in Nicaragua), or the identification of key genes and the deciphering of the physiological mechanisms involved in adaptation to abiotic stress (Alves et al. 2017). A redefinition of the Arabica coffee ideotype has been proposed, taking into account the issues of climate change and cultivation in AFS. This target ideotype helps guide future breeding programmes to increase the profitability and competitiveness of coffee-based AFS against intensive full-sun cultivation systems, which are currently more profitable but less sustainable (Breitler et al. 2022).

21.2.2 Combining Modern Breeding Methods with Traditional Knowledge for Adaptation to Climate Change: The Case of Sorghum

The sorghum improvement programme in West Africa primarily focuses on developing drought-resistant varieties, using the cost-effective method of ensuring crop maturity before the onset of severe water stress. Initial efforts concentrated on developing early varieties, suitable for shorter rainy seasons (Pixley et al. 2023). However, the adoption of these new varieties was very limited due to the high spatio-temporal variability of the seasons. Therefore, it was preferred to value the expertise of Sudan-Sahelian farmers who, over generations, have developed photoperiod-sensitive varieties that naturally adjust their maturity according to the end date of the rainy season (Fig. 21.1) (Clerget et al. 2021).

This sensitivity to photoperiod and the so-called *stay-green* phenotype related to drought tolerance (Vadez et al. 2011) have been integrated into marker-assisted selection (MAS) programmes, thus accelerating genetic gain. The combination of MAS with crop modelling allows for the consideration of strong G × E interactions, and thus the design of varieties specifically adapted to climate change (Guitton et al. 2018). Finally, the development of varieties directly in the target environment, with

---

[4] https://www.breedcafs.eu
[5] https://fundacionnicafrance.org

Fig. 21.1 African sorghums, known as "photoperiodic", naturally adjust their development according to the sowing date (June sowing on the left, July in the center and August on the right). (Credit: photo by M. Vaksmann)

the participation of farmers, has allowed for the production of varieties more adapted to local conditions (Vom Brocke et al. 2020).

21.2.3 Towards Breeding Programmes that Value Genetic Diversity for Plant Adaptation to Climate Change and its Mitigation: The Case of Rice

Rice cultivation in flooded (or anaerobic) conditions is currently controversial due to unsustainable water management and the emission of greenhouse gases (GHGs), particularly methane emissions from the plant root system (Rajendran et al. 2024). However, these anaerobic conditions yield better results than aerobic cultivation conditions and remain predominant despite the environmental controversy (Saito et al. 2018).

In this context, the selection of varieties adapted to aerobic conditions is a major challenge (Cirad-Ciat collaboration[6]) (Chatel et al. 2008). However, the effectiveness of breeding programmes is limited, as the traits associated with adaptation to aerobic conditions have not or have been little selected so far. The physiological and genetic bases of adaptation to aerobic and drought conditions must then be studied (Luquet et al. 2016). While genes or QTL (*Quantitative Trait Loci*) candidates for creating varieties adapted to these conditions have already been identified, few of them prove effective enough for marker-assisted selection. Recurrent genomic selection seems a conclusive option (Grenier et al. 2015).

The rise in temperatures is also a concern for rice cultivation. A multi-location agronomic evaluation of a panel of 50 rice lines showed a significant decrease in average yields (by about 40%) in the hottest environments (Rebolledo et al. 2023). These results justify the deployment of screening methods to identify sources of tolerance within the intra and interspecific diversity of rice species *japonica, indica* and *glaberrima* as part of breeding programmes (Cirad-Ciat collaboration).

21.3 Interspecific Grafting as a Solution for Climate Change Adaptation: The Case of Citrus

In most cultivated trees, characterised by their long biological cycle and selection periods of several decades, the identification of species more resilient to climate change and abiotic stress, and their rapid valorisation through interspecific hybridisation or grafting, should be prioritised.

Cultivated and marketed citrus fruits are the result of interspecific hybridisations within the genus *Citrus*, from four ancestral species: *C. reticulata, C. maxima, C. medica* and *C. micrantha* (Wu et al. 2018). In current citrus cultivation, trees are almost always propagated by grafting the commercial variety onto a rootstock selected for its vigour, resistance or tolerance to biotic and/or abiotic stress. More than 20 genotypes of rootstocks from the genus *Citrus* related genera (such as *Poncirus*), or even interspecific or intergeneric hybrids have been recorded worldwide, and their use depends on the production region, local practices and their agronomic performance (Bowman and Joubert 2020).

In regions where rainfed cultivation is common, such as in Latin America, water use efficiency has become a priority target for the selection of new rootstocks (Girardi et al. 2021). Thus, the analysis of orange production (*C. sinensis* [L.] Osbeck var. Valencia), grafted onto 23 genotypes from interspecific, intergeneric hybridisations or hybrid selection over a 10-year period in rainfed cultivation in Brazil (tropical savannah climate), has allowed the selection of new rootstocks

---

[6] https://alliancebioversityciat.org/fr

resistant to water deficit. On the other hand, the study of different interspecific combinations of scions and rootstocks (Santana-Vieira et al. 2016) showed that they could exhibit contrasting adaptive behaviours to water stress (avoidance *vs* tolerance) and beneficial in the case of cultivation areas with hard, dry and shallow soils.

21.4 Conclusion and Perspectives

Climate change is global, but its effects generate increasing intra and interannual variability at the local scale (report AR6, IPCC[7]) which is important to characterise in order to identify adaptation solutions for agroecosystems to these diverse, current and future contexts. The examples given in this chapter support that the mobilisation of agrobiodiversity is an essential lever for the design of diversified and sustainable climate change adaptation solutions (Fig. 21.2). However, it highlights the need to combine approaches that allow for better characterisation and consideration of the variability of constraints (climatic, agroecological and societal) and an increased mobilisation of the genetic and adaptive diversity present within and between cultivated or wild species.

Diagram.

21.4.1 Prioritising Cultivation Systems that Utilise Intra and Interspecific Diversity

Ongoing work to accelerate the understanding of these systems and their deployment on a larger scale highlights the potential of approaches combining multisite and multi-stakeholder experiments, and the modelling of cultivation systems for their analysis and for predicting their performance (Gaudio et al. 2022). In this regard, the *in situ* and long-term facilities of Cirad and its partners represent a remarkable added value, such as the DP Agroforesta[8], the Niakhar observatory (Delaunay et al. 2018). They should also rely on experiments under more controlled conditions, allowing for example the ecophysiological understanding of adaptation to climate change — Abiophen[9] and M3P (Welcker et al. 2015). An effort should also be made to better understand the ability of agrosystems to sequester carbon and maintain soil fertility (Bolero project[10]), as well as the role of plant-microbiome interactions in accessing soil resources and water.

---

[7] https://report.ipcc.ch/ar6syr/index.html

[8] https://www.cirad.fr/dans-le-monde/dispositifs-en-partenariat/agroforesta

[9] https://www.cirad.fr/collaborer-avec-nous/science-ouverte/infrastructures-ouvertes-du-cirad/abiophen-et-serres-experimentales

[10] https://www.bolero-project.eu

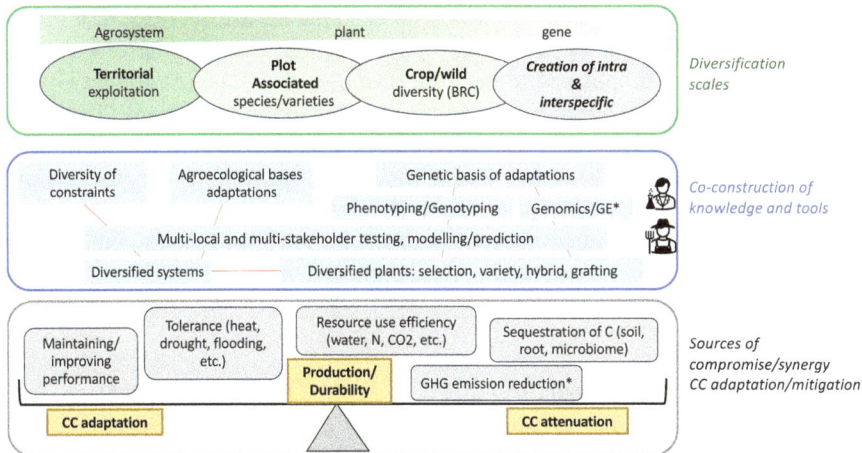

Fig. 21.2 Valorisation of biodiversity through interdisciplinary and multi-stakeholder approaches, for the sustainable adaptation of cropping systems to climate change and for its mitigation. GE: genome editing. Diagram representing the valorisation of biodiversity through interdisciplinary approaches. There are several scales of diversification, ranging from the agrosystem (territory exploitation, species and variety association), to the plant (use of cultivated and wild diversity), and to the gene (through the creation of intra and interspecific hybrids). This valorisation, multi-stakeholder, requires a co-construction of knowledge and tools: integration of agroecological and genetic bases, setting up of multi-local trials, phenotyping, genotyping, modelling and predictions in order to produce diversified agrosystems, in terms of crops, varieties, types of plants (hybrids, grafted plants). This pooling of knowledge aims at adapting to climate change, and its mitigation, while maintaining production and sustainability of systems. The targets are: maintaining or improving yields, better tolerance to abiotic stresses, increased efficiency in resource use (water, nitrogen, carbon), reduction of greenhouse gases emission, and thus a better carbon balance at the agrosystem scale

21.5 Selection Programmes Based on Multi-criteria and Multi-stakeholder Approaches

21.5.1 Towards Increasingly Massive and Multi-scale Data

This chapter highlights that, in order to valorise the genetic diversity of cultivated or related wild plants for adaptation to climate change, selection programmes must generate and manipulate increasingly massive and multi-scale data sets: from biological data (physiological, genetic and other "omics") characterising sources of adaptation among thousands of individuals composing the genetic diversity of a target species, to agroclimatic and societal data (surveys) acquired in multi-local trials representative of the sets of constraints to which the cultivated plants need to be adapted. The integrated analysis of these data represents a numerical challenge now at the heart of varietal selection programmes and the upstream research they require (see paragraph 2.1). Data analysis and modelling approaches (mechanistic statistical, data-driven learning) exist for this, but need to be optimised and better complemented to: (1) understand the complex biological systems to be improved

(Luquet et al. 2016), (2) develop "proxies" (for example the use of NIRS, near infrared spectroscopy, or imaging) allowing high-throughput phenotyping of adaptive traits that are more complicated to measure (Gano et al. 2021), (3) predict G × E interactions and varietal ideotypes (genomic and phenomic prediction) (Tong and Nikoloski 2021), and (4) assess the added value in varied and future agroclimatic contexts (Parent et al. 2018). The use of artificial intelligence (AI) becomes essential in this context. While this type of approach is gaining momentum in support of selection programmes for intensified cultivation contexts, this chapter highlights the ongoing efforts to benefit diversified tropical agro-systems.

21.5.2 Towards the Consideration of New Traits of Interest for Adaptation and Mitigation of Climate Change in Selection Programmes

A significant part of the G × E interactions explaining plant adaptation to climatic conditions occurs at the soil level, between the root system, resources and the microbiome. A small fraction of the microbes present in the rhizosphere can be beneficial for the host plant, inducing beneficial morphological and physiological modifications of the root system (Gonin et al. 2023). These interactions between roots and the associated microbiota can effectively contribute to tolerance to abiotic stresses (for example deeper rooting, better nutrient or water absorption, etc.), particularly to drought, but also to an increase in carbon sequestration in the soil (De Vries et al. 2020). Promoting the growth of these beneficial microbes through targeted strategies can improve crop health and productivity in a biological approach. Work in this area is rare and recent.

The challenge of mitigating climate change also requires addressing the reduction of greenhouse gas emissions and the sequestration of carbon by agricultural systems. This challenge is global (see paragraph 2), but particularly concerns rice cultivation in flooded conditions, leading to significant methane emissions that impact the environment. The potential to harness the genetic diversity of rice to create varieties that emit less methane in flooded conditions, or, more importantly, that can produce in aerobic conditions at yield levels approaching those of flooded conditions (Saito et al. 2018), has been highlighted in this chapter. Beyond these results, recent studies have shown the existence of genetic diversity in rice in its ability to assimilate atmospheric CO_2 through photosynthesis and thus increase production, but also carbon accumulation in the plant, including at the root level. These original results pave the way for their valorisation in varietal selection, both in terms of adaptation and mitigation of climate change (Dingkuhn et al. 2020).

21.5.3 New Challenges for Grafting

Drawing on the experience with citrus fruits, we can envisage the use of wild species in perennial plants, combined with grafting techniques, as a quick solution to counteract the effects of climate change. The genus *Coffea* includes no fewer than

130 species native to West and Central Africa, a large part of which is preserved in botanical gardens or biological resource centres (BRC in La Réunion, international collection at CATIE in Costa Rica, BRC in French Guiana). This diversity is currently being explored to identify coffee species more tolerant to heat and drought and to test them as hybrid parents or as rootstocks to maintain the production of Arabica and Robusta (Bolero project[11]).

21.6 Role of New Biotechnologies

Although a large number of candidate genes to improve adaptation to biotic and/or abiotic stresses have been identified and sequenced thanks to a deep understanding of genetic diversity in most annual and perennial cultivated species, their valorisation through genome editing (with the CRISPR-Cas9 tool among other techniques) is still in its infancy. The first edited plants proposed for commercialisation are modified for simple traits, such as quality and resistance to biotic stress (Pixley et al. 2022). In coffee, as in most tropical fruit trees where scientific communities are limited and the seed industry is absent, these innovations for breeding programmes are mainly carried out through projects involving research centres, universities, industrial partners and farmers. This is the case with the Gardens initiative of the PEPR SVA,[12] which proposes to study the potential contribution of agrobiodiversity in tropical and temperate zones to the agronomic, socio-economic and ecological performance of orchard-vegetable systems. These systems, gradually abandoned over the centuries, are still very present in tropical areas, but are neglected by intensive agriculture. The Gardens initiative will leverage genome editing tools for the design of varietal ideotypes adapted to these systems in the context of climate change and mobilising genetic diversity within interacting species.

In cereals, the effort is greater and continuous, thanks to a large community of researchers, a powerful seed industry and international centres dedicated to these plants (CGIAR[13]). Adding to this the benefit of a short generation time, we can expect in the coming years the arrival on the market of genetically edited varieties that are more resilient to climate change. An increasing number of studies using this technique are already identifying new genes of interest for adaptation and mitigation, with a view to their integration into breeding programmes (such as the Greener project[14] for rice adaptation genes to aerobic conditions).

While the rise of new technologies for genome editing is undeniable, its practical implementation is less prevalent in tropical plants of agriculture in the South. To address this gap, research partners in Montpellier (Cirad, IRD, INRAE, University

---

[11] https://www.bolero-project.eu/

[12] PEPR SVA: Advanced plant breeding in response to the climate challenge. The priority research programmes and equipment (PEPR) constitute the upstream/research aspect of the France 2030 strategies. https://www.pepr-selection-vegetale.fr

[13] CGIAR: Consultative Group on International Agricultural Research.

[14] https://umr-agap.cirad.fr/recherches/projets-de-recherche/greener

of Montpellier) are setting up a shared infrastructure specialising in the editing of genomes of tropical and Mediterranean plants (Editrop, PEPR SVA project).

21.7 Role of Biological Resource Centres

BRCs, particularly those dedicated to Mediterranean and tropical plants, have and will play a major role in the context of climate change. The genetic diversity they contain offers a unique source of adaptation to current and future agroclimatic constraints. It is necessary to preserve this diversity and accelerate its characterisation for its valorisation in breeding programmes. In this sense, BRCs must increasingly play the role of interaction platforms between research, varietal selection and actors in agricultural sectors in the North and South. In this context, the exchange of genetic resources is conducted in strict accordance with the Nagoya Protocol (2010), aimed at better protecting the planet's species and ecosystems and sharing the benefits more equitably.

References

Alves GSC, Torres LF, Déchamp E, Breitler J-C, Joët T, Gatineau F et al (2017) Differential fine-tuning of gene expression regulation in coffee leaves by CcDREB1D promoter haplotypes under water deficit. J Exp Bot 68:3017–3031. https://doi.org/10.1093/jxb/erx166

Benitez-Alfonso Y, Soanes BK, Zimba S, Sinanaj B, German L, Sharma V et al (2023) Enhancing climate change resilience in agricultural crops. Curr Biol 33:R1246–R1261. https://doi.org/10.1016/j.cub.2023.10.028

Bowman KD, Joubert J (2020) Citrus rootstocks. In: The Genus Citrus. Elsevier, pp 105–127

Bravo-Peña F, Yoder L (2024) Agrobiodiversity and smallholder resilience: a scoping review. J Environ Manag 351:119882. https://doi.org/10.1016/j.jenvman.2023.119882

Breitler J-C, Etienne H, Léran S, Marie L, Bertrand B (2022) Description of an Arabica coffee ideotype for agroforestry cropping systems: a guideline for breeding more resilient new varieties. Plants 11:2133. https://doi.org/10.3390/plants11162133

Chatel M, Ospina RY, Rodriguez F, Lozano VH, Delgado H (2008) Upland rice composite population breeding and selection of promising lines for Colombian savannah ecosystem. CGIAR Database Resource

Clerget B, Sidibe M, Bueno CS, Grenier C, Kawakata T, Domingo AJ et al (2021) Crop photoperiodism model 2.0 for the flowering time of sorghum and rice that includes daily changes in sunrise and sunset times and temperature acclimation. Ann Bot 128:97–113. https://doi.org/10.1093/aob/mcab048

Delaunay V, Desclaux A, Sokhna C (eds) (2018) Niakhar, memories and perspectives. Multidisciplinary research on change in Africa. IRD Editions and L'Harmattan Senegal, Marseille and Dakar. 535 p

De Vries FT, Griffiths RI, Knight CG, Nicolitch O, Williams A (2020) Harnessing rhizosphere microbiomes for drought-resilient crop production. Science 368:270–274. https://doi.org/10.1126/science.aaz5192

Dingkuhn M, Luquet D, Fabre D, Muller B, Yin X, Paul MJ (2020) The case for improving crop carbon sink strength or plasticity for a CO_2-rich future. Curr Opin Plant Biol 56:259–272. https://doi.org/10.1016/j.pbi.2020.05.012

Ganeme A, Kondombo CP, Raboin L-M, Dusserre J, Kabore R, Adam M et al (2022) Characterizing sorghum (*Sorghum bicolor* [L.] Moench) varieties diversity to identify those with contrasting traits of interest for intercropping systems in the Sudano-Sahelian zone of West Africa. Plant Genet Resour Charact Util 20:87–97. https://doi.org/10.1017/S1479262122000168

Gano B, Dembele JSB, Tovignan TK, Sine B, Vadez V, Diouf D et al (2021) Adaptation responses to early drought stress of West Africa sorghum varieties. Agronomy 11:443. https://doi.org/10.3390/agronomy11030443

Gaudio N, Louarn G, Barillot R, Meunier C, Vezy R, Launay M (2022) Exploring complementarities between modelling approaches that enable upscaling from plant community functioning to ecosystem services as a way to support agroecological transition. Silico Plants 4:diab037. https://doi.org/10.1093/insilicoplants/diab037

Girardi EA, Ayres AJ, Girotto LF, Peña L (2021) Tree growth and production of rainfed valencia sweet orange grafted onto trifoliate orange hybrid rootstocks under aw climate. Agronomy 11:2533. https://doi.org/10.3390/agronomy11122533

Gonin M, Salas-González I, Gopaulchan D, Frene JP, Roden S, Van De Poel B et al (2023) Plant microbiota controls an alternative root branching regulatory mechanism in plants. Proc Natl Acad Sci 120:e2301054120. https://doi.org/10.1073/pnas.2301054120

Grenier C, Cao T-V, Ospina Y, Quintero C, Châtel MH, Tohme J et al (2015) Accuracy of genomic selection in a rice synthetic population developed for recurrent selection breeding. PLoS One 10:e0136594. https://doi.org/10.1371/journal.pone.0136594

Guitton B, Théra K, Tékété ML, Pot D, Kouressy M, Témé N et al (2018) Integrating genetic analysis and crop modelling: a major QTL can finely adjust photoperiod-sensitive sorghum flowering. Field Crop Res 221:7–18. https://doi.org/10.1016/j.fcr.2018.02.007

Haggar J, Barrios M, Bolaños M, Merlo M, Moraga P, Munguia R et al (2011) Coffee agroecosystem performance under full sun, shade, conventional and organic management regimes in Central America. Agrofor Syst 82:285–301. https://doi.org/10.1007/s10457-011-9392-5

Heming NM, Schroth G, Talora DC, Faria D (2022) Cabruca agroforestry systems reduce vulnerability of cacao plantations to climate change in southern Bahia. Agron Sustain Dev 42:48. https://doi.org/10.1007/s13593-022-00780-w

Levard L (coord) (2023) Guide for the evaluation of agroecology. Method for assessing its effects and the conditions for its development Gret Editions, Quae Editions, 320 p

Luquet D, Rebolledo C, Rouan L, Soulie J-C, Dingkuhn M (2016) Heuristic exploration of theoretical margins for improving adaptation of rice through crop-model assisted phenotyping. In: Yin X, Struik PC (eds) Crop systems biology. Springer, pp 105–127. https://doi.org/10.1007/978-3-319-20562-5_5

Nerva L, Sandrini M, Moffa L, Velasco R, Balestrini R, Chitarra W (2022) Breeding towards improved ecological plant–microbiome interactions. Trends Plant Sci 27:1134–1143. https://doi.org/10.1016/j.tplants.2022.06.004

Parent B, Leclere M, Lacube S, Semenov MA, Welcker C, Martre P et al (2018) Maize yields over Europe may increase in spite of climate change, with an appropriate use of the genetic variability of flowering time. Proc Natl Acad Sci 115:10642–10647. https://doi.org/10.1073/pnas.1720716115

Pixley KV, Cairns JE, Lopez-Ridaura S, Ojiewo CO, Dawud MA, Drabo I et al (2023) Redesigning crop varieties to win the race between climate change and food security. Mol Plant 16:1590–1611. https://doi.org/10.1016/j.molp.2023.09.003

Pixley KV, Falck-Zepeda JB, Paarlberg RL, Phillips PWB, Slamet-Loedin IH, Dhugga KS et al (2022) Genome-edited crops for improved food security of smallholder farmers. Nat Genet 54:364–367. https://doi.org/10.1038/s41588-022-01046-7

Raboin L-M, Batieno BJ, Gozé E, Douzet J-M, Poda L, Koala WA et al (2023) The mung bean, *Vigna radiata* (L.), an alternative to the sorghum-cowpea association for the diversification of crops in Sudan-Sahelian conditions? Cah Agri 32:26. https://doi.org/10.1051/cagri/2023019

Rahn E, Vaast P, Läderach P, van Asten P, Jassogne L, Ghazoul J (2018) Exploring adaptation strategies of coffee production to climate change using a process-based model. Ecol Model 371:76–89. https://doi.org/10.1016/j.ecolmodel.2018.01.009

Rajendran S, Park H, Kim J, Park SJ, Shin D, Lee J-H et al (2024) Methane Emission from rice fields: necessity for molecular approach for mitigation. Rice Sci 31:159–178. https://doi.org/10.1016/j.rsci.2023.10.003

Rebolledo MC, Ranaivoson L, Falconnier G, Adam M, Ibrahin A, Mallikarjuna S et al (2023) Global Rice Field laboratory to understand rice response to climate variability. https://doi.org/10.5281/ZENODO.10399125

Rigal C, Wagner S, Nguyen MP, Jassogne L, Vaast P (2022) SHADETREEADVICE methodology: guiding tree-species selection using local knowledge. People Nat 4:1233–1248. https://doi.org/10.1002/pan3.10374

Saito K, Asai H, Zhao D, Laborte AG, Grenier C (2018) Progress in varietal improvement for increasing upland rice productivity in the tropics. Plant Prod Sci 21:145–158. https://doi.org/10.1080/1343943X.2018.1459751

Saj S, Jagoret P, Ngnogue HT, Tixier P (2023) Effect of neighbouring perennials on cocoa tree pod production in complex agroforestry systems in Cameroon. Eur J Agron 146:126810. https://doi.org/10.1016/j.eja.2023.126810

Santana-Vieira DDS, Freschi L, Almeida LADH, Moraes DHSD, Neves DM, Santos LMD et al (2016) Survival strategies of citrus rootstocks subjected to drought. Sci Rep 6:38775. https://doi.org/10.1038/srep38775

Sarzynski T, Vaast P, Rigal C, Marraccini P, Delahaie B, Georget F et al (2024) Contrasted agronomical and physiological responses of five Coffea arabica genotypes under soil water deficit in field conditions. Front Plant Sci 15:1443900. https://doi.org/10.3389/fpls.2024.1443900

Steiger L, Nagai C, Moore H, Morden W, Osgood V, Ming R (2002) AFLP analysis of genetic diversity within and among Coffea arabica cultivars. TAG Theor Appl Genet 105:209–215. https://doi.org/10.1007/s00122-002-0939-8

Tong H, Nikoloski Z (2021) Machine learning approaches for crop improvement: leveraging phenotypic and genotypic big data. J Plant Physiol 257:153354. https://doi.org/10.1016/j.jplph.2020.153354

Turreira-García N (2022) Farmers' perceptions and adoption of Coffea arabica F1 hybrids in Central America. World Dev Sustain 1:100007. https://doi.org/10.1016/j.wds.2022.100007

Vadez V, Deshpande SP, Kholova J, Hammer GL, Borrell AK, Talwar HS et al (2011) Stay-green quantitative trait loci's effects on water extraction, transpiration efficiency and seed yield depend on recipient parent background. Funct Plant Biol 38:553. https://doi.org/10.1071/FP11073

Vom Brocke K, Kondombo CP, Guillet M, Kaboré R, Sidibé A, Temple L et al (2020) Impact of participatory sorghum breeding in Burkina Faso. Agric Syst 180:102775. https://doi.org/10.1016/j.agsy.2019.102775

Welcker C, Cabrera BL, Grau A, Tardieu F, Negre V et al (2015) M3P: the "Montpellier plant phenotyping platforms". EPPN Plant Phenotyping Symposium, Barcelona, Spain

Wu GA, Terol J, Ibanez V, López-García A, Pérez-Román E, Borredá C et al (2018) Genomics of the origin and evolution of citrus. Nature 554:311–316. https://doi.org/10.1038/nature25447

Open Access This chapter is licensed under the terms of the Creative Commons Attribution-NonCommercial-NoDerivatives 4.0 International License (http://creativecommons.org/licenses/by-nc-nd/4.0/), which permits any noncommercial use, sharing, distribution and reproduction in any medium or format, as long as you give appropriate credit to the original author(s) and the source, provide a link to the Creative Commons license and indicate if you modified the licensed material. You do not have permission under this license to share adapted material derived from this chapter or parts of it.

The images or other third party material in this chapter are included in the chapter's Creative Commons license, unless indicated otherwise in a credit line to the material. If material is not included in the chapter's Creative Commons license and your intended use is not permitted by statutory regulation or exceeds the permitted use, you will need to obtain permission directly from the copyright holder.

Chapter 22
Territorializing Climate Action

Camille Jahel, Amandine Adamczewski, Jérémy Bourgoin, Guillaume Lestrelin, Ronan Mugelé, René Poccard-Chapuis, Fatma Rostom, Tiago Teixeira Da Silva Siqueira, and Elodie Valette

Abstract Climate change is a global issue but is experienced and understood differently across territories, which are spaces appropriated and governed by human groups. The territorial approach highlights how local social, political, and ecological contexts shape diverse climate impacts and responses. Territories reveal the connections between societies and climate change, making them crucial for designing relevant adaptation and mitigation strategies. This approach balances the broad global perspective with local specificities, enabling tailored solutions. The chapter explores how territories can be mobilized as key spaces to address climate change effectively.

Climate change is often presented—notably in the reports of the Intergovernmental Panel on Climate Change (IPCC)—as a "global issue" or a "planetary challenge". Indeed, it affects the entire Earth system and societies worldwide. However, from the perspective of these societies, climate changes are multiple; they are primarily

---

C. Jahel (✉)
Cirad, UMR TETIS, Montpellier, France
e-mail: camille.jahel@cirad.fr

A. Adamczewski
Cirad, UMR G-EAU, Montpellier, France

J. Bourgoin
Cirad, UMR TETIS, International Land coalition, Rome, Italy

G. Lestrelin
Cirad, UMR TETIS, National Agronomic Institute of Tunisia (INAT), Tunis, Tunisia

R. Mugelé
UMR Prodig, Aubervilliers, France

R. Poccard-Chapuis
Cirad, UMR SELMET, EMBRAPA Eastern Amazon, Belém, Brazil

F. Rostom · E. Valette
Cirad, UMR ART-Dev, Montpellier, France

T. Teixeira Da Silva Siqueira
Cirad, UMR SELMET, Saint-Pierre, La Réunion, France

understood, in their causes as well as their consequences, through the territories in which they evolve. The territory can be understood as a space appropriated (in terms of ownership and identity) and governed by an organised human group. The territorial approach to climate change emphasises that socio-environmental changes are primarily experienced by societies living "here and now". It involves anchoring and translating global change into the space-times of societies, in the North and in the South, in cities and in rural areas, according to different political, social, ecological, and economic contexts (Le Treut 2022).

The impacts of climate change are highly variable from one territory to another, whether concerning biophysical transformations (for example, rainfall records or droughts) which increase differentiated exposure to risks, or the responses of local societies given their history, their expectations, their experiences or their political struggles. All these factors also contribute to localising or specifying ways of relating to the issue of climate change according to very diverse considerations.

Because territories outline and make visible the organic links between societies and climate change, they are key spaces for reflection and for implementing relevant adaptation and mitigation actions. Again, variability of actions is high between territories, due to the diversity of resources they have, their history, their social capital, their access to information and the means of action of their populations. Halfway between the "too macro" to integrate specificities and the "too local" with limited impacts, territorial approaches propose a framework for collectively thinking, organising and developing contextualised mitigation and adaptation solutions. The aim of this chapter is to show how and why the territory, as a concept, as a place where biophysical changes are experienced, as a social entity and arena, can be mobilised to face climate change.

22.1 The Territory, an Arena for Collective Deliberation and Negotiation

Due to their increasing magnitude and intensity, floods, megafires, hurricanes, storms, droughts render individual or isolated adaptation initiatives powerless. These climate-related events call for large-scale collective commitments. This implies the coordination and collaboration of a diversity of actors, sometimes with divergent interests, in a situation of power asymmetry and incomplete knowledge. The territorial approach promotes coalitions of actors, through the sharing of a common identity, the result of history, experience, living in the same space. In the face of climate change, the inhabitants of a territory also share the same threats, the same uncertainties, although their vulnerabilities and their adaptation capacities are different. Thinking about climate change through the territory means making this threat the driving force of collective action, transforming it into a common aspiration, while avoiding that it reinforces divisions.

Territorial approaches thus propose an operational multiscale framework, aimed at transformation, and based on the principles of equity, participation, inclusion and accountability. It is about making the territory a space for negotiation, an arena for building compromises and ideas to adapt and mitigate climate change. Territorial arrangements can take various forms of observatories, living labs (Box 22.1), third places, citizen debate platforms, etc. These are arenas in which the actors of the territory mobilise to bring about adaptation and mitigation actions.

> **Box 22.1 Feedback 1. Construction of Living Labs in Senegal**
>
> The concept of *living lab*, relatively new, admits different definitions all converging towards the idea of platform anchored in a particular social and local dynamic, where citizens, residents, users are considered as key actors in research and innovation processes. This approach is supposed to stimulate collaborations between heterogeneous profiles of people with the aim of developing unexpected discoveries (Romero 2017).
>
> Around Lake Guiers in Senegal, two *living labs* have recently been established. They are part of a context of scarcity of land and water resources exacerbated by climate change and generating social and spatial tensions between different categories of actors. The ambition of the *living labs* is to improve the health of humans, agricultural production systems and the environment, and then positively affect the livelihoods of the populations of the territories. Anchored in local territorial management institutions, the two *living labs* associate local actors (farmers, breeders, fishermen, elected officials, entrepreneurs) with scientists who will accompany the experiments. After a phase of territorial diagnosis, the stakeholders chose to prioritise certain experiments. For this, two forums, one per *living lab*, bringing together all the users and scientists, were constituted and thought of as dialogue arenas allowing collective validation of choices. Then, the communities of actors, users and scientists, organised around the experiments. This process of territorial validation, at the scale of the *living lab*, through the forum, allows the community to engage in actions previously put up for debate. This governance tool, based on debate with the diversity of territory users, allows the formation of collectives capable of organising, validating, choosing and then testing. It is about anticipating the necessary adaptations that the community *living lab* will need to be able to implement, beyond the project time.

22.2 Towards Multisectoral Initiatives Based on Spatial Complementarities

From this territorial intelligence emerge contextualised initiatives for the adaptation and mitigation of the local effects of climate change, such as ideas of experiments, new synergies, solidarities and forms of governance. Due to the diversity of skills and knowledge gathered, territorial initiatives strive to break away from siloed approaches to propose a multi-sectoral approach. The spatialisation of problems and solutions is a structuring element: the links between biophysical and social realities are anchored in a geographical space. Climate adaptation or mitigation actions are thus based on a systemic conception of the territory, which values synergies and interactions between the different components of the territorial system.

The management of flood risks, increased by the effects of climate change, is interesting in this respect. Flooding is certainly the result of a meteorological hazard, but also of a vulnerability built up over time within territories. In eastern Amazonia, Brazil, most watersheds have been affected by deforestation and agricultural practices that poorly protect the soil against erosion, for five to ten decades. Year after year, the action of erosive rains silts up the hydrographic drainage network. When an event of exceptionally intense rainfall occurs, such as during La Niña episodes, the flows are no longer absorbed either by the forest vegetation, which has disappeared, or by the soils, which have become impermeable, or by the riverbeds, which are silted up. Therefore, a flood occurs, which can be dramatic as in Paragominas in 2018, this city being built in the middle of an alluvial plain. This event mobilised the public authorities, and a payment plan for environmental services has since been put in place. All farmers in the watershed receive an incentive amount, deducted from the bills of each household connected to the urban drinking water supply network. The condition to receive this payment is that farmers comply with the land use plan defined by the public authorities, after extensive consultation. This plan provides for (1) anti-erosion arrangements and agricultural practices in areas identified as suitable for agriculture, (2) forest restoration practices in areas identified as crucial for the water cycle. The entire system is therefore based on a mapping of soil aptitudes, i.e. a precise zoning sanctioned by a municipal law that defines the rules of use for each zone. This reorganisation of mixed landscapes, composed of forests and agricultural land, should make it possible to desilt rivers, reduce flood risks and improve the quality of the waters of the Uraim river, thus reducing the costs of their treatment to supply the drinking water network. This saving would then be used to increase incentives for farmers.

Another illustrative example of territorial initiatives for the adaptation to climate change is the transformation of territorial agri-food systems. These are defined as "the set of actors, networks and mechanisms which guide the way food is produced and consumed at the territorial level. They therefore include, beyond the flows of production, collection, transformation, and distribution—which correspond to the agri-food system in its economic definition—consumers, civil society, agricultural or rural development supporting services, and local authorities and public institutions" (Authors' translation of Lamine et al. 2022). Agri-food systems play a pivotal role in climate change, being at the same time (i) major emitters of greenhouse gases, (ii) particularly vulnerable, and (iii) with a significant potential for mitigation and adaptation. Thus, the reconfiguration of agri-food systems at the territorial level helps rethinking production and the circulation of products, as well as consumption patterns and waste treatment, offering a breakdown with the globalised industrial model. The establishment of short food supply chains, for example, can contribute to "reformulating commercial links in the direction of increased cooperation not only between producers, but also between producers and consumers, between consumers, between actors in the same territory or region" (Chiffoleau and Prevost

2012). Box 22.2 presents a case study of the reconfiguration of an agricultural sector to adapt to the impacts of climate change in Mediterranean territories.

Finally, because they involve multi-stakeholder consultation, territorial approaches help to improve the success of certain greenhouse gas emission reduction initiatives. This is the case, for example, with energy transition strategies towards less emitting energy sources, such as solar or wind. It has been shown that projects disconnected from local socio-economic and environmental contexts can generate impacts that are not desired by local actors, whether on landscapes, ecosystems, employment structure, energy access but also on land issues, leading to strong local opposition movements (Siamenta and Dunlap 2019) (see Chap. 19).

Box 22.2 Feedback 2. Territorialised Initiatives in the Face of Climate change in the Mediterranean Area: The Case of the Pistachio in Provence

Some agricultural productions in Mediterranean areas are heavily affected by recurrent droughts and the rising temperature (Ivits et al. 2012). Territorialised initiatives based on more resilient agricultural production systems and sectors are emerging to face these impacts, as illustrated by the pistachio sector in Vaucluse. Once very common in the area, the pistachio tree has returned to the wild. Drawing on its significant hardiness (resistance to drought and extreme temperatures), its histocam presence and its adaptation to the Mediterranean territory, certain quality-oriented value chains are re-emerging thanks to the collective work of local actors (Debolini and Siqueira 2021).

The first reflections and studies were initiated in 2016 by some local pioneers. They made trips to pistachio-producing areas and consulted departmental archives. These elements laid the first foundations for the development of the sector. In 2018, civil society actors, farmers and processors came together to create a non-profit association, called Pistachio in Provence. In parallel, some pioneer farmers planted the first plants despite the lack of local nurseries for their supply. From 2018–2019 onwards, in response to requests from several local farmers, the Vaucluse Chamber of Agriculture began conducing technical and agronomic studies to facilitate local planting. It also offered technical support to interested farmers. An initial Feader project allowed local actors from agricultural advisory services, technical and scientific research and farmers to come together. In 2021, farmers organised themselves into a union to build a project aimed at structuring of the industry together with local transformation actors. Currently, over 400 hectares have been planted, and stakeholders are working towards obtaining recognition for Provence pistachios under a quality and origin label—SIQO.[1]

---

[1] https://rd-agri.fr/detail/DOCUMENT/chambres_d%27agriculture_294233

The acceptability of installing renewable energy infrastructures therefore does not depend solely on the *not-in-my-backyard syndrome* that is empirically contested and that pejoratively plays down opposition arguments to these infrastructures. The way projects are implemented is a crucial factor influencing this acceptability. In particular, governance and the financial participation of citizens and local communities in such projects are key (Mussal and Kuik 2011) and such participation depends on the level of trust among citizens and the sharing of a local identity (Kalkbrenner and Rossen 2016), two levers at the heart of territorial approaches.

22.3 Territorial Approaches for Climate Justice

The importance of the acceptance of climate change mitigation and adaptation initiatives by the inhabitants of a territory leads to questioning the concept of climate justice. Mitigation and adaptation solutions are more easily accepted and adopted if they are fair (Newell et al. 2021). However, although justice is increasingly mobilised at national or international scales (Gupta et al. 2023), particularly on issues of north-south inequalities in the face of climate change, the application of these notions remains rare at the local level (Tsayem and Philippe 2022), at best, confined to urban territories (Bulkeley et al. 2014).

Climate justice refers to an ethical approach translated into a multidimensional legal analysis framework aimed at promoting equity in responses to the challenges posed by climate change (Schlosberg and Collins 2014). To do so, it posits the following fundamental principles:

- Social equity: ensuring that the most vulnerable and marginalised populations do not disproportionately bear the consequences of climate change and that they also benefit from the solutions;
- Environmental equity: avoiding the concentration of the harmful effects of climate change in certain regions and guaranteeing equitable access for populations to natural resources;
- Responsibility: recognising the differentiated responsibilities (of countries, social groups, etc.) in contributing to greenhouse gas emissions, in mitigating the impacts of climate change, and the adaptation strategies developed;
- Participation and inclusion: involving a wide diversity of stakeholders in climate-related decision-making processes, with an emphasis on the representation of affected populations;
- Fair solutions: developing mitigation and adaptation policies and measures that take into account the diverse needs and realities of populations and seek to reduce existing inequalities.

Through these distributive, procedural and recognition issues, climate justice emphasises the relationships, institutions and social structures that are both the origine of climate change and contribute to shaping responses to it. By adopting a more transformative perspective, questions of inequality and power relations lie at the heart of climate justice (Newell et al. 2021).

Territorial approaches suggest that heterogeneity, synergies and spatial complementarities may provide a relevant entry point for addressing the challenges posed by climate change through consultation processes, decision-making and collective action (see paragraph 1). Indeed, they are grounded on the question of inter and intra-territorial inequalities—inequalities in terms of endowment and access to goods and resources, but also in terms of exposure to risks, capacity to adapt, contribution to environmental issues, etc. (Chaumel and La Branche 2008). By fostering interactions and collaborations between actors, promoting participation and inclusion of local populations in the design, planning and implementation of policies, generating collective learning, and establishing new modes of governance, they can also contribute to reducing inequalities and changing the power relations that govern territories, in the direction of a consolidated climate justice, as illustrated in Box 22.3. This case

> **Box 22.3 Case Study 3. Territorial Platforms for Climate Justice in Tunisia**
>
> Since 2018, the Programme for Climate Change Adaptation in Vulnerable Rural Areas (PACTE) in Tunisia has been piloting the establishment and facilitation of territorial planning platforms in six rural regions. This multi-stakeholder consultation mechanism aims to:
>
> - contribute to reducing inter and intra-territorial inequalities through the co-design of integrated development plans and the implementation of public investments for climate change adaptation targeting highly marginal rural areas that have historically neglected by territorial planning policies;
> - alter existing power relations by establishing local governance bodies for territorial diagnosis, planning, and monitoring-evaluation of public investments. These bodies are composed of elected representatives from different local communities, institutional actors, and civil society. They are engaged in a consultation process designed to address territorial inequalities within the intervention regions themselves.
>
> The mechanism has generated massive participation at the local level, with notably 4300 direct participants in the diagnosis and definition of priority development issues and 11,500 collective action proposals collected from citizens (Braiki et al. 2022). Based on these inputs, development plans have been drawn up, taking into account spatial heterogeneity (in terms of access to natural resources, labour, land, etc.) and aiming to reduce territorial inequalities.
>
> Since 2023, these plans have guided public investments across numerous sectors. While networks and social structures conducive to emancipation and the empowerment of collective action have also spontaneously emerged, the interim results remain nuanced. For example, techno-administrative norms have hindered the programme's ambition to benefit the most marginal communities. Pre-existing power relations, between local notables and ordinary citizens, have also been partially reproduced in the new bodies, generating new distributive inequalities.

study also highlights the limitations of territorial approaches, which sometimes fail to achieve the desired changes, particularly when they encounter institutional lock-ins at regional or national scales, such as rigid standards incompatible with change or centralised power structures. This underscores the importance of developing territorial approaches capable of articulating different scales of action.

22.4 Territorial Approaches at the Intersection of Scales of Action

The ambition of some territorial initiatives for climate change mitigation and adaptation can conflict with national and regional sectoral public policies if they have not been thoughtfully articulated. For example, in the case of the transformation of territorial agri-food systems presented above, short supply chains can have negative impacts if they have been designed without coordination or scale articulation. It is therefore not automatic to significantly reduce the carbon footprint of transport through short supply chains if there is no adapted logistics (Ademe 2017). Territorial agri-food systems and associated territorial innovations therefore require specific local (Box 22.4) and national public policies, highlighting the relevance of territorial approaches articulated with other scales. The design of new multi-level governance mechanisms, integrating the interdependencies between different scales of action, and new institutional practices is necessary to enable integrated territorial approaches.

In practice, territorial development initiatives often have to accommodate the existing institutional framework, advocating for increased permeability of institutional and sectoral boundaries (Rhodes 1996) within the framework of projects. Conversely, when a successful territorial experience has demonstrated the relevance of reshaping the institutional framework (Box 22.4), it is sometimes difficult to transfer this successful experience to the specificities of other territories in a country. It is unlikely that the "large-scale impacts" sought by development agencies to meet the challenges of the 2030 agenda will result from the replication of local successes (Caron et al. 2017). Successful experiences here are rarely reproducible and extrapolatable elsewhere, due to their specificities, the volume of resources to be invested, and the need to act at different scales to induce changes. Scaling up involves a more complex process than simply diffusing or propagating a product or model (Moore et al. 2015). Tunisia provides us with an example of building an institutional framework at the national level, taking into account territorial specificities. After several years of experimentation, the Tunisian Ministry of Agriculture developed an innovative approach in 2016 for the implementation of its new water and soil conservation strategy, proposing a new action framework, based on a consultation process between local actors and regional service experts and on the integration of sectoral approaches at the watershed level (Chevrillon et al. 2017).

Box 22.4 Case Study 4. Paragoclima, or the Commitment of an Amazonian Territory to a Low-Carbon Development Trajectory

The Amazon is severely affected by the impacts of climate change: fires, droughts, and floods strike the region with alarming frequency. At the same time, the Amazon is also internationally held accountable, being a "climate killer" due to the massive deforestation affecting the global climate. This dual status of victim and perpetrator of climate change exposes the Amazon to a major territorial risk, hence the need for a territorialised response.

Paragominas is known as the first Amazonian municipality to have defeated the spectre of illegal deforestation, 15 years ago. In early 2023, it centred its new development strategy, called Paragoclima,[2] on the exceptional virtues of the Amazonian climate, to transform atmospheric carbon into agricultural and forestry biomasses of interest. Heat, rainfall, and solar radiation naturally accelerate the metabolism of plant and animal organisms, which, if managed within a framework of good agricultural practices, restore soil fertility, increase the productivity of plantations, pastures, forests, and agroforests. These good practices are cost-effective in terms of inputs, investments, and mechanised work, thus being socially inclusive. Moreover, the new environmental code organises land use according to their aptitudes, thereby recomposing mixed landscapes of forests, pastures, and agricultural lands, capable of generating income while regulating the cycles of water and carbon, in a habitat favourable to biodiversity. Monitored and certified using a comprehensive set of indicators, this development strategy reassures investors and gains the support of local stakeholders, regardless of their social group, as evidenced by the massive support for Paragoclima, formalised by a territorial pact in May 2023 (65 local institutions have committed to it).

Bottom-up approaches play a key role in research for development. It seems important that top-down approaches are not neglected and that support for their practices is developed in parallel. Often described as exogenous, out-of-place and failing in terms of transparency and accountability, it is important to invest in political dialogue and reconsider these top-down and institutional approaches in connection with bottom-up ones, in order to provide grounding for public action and opportunities for learning for all parties involved in multi-level governance. Box 22.5 develops an example of back-and-forth between territories and the national level in Congo for the development of public policies.

---

[2] For more information: https://paragoclima.com.br; https://www.terramaz.org

> **Box 22.5 Feedback 5. Back-and-Forth Between Public Policy and Territory in Congo**
>
> The forest territory of Northern Congo is mainly oriented towards the production of timber (7.7 Mha, nineteen concessions) and the maintenance of exceptional biodiversity through a significant network of protected areas (1.8 Mha, six protected areas). The forests of Northern Congo play a central role in mitigating climate change.
>
> This territory is experiencing land pressure resulting from population growth and migrations from neighbouring countries, but also from practices of often informal and extensive extractive operations, leading to degradation of the forest cover, deforestation, or pollution. The expansion of urban centers is also associated with the progressive opening up of this territory. Indigenous and autochthonous populations, who constitute the majority on the territory's inhabitants, are poorly represented formally in censuses and land allocation plans.
>
> As part of the Northern Congo Forest Landscape project (PPFNC), Cirad supports the establishment of a multi-sectoral and multi-stakeholder platform. This is intended to guide future political choices regarding land-use planning at the level of the Northern Congo forest landscape. A first phase of three workshops held in 2023 resulted in the establishment of a shared and concerted diagnosis of the territory among sectors and actors. Within this platform, eight central administrations, three private companies, two conservation NGOs and six civil society structures have developed a shared vision of the territory. Over a period of eight months, and iteratively, stakeholders have built a model of territorial representation from initial sectoral visions, progressively supported by quantitative data and field feedback. This sharing has led to the emergence of divergent postures and interests and to collectively develop a roadmap for the future activities of this platform, between territorial foresight and multi-actor advocacy.

22.5 Conclusion

Although often relegated to the background in debates on adaptation and mitigation, territorial approaches play a crucial role in the fight against climate change. This chapter outlines the general principles on which these approaches are based to address climate issues—collaboration, inclusivity, trust, justice, spatialisation of problems and responses, multi-sectorality and multi-scalarity—while highlighting the diversity of forms they can take depending on contexts and specificities. Territorial approaches are therefore not ready-made tools, they offer frameworks for thinking and building collective action, which must be reinvented and readapted to each territory and each issue. These frameworks do not replace more global or local approaches, rather, they aim to complement, articulate, strengthen and contribute to reshaping them. This is a major challenge of territorial approaches, as their development can be sometimes limited by their short temporal horizon and the locked institutional frameworks in which they are often embedded.

The focus on the territory, its internal dynamics, and the local impacts of climate change, in connection with other scales of action, should not overshadow the interterritorial dimension of adaptation and mitigation strategies. We have demonstrated how the biophysical manifestations of climate change and the vulnerability of populations vary from one territory to another. These inequalities are even more pronounced between territories in the North and those in the South, with the North historically being the origin of climate change and the poorest countries in the South often facing the most severe impacts and damage. In addition to the essential questions of interterritorial climate justice raised here, these inequalities also reflect significant knowledge gap issues: the potential or actual effects of climate change are more or less documented depending on the territory. This is a fundamental challenge for research which should also be more inclusive and "fair", particularly between certain territories in the South that are less accessible or less equipped and others in the North where common concerns about climate change are perceived as legitimate, but not very audible.

References

Ademe, 2017. Avis de l'Ademe - Alimentation - Les circuits courts de proximité. 8 p

Braiki H, Hassenforder E, Lestrelin G, Morardet S, Faysse N, Younsi S et al (2022) Large-scale participation in policy design: citizen proposals for rural development in Tunisia. EURO J Decis Process 10:100020. https://doi.org/10.1016/j.ejdp.2022.100020

Bulkeley H, Edwards GAS, Fuller S (2014) Contesting climate justice in the city: examining politics and practice in urban climate change experiments. Glob Environ Chang 25:31–40. https://doi.org/10.1016/j.gloenvcha.2014.01.009

Caron P, Valette É, Wassenaar T, Coppens DE, Papazian V (2017) Living territories to transform the world. Versailles, Quæ editions, p 274

Chaumel M, La Branche S (2008) Inégalités écologiques: vers quelle définition? 1:101–110. https://doi.org/10.4000/eps.2418

Chevrillon A, Haha NB, Burte J (2017) Towards a territorialisation of rural policies in Tunisia: the example of water and soil conservation policies. In: Living territories to transform, p 167

Chiffoleau Y, Prevost B (2012) Les circuits courts, des innovations sociales pour une alimentation durable dans les territoires. Norois, 7–20. https://doi.org/10.4000/norois.4245

Debolini M, Siqueira TTS (2021) Crop diversification as a strategy for adaptation to climate change: potential and limits on the case of pistachio in the French Mediterranean region. In: Landscape conference – diversity for sustainable and resilient agriculture, p 109

Mussal FD, Kuik O (2011) Local acceptance of renewable energy—A case study from southeast Germany. Energy Policy 39(6):3252–3260. https://doi.org/10.1016/j.enpol.2011.03.017

Gupta J, Prodani K, Bai X, Gifford L, Lenton TM, Otto I et al (2023) Earth system boundaries and earth system justice: sharing the ecospace. Environ Polit:1–21. https://doi.org/10.1080/09644016.2023.2234794

Ivits E, Cherlet M, Tóth G, Sommer S, Mehl W, Vogt J, Micale F (2012) Combining satellite derived phenology with climate data for climate change impact assessment. Glob Planet Chang 88-89:85–97. https://doi.org/10.1016/j.gloplacha.2012.03.010

Kalkbrenner BJ, Roosen J (2016) Citizens' willingness to participate in local renewable energy projects: the role of community and trust in Germany. Energy Res Soc Sci 13:60–70. https://doi.org/10.1016/j.erss.2015.12.006

Lamine C, Dodet F, Demené C, Rotival D, Latré L, Sabot N et al (2022) Transformations du système agri-alimentaire territorial en sud Ardèche: co-construire une périodisation du passé... qui fasse sens pour l'avenir. Géocarrefour 96(96/3). https://doi.org/10.4000/geocarrefour.20864

Le Treut H (2022) Les territoires; espaces de solutions concrètes, d'expérimentation et d'éducation. In: Climate and civilisation. Érès, Toulouse, pp 149–156

Moore ML, Riddell D, Vocisano D (2015) Scaling out, scaling up, scaling deep: strategies of non-profits in advancing systemic social innovation. J Corp Citizsh 58:67–84

Newell P, Srivastava S, Naess LO, Torres Contreras GA, Price R (2021) Toward transformative climate justice: an emerging research agenda. Wiley Interdiscip Rev Clim Chang 12(6):e733. https://doi.org/10.1002/wcc.733

Rhodes RAW (1996) The new governance: governing without government. Polit Stud 44(4):652–667

Romero HN (2017) The emergence of living lab methods. In: Keyson DV, Guerra-Santin O, Lockton D (eds) Living labs: design and assessment of sustainable living. Springer International Publishing, pp 9–22. https://doi.org/10.1007/978-3-319-33527-8_2

Schlosberg D, Collins LB (2014) From environmental to climate justice: climate change and the discourse of environmental justice. Wiley Interdiscip Rev Clim Chang 5(3):359–374. https://doi.org/10.1002/wcc.275

Siamanta, Zoi Christina et DUNLAP, Alexander (2019) Accumulation by wind energy: wind energy development as a capitalist Trojan horse in Crete, Greece and Oaxaca, Mexico. ACME: An Int J for Critical Geographies 18(4):925–955

Tsayem DM, Philippe C (2022) Landmarks and epistemic characteristics of climate justice. Nat Sci Soc 30:14–30. https://doi.org/10.1051/nss/2022016

Open Access This chapter is licensed under the terms of the Creative Commons Attribution-NonCommercial-NoDerivatives 4.0 International License (http://creativecommons.org/licenses/by-nc-nd/4.0/), which permits any noncommercial use, sharing, distribution and reproduction in any medium or format, as long as you give appropriate credit to the original author(s) and the source, provide a link to the Creative Commons license and indicate if you modified the licensed material. You do not have permission under this license to share adapted material derived from this chapter or parts of it.

The images or other third party material in this chapter are included in the chapter's Creative Commons license, unless indicated otherwise in a credit line to the material. If material is not included in the chapter's Creative Commons license and your intended use is not permitted by statutory regulation or exceeds the permitted use, you will need to obtain permission directly from the copyright holder.

Chapter 23
Food Systems and Climate Change: Mitigation and Adaptation in Agri-Food Chains and Consumption

Marie Walser, Carine Barbier, Nicolas Bricas, and Patrice Dumas

Abstract While climate change mitigation and adaptation in agricultural production have been extensively studied, the post-production stages of the food system—including storage, processing, transportation, distribution, and consumption—have received comparatively limited attention. Yet, these stages face significant climate-related challenges, such as disruptions to supply chains, and are responsible for about one-third of global food system greenhouse gas emissions. The impacts and appropriate responses differ between industrialized countries, with more processed diets and complex logistics, and less industrialized countries, where food systems tend to be simpler. This chapter highlights key mitigation and adaptation strategies that can be implemented in the post-production food system stages and food consumption. It concludes with policy recommendations to support the transformation of food system transformation in the face of climate change.

The scientific literature has extensively documented the pathways for mitigation and adaptation to climate change in the context of agricultural production (see, for example, Chaps. 13, 14, and 15). In contrast, relatively less attention has been devoted to the post-production stages—namely storage, processing, transportation, distribution, and food service—as well as to consumption. Yet, the downstream segments of food systems are increasingly confronted with challenges related to climate change:

M. Walser (✉)
Unesco Chair in World Food Systems, Cirad, Institut Agro Montpellier, Montpellier, France

C. Barbier
UMR International Centre for Research on Environment and Development, CNRS, ENPC, Cirad, AgroParisTech, EHESS, Montpellier, France

N. Bricas
Cirad, UMR MoISA, Montpellier, France

MoISA, University of Montpellier, CIHEAM, Cirad, INRAE, Institut Agro Montpellier, IRD, Montpellier, France

P. Dumas
Cirad, UMR Cired, Montpellier, France

- On the one hand, climate-induced disruptions—such as poor harvests, infrastructure degradation, water scarcity, the proliferation of pests and pathogens, and heightened price volatility for raw materials—pose serious threats to the functioning of supply chains and, consequently, to food security;
- On the other hand, these post-production stages significantly contribute to climate change themselves, accounting for approximately one-third of global food system greenhouse gas (GHG) emissions (Crippa et al. 2021). However, distinctions must be made between industrialised countries, where diets are typically more processed and meat-intensive and rely heavily on both domestic and international transport and refrigeration, and less industrialised countries, where food processing, distribution networks, cold chain infrastructure, and logistics are far less developed. Consequently, many of the mitigation strategies outlined in this chapter are particularly relevant for high-emitting, industrialised nations.

Without claiming to be exhaustive or quantifying the full mitigation and adaptation potential, this chapter aims to identify key levers for climate action in the post-production stages of food systems. It begins by examining measures available to economic actors in the domains of food processing, storage, transportation, and distribution. It then turns to changes in consumer behaviour at the consumption stage. The chapter concludes with a set of policy proposals designed to foster enabling environments for food system transformation in the face of climate change.

23.1 Mitigation and Adaptation in Processing, Storage, Transport, and Distribution

23.1.1 Mitigation Strategies in Post-Production Stages

Within the food processing sector, GHG emissions are primarily associated with energy-intensive manufacturing processes. Certain industries—notably those dealing with sugar, dairy, and starchy products—are especially energy-demanding and therefore particularly emissive (HCC 2024).

Food processing facilities may pursue several mitigation strategies:

- Conduct energy audits and diagnostics to enhance process efficiency (HCC 2024);
- Expand the use of renewable energy sources, a viable option given that many food processing operations require only low to moderate heat (Crippa et al. 2021). In regions with abundant solar radiation, solar energy can significantly increase the share of renewables in the energy mix;
- Employ biomass derived from agricultural and forestry by-products (e.g., bagasse, kernels, husks, peels) (see Chap. 19), offering a cost-effective, reliable, and sustainable energy solution particularly suited to the needs of processors in the Global South, such as in West Africa (Rivier 2017; Graefe et al. 2011);

- Promote fermentation processes where appropriate, as these methods both preserve food and offer nutritional benefits while consuming minimal energy (Rastogi et al. 2022).

Although less directly related to energy consumption, food companies can also mitigate emissions by sourcing raw materials and products from agricultural systems that adopt climate-friendly practices (HCC 2024).

Packaging decisions constitute another significant mitigation pathway. Packaging production, usage, and disposal collectively account for 5.4% of global food system emissions (Crippa et al. 2021). The most effective strategy remains upstream reduction—minimising over-packaging, promoting bulk sales, expanding the use of large formats (especially for non-perishable goods), and transitioning to reusable containers. The choice of materials is also crucial. Life Cycle Assessments (LCAs), which evaluate environmental impacts across the entire packaging value chain, serve as essential decision-support tools. These assessments must also account for potential trade-offs, such as increased food loss due to reduced packaging, which itself carries a carbon footprint.

Storage, particularly within cold chains, is another high-emission segment. It constitutes 43% of the energy consumption in the food distribution sector (Behfar et al. 2018). Furthermore, the refrigeration process can lead to emissions of high Global Warming Potential (GWP) gases, such as hydrofluorocarbons (HFCs), especially when equipment is poorly maintained or improperly disposed of. While these emissions currently stem primarily from industrialised countries with well-developed cold chains (Crippa et al. 2021), both rising global incomes and increasing temperatures are expected to drive significant growth in refrigeration demand worldwide (UNEP 2023). Though refrigeration enhances food preservation and reduces losses, there is a growing need to incorporate spatial considerations—such as the geographic proximity of processing facilities to markets—into LCAs, to minimise transportation and the associated need for cold storage. Moreover, phasing out high-GWP refrigerants in favour of low-emission alternatives is urgent. Tackling the illicit trade in environmentally harmful refrigerants is likewise imperative.[1]

Among the post-production phases, transport represents a significant source of GHG emissions, accounting for between 4.8% and 19.0% of total emissions within food systems, depending on the system boundaries applied (Crippa et al. 2021; Li et al. 2022). Emissions from transport are 2.7 times higher in industrialized countries—specifically North America, Europe, and Oceania—than in other regions worldwide (Li et al. 2022), underscoring the disproportionate responsibility of these regions in mitigating transport-related emissions.

The ongoing urbanization and expansion of international trade have substantially extended food supply chains globally. A critical mitigation strategy involves shortening supply distances across all food chains and moving away from highly

---

[1] Note that the emissions of fluorinated gases are regulated by the Kigali Amendment to the Montreal Protocol, which came into effect in 2019 and relates to the protection of the ozone layer, and by the Paris Agreement on climate change (2015)

globalized, long-distance supply networks. Three principal levers can be employed to achieve this:

- Increasing transportation costs to discourage extensive commercial exchanges;
- Relocating production at scales appropriate to demand. Urban centers, even when considering their surrounding hinterlands, are largely insufficiently supplied. While promoting peri-urban vegetable cultivation is necessary, it remains inadequate. Agricultural sectors that require substantial land areas must be planned at broader territorial scales;
- Reducing the meat content of diets diminishes the demand for animal feed transport. Notably, oilcakes—often imported—constitute up to a quarter of agricultural product traffic (Barbier et al. 2019).

However, this intuitive preference for localizing food sectors, driven by presumed environmental benefits from reduced transport distances, requires thorough evaluation. Differing production methods may lead to varying environmental performances. When factoring in carbon emissions associated with land-use change, agricultural production remains the most climate-impacting phase due to its overall GHG emissions balance, despite the increasing significance of post-production emissions. Food distribution can facilitate the integration of local products into retail assortments by optimizing logistics through the establishment of territorial hubs, collaborative pooling, energy-efficient fleet renewal, the adoption of low-carbon energy sources, and the development of rail and inland waterway freight transport (HCC 2024).

These measures should be contextualized alongside efforts to reduce food losses, which are mitigated by processing, refrigeration, and packaging. Food loss and waste account for approximately 9% of total GHG emissions from the food system (Crippa et al. 2021), contributing to the greenhouse effect both through emissions from unused production and the methane released during landfill decomposition (IPCC 2021). Numerous mitigation strategies exist at all points along the food chain (Hanson and Mitchell 2017).

The scale and nature of losses vary by country. In less industrialized nations, post-harvest losses predominantly result from product deterioration during storage. To address this, Kitinoja (2013) advocates for the deployment of "fresh chains" (12–18 °C), which are particularly suitable for tropical and subtropical products such as fruits, vegetables, roots, and tubers. Unlike the cold chain (0–4 °C), which it complements, the fresh chain relies on low-cost, energy-efficient technologies often made from local materials and easy to maintain—such as evaporative cooling systems or night-ventilated refrigerated storage facilities. Furthermore, training technicians in proper equipment use and end-of-life management represents a key lever for reducing losses (Yahia 2010).

Conversely, in industrialized countries, food loss and waste predominantly occur at the terminal stages of the food system. Several mitigation avenues exist; a few notable examples include:

- Processors and distributors striving to align their orders closely with production capacities or sales forecasts. Although the redistribution of unsold food to charitable organizations is often presented as a dual solution to food waste and insecurity, it is important to recognize the limitations of this approach, which does not fully resolve either issue but rather accepts their coexistence (Bricas and Scherer 2021);
- Retailers educating consumers about the distinction between various expiry dates to prevent confusion that leads to premature disposal of food at home;
- Restaurateurs offering smaller portion sizes, with the option for customers to request additional servings at extra cost or not, thereby allowing them to better match their intake to their appetite. Encouraging customers to take leftovers home, supported by the provision of appropriate packaging, further helps reduce waste (Gunders 2012).

23.1.2 Adaptation Strategies in Post-Production Stages

While climate changes severely affects agricultural production, they also threaten downstream segments of food systems. Several adaptation pathways are emerging, particularly for food processing actors.

Global warming increases pressure on water resources, which are essential to processing operations (washing, cooking, cooling, etc.). Valta et al. (2015) list a range of adaptation measures to address water scarcity, including water audits, process optimisation to reduce usage, and safe water recycling in accordance with health regulations.

Rising temperatures also foster the growth of pathogens in agricultural and food products, raising safety concerns and contributing to losses. This affects fresh products such as fruits and vegetables or animal products, but also grains and even some processed foods (Misiou and Koutsoumanis 2022). Better control of storage conditions is therefore essential.

Furthermore, climate change heightens the uncertainty of raw material supply. Since the twentieth century, industrial food processing has evolved toward automation, treating large volumes of products in a standardised way and using standardised raw materials (Bricas et al. 2021). While diversification and localisation of agricultural production are resilience levers, the food industry must also adapt by developing more flexible processing systems capable of accommodating raw materials with variable characteristics (HCC 2024).

Climate change may also intensify food and water insecurity in vulnerable regions (Roberts 2008; Awuor et al. 2008). Although the use of strategic food reserves has diminished due to economic and logistical reasons, their revival—at household, national, or regional scales—is now being reconsidered (IFPRI 2023). Such stocks would buffer the effects of production volatility, though they depend on reliable transport infrastructures for their operation.

23.2 Mitigation and Adaptation to Climate Change Through Consumption

23.2.1 Mitigation Through Consumption Practices

At the household level, reducing GHG emissions requires a combination of transformations of food practices whilst adopting an energy sobriety perspective.

When considering the procurement of food supplies, the decarbonisation of "last-mile" logistics is essential. For households located in well-served urban areas, low-emission transport modes for shopping—such as walking, cycling, or public transport—should be prioritised and facilitated. In rural or peri-urban areas where such access is limited, pooling food supply trips and integrating quality food retail within territorial planning policies become critical. The use of reusable containers for bulk purchases also contributes to a reduction in the volume of packaging and associated emissions.

Domestic refrigeration is another source of emissions, particularly through over-equipment, inefficient appliances, or poor recycling of cooling systems, which may leak fluorinated gases. Efforts should thus focus on avoiding over-equipment, promoting energy-efficient devices and raising awareness about best practices and recycling at end-of-life. These issues are not exclusive to industrialised countries: even in off-grid rural areas, refrigeration is often ensured by gas-powered appliances, making this a global concern.

Food preparation, particularly cooking, also represents a lever for GHG emission reduction. Impacts vary depending on the cooking method, energy source, equipment efficiency, and user behaviour (ZhongYue et al. 2015). In less industrialised countries, biomass-based cooking methods (wood, charcoal) are still widespread, and associated with high emissions. MacCarty et al. (2008) recommend transitioning to "improved" cookstoves that reduce GHG emissions while maintaining usability. In industrialised contexts, efforts should focus on reducing energy consumption from cooking through (Frankowska et al. 2020):

- wider adoption of low-emission cooking methods (microwave, pressure cooker, slow cooker) and cleaner energy sources (e.g. biogas);
- public information campaigns on energy-efficient practices (e.g. using lids, adapting heating levels);
- promotion of batch cooking, which spreads energy use over several meals, increasing overall efficiency (Frankowska et al. 2020).

According to the IPCC's Sixth Assessment Report (2023), dietary change is among the most effective levers for reducing GHG emissions globally. The upstream stages of food systems—and their associated emissions—are strongly influenced by consumption patterns. Three key dietary shifts stand out:

1. Reducing overconsumption: in many industrialised countries, the overabundance of food and marketing-driven incentives contribute to both food waste and excessive intake—especially of fats, sugars, and animal proteins. Curtailing this overconsumption is not only vital for climate mitigation but also for public health, as rates of obesity are rising dramatically worldwide (Micha et al. 2022).
2. Transitioning to more plant-based diets: in contexts of high animal product consumption, shifting toward diets richer in plant-based foods is widely recognised as an effective way to lower emissions and reduce diet-related health risks (Moore 1971; Poore and Nemecek 2018; Steinfeld et al. 2006; Searchinger et al. 2019; Westhoek et al. 2011) (see Chap. 10). Fouillet et al. (2023) show that diets in which 70–80% of protein intake comes from plant sources can remain nutritionally balanced.
3. Limiting ultra-processed foods: these products are often associated with both higher emissions and caloric overconsumption (Hendrie et al. 2016). Favouring local, seasonal, and sustainably sourced products over ultra-processed ones contributes to both mitigation and resilience.

Regardless of the dietary shift—whether toward frugality, plant-based eating, or reduced processing—it is essential to consider the wide diversity of contexts: between countries, within countries (urban vs. rural), and across population groups. These differences arise from geographic, cultural, economic, and historical factors. For instance, while meat consumption is high in China, it is much lower in India, partly due to cultural and religious reasons (Schönfeldt and Gibson 2012). In rapidly developing regions, dietary aspirations may still reflect patterns shaped by past scarcity. Therefore, dietary transitions must be adapted to local realities and should primarily target populations that overconsume. Promoting one-size-fits-all "optimal" diets based on environmental and health averages carries the risk of low public acceptance if local specificities are ignored.

23.2.2 Adaptation at the Consumption Stage

At the level of consumption, adaptation to climate change primarily entails enhancing household resilience in the face of temporary shortages, increased variability in the quality and quantity of available foodstuffs, and heightened price volatility. In rural communities exposed to climatic hazards, where food security relies only partially on market mechanisms, long-established local practices have developed to prevent shortages. These practices encompass a range of adaptive strategies that vary between communities: diversification of food production, cultivation of climate-resilient crops (such as cassava), food storage, and the accumulation of assets that can be mobilized in times of need (e.g., livestock, jewelry). Social and familial solidarity networks, recourse to foraging or hunting during acute agricultural crises, as well as temporary or permanent migration to reduce food demand, further exemplify these adaptive responses. In more market-integrated areas, where self-production has diminished but price fluctuations remain significant, similar strategies prevail—including diversification, storage, solidarity mechanisms, risk pooling, mobility, and even a resurgence of subsistence production activities.

23.3 Supporting Actors and Consumers: Political Responses to Climate Change

The mitigation and adaptation strategies discussed herein delineate the principal levers available within post-production stages to address the climate-related challenges confronting food systems. Their efficacy is enhanced when deployed in concert across multiple stages and territorial scales. Crucially, these strategies necessitate robust public support; the transformation of food systems cannot rely solely on voluntary private sector initiatives or shifts in individual consumer behavior. Indeed, a substantial body of research underscores the limitations of the notion of the "enlightened" consumer, whose awareness alone is deemed sufficient to drive behavioral change (Lahlou 2017).

The construction and institutionalization of a new social food contract fundamentally presuppose the establishment of enabling environments that facilitate behavioral shifts and, importantly, regulate private actors—especially those operating within oligopolistic market structures and bearing significant responsibility for food system GHG emissions. Such enabling environments can be categorized as follows:

- Cognitive Environments: these shape and influence preferences and attitudes, both consciously (through education, information dissemination, awareness campaigns, and advisory services) and unconsciously (via advertising effects, rumors, and message repetition). For economic actors, cognitive environments encompass institutions dedicated to training, information, research, etc. Strengthening institutions tasked with guiding actors through transitions in production models is imperative. For consumers, beyond communication campaigns, regulatory measures targeting advertising and marketing practices—particularly those employed by large retail chains—are vital.
- Material Environments: these either facilitate or constrain access to and utilization of food. This dimension includes the physical environment—proximity of retail outlets, availability of storage facilities, kitchen and waste management infrastructure, and dining amenities. Urban commercial planning that strategically locates food retail points, improves their functionality, and enhances their appeal is a significant lever to ensure access to quality food, prevent "food deserts" or "food swamps," and elevate the profile of certain products. Several African cities, for example, have initiated efforts to improve venues and conditions for local artisanal food production and sale, countering the expansion of large-scale retailing that favors environmentally and health-wise problematic

industrial products (see the AfriFOODlinks project).[2] Furthermore, by incorporating environmental standards into trade agreements, states can leverage policy to enhance the quality of imported foods. Equally important are food prices and payment conditions, which constitute core elements of the material environment. Rather than simply reflecting supply-demand equilibrium, prices should be viewed as instruments to incentivize or dissuade consumption patterns. Contrary to purely liberal approaches that reject subsidies or taxes, price interventions—through regulation, public support, and taxation—represent critical tools to guide stakeholder behavior. Given that climate change will impact food quality, accessibility, and may exacerbate pre-existing inequalities, public authorities must be prepared to support vulnerable populations to avert deepening food insecurity (Fanzo et al. 2018).

- Social Environments: food consumption is deeply embedded in social practices and symbolic meanings, transmitted culturally, intergenerationally, and shaped by immediate social contexts. Reductions in animal product consumption or portion sizes, for instance, may conflict with social norms that valorize such products or abundance, particularly in hospitality contexts. Modifying these social norms involves leveraging the influence of lifestyle exemplars—celebrities, religious figures, athletes, influencers—as well as representing alternative practices through media and cultural narratives, such as television series.

Transforming the cognitive, material, and social environments underpinning food systems is indispensable for confronting climate challenges and advancing broader sustainability objectives. Achieving this necessitates a recalibration of governance power relations in favor of public authorities and, crucially, citizens—anchored in the principles of food democracy. While citizen engagement, whether direct or through representative institutions, tends to be most effective at local scales (neighborhoods, villages, cities), policy incentives must be coherently coordinated across multiple levels, ranging from the local to the global.

1. www.fao.org/agroecology/fr; agroecology@fao.org.
2. https://www.cocoa4future.org/.
3. https://www.fondationavril.org/projects/cirad-oracle/.
4. https://www.breedcafs.eu.
5. https://fundacionnicafrance.org.
6. https://alliancebioversityciat.org/fr.
7. https://report.ipcc.ch/ar6syr/index.html.
8. https://www.cirad.fr/dans-le-monde/dispositifs-en-partenariat/agroforesta.
9. https://www.cirad.fr/collaborer-avec-nous/science-ouverte/infrastructures-ouvertes-du-cirad/abiophen-et-serres-experimentales.
10. https://www.bolero-project.eu.

---

[2] https://afrifoodlinks.org/

11. PEPR SVA: Advanced plant breeding in response to the climate challenge. The priority research programmes and equipment (PEPR) constitute the upstream/ research aspect of the France 2030 strategies. https://www.pepr-selection-vegetale.fr.
12. CGIAR: Consultative Group on International Agricultural Research.
13. https://umr-agap.cirad.fr/recherches/projets-de-recherche/greener.
14. https://rd-agri.fr/detail/DOCUMENT/chambres_d%27agriculture_294233.
15. For more information: https://paragoclima.com.br; https://www.terramaz.org.
16. Note that the emissions of fluorinated gases are regulated by the Kigali Amendment to the Montreal Protocol, which came into effect in 2019 and relates to the protection of the ozone layer, and by the Paris Agreement on climate change (2015).
17. https://afrifoodlinks.org/.

References

Awuor CB, Orindi VA, Ochieng AA (2008) Climate change and coastal cities: the case of Mombasa, Kenya. Environ Urbanization 20(1):231–242. https://doi.org/10.1177/0956247808089158

Barbier C, Couturier C, Pourouchottamin P, Cayla J-M, Sylvestre M, Pharabod I et al (2019) The energy and carbon footprint of food in France. CIRED, Paris, p 24p

Behfar A, Yuill D, Yu Y (2018) Supermarket system characteristics and operating faults (RP-1615). Sci Technol Built Environ 24(10):1104–1113. https://doi.org/10.1080/23744731.2018.1479614

Bricas N, Scherer P (2021) Fighting against insecurity through food aid? In: Bricas N, Conaré D, Walser M (eds) An ecology of food. Quae editions, Versailles, pp 205–214

Bricas N, Conaré D, Walser M (2021) The industrialisation of the food supply. In: Bricas N, Conaré D, Walser M (eds) An ecology of food. Quae editions, Versailles, pp 81–93

Crippa M, Solazzo E, Guizzardi D et al (2021) Food systems are responsible for a third of global anthropogenic GHG emissions. Nat Food 2:198–209. https://doi.org/10.1038/s43016-021-00225-9

Fanzo J, Davis C, McLaren R, Choufani J (2018) The effect of climate change across food systems: implications for nutrition outcomes. Glob Food Sec 18:12–19. https://doi.org/10.1016/j.gfs.2018.06.001

Fouillet H, Dussiot A, Perraud E et al (2023) Plant to animal protein ratio in the diet: nutrient adequacy, long-term health and environmental pressure. Front Nutr 10:1178121. https://doi.org/10.3389/fnut.2023.1178121

Frankowska A, Rivera XS, Bridle S et al (2020) Impacts of home cooking methods and appliances on the GHG emissions of food. Nat Food 1:787–791. https://doi.org/10.1038/s43016-020-00200-w

Graefe S, Dufour D, Giraldo A et al (2011) Energy and carbon footprints of ethanol production using banana and cooking banana discard: a case study from Costa Rica and Ecuador. Biomass Bioenergy 35(7):2640–2649. https://doi.org/10.1016/j.biombioe.2011.02.051

Gunders D (2012) Wasted: how America is losing up to 40 percent of its food from farm to fork to landfill. NRDC, 58 p. https://www.nrdc.org/sites/default/files/wasted-2017-report.pdf

Hanson C, Mitchell P (2017) The business case for reducing food loss and waste. Champions 12.3, Washington, DC, 14 p. https://champions123.org/sites/default/files/2020-08/business-case-for-reducing-food-loss-and-waste.pdf

HCC (2024) Accelerating the climate transition with a low-carbon, resilient and fair food system, 167 p. https://www.hautconseilclimat.fr/wp-content/uploads/2024/01/2024_HCC_Alimentation_Agriculture_25_01_webc_vdef_c.pdf
Hendrie GA, Baird D, Ridoutt B et al (2016) Overconsumption of energy and excessive discretionary food intake inflates dietary greenhouse gas emissions in Australia. Nutrients 8(11):690. https://doi.org/10.3390/nu8110690
IFPRI (2023) Global food policy report 2023: rethinking food crisis responses. IFPRI, Washington, p 140. https://cgspace.cgiar.org/bitstreams/a935566d-ddea-49cc-a2a7-a0c60743e6cf/download
IPCC (2021) Climate change 2021: the physical science basis. https://www.ipcc.ch/report/ar6/wg1/
IPCC (2023) Climate change 2023, synthesis report. Contribution of working groups I, II and III to the sixth assessment report of the intergovernmental panel on climate change, 184 p. https://doi.org/10.59327/IPCC/AR6-9789291691647
Kitinoja L (2013) Use of cold chains for reducing food losses in developing countries, PEF White Paper 13-03, 16 p
Lahlou S (2017) Installation theory. The societal construction and regulation of behaviour. Cambridge University Press, Cambridge, p 510
Li M, Jai N, Lenzen M, Malik A et al (2022) Global food-miles account for nearly 20% of total food-systems emissions. Nat Food 3:445–453. https://doi.org/10.1038/s43016-022-00531-w
MacCarty N, Ogle D, Still D et al (2008) A laboratory comparison of the global warming impact of five major types of biomass cooking stoves. Energy Sustain Dev 12(2):56–65. https://doi.org/10.1016/S0973-0826(08)60429-9
Micha R, Di Cesare M, Zanello G (2022) Global nutrition report 2022, 140 p. https://globalnutritionreport.org/reports/2022-global-nutrition-report/
Misiou O, Koutsoumanis K (2022) Climate change and its implications for food safety and spoilage. Trends Food Sci Technol 126:142–152. https://doi.org/10.1016/j.tifs.2021.03.031
Moore LF (1971) Diet for a small planet. Ballantines Books, New York, p 301
Poore J, Nemecek T (2018) Reducing food's environmental impacts through producers and consumers. Science 360:987–992. https://doi.org/10.1126/science.aaq0216
Rastogi YR, Thakur R, Thakur P et al (2022) Food fermentation—significance to public health and sustainability challenges of modern diet and food systems. Int J Food Microbiol 371:109666. https://doi.org/10.1016/j.ijfoodmicro.2022.109666
Rivier M (2017) Analysis and multicriteria optimisation of a heat transfer and drying process for an application in West Africa, Doctoral thesis, National Institute of Higher Agricultural Studies of Montpellier, 176 p
Roberts D (2008) Thinking globally, acting locally—Institutionalising climate change at the local government level in Durban, South Africa. Environ Urban 20(2):521–537. https://doi.org/10.1177/0956247808096126
Schönfeldt HC, Gibson HN (2012) Dietary protein quality and malnutrition in Africa. Br J Nutr 108(S2):S69–S76. https://doi.org/10.1017/S0007114512002553
Searchinger T, Waite R, Hanson C, Ranganathan J, Dumas P, Matthews E (2019) Creating a sustainable food future: a menu of solutions to feed nearly 10 billion people by 2050 – synthesis report, World Resources Institute, 556 p
Steinfeld H, Gerber P, Wassenaar T, Castel V, Rosales M, De Haan C (2006) Livestock's long shadow: environmental issues and options. FAO, Rome, p 414
UNEP (2023) Emissions gap report 2023: broken record—temperatures hit new highs, yet world fails to cut emissions (again), Nairobi, 80 p. https://doi.org/10.59117/20.500.11822/43922
Valta K, Moustakas K, Sotiropoulos A et al (2015) Adaptation measures for the food and beverage industry to the impact of climate change on water availability. Desalin Water Treat 57(5):2336–2343. https://doi.org/10.1080/19443994.2015.1049407
Westhoek H, Rood T, van den Berg M et al. (2011) The protein puzzle: the consumption and production of meat, Dairy and Fish in the European Union. PBL Netherlands Environmental Assessment Agency, The Hague, 218 p

Yahia E (2010) Development and challenges of the cold chain in the developing world. Acta Horticult 877:127–132. https://doi.org/10.17660/ActaHortic.2010.877.9

ZhongYue X, DaWen S, Zi Z et al (2015) Research developments in methods to reduce the carbon footprint of cooking operations: a review. Trends Food Sci Technol 44(1):49–57. https://doi.org/10.1016/j.tifs.2015.03.004

Open Access This chapter is licensed under the terms of the Creative Commons Attribution-NonCommercial-NoDerivatives 4.0 International License (http://creativecommons.org/licenses/by-nc-nd/4.0/), which permits any noncommercial use, sharing, distribution and reproduction in any medium or format, as long as you give appropriate credit to the original author(s) and the source, provide a link to the Creative Commons license and indicate if you modified the licensed material. You do not have permission under this license to share adapted material derived from this chapter or parts of it.

The images or other third party material in this chapter are included in the chapter's Creative Commons license, unless indicated otherwise in a credit line to the material. If material is not included in the chapter's Creative Commons license and your intended use is not permitted by statutory regulation or exceeds the permitted use, you will need to obtain permission directly from the copyright holder.

Chapter 24
Agricultural Methane: A Lever for Reducing Greenhouse Gas Emissions to Comply with the Paris Agreement

Rémi Prudhomme, Myriam Adam, and Mohamed Habibou Assouma

Abstract Methane is the second most significant anthropogenic greenhouse gas after CO_2, with a strong short-term warming effect. Agriculture accounts for about 40% of anthropogenic methane emissions, mainly from ruminant enteric fermentation, manure management, and rice cultivation. This chapter reviews global and national policies to reduce methane emissions, such as the Global Methane Pledge, which targets a 30% reduction by 2030. It explores some specific mitigation strategies in the agricultural sector: improved livestock feeding, livestock genetic selection, water management in rice systems (e.g., alternate wetting and drying), and manure treatment technologies. Reducing methane has a rapid climate benefit and is crucial for meeting Paris Agreement goals while balancing food security and rural livelihoods.

Methane emissions are the second source of greenhouse gas (GHG) emissions most influenced by human activity and the most climate-forcing after carbon dioxide emissions. Over the past two centuries, methane (CH_4) emissions have almost doubled since 1900, primarily due to human activities (IPCC 2022). Methane is

---

R. Prudhomme (✉)
Cirad, UMR CIRED, Nogent-sur-Marne, France
e-mail: remi.prudhomme@cirad.fr

M. Adam
Cirad, UMR AGAP Institute, Montpellier, France

UMR AGAP Institute, Univ Montpellier, Cirad, INRAE, Institut Agro, Montpellier, France

Faculty of Agriculture and Food Processing, National University of Battambang, Battambang, Cambodia

M. H. Assouma
UMR SELMET, University of Montpellier, Cirad, INRAE, Montpellier Institut Agro, Montpellier, France

Cirad, UMR SELMET, dP ASAP (Agro-sylvo-pastoral systems in West Africa), Bobo Dioulasso, Burkina Faso

International Research-Development Centre on Livestock in Subhumid Zone (CIRDES), Bobo-Dioulasso, Burkina Faso

responsible for 30% of global warming from the pre-industrial period to the present day. It differs from CO_2 due to its short lifespan (11.8 years) (IPCC 2022), but its high global warming potential (GWP), i.e., its ability to trap heat in the atmosphere, is 28 times more warming over a hundred-year period. Methane thus has a significant direct effect on global warming, but also through atmospheric chemical reactions that lead to the production of ozone (O_3) and water vapour, which are themselves GHGs, adding to radiative forcing.[1]

Methane emissions can be of anthropogenic origin (359 $MtCH_4$/year over the period 2008–2017, or about 60% of methane emissions) or natural (217 $MtCH_4$/year, or about 40% of emissions) (Saunois et al. 2020; summarised in Fig. 24.1). Among these anthropogenic emissions, methane emissions come from agriculture for 40% (141 $MtCH_4$/year over the period 2008–2017), from fossil fuel extraction for 33% (114–115 $MtCH_4$/year over the period 2008–2017), from waste management and landfills for 18% (64 $MtCH_4$/year over the period 2008–2017) and from biomass combustion for 9% (32 $MtCH_4$/year over the period 2008–2017) (Saunois et al. 2020; summarised in Fig. 24.1). These total methane emissions are steadily increasing at a constant rate (Tollefson 2022). Agricultural methane comes from the degradation of organic matter by methanogenic bacteria, in an oxygen-poor environment.

Although methane emissions associated with fossil fuel production represent a large part of anthropogenic methane emissions and therefore a significant lever for reducing methane emissions (a 30% reduction in anthropogenic methane emissions can be achieved through a drastic reduction in the intensity of emissions associated with oil and gas activities), we will focus in this chapter on agricultural methane emissions, which are the subject of this book. Methane emissions in agriculture can come from (1) livestock production, which is the largest source of agricultural methane emissions (109 $MtCH_4$/year on average over the period 2008–2017), largely from enteric fermentation[2] (90% of this methane comes from direct emissions from ruminants) and from manure management for the rest, and (2) emissions from rice cultivation (31 $MtCH_4$/year on average over the period 2008–2017) (Saunois et al. 2020, summarised in Fig. 24.1). Agriculture also influences some of the methane emissions from biomass combustion or biofuel (30 $MtCH_4$/year over the period 2008–2017) through practices such as slash-and-burn agriculture, with a predominance of emissions observed in tropical areas (about 65% of emissions at latitudes below 30 °N) (Saunois et al. 2020). Uncertainties about current estimates of natural methane emissions linked to permafrost thaw or methane emissions associated with fossil fuels suggest a clear underestimation of these methane emissions, which contributes to putting methane in the spotlight.

---

[1] Radiative forcing: the difference between the radiative energy received and the radiative energy emitted by a given climate system.

[2] Enteric fermentation: a digestive process in which microorganisms break down substrates (especially carbohydrates) into simpler molecules, allowing their absorption into an animal's bloodstream.

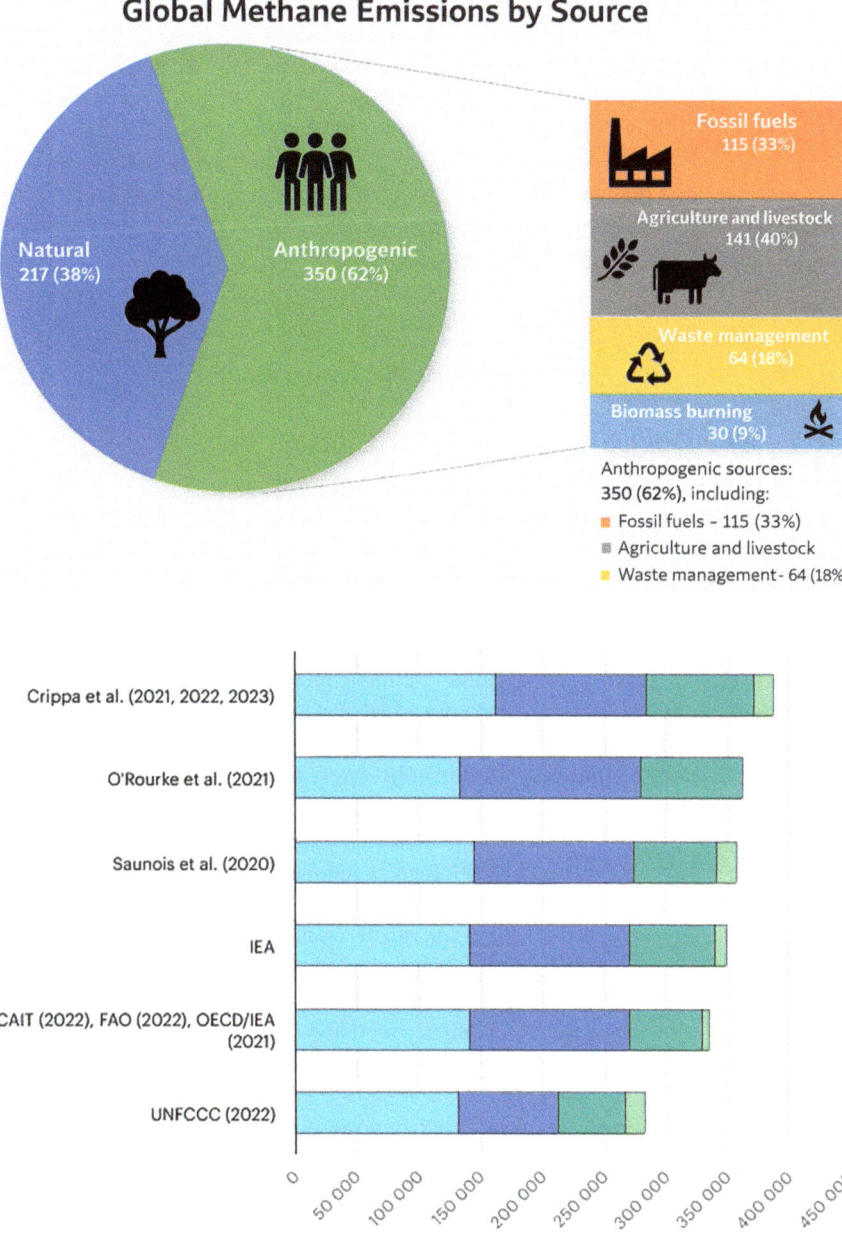

Fig. 24.1 Distribution of methane emissions (**a**) between natural and anthropogenic methane emissions (annual methane emission sources between 2007 and 2018 in MtCH$_4$/year), and (**b**), among anthropogenic methane emissions, between emission sources from agriculture, fossil fuel extraction, biomass combustion or waste management (annual anthropogenic methane emission sources in 2023 in ktCH$_4$/year). (Sources: Saunois et al. 2020; International Energy Agency, 2023. Methane Tracker Database, IEA, Paris. License: Creative Commons Attribution CC BY 4.0. https://www.iea.org/data-and-statistics/data-tools/methane-tracker-data-explorer)

The rest of this chapter generally addresses scenarios and policies for reducing methane emissions at the international and national level. Then the methane emissions directly associated with the agricultural sector, namely emissions from the livestock sector and rice cultivation, will be analysed, especially in light of methane reduction policies, and by proposing solutions or mitigation options for these emissions. Although natural methane emissions (38% of methane emissions mainly from wetlands) and methane emissions associated with fossil fuel extraction (33% of anthropogenic emissions) are significant sources, we will only briefly discuss them in this chapter to focus on agricultural emissions.

24.1 Policies for Reducing Methane Emissions

24.1.1 The Political Objectives for Reducing Methane

In the Kyoto Protocol established in 1997, methane was included as a GHG in the international climate policies of the United Nations Framework Convention on Climate Change (UNFCCC), as a significant contributor to climate change. But this led to few methane reduction measures. This lack of interest in methane emissions has recently transformed into a particular focus, materialised by the launch at COP26 in Glasgow in 2021 of a global methane agreement (the Global Methane Pledge) by the United States and Europe, which has currently been joined by 150 countries representing just over 50% of current methane emissions (Crippa et al. 2023). The signatory countries commit to collectively reducing their methane emissions by 30% by 2030 compared to 2020, focusing on a reduction in the energy and waste sectors. However, a reduction of methane in the agricultural sector is mentioned in the agreement through technological innovations and partnerships with farmers.

In this growing interest in reducing methane emissions, 476 specific methane reduction measures are present in 168 nationally determined contributions (Malley et al. 2023). These mitigation measures (for example, optimising animal feed management, genetic selection to reduce enteric methane emissions from ruminants or implementing water-saving technology, such as Alternate Wetting and Drying (AWD) in rice cropping systemsg) cover 40% of current methane emissions (for example, the Nationally Determined Contributions, NDCs,[3] of Senegal; Box 24.1) and, if implemented to their maximum technical potential, they would reduce methane emissions by 31%. Despite a significant technical potential for reducing

---

[3] Nationally Determined Contributions or NDCs: national climate commitments defined within the framework of the Paris Agreement.

methane emissions in the agricultural sector, these reductions in this sector remain low in these commitments (Malley et al. 2023). If we take the example of Cambodia (Box 24.2), the NDCs mainly target the forest sector; although agriculture is mentioned, funding for reducing GHG emissions, including methane, remains low.

Box 24.1 Emission Reduction Measures in the Agricultural Sector in Senegal's Nationally Determined Contribution

Senegal is currently seeking to increase its agricultural production (peanuts, paddy rice or fruits and vegetables) with the deployment of the second phase of the Programme to Boost and Acceleratef the Growthof Senegalese agriculture (Pracas2, 2019–2023), the agricultural component of the Emerging Senegal Plan (PSE). Senegal also aims to increase animal production through the modernisation of farming practices and support for meat, poultry and dairy sectors, in accordance with the Livestock Development Policy Letter (2017–2021).

The strategic actions of the NDC* are:

- the conversion of 99,621 ha of agricultural land to support natural regeneration (RNA) practices and 4500 ha where compost will be applied by 2030, and the provision of organic manure and improved compost with biogas production;
- the conversion of 28,500 ha of irrigated rice to an intensive rice cultivation system (SRI) that reduces both water volumes used and methane emissions, and the transition to 498,105 ha for RNA and 14,400 ha for compost.

* Nationally Determined Contributions in Senegal (2020), https://unfccc.int/sites/default/files/NDC/2022-06/CDNSenegal%20approuv%C3%A9e-pdf-.pdf

---

24.1.2 The Role of Methane Emissions in Climate Scenarios

To assess the potential for reducing methane emissions in the future, it is essential to consider it in the development of climate scenarios. The importance of methane in the climate scenarios of the scientific literature is increasing for several reasons: (1) in more complex climate scenarios, emissions other than CO_2, such as methane emissions, play a role as important as CO_2 emissions, and (2) the short-term effect of a reduction in methane emissions is of interest.

Box 24.2 Nationally Determined Contributions in Cambodia (2020–2025)

The NDC aims to reduce GHG emissions by 41.7% by 2030, compared to the 2016 business as usual (BAU) scenario for the FOLU sectors (Food and Land Use, i.e., agriculture, energy, industry and waste). Agriculture accounts for only 0.9% of the total funding allocated to mitigation measures, while LULUCF (Land Use, Land-Use Change and Forestry) accounts for 60% of the total mitigation funding (out of a total of 5.7 billion dollars). Regarding adaptation measures, although agriculture is the sector with the most priority actions (17 actions), it only represents 15% of the total adaptation funding (out of a total of two billion dollars).

Agriculture is highlighted in the priority actions for adaptation, and not for mitigation, while the opposite is true for the FOLU sector. The estimated funds needed are substantial for the FOLU sector in its mitigation actions, while they are less for the adaptation actions of agriculture. Soil conservation is well taken into account, particularly in the adaptation action aimed at increasing the sustainability and efficiency of agricultural land management techniques through conservation agriculture (written objective: protect soils against erosion; increase soil organic carbon, etc.; budget: about 25 million US dollars).

In ambitious climate scenarios limiting global warming to +1.5 °C compared to the pre-industrial era, a reduction in biogenic methane emissions[4] of 34% and 42% by 2050 compared to 2010 is estimated (IPCC 2018). These scenarios indeed have in common the goal of achieving "net zero emissions" (all gases combined) around 2050 and therefore explicitly tackle the question of the place of emissions other than CO_2 and in particular methane in climate scenarios compatible with the commitments of the Paris Agreement. This net zero emission goal therefore directly links the methane reduction target with the carbon dioxide reduction target. To achieve this goal (by summing GHG emissions and GHG sinks), the ambition to reduce methane emissions therefore influences the available carbon budget (IPCC 2018).

But methane has an essential role in the net zero emission targets due to its short atmospheric lifespan: (1) the question of equivalence between methane and a long-lived GHG like CO_2 is particularly important (Box 24.3), (2) a reduction in methane emissions has a significant short-term effect on the global warming trajectory. Regarding this second aspect, in ambitious climate scenarios of emission reduction with a low overshoot of the climate target in the middle of the century, between 30%

---

[4] Biogenic methane emissions: methane emissions resulting from the decomposition of organic matter by methanogenic archaea in an anaerobic environment, primarily occurring in agriculture and waste management.

and 60% of the reduction in global warming in the coming decades (2030–2040) systematically includes reductions in methane emissions (Ocko et al. 2021). This reduction in short-term warming plays a key role in stabilising temperatures at +1.5 °C or + 2 °C compared to the pre-industrial era, by avoiding a significant increase in the global average temperature by mid-century, which would in turn disrupt the methane cycle, notably causing natural methane emissions from wetlands (Peng et al. 2022). Even though agricultural emissions only account for a portion of total methane emissions (40% of anthropogenic methane emissions), they therefore represent a significant lever for limiting natural methane emissions and thus significantly contributing to the fight against global warming. This is particularly reflected in scientific and political debates around the use of equivalence metrics between methane and longer-lived greenhouse gases (GHGs) such as CO_2.

> **Box 24.3 Methane Equivalence Metrics**
>
> GHG reduction strategies can be set with a common emission reduction goal or with independent goals for each gas. In the case of a common goal, the carbon dioxide equivalent or CO_2 equivalent, abbreviated as $eqCO_2$, is used as an equivalence metric to compare the effect of emissions from various GHGs based on their GWP, by converting the quantities of other gases into an equivalent quantity of carbon dioxide with the same GWP over a given period. For example, the GWP100 (GWP over 100 years) of methane is 28 while from nitrous oxide (N_2O), it is 273. The peculiarity of methane is its short lifespan compared to other GHGs like CO_2 and N_2O. The choice of equivalence metric is then delicate, as the metric usually used in IPCC reports (the GWP100) underestimates the importance of methane in the short term. The GWP* was proposed to address this lack by proposing a metric that takes the flow and not the stock of methane in the atmosphere as the source of methane's radiative activity, but it raises questions of fairness in its use (Prudhomme et al. 2021). In this context, it is no longer an accumulation of GHGs in the atmosphere that causes warming, but the emission flow. A neutral effect on the climate is therefore no longer achieved by reducing emissions to zero (as is the case for CO_2), but by stabilisation, or even a slight decrease, taking into account its long term warming effect. To avoid exacerbating debates associated with equity issues, particularly between Northern and Southern countries with the use of the GWP* indicator, some authors propose to construct explicit methane reduction targets, in addition to CO_2 targets, and to avoid converting methane emissions into $eqCO_2$ (Prudhomme et al. 2021).

24.2 Methane Reduction Technologies in Agriculture

24.2.1 Rice Production

Rice cropping systems (see Chap. 13) store up to 108 t/ha of organic carbon in the soil to a depth of one meter (about 10% more than global soils in general) (Liu et al. 2021) and contribute to local and global food security. However, about 30% of methane (CH_4) and 11% of nitrous oxide (N_2O) from global agriculture are emitted by rice paddies (Gupta et al. 2021). Indeed, these are often intensively managed (i.e., they use a lot of synthetic inputs: fertilizers, pesticides), and about 50% of the 180 Mha of rice grown worldwide are in flooding conditions for at least part of the year (Bouman et al. 2006). This results in anoxic and reducing soil conditions, in which the decomposition of organic matter contributes to the active process of the carbon and nitrogen (N) cycle, leading to GHG emissions, particularly CH_4.

Recently, Qian et al. (2023) examined the effect of management practices (continuous flooding, addition of organic matter) and showed that management strategies have a potential for mitigating GHG emissions. The transition from intensive rice cropping systems with two to three rice cycles per year to more diversified rice cropping systems will result in better water and nutrient management and potentially a reduction in GHG emissions. Options such as integrating rainfed crops other than rice into the cropping scheme or alternating wet and dry periods for rice through irrigation are known to reduce CH_4 emissions, but N_2O emissions can also increase significantly, as aerobic conditions are more conducive to the formation of large amounts of N_2O. These type of agroecological practices (see Chap. 21) allow for the production of a more diverse biomass, potentially leading to a greater contribution of organic matter in cropping systems (Shumba et al. 2023). However, unlike in rainfed conditions, the internal dynamics of soil carbon (C) and nitrogen (N) induced by agri-environmental practices within various water management regimes are not well understood, with numerous feedback loops between the carbon and nitrogen cycles. In rice cropping systems, typically characterised by partial water control in many regions of the world, soil carbon storage is often offset by increased emissions of methane (CH_4) and nitrous oxide (N_2O) (Liu et al. 2021). Another strategy for reducing GHG emissions is the development of rainfed rice. Flooding, which promotes the presence of methanogenic bacteria, is not strictly necessary for rice growth. This type of rice can be grown under strict rainfed conditions on lands sheltered from temporary submersions (as with wheat). Thus, high-performing rainfed rice could allow for crop rotation diversification to reduce pressure on freshwater resources and contribute to reducing methane emissions (see Chap. 5 for an explanation of the development potential of rice varieties that could be grown under rainfed conditions). Finally, the Alternate Wetting and Drying (AWD) technology is well-known, and can reduce water consumption in rice production by 30%, offering farmers the opportunity to lower their production costs without yield loss, while also reducing GHG emissions. AWD involves periodically draining the field to a certain threshold and then flooding it again. It has been proven that AWD technology can reduce GHG emissions, particularly methane (CH_4), associated with rice production by 30% to 70%, without causing a decrease in yield. During the dry phases, methane-producing bacteria are inhibited, reducing GHG emissions. However, implementing this technique remains challenging in rural areas, where managing timely field drainage can be difficult. Depending on local conditions, the number of drainages and the number of days the field is not flooded vary.

24.2.2 Ruminant Livestock Farming

Approximately 32% of global methane emissions related to human activity come from the livestock sector, specifically from microbial processes related to enteric fermentation in animals, particularly ruminants, from the treatment of livestock effluents, and from the production of animal feed. Methane emissions from the livestock sector were estimated in 2015 at 3.4 Gt CO_2 eq, accounting for 54% of global GHG emissions from the livestock sector: this makes methane the most significant GHG in this sector (FAO 2023a). According to various authors, Ruminant livestock farming contributes between 70% (Gerber et al. 2013) and 90% (Scholtz et al. 2020) of the livestock sector's methane emissions worldwide, with enteric methane being the main source. Enteric methane comes from the fermentation process resulting from the decomposition of plant matter by microorganisms in the rumen of ruminants (cattle, buffalo, goats, and sheep). Several reviews on methane emissions from the livestock sector are available and detail the different sources (Tedeschi et al. 2022). According to the latest IPCC report (2022), reducing enteric CH_4 emissions from ruminant production is a key element of strategies designed to limit the increase in global temperature to +1.5 °C, as mentioned in the Paris Climate Agreement. Research in the field of enteric methane mitigation accelerated over the past two decades, and various strategies for reducing enteric methane have been studied: production intensification, feed management (including the use of concentrates, management of herbaceous and woody forages and pastures, and the use of agricultural by-products), rumen manipulation, genetic improvement to enhance animal productivity and health, and the selection of animals with low enteric methane emissions. Several reviews exist in the literature on strategies for reducing enteric methane emissions (FAO 2023b). The effectiveness of these strategies varies depending on the farming system (extensive or intensive) and the region (temperate or tropical). It is therefore important to develop research on methane emission measurement and inclusion in methane emission inventories in regions with very few references, such as sub-Saharan Africa (Assouma et al. 2018; Blanfort et al. 2023). This is the challenge that the Cassecs project in Sahelian countries is trying to address, as described in Box 24.4 (see also Box 16.1). Sahelian countries emit very little GHGs and contribute to less than 0.2% of global GHG emissions from human activities. This project is an opportunity for this region, as it allows for the production of references adapted to the extensive livestock systems of this region, as well as exploring the main avenues for reducing enteric methane emissions.

> **Box 24.4 Calculation of Methane Emission Factors in the Cassecs Project**
>
> As part of the Cassecs project*, feeding practices were tested to assess their impacts on enteric methane emissions of local cattle breeds. The project funded the establishment of a platform for measuring and monitoring enteric methane emissions from animals equipped with a GrenFeed device (a mobile system for measuring respiratory gases emitted by ruminants) in Burkina Faso. In the extensive livestock systems in the Sahel, fodder from leguminous shrubs is used by farmers to feed ruminants. A study was initiated in cattle supplemented with two main species of shrub forages cultivated by farmers in the form of a fodder bank (Leucena leucocephala and Gliricidia sepium). The results of this study show that supplementation with leguminous shrub forages improved daily dry matter intake (DMI), digestibility and average daily live weight gain, and reduced enteric methane emissions by about 25% per kilogram of DMI. These results support for the promotion of the establishment of fodder tree banks as an action for mitigating CH_4 emissions in sub-Saharan Africa. This type of reference is very useful to support countries in drafting their NDCs (Nationally Determined Contributions). These data indeed contribute to better taking into account the livestock sector in mitigation strategies by reducing methane emissions from the livestock sector in synergy with the role of extensive livestock farming in carbon sequestration in pastoral ecosystems.
>
> * https://www.cassecs.org/.

24.2.3 Manure Management

Methane is also produced from manure management, as it is produced under anaerobic conditions (without oxygen in the environment) by archaea bacteria that use organic matter. Methane emissions related to manure management represent 7.8% of total emissions from the livestock sector worldwide (FAO 2023a). These emissions vary depending on the livestock systems and breeds. High methane emissions related to manure management come from pig farming. Syntheses in the literature describe these forms of methane emissions (FAO 2023a). Some liquid slurry storage systems such as ponds promote more methane production than open systems although they emit more N_2O by volatilisation.

Numerous strategies have been proposed to reduce CH_4 emissions from manure management. These strategies are summarised in Table 24.1.

Table 24.1 Strategies for reducing methane emissions related to livestock manure management

	Emission reduction strategies
Valorisation of methane as bioenergy	Collection and capture of biogas
	Manure management with biodigesters
Management and storage of manure	Frequent removal of manure from livestock buildings or storage
	Separation of solids, use of filters and purifiers
Actions on manure	Cooling of manure
	Acidification of manure
	Addition of amendments that inhibit the production of CH_4
Valorisation of manure in plant production	Spreading of manure on pastures and agricultural soils
Specific arrangements	Management of manure under aerobic conditions
	Use of filters and purifiers

24.3 Conclusion: The Multidimensional Aspect of Methane

Since the Paris Agreement and the Global Methane Pledge, methane emissions have been under scrutiny because (1) reducing emissions of this short-lived greenhouse gas (GHG) can limit warming in the short term and (2) they are intrinsically linked to long-lived GHG emissions such as CO_2 and N_2O.

The management of methane emissions directly affects CO_2 emissions through several channels: (1) methane emissions are often associated with CO_2 emissions (livestock emissions) or N_2O emissions (rice emissions) through joint emission processes, (2) methane is part of general GHG emission reduction strategies, which include CO_2 reduction targets. The final effect on global warming of these methane emission reduction practices must therefore include the effects on CO_2 emissions, also considering equivalence metrics (Box 24.3), and (3) these methane emissions are closely linked to several dimensions of sustainable development.

Indeed, methane emissions are intimately associated with aspects of food security and poverty reduction. In fact, rice production is central to the food security strategies of some countries, such as India, and livestock farming is at the heart of many farmers' strategies in developing countries to ensure their food security and nutritional intake (see Chap. 16). It also helps to fight poverty in these rural areas by providing income for farmers. Although not discussed in this chapter, the management of methane emissions associated with wetlands would also affect the biodiversity of these areas.

Regarding climate, recognising and taking into account this multidimensional aspect of methane emissions would help to better understand the difficulties encountered in international negotiation arenas, such as climate negotiations.

References

Assouma MH, Lecomte P, Hiernaux P, Ickowicz A, Corniaux C, Decruyenaere V et al (2018) How to better account for livestock diversity and fodder seasonality in assessing the fodder intake of livestock grazing semi-arid sub-Saharan Africa rangelands. Livest Sci 216:16–23. https://doi.org/10.1016/j.livsci.2018.07.002

Blanfort V, Assouma MH, Bois B, Edouard-Rambaut L-A, Vayssières J, Vigne M (2023) Efficiency to account for the complexity of the contributions of livestock grazing systems to climate change. In: Ickowicz A, Moulin C-H (eds) Livestock grazing systems and sustainable development in the Mediterranean and tropical areas. Recent knowledge on their strengths and weaknesses, Versailles, Quæ editions, pp 82–99

Bouman BAM, Yang X, Wang H, Wang Z, Zhao J, Chen B (2006) Performance of aerobic rice varieties under irrigated conditions in North China. Field Crop Res 97:53–65. https://doi.org/10.1016/j.fcr.2005.08.015

Crippa M, Guizzardi D, Solazzo E, Muntean M, Schaaf E, Monforti-Ferrario F et al (2023) GHG emissions of all world countries. Publications Office of the European Union. https://doi.org/10.2760/235266

Crippa M, Solazzo E, Guizzardi D, Monforti-Ferrario F, Tubiello FN, Leip A (2021) Food systems are responsible for a third of global anthropogenic GHG emissions. Nature Food 2(3): 198–209. https://doi.org/10.1038/s43016-021-00225-9

Crippa, Monica, Diego Guizzardi, Manjola Banja, et al (2022) CO2 emissions of all world countries. *JRC Science for Policy Report, European Commission*, EUR 31182.

Crippa, Monica, Diego Guizzardi, Efisio Solazzo, et al (2023) GHG emissions of all world countries. Publications Office of the European Union, publication en ligne anticipée.

FAO (2023a) Methane emissions in livestock and rice systems – sources, quantification, mitigation and metrics. FAO, Rome. https://doi.org/10.4060/cc7607en

FAO (2023b) Pathways towards lower emissions – a global assessment of the greenhouse gas emissions and mitigation options from livestock agrifood systems. FAO, Rome. https://doi.org/10.4060/cc9029en

Gerber PJ, Steinfeld H, Henderson B, Mottet A, Opio C, Dijkman J et al (2013) Tackling climate change through livestock—a global assessment of emissions and mitigation opportunities. Food and Agriculture Organization of the United Nations (FAO), Rome, p 115

Gupta K, Kumar R, Baruah KK, Hazarika S, Karmakar S, Bordoloi N (2021) Greenhouse gas emission from rice fields: a review from Indian context. Environ Sci Pollut Res 28:30551–30572. https://doi.org/10.1007/s11356-021-13935-1

International Energy Agency (2023) Methane Tracker Database, IEA, Paris. https://www.iea.org/data-andstatistics/data-tools/methane-tracker-data-explorer

IPCC (2018) Global warming of 1.5°C. An IPCC Special Report on the impacts of global warming of 1.5°C above pre-industrial levels and related global greenhouse gas emission pathways, in the context of strengthening the global response to the threat of climate change, sustainable development, and efforts to eradicate poverty. https://doi.org/10.1017/9781009157940

IPCC (2022) Climate change 2022: mitigation of climate change. In: Contribution of working group III to the sixth assessment report of the intergovernmental panel on climate change. Cambridge University Press. https://doi.org/10.1017/9781009157926

Liu Y, Ge T, van Groenigen KJ, Yang Y, Wang P, Cheng K et al (2021) Rice paddy soils are a quantitatively significant carbon store according to a global synthesis. Commun Earth Environ 2:154. https://doi.org/10.1038/s43247-021-00229-0

Malley CS, Borgford-Parnell N, Haeussling S, Howard IC, Lefèvre EN, Kuylenstierna JCI (2023) A roadmap to achieve the global methane pledge. Environm Res Clim 2(1):011003. https://doi.org/10.1088/2752-5295/acb4b4

Ocko IB, Sun T, Shindell D, Oppenheimer M, Hristov AN, Pacala SW et al (2021) Acting rapidly to deploy readily available methane mitigation measures by sector can immediately slow global warming. Environ Res Lett 16(5):054042. https://doi.org/10.1088/1748-9326/abf9c8

Peng S, Lin X, Thompson RL, Xi Y, Liu G, Hauglustaine D, Lan X et al (2022) Wetland emission and atmospheric sink changes explain methane growth in 2020. Nature 612(7940):477–482. https://doi.org/10.1038/s41586-022-05447-w

Prudhomme R, O'Donoghue C, Ryan M, Styles D (2021) Defining national biogenic methane targets: implications for national food production & climate neutrality objectives. J Environ Manag 295:113058. https://doi.org/10.1016/j.jenvman.2021.113058

Qian H, Zhu X, Huang S, Linquist B, Kuzyakov Y, Wassmann R et al (2023) Greenhouse gas emissions and mitigation in rice agriculture. Nat Rev Earth Environ 4:716–732. https://doi.org/10.1038/s43017-023-00482-1

Saunois M, Stavert AR, Poulter B, Bousquet P, Canadell JG, Jackson RB et al (2020) The global methane budget 2000–2017. Earth Syst Sci Data 12(3):1561–1623. https://doi.org/10.5194/essd-12-1561-2020

Scholtz MM, Neser FWC, Makgahlela ML (2020) A balanced perspective on the importance of extensive ruminant production for human nutrition and livelihoods and its contribution to greenhouse gas emissions. S Afr J Sci 116(9/10). https://doi.org/10.17159/sajs.2020/8192

Shumba A, Chikowo R, Corbeels M, Six J, Thierfelder C, Cardinael R (2023) Long-term tillage, residue management and crop rotation impacts on N_2O and CH_4 emissions from two contrasting soils in sub-humid Zimbabwe. Agric Ecosyst Environ 341:108207. https://doi.org/10.1016/j.agee.2022.108207

Tedeschi LO, Abdalla AL, Álvarez C, Anuga SW, Arango J, Beauchemin KA et al (2022) Quantification of methane emitted by ruminants: a review of methods. J Anim Sci 100(7):skac197

Tollefson J (2022) Scientists raise alarm over 'dangerously fast' growth in atmospheric methane. Nature. https://doi.org/10.1038/d41586-022-00312-2

United Nations Framework Convention on Climate Change (UNFCCC) (2022) Greenhouse Gas Data Interface. Available at: https://di.unfccc.int/

Open Access   This chapter is licensed under the terms of the Creative Commons Attribution-NonCommercial-NoDerivatives 4.0 International License (http://creativecommons.org/licenses/by-nc-nd/4.0/), which permits any noncommercial use, sharing, distribution and reproduction in any medium or format, as long as you give appropriate credit to the original author(s) and the source, provide a link to the Creative Commons license and indicate if you modified the licensed material. You do not have permission under this license to share adapted material derived from this chapter or parts of it.

The images or other third party material in this chapter are included in the chapter's Creative Commons license, unless indicated otherwise in a credit line to the material. If material is not included in the chapter's Creative Commons license and your intended use is not permitted by statutory regulation or exceeds the permitted use, you will need to obtain permission directly from the copyright holder.

Chapter 25
The Heterogeneity of Institutionalization Pathways for Climate Policies and Instruments in Agriculture: A Comparative Analysis of Senegal, Colombia, Brazil, and France

Marie Hrabanski, Jean François Le Coq, Gilles Massardier, Carolina Milhorance, Yves Montouroy, and Eric Sabourin

Abstract Climate change mitigation and adaptation have become key public policy challenges worldwide, with national plans emerging since the early 2000s and most countries including agriculture in their climate commitments. By 2020, over 90% of nationally determined contributions (NDCs) addressed agricultural adaptation, but many developed countries still neglect food systems in their climate policies. This chapter examines how institutionalization paths of climate policies in Colombia, Brazil, France, and Senegal influence the implementation of agricultural climate actions. It analyzes factors like policy framing, proactivity, sector integration, policy instruments, and innovation. The study highlights how these national differences affect each country's ability to transition towards climate-smart agriculture.

25.1 Introduction

In response to the urgency, the issues of mitigation and adaptation of our societies to climate change (CC) have now become major challenges for public policies in both the Global North and South. The first national climate change adaptation plans emerged in the early 2000s (in 2004 in France, 2006 in Senegal, and 2011 in Colombia and Brazil), some of which included agriculture. Subsequently, the Paris

---

M. Hrabanski (✉) · J. F. Le Coq · G. Massardier · C. Milhorance · E. Sabourin
Cirad, UMR Actors Resources Territories & Development (ART-Dev), Montpellier, France
e-mail: marie.hrabanski@cirad.fr

Y. Montouroy
LC2S, University of the Antilles, Guadeloupe, Martinique, France

© The Author(s) 2026
V. Blanfort et al. (eds.), *Climate Impacts and Challenges in Agriculture, Forests and Food Systems*, https://doi.org/10.1007/978-3-032-04331-3_25

Agreement in 2015 required signatory countries to submit their nationally determined contributions (NDCs) to reduce global emissions. Thus, the integration of agricultural-climate issues has primarily been promoted at the national level in countries of the North and South. By 2020, over 90% of these NDCs included the adaptation of agriculture to climate change, and approximately 80% identified mitigation objectives in the agricultural sector. However, it is noteworthy that in 2023, the majority of developed countries (62%) did not present any measures related to food systems in their NDCs, with the term not even being mentioned. Despite these efforts, the vulnerability of agriculture to climate change is increasing (OECD 2023): mechanisms to manage these vulnerabilities must be developed and improved.

This chapter analyzes how the pathways of institutionalization of climate policies, specific to each country, explain the institutional blockages that hinder the concrete implementation of public action (Hrabanski and Montouroy 2022). These pathways of institutionalization will be analyzed in four countries (Colombia, Brazil, France, and Senegal) and compared according to the following five variables: (i) the type of framing of the climate issue in agriculture (the place of mitigation vs. adaptation, main concepts used in policy documents on the adaptation of agriculture to climate change (AAACC), hierarchization of public problems); (ii) the degree of proactivity towards climate change (proactive countries vs. reactive/followers of the international climate agenda) and the dynamics of extraversion (international influences) or endogenization (internal logics) in the implementation of AAACC policies; (iii) the (intersectoral) integration of mechanisms or, conversely, their sectoral nature (solely agricultural); (iv) the types of available instruments (regulatory, incentive-based, informational (Vedung 1998), or combining different types of instruments (Pacheco-Vega 2020)), and their degree of innovation.

This chapter analyzes the heterogeneity of the pathways of institutionalization of climate issues for agriculture in these four national contexts of the South and North. It is structured into three parts. The first part will analyze, for the four countries, the following three variables: framing, extraversion, and integration. The second part will examine the last two variables: public policy instruments and innovation. Finally, the third part will discuss the results to understand the link between the national variety of institutionalization pathways and the countries' capacity for transition towards AAACC. While addressing both issues related to the institutionalization of adaptation policies and mitigation policies, the chapter focuses more on the former.

25.2 Institutionalizing Agricultural Transition Policies—Framing, Degree of Extraversion, and Integration

25.2.1 In Colombia: Uncoordinated and Extraverted Institutionalization

In Colombia, a first climate policy instrument dedicated to adaptation was created in 2011: the National Adaptation Fund (Milhorance et al. 2022b), followed by the formulation and launch of the National Adaptation Plan (NAP) in the same year. The government aimed to provide a multisectoral and integrated governance structure at a high level: the responsibility for climate policymaking was transferred from the Ministry of Environment to the National Planning Department (DNP), and the 2010–2014 National Development Plan was reformulated to incorporate climate change, with the aim of building a cross-cutting, coordinated climate response across ministries.

However, the agroecological and climate transition was soon steered by a poorly coordinated extraverted dynamic. External actors played a significant role in "climatizing" existing sectoral policies: the International Center for Tropical Agriculture (CIAT) promoted Climate-Smart Agriculture (CSA) to the DNP as early as the 2010s; the FAO simultaneously advocated for climate resilience within the Ministry of Agriculture; and USAID contributed by encouraging the adoption of resilience frameworks across the Ministries of Mines and Energy, Water, and Housing, applying the concept to diverse policy areas. At the same time, NGOs such as WWF and The Nature Conservancy promoted Ecosystem-based Adaptation to the Ministry of Environment, highlighting the importance of biodiversity conservation. Agroecology, however, is not a core concept in Colombia's NAP. While formally integrated into planning documents, the absence of concrete implementation measures suggests these concepts were employed more symbolically than effectively (Milhorance et al. 2022b).

Despite these conceptual advancements, the 2011 Colombian NAP was not designed as an operational plan and lacked coherence. It comprised various documents that encouraged the development of sectoral and territorial policies by providing methodological tools for informed decision-making. Yet, this proliferation of sectoral documents led to fragmented dynamics shaped by divergent stakeholder interests. Moreover, the NAP's formulation was bureaucratic, non-participatory (Milhorance et al. 2022b), and extraverted—meaning it emerged from dynamics external to Colombia. International development actors such as the Inter-American Development Bank, the World Bank, the United Nations Development Programme (UNDP), and the FAO played a key role by offering technical and financial support. The Colombian government used climate policy opportunistically—to support territorial implementation of peace accords, attract foreign investment, and strengthen international dialogue (Bustos 2018).

Ultimately, the implementation of climate policy in Colombia reveals a disconnect between national-level planning (often based on decontextualized international recommendations; Mariño 2011) and local realities. Despite the existence of multiple ad hoc committees, coordination between sectoral and territorial institutions remains weak. Moreover, implementation is hindered by limited political commitment and the absence of dedicated funding.

25.2.2 In Senegal: Climate Institutionalization in Competition with Food Security

Following its ratification of the UNFCCC in 1992, Senegal—operating under an aid regime (Lavigne 2017)—began to structure its climate agenda in alignment with international recommendations (Boutroue et al. 2022). This approach materialized in 2006 with the drafting of a National Adaptation Programme of Action (NAPA), and was furthered by the launch of a National Adaptation Plan (NAP) process beginning in 2015. The integration of climate concerns into the agricultural sector initially relied on the mobilization of the Climate-Smart Agriculture (CSA) concept, later followed by a policy shift toward agroecology (Milhorance et al. 2022a). These two frameworks contributed to the institutionalization of climate issues within Senegal's agricultural policies (Boutroue et al. 2022), reflecting a broader ambition to link climate adaptation and mitigation with food security objectives. Nevertheless, the strategies advanced have remained primarily focused on agricultural productivity and food security, with limited substantive integration of adaptation goals.

From the perspective of extraversion, Senegal's adaptation planning has involved numerous international technical and financial partners. USAID supported and funded the Ministry of Fisheries between 2016 and 2018 in the development of its sectoral NAP; similarly, the French Development Agency (AFD) has provided assistance since 2018 via the Adapt'Action initiative. Other efforts—led by the German Agency for International Cooperation (GIZ), the FAO, the UNDP, and the Green Climate Fund—underscore Senegal's significant dependence on extranational partners.

At the institutional level, Senegal's NAP adopted a sectoral and participatory approach under the coordination of the Ministry of Environment and Sustainable Development. However, the institutional integration of climate concerns is shaped by inter-agency competition. On one side stands the Centre for Ecological Monitoring (CSE), an operational arm of the Ministry of Environment advocating for an environmental framing of climate issues. On the other side is ANACIM (National Agency of Civil Aviation and Meteorology), which interprets adaptation primarily through the lens of food security—reflecting its institutional mandates and technical expertise (Boutroue et al. 2022).

25.2.3 In Brazil: Institutionalizing a New Issue Within Routinely Structured Sectors

Brazil's climate policy trajectory, shaped by the international agenda, began with its ratification of the UNFCCC in 1992. Brazilian diplomacy played a proactive role from the outset, emphasizing the responsibility of developed countries in reducing emissions. At the same time, Brazil resisted mechanisms such as REDD+ (Reducing Emissions from Deforestation and Forest Degradation) that would assign monetary value to avoided deforestation. This position reflects internal tensions: the agribusiness sector seeks to avoid restrictions that could limit the conversion of forested areas into agricultural land, thereby supporting the expansion of soy cultivation and cattle ranching (Friberg 2009; Lahsen 2009).

Under President Lula's administration, a series of climate-related plans were launched. The National Climate Change Plan (2008) aimed to curb emissions, particularly in the Amazon region (Gallo and Albrecht 2019). The Low-Carbon Agriculture Plan (Plan ABC) focused primarily on mitigation, while the National Adaptation Plan (NAP) was introduced in 2016. Although these policies adopted a formally intersectoral approach, their implementation suffered from a lack of coordination and integration across their different dimensions. Land use conflicts further undermined these efforts (Zellhuber 2016).

The Brazilian NAP addressed contradictory concerns and regional specificities (Milhorance et al. 2022a, b). For instance, Plan ABC concentrated on agribusiness in the Cerrado (Central-West region), initially drawing on the principles of Climate-Smart Agriculture (CSA), which combines adaptation, mitigation, and productivity goals (Lipper et al. 2014). This sector has prioritized technological innovation to boost its competitiveness in global markets, in stark contrast to the financial and institutional limitations faced by agroecological and conservation-oriented initiatives (see Chapter on Concepts). In the Northeast, adaptation policies emphasized agroecology and improved water access for family farms (Sabourin et al. 2022), aligned with a socio-ecological resilience framework (Brazil 2016; Monteiro et al. 2022). In the North (Amazon region), efforts focused on preserving ecosystem services provided by tropical forests—critical for Indigenous populations (Brazil 2016).

Thus, the framing of climate policy in Brazil emerges from a complex political process characterized by negotiations among competing sectoral interests and divergent development models (Milhorance et al. 2022b).

25.2.4 In France: A Multi-level and Negotiated Institutionalization at National and Territorial Scales

In France, the integration of climate objectives results from complex multi-level processes involving European, national, and territorial levels. Most recently, the new Common Agricultural Policy (CAP) for the 2023–2027 period has reaffirmed

EU-level climate goals within the broader framework of the European Green Deal, which aims to "reduce the climate footprint" (Hrabanski et al. 2025). In practice, however, this renewed climate ambition is constrained by the persistence of the CAP's foundational triptych—growth, productivity, and competitiveness (Bodiguel 2014). Other environmental concerns (such as biodiversity, water management, and pesticide reduction), as well as non-environmental objectives (territorial cohesion, fiscal balance), often compete with climate objectives for policy attention (Sabourin et al. 2018; Biabiany et al. 2022; Montouroy et al. 2022).

At the national level, climate change entered the agricultural policy agenda slowly and belatedly. The first National Adaptation Plan to Climate Change (PNACC) was adopted in 2011. Following the 2015 Paris Agreement, both adaptation and mitigation in agriculture gained political visibility—illustrated by the launch of the "4 per 1000" initiative (Demenois et al. 2022). Among these, adaptation to climate change (AACC) has garnered increasing attention from policymakers due to its implications for economic, political, food, and environmental security. This shift has also been driven by mobilization within the professional agricultural sector, which has been increasingly affected by extreme weather events (e.g., droughts, hailstorms) since the late 2010s (Dantec and Roux 2019).

A second PNACC (2018–2022) was adopted, and multiple parliamentary reports have since maintained agriculture's adaptation as a high-priority item on the political agenda. Within this context—and amid growing concerns over water scarcity—the government launched the "Varenne agricole de l'eau et de l'adaptation au changement climatique" in 2022.

The direct consequence of climate change on agriculture and rural territories has been a growing complexity in governance. Three main, often parallel or loosely coordinated, logics now shape the climate transition in agriculture, revealing a rather baroque multi-level architecture (Hrabanski et al. 2025): Conditional EU aid; National and territorial planning frameworks that aim to "de-sectoralize" climate governance by integrating cross-cutting concerns; Conversely, siloed sectoral subsidies—often organized by commodity chains and co-produced through neo-corporatist arrangements deeply embedded in French agricultural governance (notably through the FNSEA) (Montouroy et al. 2022).

25.3 The Political Institutionalization of the Climate Challenge Through Policy Instruments and Their Degree of Innovation

An inventory of public instruments for adapting agriculture to climate change was conducted in 2019 across the four countries examined. Although mitigation instruments were not included, this inventory enables an empirical assessment of the tools available for adaptation and, more broadly, of each state's capacity to respond effectively to the challenges posed by climate change.

25.3.1 In Colombia: Numerous Innovations, But Fragmented Local Implementation and Reliance on Extranational Donors

Currently, Colombia's National Adaptation Plan (NAP) is largely implemented by domestic NGOs with funding from the Green Climate Fund, while the national government plays a coordinating role. The state had already acknowledged its limited financial and administrative capacity to meet climate goals, prompting national and local actors to seek partnerships with the private sector and international donors to secure implementation funding.

The inventory identified around thirty adaptation instruments in the Colombian agricultural sector. These instruments demonstrate significant capacity for policy innovation: 90% were introduced after 2009 (Noblanc 2019). Most do not rely on regulation or coercion, but are instead incentive-based or informational. Examples include multi-stakeholder platforms for dialogue and climate information dissemination in the Cauca department (Martinez Tintinago 2019; Osorio-García et al. 2019). Interestingly—despite Colombia's largely extraverted policy trajectory—many of these instruments have proven relatively stable over time, with the majority being institutionalized beyond short-term funding cycles.

The Colombian NAP incorporated international strategies aligned with the CSA paradigm and introduced new instruments, particularly deliberative platforms. However, despite this apparent innovation, the plan lacks concrete implementation mechanisms. It was designed primarily at the national level by technical staff, with limited participation from civil society or key sectoral policymakers who could have assigned dedicated funding. International consultants provided technical input but were mainly involved in drafting policy documents and conducting studies.

Officials from the National Planning Department and the Ministry of Agriculture, as well as international consultants, noted the country's tendency to focus on normative frameworks rather than on actual implementation. As one Ministry of Agriculture official acknowledged, "Colombia looks exemplary on paper, with strong laws—but implementation fails due to lack of funding." Ultimately, despite frameworks such as the 2018 Agricultural Frontier Law and the Green Growth Strategy (CONPES 3934), government budgets allocated to climate action remain limited. In parallel, local-level implementation of adaptation policies is often interwoven with the post-conflict agenda. This dual focus has revealed a clear gap between national-level policy formulation and its local application. Moreover, the instruments deployed on the ground are frequently fragmented, intermittent, and shaped by power struggles among local development actors competing for influence (Howland 2022a, b).

25.3.2 In Senegal: Limited Innovations Due to a Project-Based Governance Model

The inventory of adaptation instruments in Senegal reveals a significant number of projects (about thirty between 2006 and 2022) formally dedicated to agricultural adaptation to climate change, funded by extranational donors. However, none of these instruments relies on regulation or coercion. Rather, they are informational tools, based on the dissemination and exchange of knowledge about climate change—such as Climate Smart Agriculture platforms (C-CASA) and Integrated Territorial Climate Plans (PCTI). Incentive-based instruments are also in place, for example, the FAO's "One Million Cisterns for the Sahel" or the "Land and Ecosystem Management Strengthening Project" (PRGTE) in the Niayes and Casamance regions. Some instruments combine both types.

Most of the adaptation instruments identified were created after the first national adaptation plan (2006), indicating a degree of innovation. However, this innovation is tempered by the country's project-based governance model, which characterizes many aid-dependent states. Except for indexed insurance schemes, these instruments are integrated into development projects—types of public policy in themselves (Lavigne 2017). The predominance of extranational actors and a project-based governance framework hinder coordination and integration, often resulting in temporally unstable, short-term, and territorially unequal initiatives. Despite the proliferation of projects, empirical research reveals that implementation often falls short, as the projects frequently fail to meet the actual needs and expectations of farmers.

25.3.3 In Brazil: Few New Instruments and a Repackaging of Existing Policies

The analysis of Brazil's adaptation instruments highlights two main findings. First, as in the other countries, there is a strong emphasis on incentive-based tools, particularly credit lines for low-carbon agriculture (ABC Credit). However, this credit represented only 1% of total rural credit between 2010 and 2018. With high interest rates (8.5% in 2016/17, reduced to 6% in 2018/19), it remains less attractive than conventional agricultural loans, which are also less bureaucratic. Conversely, smallholders and traditional communities lack access to tailored financing for sustainable practices.

Second, and more notably, Brazil's climate policy shows limited innovation and heavy reliance on repackaged existing tools. The government layered new climate objectives onto pre-existing policies without altering their core structures (Milhorance 2020). As such, financial incentives to agribusiness were rebranded to align with the new climate agenda. For example, measures like technological upgrades for low-carbon agriculture and the rehabilitation of degraded lands essentially perpetuate productivity-oriented agricultural practices, rather than triggering a systemic shift. In Brazil's NAP, these policies promote agro-industrial production dependent on chemical inputs, mechanization, and fossil energy—even when reframed as "precision" agriculture.

25.3.4 In France: Some Innovation, But Many Recycled Instruments

France's adaptation tools display three key features. First, there is a clear preference for incentive-based and informational instruments, often used in combination (Pacheco-Vega 2020). Incentive-based instruments include financial supports such as Common Market Organization (CMO) mechanisms or climate risk insurance. These are designed at the European and regional levels (for CAP aids) and at the national level by the Ministry of Agriculture and public institutions. Informational instruments include observatories (e.g., ORACLE, Clima-XXI, Viticultural Observatory) and tools intended to influence farmers' perceptions of climate issues.

Second, adaptation instruments are mostly sectoral. The agricultural sector's long-standing organizational structure—agricultural chambers, unions, producer associations—facilitates the rapid development of sectoral instruments. However, this strong structuring also impedes intersectoral governance, complicating efforts to address non-sectoral issues. While this organization allows climate concerns to be rhetorically embedded in existing regulatory frameworks, it does not necessarily translate into cross-sectoral action.

Third, innovation in France must be qualified: many adaptation instruments are in fact repurposed tools. Existing mechanisms were rebranded for climate goals, without fundamental changes. For instance, the Agro-Environmental Measures (MAET) became Agro-Environmental and Climate Measures (MAEC) in 2014. Yet, professionals remain skeptical about their real climate impact. These symbolic changes are not always backed by tangible adaptation outcomes. In some cases, local negotiation over EU measure content enables collectively organized commodity sectors to secure subsidies that support economic viability, but not climate transition per se (Montouroy et al. 2022).

25.4 Key Findings and Pathways Toward Transformational Policies

Under the influence of international recommendations, the four countries studied have significantly integrated adaptation and mitigation into national political agendas, adopting new norms and objectives for agricultural transition. However, these agendas remain embedded in sectoral logic, despite international calls for multisectoral governance—a necessary condition for adaptive governance (Chaffin and Gunderson 2016; Koontz et al. 2015). Four main findings emerge:

Climate objectives have now been institutionalized across all four countries.

Yet, this institutionalization remains largely symbolic or rhetorical in several cases.

Institutional pathways vary across countries, limiting transformative capacity in different ways.

In Senegal and Colombia, extraverted and project-based models—described as a "mix of opportunism and coercion characterizing compliance with international public policy frameworks" (Siméant-Germanos 2019)—lead to temporary and spatially unequal funding. While many new instruments have been promoted, their institutionalization in ministries and official documents does not translate into effective change on the ground.

In France, the process is more homogeneous and sectorally embedded, but largely incremental and dependent on existing power structures. In Brazil, the bundling of pre-existing multisectoral plans allows the government to delegate implementation to local actors, often resulting in territorially uneven and institutionally fragmented outcomes.

Current instruments rarely challenge national agricultural models.

In Brazil, climate objectives coexist with both agroecological and low-carbon agribusiness models. However, real transformation requires systemic reform of dominant models. In France, although transition policies echo the "greening" of the CAP, the prevailing model—technologically modernist and export-oriented—hinders transformative adaptation, as urged by the IPCC (2022).

Climate issues are politically ranked relative to other public problems.

Competing priorities such as food security, productivity, biodiversity, water, energy, and climate justice create institutional tensions and inter-sectoral competition (Hrabanski and Montouroy 2022).

25.4.1 Toward Transformational Policies

Achieving transformational policies requires complex trade-offs. Success will depend on a country's ability to strike a balance between: Local governance and centralized planning to avoid territorial inequalities (Brazil/Colombia/Senegal); Increasing funding and designing durable instruments (Colombia/Senegal); Multisectoral planning and over-reliance on agricultural actors (France); Technocratic design and politicization of climate adaptation instruments. Thus, adaptive agriculture and agricultural climate transition cannot be addressed solely through technological and financial variables. They are also deeply conditioned by the governance mechanisms shaping public support policies.

References

Biabiany O, Massardier G, Montouroy Y (2022) The implementation process of agriculture adaptation instruments to climate change. The invisibilization of European climate policy goals in French West Indies' banana Chain. Int J Agric Sustain 20(6):1181–1193

Bodiguel L (2014) Combating climate change: the new leitmotif of the Common Agricultural Policy. Rev Eur Union 580:414–426

Boutroue B, Hrabanski M, Diao CA (2022) Governing the adaptation of agriculture to climate change by project? The limits of the climatisation of agricultural policies in Senegal. Govern Public Action 3:99–125

Brazil (2016) National adaptation plan to climate change: sectoral and thematic strategies – volume II. Ministry of Environment

Bustos MC (2018) What shapes Colombia's foreign position on climate change? Colombia Internacional 94:27–51. https://doi.org/10.7440/colombiaint94.2018.02

Chaffin BC, Gunderson LH (2016) Emergence, institutionalization and renewal: rhythms of adaptive governance in complex social-ecological systems. J Environ Manag 165:81–87. https://doi.org/10.1016/j.jenvman.2015.09.003

Dantec R, Roux JY (2019) Adapting France to climate disruptions by 2050: declared emergency. Information Report-Senate No 511

Demenois J, Dayet A, Karsenty A (2022) Surviving the jungle of soil organic carbon certification standards: an analytic and critical review. Mitig Adapt Strateg Glob Chang 27(1):1

Friberg L (2009) Varieties of carbon governance: the clean development mechanism in Brazil—a success story challenged. J Environ Dev 18(4):395–424

Gallo P, Albrecht E (2019) Brazil and the Paris agreement: REDD+ as an instrument of Brazil's nationally determined contribution compliance. Int Environ Agreem: Politics Law Econ 19:123–144

Howland F (2022a) Local climate change policy and rural development in Colombia's post-peace agreements context. Int J Agric Sustain 20(7):1260–1127

Howland F (2022b) Disaster risk management, or adaptation to climate change? The elaboration of climate policies related to agriculture in Colombia. Geoforum 131:163–172. https://doi.org/10.1016/j.geoforum.2022.02.012

Hrabanski M, Boutroue B, Massardier G, Thomas A (2025) Public policies for mitigation and adaptation of French agriculture to climate change: between late agenda setting and complex institutional marquetry. In: Debaeke P, Graveline N, Lacor B, Pellerin S, Renaudeau D, Sauquet E (eds) Agriculture and climate change. Impacts, adaptation and mitigation. Quæ editions, Versailles, 398 p

Hrabanski M, Montouroy Y (2022) The differentiated 'climatisations' of public action: Standardising the study of the 'climate change' problem. Govern Public Action 3:9–31

IPCC (2022) Summary for policymakers. In: Climate change 2022: impacts, adaptation and vulnerability. https://www.ipcc.ch/report/ar6/wg2/chapter/summary-for-policymakers/

Koontz TM, Gupta D, Mudliar P, Ranjan P (2015) Adaptive institutions in social-ecological systems governance: a synthesis framework. Environ Sci Pol 53(Part B):139–151. https://doi.org/10.1016/j.envsci.2015.01.003

Lahsen M (2009) A science-policy interface in the global south: the politics of carbon sinks and science in Brazil. Clim Chang 97(3):339–372

Lavigne Delville P (2017) For a socio-anthropology of public action in 'aid regime' countries. Anthropol Dev 45:33–64. https://doi.org/10.4000/anthropodev.542

Lipper L, Thornton P, Campbell BM, Baedeker T, Braimoh A, Bwalya M, Torquebiau EF (2014) Climate-smart agriculture for food security. Nat Clim Chang 4(12):1068–1072

Mariño N (2011) Reflections on the cultural perspective in climate change policies in Colombia: an approach to the cultural and spatial analysis of public policies. In: Ulloa A (ed) Cultural perspectives of climate. National University of Colombia, Bogota, pp 495–528

Martinez Tintinago MH (2019) The agroclimatic technical tables in the Cauca region [Colombia]: the limits of a project for an adaptation instrument of agriculture to climate change. Master's thesis 2. Univ. Paris 1 Panthéon Sorbonne, Iedes, Paris, 108 p

Milhorance C (2020) Diffusion of Brazil's food policies in international organisations: assessing the processes of knowledge framing. Polic Soc 39(1):36–52

Milhorance C, Camara AD, Sourisseau JM, Piraux M, Assembène MC, Sirdey N et al (2022a) The integration of agroecology into public policies in Senegal. ISRA

Milhorance C, Howland F, Sabourin E, Le Coq J-F (2022b) Tackling the implementation gap of climate adaptation strategies: understanding policy translation in Brazil and Colombia. Clim Pol:1–17. https://doi.org/10.1080/14693062.2022.2085650

Monteiro D, Silveira L, Petersen P (2022) "Fartura tem de montão": public policies and socio-ecological resilience in traditional communities of pastureland in the Sertão do São Francisco, Bahia. In: Sabourin E et al (eds) Public policies for adapting agriculture to climate change in semi-arid Northeast Brazil. E-papers, Rio de Janeiro, pp 183–210

Montouroy Y, Biabiany O, Massardier G (2022) The local implementation of instruments as a vector for the de-climatisation of public policies: the case of agricultural policy and the banana sector in Guadeloupe. Gov Public Action 3:127–152

Noblanc M (2019) Mapping of political instruments for adapting agriculture in Colombia (mango and market gardening sector), Master's thesis

OECD (2023) https://www.oecd.org/publications/climate-change-adaptation-policies-to-foster-resilience-inagriculture-5fa2c770-en.htm

Osorio-García AM, Paz L, Howland F, Ortega LA, Acosta-Alba I, Arenas L et al (2019) Can an innovation platform support a local process of climate-smart agriculture implementation? A case study in Cauca, Colombia. Agroecol Sustain Food Syst 44(3):378–411. https://doi.org/10.1080/21683565.2019.1629373

Pacheco-Vega R (2020) Environmental regulation, governance, and policy instruments, 20 years after the stick, carrot, and sermon typology. J Environ Policy Plan 22(5):620–635

Sabourin E, Le Coq JF, Fréguin-Gresh S, Marzin J, Bonin M, Patrouilleau MM et al (2018) Public policies to support agroecology in Latin America and the Caribbean. Perspective 45:1–4

Sabourin E, Oliveira LMR, Goulet F, Martins ES (2022) Public policies for adapting agriculture to climate change in semi-arid Northeast Brazil. E-papers, Rio de Janeiro, 236 p

Siméant-Germanos J (2019) Thinking about environmental engineering in Africa in light of the social sciences of development. Zilsel 6(2):281–313

Vedung E (1998) Policy instruments: typologies and theories. In: Bemelmans-Videc M-L, Rist RC, Vedung E (eds) Carrots, sticks, and sermons: policy instruments and their evaluation. Transaction Publishers, pp 21–58

Zellhuber A (2016) Environmental policy in Brazil. Tensions between conservation and the ideology of growth. In: The political system of Brazil. Springer Berlin, Berlin/Heidelberg, pp 329–350

Open Access This chapter is licensed under the terms of the Creative Commons Attribution-NonCommercial-NoDerivatives 4.0 International License (http://creativecommons.org/licenses/by-nc-nd/4.0/), which permits any noncommercial use, sharing, distribution and reproduction in any medium or format, as long as you give appropriate credit to the original author(s) and the source, provide a link to the Creative Commons license and indicate if you modified the licensed material. You do not have permission under this license to share adapted material derived from this chapter or parts of it.

The images or other third party material in this chapter are included in the chapter's Creative Commons license, unless indicated otherwise in a credit line to the material. If material is not included in the chapter's Creative Commons license and your intended use is not permitted by statutory regulation or exceeds the permitted use, you will need to obtain permission directly from the copyright holder.

Chapter 26
Finance, Agriculture and Climate

Antoine Ducastel

Abstract Since 2020, many policies and studies have highlighted a significant funding gap for climate adaptation and mitigation in agriculture. This chapter examines how different actors frame the finance-agriculture-climate nexus and the public instruments proposed to address funding shortages. It reviews recent efforts to mobilize climate finance, focusing on innovative public-private financial tools and repurposing existing public funds toward "sustainable" agriculture. The financialization of agricultural climate policies emphasizes risk-return dimensions, shaping both the volume and the flows of funding. This often limits support for climate-vulnerable groups and countries.

Since 2020, a multitude of policies (international and national) and studies have focused on the relationships between finance, agriculture (or more broadly agrifood systems) and climate. All these initiatives highlight the deficit of funding, both public and private, allocated to adaptation and mitigation policies in agriculture, and, propose solutions to bridge this deficit.

This chapter provides a brief overview of the relationships between finance, climate and agriculture, focusing on i) how they are problematised by different and competing groups of actors and ii) what are the policies promoted to tackle these challenges?

We will first review a few recent initiatives evaluating climate finance for agriculture. These efforts are part of a financialised approach for agricultural development (Chiapello et al. 2023; Ducastel et al. 2023) and climate change policies (Bracking and Leffel 2021; Carton and Bigger 2020; Chiapello 2020): focus on the mobilisation of (private) funding. To tackle the chronic funding deficit, two paths are primarily explored: the search for innovative public-private financial instruments; and the repurposing of existing public subsidies towards more "sustainable" activities. We will detail these two areas through concrete examples implemented in both the Global North and South.

A. Ducastel (✉)
Cirad, UMR Actors Resources Territories & Development (ART-Dev), FAO, Rome, Italy
e-mail: antoine.ducastel@cirad.fr

In a nutshell, the influence of (public and private) financial actors over the design and implementation of climate finance policies for agriculture frames and limits *de facto* the distributive outcomes of these policies — both on the quantitative (i.e., the volume of climate finance allocated to agriculture) and the qualitative levels (i.e., the access of the most climate-vulnerable countries and groups, e.g., smallholder farmers, Indigenous people, women).

26.1 Measuring the *Financial Gap* to Better Bridge It?

Since 2020, a few studies have assessed the volume of climate finance flowing to agriculture and agrifood systems (Ifad and CPI 2020; FAO 2022; CPI 2023; ROPPA 2023). These sectoral studies draw inspiration from existing landscapes of climate finance carried out at international and national scales (CPI 2023; I4CE 2023a), in the wake of the Paris Agreement (2015).[1] While they faced the same methodological limitations — i.e., the lack of data, the absence of "sustainable" or "green" standards (Chiapello 2020; Jachnik et al. 2019) — these works nonetheless identify three general trends:

- The proportion of climate finance flowing to agriculture is limited and decreasing;[2]
- The distribution of this funding into agrifood systems is highly uneven as the most vulnerable groups to climate change i.e., small-scale family farmer, Indigenous communities and women- are largely overlooked.[3]
- Public funding represents the bulk of these flows.[4]

Beyond climate finance, the underfunding of agriculture by development finance is a long-term and structural dynamic, since the structural adjustment policies of the 1980s and 1990s (Ducastel et al. 2023; Gabas et al. 2017). This chronic underfunding is often contrasted with the financial needs required to achieve the Sustainable Development Goals (SDGs) and carbon neutrality. For instance, the Food and Land Use Coalition estimates that "*annual investments of around 300 to 350 billion dollars are needed to ensure the transition to sustainable food systems and land uses, and to tackle climate change-related issues*" (FOLU 2019).[5]

---

[1] Article 2.1c.

[2] "Cumulative financing for agriculture, forestry and land use was only 20 billion dollars per year in 2017-2018, representing 3% of total global climate financing for the period" (Ifad and CPI 2020).

[3] "Only 2% of international public climate financing—2 billion dollars—was dedicated to small family farmers and rural communities in 2021" (ROPPA 2023).

[4] "Climate financing for small-scale agriculture comes 95% from the public sector, notably from government grants, and from multilateral and bilateral development financing institutions" (Ifad and CPI 2020).

[5] There are many other estimates of these needs, stated for example Shakhovskoy et al., Thornton et al. or the World Bank: by or the: "300 to 400 billion dollars of additional investments per year, less than 0.5% of global GDP".

This "financial gap" between existing climate finance flows and the financial requirements to archive climate and environmental objectives has over the years become a major issue for climate change and agriculture policies. This framing of climate policies as mainly a financial issue, i.e., the financialisation of climate policies,[6] is now the dominant paradigm in climate governance (Bracking and Leffel 2021; Carton and Bigger 2020; Chiapello 2020).

To address this funding gap, a specialized community of public and private actors is emerging, bringing together international organisations specialising in agriculture and the environment (FAO, Ifad, World Bank, Unep), who are seeking to finance their action programme in a constrained fiscal context; development finance institutions either multinational (Green Climate Fund), regional (African Development Bank) or national (French Development Agency)- often at the forefront of financial innovation (Ducastel et al. 2023); institutional investors (pension funds, sovereign funds) and investment companies (Phatisa, an African private equity fund manager); specialised consulting firms (Dalberg, Food System for the Future) and service providers; and finally, a few think tanks (Climate Policy Initiative, IFPRI).

They established coalitions, such as the Good Food Finance Network,[7] launched at the UN Food system summits by a few agri-food companies (Yara), investors (Rabobank, Green Climate Fund) and international organisations (Unep, FI, World Bank) to promote a reform programme for the architecture of "agri-food finance".. In parallel, other initiatives have been set up within the UNFCCC framework. For instance, at the COP27 in Sharm el-Sheikh, the Egyptian presidency launched the FAST initiative with the stated objective to improve both the quantity and the quality of climate finance for agrifood systems.[8]

These initiatives rely on three main assumptions. First, in the current context of austerity policies, public funding is insufficient to bridge the *financial gap*. Second, private financial institutions hold a large fraction of global savings and are always looking for new investment opportunities.[9] Third, public funding should primarily have a "leverage effect" on private finance, particularly by "de-risking" (Gabor 2023) investments in "sustainable" agriculture.

---

[6] According to Eve Chiapello, these are "public policies that seek to capture the forces of private finance, to engage its actors, and which also rely on its techniques, and its forms of reasoning" (Chiapello 2020).

[7] https://goodfood.finance/

[8] https://www.fao.org/climate-change/action-areas/access-to-climate-finance/fast/en

[9] "Public financial resources, including official development assistance, are not sufficient. Private financing is essential to bridge the financing gap and support the rapid transition of food systems" (Unep 2023).

26.2 Promoting Innovative Financial Instruments to leverage Private Capital

Recently, a few financial innovations have been promoted to address the funding gap for agrifood system. Despite their diversity, these policies are:

- public-private, for both their design and implementation;
- incentive-based and voluntary (Chiapello 2020);
- market-based, targeting the so-called market failures;
- often imported from other economic sectors.

Without being exhaustive,[10] we focus on two particular innovations: blended finance, and carbon offset projects.

26.2.1 Blended Finance

In the last decade, *blended finance*[11] became the new reference paradigm in development finance (Christiansen 2021; Mah 2023; Mawdsley 2018). Blended finance encompasses a wide range of instruments whose common goal is to "de-risk" investments: concessional loans, private equity funds (Ducastel 2016), technical assistance, or guarantee mechanisms.

This *blended finance* was relatively late in being imported into the agricultural sector. However, in recent years, several initiatives have been launched (Box 26.1).

Box 26.1 A Proliferation of Blended Funds for Agriculture

The European Union (EU) was a pioneer in blending finance for development by launching the EU-Africa Infrastructure Trust Fund (EU-AITF) in 2007. Then, several sectorial or regional blending windows have been set up, such as the ElectriFI initiative for the development of renewable energies in the Global South. From 2018, the EU moved to agriculture acquiring stakes in two agrifood investment funds, i.e., African Agriculture Fund and Trade managed by the Deutsch Bank and the Agri-Business Capital Fund managed by IFAD. In addition, the EU set up its own blending vehicle targeting agrifood systems, i.e., the AgriFI facility.

UNEP and Rabobank also set up an agrifood blending fund, i.e., the Agri3 fund with the aim of mobilising a billion dollars for forest protection and restoration projects by providing guarantees and technical assistance to support project development.

The Ugandan Development Bank, in partnership with the FAO Investment Centre, launched its AgrInvest initiative to strengthen the bank's capacity for investing in agriculture and to enhance its environmental assessment procedures (FAO 2023).

26.2.2 Carbon Offset Projects

The development of carbon credit markets, particularly voluntary markets, is another promise for funding "sustainable" agriculture. To offset their CO_2 emissions, companies (such as Microsoft, Danone, Unilever, CMA CGM) targeting carbon neutrality would acquire carbon credits. These credits are equivalent to the reduction or removal of one metric tonne of CO_2 or its equivalent thanks to "nature-based solutions", such as tree planting projects, agroforestry or regenerative agriculture. These payments for results are certified by third-party organisations that ensure compliance with environmental and social standards.[12] Despite recent controversies on the one hand, the current instability and narrowness of voluntary carbon credit markets on the other hand, many operators and financial intermediaries anticipate a demand surge for these credits [13] and a few entrepreneurs have launched dedicated carbon funds, such as the Dryada fund.

Despite the public resources invested, these innovative financial instruments have not closed the financing gap so far. Conversely, studies observe the persistence or even increase in 'agrifood systems' underfunding (FAO 2022, 2023). These studies also note a recurring uneven distribution in agrifood systems. On one hand, financial institutions and AgTech start-ups in climate-smart agriculture (Dey and Mishra 2022) or carbon sequestration, have the resources to capture these funds (Climate Policy Initiative 2023). As such, climate finance largely benefits to "*a new investment chain with its specialised actors (evaluators, auditors, investment funds) and its ecosystem*" (Chiapello 2020). On the other hand, (small) family farmers face high barriers in accessing climate finance.

> Across global and local agrifood systems, sustainable finance can support decarbonization efforts; still many actors are left out in the short term. Such dynamics are not exclusive to sustainable finance but rather finance in general: smallholder farmers, agrifood SMEs and other actors particularly in developing countries do not stand to benefit immediately from the developments in sustainable finance (FAO 2023).

Finally, the "greeness" of the projects supported by climate finance is regularly criticised by researchers (van Veelen 2020) and environmental NGOs (ONE France 2023), as well as their outcomes in terms of social and environmental justice (Franco and Borras 2019).

While climate finance practitioners often complain about the lack of recognized standards, the persistence of underfunding and inequalities is due to the exclusive focus on financial risk and return in a sector perceived as too risky (with high transaction costs, frequent climate hazards, etc.) and delivering scarce returns compared

---

[12] For example, the "Climate, Community and Biodiversity" standards developed and promoted by the certification body Verra.

[13] "Although these markets remain limited to about 2 billion dollars, the global demand for carbon credits is expected to increase by 100 times by 2050" (Climate Policy Initiative 2023).

to the levels expected by investors.[14] The financialisation of agricultural development and climate change policies inevitably leads to a disinvestment in the sector. Paradoxically, it is still the financial industry which is called to the rescue.

In addition, these policies also raise two other issues. First, the climate finance landscape is currently characterised by a proliferation and coexistence of initiatives without coordination. This proliferation, caused by institutional competition and bureaucratic politics, is an important limit for the planning of climate change mitigation and adaptation in agrifood systems. Second, while they focus on the supply-side of funding (i.e., the how) these policies leave aside the question of the "sustainable" practices to support (i.e., the what) e.g., promoting agroecology or strengthening the efficiency of conventional agriculture?

26.3 "Greening" Existing Public Financing

While the quest for innovative climate finance continues, another front has been recently opened up: the repurposing of existing public subsidies from "harmful" to "sustainable" agricultural activities. Indeed, several recent reports highlighted the potential of repurposing. In 2020, a World Bank report sought to assess the volume of public funding allocated to agriculture worldwide and the proportion dedicated to mitigating or adapting to climate change:

> Only a modest portion of current agricultural support has the potential to help mitigate emissions or even to increase production efficiencies generally. The roughly US$300 billion in market price supports boost prices to some farmers but at costs to others. Of the US$300 billion in direct spending, roughly 43 percent is designed to support farmer income and another 30 percent supports production. Only 9 percent of direct spending explicitly supports conservation, while another 12 percent supports research and technical assistance (Malins et al. 2020).

The authors of the report therefore propose repurposing a significant portion of these public subsidies, particularly those linked to production and inputs, towards mitigation and adaptation to climate change. Subsequently, several reports and studies have explored this proposal, detailing its advantages and feasibility conditions (Scown et al. 2020; Malins et al. 2020; IFPRI 2022; World Bank 2023; CAWR 2023; Climate Policy Initiative 2023; UNEP 2023). Two issues are regularly highlighted.

---

[14] "The risk-return profile of certain agri-food investments makes it difficult to attract private capital. For projects in primary agriculture, it is more difficult to design a standardised risk-return model, attractive to private investors and easily replicable. Most private investors still associate the agricultural sector with a higher risk, more complicated project management and less predictability. Moreover, most private investment companies are often not able to bear the operating and transaction costs of many agri-food projects, particularly those of small size. In order to reduce transaction costs and maximise returns, large institutional investors such as sovereign funds prefer to invest on a larger scale. Even the smaller impact funds struggle to invest in small-scale projects without a certain level of subsidy for transaction costs" (FAO 2023).

First, as illustrated by the title of the FAO and UNEP report on the topic ("A multi-billion dollar opportunity", in 2021), the rapid success of the repurposing narrative is primarily explained by the "discovery" of a new source of funding in a structurally constrained fiscal context: rather than inventing 'new' financial mechanisms from scratch, governments should make more efficient use of the existing ones. Indeed, the promoters of repurposing develop a virulent critique of the existing public subsidies and their negative impacts, whether on the economic (lack of efficiency), social (a very unequal distribution) and ecological (destruction of ecosystems) levels. They therefore propose to "rationalise" (or "detox") public subsidies in the name of a triple efficiency (economic, social and ecological).

> Government subsidies today make up an enormous share of public budgets worldwide, perhaps larger than at any point in human history. In many countries, the magnitude of explicit subsidies in the natural resource sectors exceeds that of investments in important public goods such as health and education. This report identifies and quantifies known as well as new channels through which poorly designed subsidies in natural resource sectors, though often well intentioned, deepen inequality, diminish productivity, and drive the destruction of ecosystems. Especially in an era of fiscal constraints and degrading natural capital, reform and repurposing of perverse and harmful subsidies offer an opportunity to promote greater sustainability, inclusion, and shared prosperity (World Bank 2023).

Second, the social impacts of repurposing are taken seriously. Indeed, promoters aim to anticipate, and often compensate, the actors and territories affected (or "sacrificed") by these policies. The FAO and UNEP report (2021) proposes a six-step protocol to evaluate *ex ante* and *ex post* the impacts of these policies. In the same line, the World Bank emphasises communication and compensation instruments to successfully implement these reforms.

> The reversal of subsidy reforms across the world points to the risks of poorly designed strategies that neglect distributional consequences, the magnitude of resistance, and the need for building a strong coalition in favor of change (World Bank 2023).

The political project of repurposing public subsidies addresses what has long been a blind spot in climate finance policies:[15] public subsidies supporting "harmful" activities emitting a lot of CO_2 or pollutants. Indeed, while "green" financial flows have been widely debated, these "bown" flows have long remained under the radar. However, these top-down proposals are largely disconnected from agricultural spheres, and the repurposing agenda covers a wide range of political agendas.

On the one hand, repurposing public subsidies is to date a top-down wish with very few concrete experiences. These reports often highlight the very same "good" practices: European agri-environmental measures (see Chap. 25), the Swiss agricultural policy, and public credits to Brazilian farmers conditioned on sustainable forest management practices. It is therefore difficult to analyse and compare empirical evidences underlying the successful implementation of these policies. These studies tend instead to downplay the political dimensions of these reforms, either during the

---

[15] In the energy sector, the first measures to eliminate public subsidies for carbon-based energies date back to 2009, yet even today they are not really implemented.

design phase, when different social groups (e.g., farmers' unions, agro-industrial companies, administrative and political elites) compete with each other around financial arbitrations (Iddri 2023), or during the implementation, when conflicts may undermine or disrupt negotiated measures. However, the recent failed attempts to remove tax exemptions on diesel for farmers in France or Germany, as well as the protests against the Nitrogen Plan in the Netherlands, highlighted the conflictual dimension of this agenda.

On the other hand, the repurposing of agricultural public subsidies covers antagonistic conceptions and objectives. On one side, a few "reformers" questions the size, and sometimes even the future, of certain agrifood activities which are particularly high emitters in CO_2 or nitrogen -e.g., livestock farming, certain cash crops such as sugar. These reformers recommend to disinvest from these activities in order to finance a decarbonised agrifood policy (CAWR 2023; I4CE 2023b). On the other side, "optimisers" call for improving the efficiency of these activities to turn them sustainable. By improving agricultural yields, it would be possible, for instance, to reduce deforestation, one of the main causes of emissions in the agricultural sector (Malins et al. 2020).

26.4 Conclusion

The current interest in climate finance for agrifood systems reflect a broad awareness of the structural lack of resources that the sector suffers from, particularly the most climate-vulnerable groups (small-scale family farmers, Indigenous Peoples, and women). This policy-making process is mainly driven by a coalition of public-private actors, gathering dominant agro-industrial companies, financial institutions, and major international organisations. Consequently, these initiatives mostly deepens the privatisation and financialisation of agricultural development policies.

On the one hand, the promotion of financial innovations and the repurposing of agricultural subsidies are often a vector for rationalising public administrations in a market-friendly framework. In the name of efficiency (economic, environmental and social), the resources managed by public institutions are reduced and transferred to private operators.

On the other hand, the problematisation of climate change through the lens of the *financial gap* clearly illustrates the financialisation of the sector and the primacy given to financial logics, actors, and infrastructures to address it. This mainstream framing has significant social and political outcomes by (de)legitimising political options and futures. By considering the fiscal constraints as insurmountable, this framing excludes *de facto* fiscal or monetary alternatives, such as green taxes or green monetary policies (CAWR 2023). Finally, these climate finance policies are technocratic and top-down with a very limited participation from the "beneficiaries", raising important issues for procedural and recognition justice (Colenbrander et al. 2018).

References

Angeli Aguiton S (2021) A market infrastructure for environmental intangibles: the materiality and challenges of index insurance for agriculture in Senegal. J Cult Econ 14(5):580–595. https://doi.org/10.1080/17530350.2020.1846590

Barral S (2023) Risk management in the Common Agricultural Policy: the promises of data and finance in the face of increasing hazards. Rev Agric Food Environ Stud 104(1):67–76

Bracking S, Leffel B (2021) Climate finance governance: fit for purpose? WIREs Climate Change 12(4):e709. https://doi.org/10.1002/wcc.709

Carton W, Bigger P (2020) Finance and climate change. In: The Routledge handbook of financial geography. Routledge

CAWR (2023) Financing Agroecological transformations for climate repair. CAWR

Chiapello E (2020) The financialisation of climate policy in a deadlock. In: Making the economy of the environment. Presses de l'école des mines, Paris

Chiapello È, Engels A, Gresse EG (2023) Financializations of development. Routledge, New York

Christiansen J (2021) Fixing fictions through blended finance: the entrepreneurial ensemble and risk interpretation in the blue economy. Geoforum 120:93–102. https://doi.org/10.1016/j.geoforum.2021.01.013

Climate Policy Initiative (2023) Landscape of climate finance for agrifood systems

Colenbrander S, Dodman D, Mitlin D (2018) Using climate finance to advance climate justice: the politics and practice of channelling resources to the local level. Clim Pol 18(7):902–915. https://doi.org/10.1080/14693062.2017.1388212

Dey K, Mishra PK (2022) Mainstreaming blended finance in climate-smart agriculture: complementarity, modality, and proximity. J Rural Stud 92:342–353. https://doi.org/10.1016/j.jrurstud.2022.04.011

Ducastel A (2016) Private equity as a tool of public action or the financialisation of development in Sub-Saharan Africa. Afr Polit 144(4):135–155. https://doi.org/10.3917/polaf.144.0135

Ducastel A, Bourblanc M, Adelle C (2023) Why development finance institutions are reluctant to invest in agriculture… and why they keep trying. The financialization of development policies as an obstacle to invest in agriculture. In: Financializations of development global games and local experiments. Routledge

FAO (2022) Climate finance in the agriculture and land use sector—global and regional trends between 2000 and 2020. FAO

FAO (2023) Investing in carbon neutrality: utopia or the new green wave? Challenges and opportunities for agrifood systems. FAO

FAO, UNEP (2021) A multi-billion dollar opportunity. Repurposing agricultural support to transform food systems. FAO

FOLU (2019) Growing better. Ten critical transitions to transform food and land use. FOLU

Franco JC, Borras SM (2019) Grey areas in green grabbing: subtle and indirect interconnections between climate change politics and land grabs and their implications for research. Land Use Policy 84:192–199. https://doi.org/10.1016/j.landusepol.2019.03.013

Gabas J-J, Ribier V, Vernières M (2017) Presentation. Financing or financialisation of development? A question under debate. Worlds Dev 178:7–22

Gabor D (2023) The (European) Derisking state. SocArXiv hpbj2, Center for Open Science. https://doi.org/10.31219/osf.io/hpbj2

I4CE (2023a) Overview of climate finance. 2023 Edition

I4CE (2023b) Transition of livestock farming: managing past investments and rethinking those to come

IDDRI (2023) Greening the agrifood system through the EU budget: can we repurpose agricultural subsidies?

IFAD, CPI (2020) Examining the climate finance gap for small-scale agriculture

IFPRI (2022) Repurposing agricultural support: creating food systems incentives to address climate change. Global food policy report

Jachnik R, Mirabile M, Dobrinevski A (2019) Tracking finance flows towards assessing their consistency with climate objectives. OECD, Paris

Johnson L (2021) Rescaling index insurance for climate and development in Africa. Econ Soc 50(2):248–274. https://doi.org/10.1080/03085147.2020.1853364

Mah L (2023) The financialization of EU development policy: blended finance and strategic interests (2007–2020). In: Financializations of development. Routledge

Malins C, Searchinger TD, Dumas P, Baldock D, Glauber J, Jayne T et al (2020) Revising public agricultural support to mitigate climate change. In: Development knowledge and learning, World Bank. http://hdl.handle.net/10986/33677

Mawdsley E (2018) 'From billions to trillions': financing the SDGs in a world 'beyond aid'. Dialogues Hum Geogr 8(2):191–195. https://doi.org/10.1177/2043820618780789

ONE France (2023). The "climate finance files": the wild west of climate finance

ROPPA (2023) Untapped potential – an analysis of international public climate finance flows to sustainable agriculture and family farmers

Scown MW, Brady MV, Nicholas KA (2020) Billions in misspent EU agricultural subsidies could support the sustainable development goals. One Earth 3(2):237–225

UNEP (2023) Driving finance for sustainable food systems. A Roadmap to Implementation for Financial Institutions and Policy Makers

van Veelen B (2020) Cash cows? Assembling low-carbon agriculture through green finance. Geoforum 118. https://doi.org/10.1016/j.geoforum.2020.12.008

World Bank (2023) Detox development. Repurposing environmentally harmful subsidies. World Bank

Open Access This chapter is licensed under the terms of the Creative Commons Attribution-NonCommercial-NoDerivatives 4.0 International License (http://creativecommons.org/licenses/by-nc-nd/4.0/), which permits any noncommercial use, sharing, distribution and reproduction in any medium or format, as long as you give appropriate credit to the original author(s) and the source, provide a link to the Creative Commons license and indicate if you modified the licensed material. You do not have permission under this license to share adapted material derived from this chapter or parts of it.

The images or other third party material in this chapter are included in the chapter's Creative Commons license, unless indicated otherwise in a credit line to the material. If material is not included in the chapter's Creative Commons license and your intended use is not permitted by statutory regulation or exceeds the permitted use, you will need to obtain permission directly from the copyright holder.

Chapter 27
Science–Policy Interfaces in the Face of Climate Change

Carolina Milhorance, Antoine Perrier, Julien Demenois, Vincent Freycon, Camille Piponiot, Paul Luu, Adèle Gaveau, Marie Hrabanski, and Sélim Louafi

Abstract The Covid-19 pandemic and climate-related crises have heightened demands to integrate scientific knowledge into policymaking, often through science-policy interfaces (SPIs) that facilitate dialogue between scientists and decision-makers. These complex issues require collaboration due to uncertainties and conflicting values. While evidence-based policymaking assumes science guides action, in reality political decisions are influenced by emotions, interests, and diverse knowledges. This chapter explores the non-linear nature of science-policy interactions, analyzing decision-making tensions and the role of SPIs in fostering collective learning on climate challenges.

C. Milhorance (✉) · A. Perrier
Cirad, ART-Dev Research Unit, Montpellier, France
e-mail: carolina.milhorance@cirad.fr

J. Demenois
CIRAD, UPR Agroecology and sustainable intensification of annual crops (Aïda), Montpellier, France

Aïda, Univ. Montpellier, CIRAD, Montpellier, France

V. Freycon
Cirad, UPR Forests and Societies, Montpellier, France

C. Piponiot
Cirad, UPR Forests and Societies, Montpellier, France

P. Luu
International Initiative "4 per 1000", Montpellier, France

A. Gaveau
University of Lausanne, Lausanne, Vaud, Switzerland

M. Hrabanski
Cirad, ART-Dev Research Unity, Montpellier, France

S. Louafi
Cirad, DGD-RS, Montpellier, France

© The Author(s) 2026
V. Blanfort et al. (eds.), *Climate Impacts and Challenges in Agriculture, Forests and Food Systems*, https://doi.org/10.1007/978-3-032-04331-3_27

The Covid-19 pandemic, natural disasters exacerbated by climate change, and recurring food crises have reinforced societal demands to integrate scientific knowledge into political decision-making (Cairney et al. 2016; Weible et al. 2020). Moreover, numerous international conventions rely on scientific assessments, notably those conducted by the Intergovernmental Panel on Climate Change (IPCC) and the Intergovernmental Science-Policy Platform on Biodiversity and Ecosystem Services (IPBES).

The term "science-policy interface" (SPI) carries multiple definitions within the literature, but it is generally understood as an institutional arrangement, forum, process, or organization designed to facilitate dialogue between scientific advice and political decision-making. The establishment of these interfaces rests, among other things, on the hypothesis that certain particularly complex issues—such as global food security, climate change, or biodiversity loss—require close and dynamic interaction between scientists and policymakers. This necessity arises from the cross-sectoral nature of these challenges, the high degree of uncertainty surrounding the state of knowledge, and pronounced divergences regarding the values to be prioritized in addressing them (solidarity, efficiency, equity, responsibility, etc.) (Head 2022). Within this context, scientists strive to illuminate the multiple causes, values, and bodies of knowledge at stake, while establishing links between problems and potential solutions and providing analyses and interpretations capable of influencing political strategies.

However, the interactions between scientific research and political decision-making are not a novel subject. For over 40 years, the social sciences and humanities have studied these interactions, moving beyond the traditional linear model. This model rests on the assumption that knowledge precedes action and that the public—often perceived as insufficiently informed—can be enlightened by reliable information. The evidence-based policymaking approach exemplifies this model by postulating that policies should be grounded in scientific evidence to address societal challenges. Yet, this ideal encounters reality: scientific evidence is frequently contested, and the political process involves a diversity of actors with varied interests. In practice, policymakers mobilize a mixture of emotions, diverse knowledges, and cognitive shortcuts to advance their policies (Cairney 2013).

Thus, the interface between science and political decision-making, far from being linear or straightforward, raises fundamental questions about the nature and modalities of these interactions. This chapter is organized into three parts to explore these questions. First, it proposes an analytical framework for interactions between science and political decision-making, drawing on technocratic, decisionist, and pragmatist models. Second, it examines decision-making processes within SPIs, revealing the tensions, compromises, and power asymmetries that influence these interactions and the construction of public policies. Finally, it discusses the potential of SPIs to foster collective learning in the face of climate challenges, while addressing the issues related to their implementation and sustainability.

27.1 Revisiting Science–Policy Relations

This section provides a brief presentation of different analytical frameworks and concrete examples to explore the interactions between scientists, policymakers, and other actors in the climate and environmental fields (Boxes 27.1 and 27.2). It highlights the complex dynamics of these interactions, both in the formulation and implementation of public policies. The analysis invites a deeper understanding of political processes to better grasp the tensions and compromises that emerge from them.

27.1.1 Key Conceptual Models

Three key models illustrate contrasting perspectives on the interface between science and policy-making, depending on the relative importance given to science or politics in terms of control and authority. These models, ranging from political predominance (decisionist model) to scientific predominance (technocratic model), are part of an academic conceptualisation developed over time thanks to the contributions of many thinkers in political science, philosophy, and sociology (including Jürgen Habermas, Sheila Jasanoff, Bruno Latour, Roger Pielke, Laurens Hessels, David Guston, John Dewey, among others). These models, referred to as ideal types, aim to illuminate the dynamics of interaction between science and public decision-making. They do not reflect the full richness of this literature and should not be interpreted as opposing frameworks or as reflecting the positions defended by specific actors (Dressel 2022).

The technocratic approach conceives political practice as the execution of directives defined by a scientific and technical elite. Rooted in a positivist vision of science, it seeks to guide public policies through "optimal" solutions, based on available knowledge. This approach is based on the idea that scientific expertise is sufficient to solve public problems, minimising the influence of power relations, values, and beliefs in decision-making. It suggests an intrinsic superiority of scientific solutions over those derived from a democratic and participatory debate based on societal and economic values. As a result, prescriptive science leaves little room for critical dialogue and negotiation between technical expertise and societal values.

The decisionist approach clearly distinguishes the roles of experts and policy makers, attributing to the latter the primary authority to define political objectives. In this model, science is limited to providing facts and technical options, while the final choices are made by political actors, according to their priorities and preferences. Once the objectives are established, science objectively determines the means to achieve them, thus preserving value-free research and a neutral role of science in public debates. This approach sees science as a reliable source of knowledge, capable of guiding concrete actions without interfering in the definition of values and objectives. However, by confining science to an instrumental role, this approach neglects the need for collective reflection on complex issues that often require multi-actor negotiation or dialogue. It also raises questions about the real possibility of separating values from facts in scientific research and the ability of science to maintain its neutrality in political or social contexts imbued with values.

Critics of the approaches mentioned above highlight the deeply dynamic interactions between science and political decision-making, where values and political objectives influence not only scientific research, but also its application (Dressel 2022). The IPCC illustrates well how some individuals take on both scientific and political roles during their career, or even simultaneously, depending on the situation. Although IPCC reports are based on rigorous scientific work, they also formulate political diagnoses concerning the evolution of societies and propose strategic visions for the management of forests, soils and other ecosystems on a global scale (Louafi 2021).

The pragmatist approach considers the scientific and political spheres as interdependent, rather than separate. According to van den Hove (2007), this interdependence fosters constructive interactions, while encouraging dialogue and negotiation. This approach emphasises reflexivity, allowing scientific and political actors to constantly reassess and adapt their methods and perspectives based on feedback and contextual changes. This approach insists on the need to recognise and communicate the uncertainties inherent in science, thus enhancing transparency and collaboration. However, it presents challenges, including the risk of leading to epistemological relativism due to the social and historical construction of scientific knowledge, and the complexity of its practical application. Moreover, by explicitly admitting the social and politico-economic influences on scientific production, this approach may inadvertently encourage excessive politicisation of science.

The idea of co-production of scientific knowledge and social contexts dates back to the earliest research on science-policy interfaces. Yet, many international initiatives and some of the literature continue to rely on simplistic linear models, centred on science directly feeding political decisions. These approaches mask the complexity of the relationships between research and policy as well as the processes of formulating and evolving public policies.

Box 27.1 Agroforestry in Ivory Coast: Is Science an Arbiter of Political Decisions?

In Ivory Coast, although agroforestry is an ancient practice, it has seen a resurgence of interest since the mid-2010s, particularly as a response to increasing deforestation and to ensure the sustainability of cocoa production. Integrated into political discourse and sustainability standards, agroforestry is now seen as an increasingly consensual solution for the sustainability of cocoa. However, the decision regarding which agroforestry system to

(continued)

Box 27.1 (continued)

support involves political choices at the confluence of sometimes contradictory objectives. For example, agroforestry is practised both in areas previously cleared with the aim of reforestation and in preserved forests, leading to their deforestation or degradation (Zo-Bi and Hérault 2023).

In this context, some decision-makers see science as a legitimate arbiter for establishing the standards for implementing agroforestry systems favoured by public policies and international projects. However, this recognition can expose scientists to risks of instrumentalisation. They are regularly solicited by institutional actors and private companies to establish normative criteria concerning, for example, the number of trees per hectare, the species to plant, the necessary level of shade and the role of farmers in these systems. While these criteria and indicators may initially seem to be technical considerations, they are actually the result of political compromises and negotiations. Their application in rural areas and in the classified forests of Ivory Coast is not limited to a scientific approach, but also involves a delicate balance between economic, environmental and political imperatives (Di Roberto et al. 2023).

Box 27.2 The Role of Expertise in the Great Green Wall

The Great Green Wall project, launched in 2007 by eleven African countries, aims to restore and green the Sahelian lands over 7000 km, from the Atlantic to the Red Sea. However, despite strong political will and significant funding, its implementation remains uneven and the results achieved are mixed.

Several factors explain this partial failure. The policies of planting, largely based on biophysical scientific knowledge and professional and managerial approaches, has overlooked social and economic dimensions. This top-down designed project encountered various obstacles, including low participation of local populations in the restoration process (Cesaro et al. 2022). This approach generated tensions with pastoral communities, who saw their transhumance practices hindered. Moreover, local resource governance did not always respect established rules, exacerbating difficulties (Mugelé 2018).

In the face of these challenges, many researchers and technicians advocate for a better consideration of the sociopolitical specificities of each context. That's why, in recent years, various initiatives have been launched to encourage dialogues between scientists from different fields (social sciences, agronomy, environment, geography…) and political actors (Mugelé 2022).

27.1.2 Decision-Making Processes in Public Policy Design

In a context where the boundaries between science and political decision-making are fluid, it is essential to better understand the decision-making processes and the formulation of public policies. This invites a reassessment of certain preconceived ideas, particularly the one that science would provide objective and decisive knowledge for political decisions (Louafi 2021). This analysis also invites us to move beyond the simplistic vision of a unitary state, often reduced to the sole actions of ministers and Parliament. Three key messages emerge from this reflection:

Public policies go beyond state interventions alone and mobilise a multitude of actors at different levels, often in an uncoordinated manner. The boundaries between the public and private sectors become blurred, paving the way for a diversity of actors at the interface of these two spheres. This phenomenon, amplified by liberalisation, privatisation and globalisation, has fostered the development of various partnerships, the integration of private standards in the public sector, and the increasing involvement of NGOs, international agencies and private entities in the development and implementation of policies (Hassenteufel 2011).

Contrary to a rational vision centred on problem-solving and informed debates, the political process is often unpredictable and discontinuous. This process is shaped by the way decision-makers set priorities, establish objectives and choose specific instruments. A variety of public and private actors participate in it, sometimes forming coalitions around common paradigms—values, worldviews—to support or challenge certain decisions (Jenkins-Smith et al. 2014). The resulting policy is not necessarily dictated by science, but is rather influenced by factors such as the role of key actors, crises, economic contexts, negotiation frameworks, institutional changes and the support of other stakeholders.

Political decisions, taken at different levels, are accompanied by implementation processes often marked by vague objectives, conflicting interests and asymmetrical means. This opens the way for interpretations and decentralised decisions by the actors involved. Implementation reveals not only how a public programme is applied, but also how it is appropriated, sometimes deviating from its initial design (Lascoumes and Le Galès 2018).

These dynamics underline the importance of understanding the public action paradigms that intersect in the formulation of public policies as well as the tensions, power relations, bureaucratic routines and institutional factors that accompany them.

Climate, forestry and agricultural policies reveal tensions and synergies between various objectives: environmental conservation, food security and socio-economic development. These objectives, often interconnected but sometimes contradictory, are particularly evident in the management of landscapes modified by human activity. In West Africa, particularly in Ghana and Côte d'Ivoire, the cultivation of cocoa, coffee, rubber and oil palm, established on former forest landscapes, illustrates the effects of the disappearance of more than 80% of forests since 1900 (Aleman et al. 2018). These territories, mixing agricultural lands, areas abandoned due to lack of fertility and residual forest fragments, embody the tensions between socio-economic imperatives, biodiversity conservation, indigenous peoples' rights and pressures from global agricultural markets. The construction of public policies for these spaces requires an integrated approach, taking into account varied spatial and temporal scales as well as complex interactions and conflicts between sectors.

Finally, in the current context, marked by populism and disinformation, the complexity of interactions between science and politics becomes even more apparent. The administration of Donald Trump in the United States illustrates how scientific and democratic issues can be reduced to oversimplified narratives, leading to polarised decisions and reductive slogans. The rise of a "post-truth era" questions the authority of scientific work in public debate, weakening not only trust in science, but also certain democratic foundations (Soneryd and Sundqvist 2023). However, many works agree that science cannot be dissociated from society. Scientists, far from reducing reality to simplifications, enrich it and play an active role in social transformation processes.

27.2 Science–Policy Interfaces at the Nexus of Climate, Food Systems, and Ecology

This section explores the diversity of Science-Policy Interfaces (SPIs) in the climate field and their interconnections with food and ecological issues, examining their operation at different scales and in various contexts. Drawing from international, national, and territorial examples, it highlights their objectives, whether it be to contribute to the construction of public policies or to promote collective learning. It also analyses the structural and contextual challenges related to the production and application of scientific knowledge, incorporating North-South asymmetries and cross-sectoral issues.

27.2.1 How Science–Policy Interfaces Operate from Global to Local Levels

SPIs materialise in multiple ways and operate at different levels, notably through informal consultations, seminars, networks, projects, interaction platforms, and other spaces (Wagner et al. 2023). These spaces pursue various objectives: the synthesis and dissemination of knowledge, capacity building, support for policy implementation and monitoring, and the establishment of partnerships.

In the fields of climate change, agriculture, and forests, international initiatives such as the IPCC for the United Nations Framework Convention on Climate Change, the IPBES for the Convention on Biological Diversity, the science-policy interface of the United Nations Convention to Combat Desertification and Land Degradation, the High-Level Panel of Experts on Food Security and Nutrition, and the "4 per 1000" Initiative illustrate significant examples (Box 27.3). For instance, with the publication of six assessment reports and numerous special reports since 1990, the IPCC has accumulated over 35 years of dialogue between climate scientists and government representatives.

At the national level, numerous more or less formal bodies exist, such as multi-stakeholder committees responsible for formulating and monitoring national climate change policies. Among these examples are Brazil's expert group on climate change, the national REDD+ commission (Reducing Emissions from Deforestation and Forest Degradation) of Côte d'Ivoire, the national committee on climate change in Senegal, and the High Council for Climate in France. At a territorial level, committees are emerging to monitor the implementation of public policies and to manage concrete issues such as the impacts of climate change on the intensification of water crises.

Furthermore, research and development projects are increasingly adopting innovative approaches, such as territorial observatories. These structures specialise in the systematic collection and analysis of data to monitor and understand specific phenomena. For example, the Observatory of Socio-Environmental Dynamics (Odyssea) examines land use transformation in certain Amazonian territories by gathering and producing data from satellite monitoring, climatic and hydrological surveys, pollution analyses, and studies of socio-political dynamics. The aim is to put this data at the service of society and stimulate "citizen science", by mobilising rural unions and facilitating political change processes (European Commission 2020).

Box 27.3 The International Initiative "4 per 1000: Soils for Food Security and Climate"

Launched by France at COP21 in 2015, the International Initiative "4 per 1000: Soils for Food Security and Climate" promotes the idea that an annual increase of 0.4% in soil carbon stocks through adapted agricultural or forestry practices could significantly contribute to combating climate change and its effects. It reflects the long-standing efforts of science to integrate this solution into the political agenda (Kon Kam King et al. 2018). The French Minister of Agriculture at the time played a role in promoting this agenda at the international level. Its political adoption is also linked to the promise of reconciling adaptation and mitigation of climate change, an issue often contentious between countries of the Global South and Global North, and which was central in the negotiations of the Paris Agreement. It is also a multi-stakeholder platform facilitating dialogue between various stakeholders from the political, scientific, entrepreneurial, production, and civil society worlds.

The "4 per 1000" Initiative, along with other initiatives and coalitions (Global Soil Partnership, Adaptation of African Agriculture, Climate Action for Sustainable and Healthy Soils, Coalition for Agroecology), has contributed to bringing the issue of soils and their health to the forefront of the international agenda, thanks to ongoing dialogue among all actors, including decision-makers and scientists, not to mention producers. Indeed, with its more than 800 partners worldwide, it remains to this day one of the few global multi-stakeholder partnerships working on the issues of soil carbon, soil health for the benefit of combating climate change, desertification, biodiversity erosion, and food insecurity.

For more information, see: https://4p1000.org/

27.2.2 Embedding Political and Learning Processes of Science–Policy Interfaces

Political processes, as well as the tensions and compromises inherent in the construction of public policies, are also manifested within the ISPs, where negotiations are developed at multiple levels.

A striking example is the International Assessment of Agricultural Knowledge, Science and Technology for Development (IAASTD), conducted from 2003 to 2008. This process aimed to assess the impact of agricultural technologies on reducing hunger and poverty, while raising awareness within the IPCC about the importance of agricultural issues in the context of climate change. It highlighted the increasing interconnection between agriculture and climate, as well as the need for interdisciplinary and inclusive approaches to address global challenges. However, the IAASTD also had to reconcile scientific expertise and the inclusion of diverse perspectives. The quantitative models used for future scenarios were criticised for their reductive nature, excluding certain solutions, particularly in Africa. Moreover, the debate on genetically modified crops exacerbated divisions between industrial and sustainable agriculture, leading to the withdrawal of representatives from the biotechnology industry (Scoones 2009).

Similar tensions appear in the work of the IPBES, particularly regarding the links between biodiversity and agriculture. Its first report, published in 2015 and focusing on pollination, highlighted controversies related to the use of chemical inputs and their effects on biodiversity (Duperray et al. 2017). In the field of food security, disagreements over the rise in food prices in 2007–2008 led to the creation, in 2009, of the High-Level Panel of Experts on Food Security and Nutrition (HLPE) within the United Nations Committee on World Food Security. Since 2021, this issue has been broadened to include dimensions of equity, access to healthy food and underlying agricultural models.

These sociopolitical tensions vary according to contexts. Since the late 2000s, numerous scientific observatory projects have emerged to monitor, quantify and document large land acquisitions. However, the production of large-scale data involves trade-offs influenced by stakeholders, local realities and available resources. For example, in Cameroon, internal divergences have hindered the dissemination of data on land acquisitions, while in Madagascar the Land Observatory limits the disclosure of sensitive information due to its ministerial affiliation. Moreover, data production is faced with technical and logistical constraints, requiring qualified human resources. Their use must also meet the needs of target actors. In Senegal, for example, a mismatch between the expectations of civil society organisations and the format of the produced data has limited their impact (Grislain et al. 2023).

In parallel, the production of research and expertise within the framework of international cooperation and North-South relations raises crucial questions. These relations highlight power asymmetries and dynamics that directly influence the impact of scientific knowledge on policies.

The IPCC illustrates these issues well. Its work, divided into three groups, addresses different aspects of climate change: Group I assesses scientific knowledge; Group II analyses the impacts and possibilities for adaptation; and Group III explores mitigation solutions. This latter group, directly linked to political considerations, plays a key role in diplomatic discussions, particularly during the approval of summaries for policymakers. Since the adoption of the Paris Agreement in 2015, nationally determined contributions have strengthened the importance of SPIs at the national level, with the IPCC playing an increasing role in the development of climate strategies. Experts sit on scientific committees, advise executive members or directly participate in decision-making bodies such as the Ministries of Environment or Science (Pryck and Gaveau 2023). However, despite an increase in the number of experts from the South since the 1990s, these represented only 39% of participants in the 6th report (AR6) (Pryck 2022). This structural imbalance reflects the dominance of Northern institutions, better equipped in terms of infrastructure (such as supercomputers) and having stronger institutional networks. Other factors contribute, such as insufficient selection of authors by certain national focal points in the South, the absence of author remuneration, and limited interactions between scientific institutions, states, civil society and the private sector, as well as sometimes unsuitable communication resources (Yamineva 2017).

The science-policy dialogue can also be envisaged as a process of collective learning, drawing on various forms of knowledge, sectors and disciplines, and going beyond scientific knowledge (Cairney et al. 2016). This process relies on the collaboration of multiple actors and on the integration of different knowledge systems. It thus contributes to producing actionable knowledge, essential for facilitating the implementation of collective actions.

However, the limited interaction of the IPCC with other SPIs at the international level hinders the possibilities for intersectoral learning on common challenges. In response, an informal process was launched in 2023 in Montpellier, France, to pool the collective intelligence of expert groups working on climate, biodiversity, health and food systems. The aim is to strengthen connectivity between local, national and global scales, and to structure a community of practice for SPIs. This project emphasises the transformation of agri-food systems as a key lever for addressing the interdependent challenges of sustainable development (Caron et al. 2022). It aims to promote cross-disciplinary collaboration among SPIs, to overcome compartmentalised thinking and practices, to improve stakeholder representation, and to better connect scales, from local to global.

27.3 Conclusion

This chapter has analyzed the role of SPIs in addressing the interconnected challenges of climate change, food issues, and ecological problems. It has presented interpretive frameworks to better understand these interfaces, while highlighting the diversity of action scales and contexts in which they operate.

Drawing on work that emphasizes the deep interweaving of science and society, this chapter has illuminated the active role of scientists in processes of social transformation. Far from reducing reality to simplifications, science enriches debates and informs collective choices. A focus on political processes has allowed for the exploration of the tensions and compromises inherent in interactions between scientists, policymakers, and other actors in the formulation and implementation of public policies. This approach goes beyond linear visions that see science as a purely technical influence on political decisions. The examples studied show that SPIs are negotiation spaces where divergent interests, power asymmetries, and legitimacy issues converge.

Furthermore, SPIs offer significant potential to promote collective learning processes involving a diversity of actors: scientists, governments, civil society, and populations. These processes allow for broadening perspectives on climate challenges, questioning paradigms that hinder political change processes, and redefining action approaches.

However, major challenges persist, particularly in terms of co-production of knowledge and consideration of scientific uncertainties, in a context where trust in scientific work is increasingly questioned in public debates. In addition, the weak integration and coordination among the many existing SPIs at the global scale limit their effectiveness. A careful contextualisation of these spaces is essential, taking into account the needs for resources, skills, and funding to ensure their sustainability.

Beyond financial or coordination challenges, it is essential to understand local power dynamics, rivalries between interest groups, and to contributions from research and other forms of knowledge. These dimensions influence the quality and nature of dialogues in various institutional and territorial contexts. Moreover, the establishment of coordination platforms must avoid creating ephemeral structures, often dependent on short-term international development projects. Geopolitical issues, as well as the influence of international organisations and development agencies on public policies in Southern countries, also deserve particular attention.

Finally, research should not be limited to a role of alert or technical expertise aimed at guiding institutional reforms. A reflective and dynamic approach is essential to make science a lever of transformation in the face of global challenges.

This study was carried out within the framework of the Terri4Sol project, with financial support from the French Fund for the Global Environment: https://www.terri4sol.org/

References

Aleman JC, Jarzyna MA, Staver AC (2018) Forest extent and deforestation in tropical Africa since 1900. Nat Ecol Evol 2(1):Article 1. https://doi.org/10.1038/s41559-017-0406-1

Cairney P (2013) What is evolutionary theory and how does it inform policy studies? Policy & Politics 41(2):279–298

Cairney P, Oliver K, Wellstead A (2016) To bridge the divide between evidence and policy: reduce ambiguity as much as uncertainty. Public Adm Rev 76(3):399–402. https://doi.org/10.1111/puar.12555

Caron P, Ferrero de Loma-Osorio G, Ferroni M, Lehmann B, Mettenleiter TC, Sokona Y (2022) Global food security: Pool collective intelligence. Nature 612(7941):631–631. https://doi.org/10.1038/d41586-022-04471-0

Cesaro D, Touré I, Taugourdeau S, Delay E, Diouf A, Ba M, Diop D, Ferrari S (2022) Reforestation and pastoralism in the Sahel: (Re)conciling uses in territories for a relaunch of the Great Green Wall. In: Conference of the parties (COP15) of the United Nations convention to combat desertification. https://agritrop.cirad.fr/600970/1/POSTER_PPZS_GMV.pdf

Di Roberto H, Milhorance C, Sokhna DN, Sanial E (2023) Agroforestry in a post-forest context: perspectives and controversies of a political agenda setting in Côte D'ivoire. Bois et Forêts des Tropiques 356:81–91. https://doi.org/10.19182/bft2023.356.a37121

Dressel M (2022) Models of science and society: transcending the antagonism. Hum Soc Sci Commun 9(1):241. https://doi.org/10.1057/s41599-022-01261-x

Duperray F, Hrabanski M, Oubenal M (2017) First thematic assessment on pollination: between the legitimisation of IPBES and tensions regarding the selection of knowledge and experts. In: The intergovernmental platform on biodiversity and ecosystem services (IPBES). Routledge

European Commission (2020) Local stakeholders monitor social and environmental dynamics in the Amazon. CORDIS I European Commission. https://cordis.europa.eu/article/id/422428-involving-local-stakeholders-in-monitoring-social-and-environmental-dynamics-in-the-amazon/fr

Grislain Q, Burnod P, Bourgoin J, Anseeuw W (2023) Producing, sharing and strengthening the use of data: challenges and lessons learned from observatories of large land acquisitions. Cahiers Agric 32:30. https://doi.org/10.1051/cagri/2023024

Hassenteufel P (2011) Political sociology: public action. Armand Colin

Head BW (2022) Coping with wicked problems in policy design. In: Research handbook of policy design. Edward Elgar Publishing, pp 155–175. https://www.elgaronline.com/display/edcoll/9781839106590/9781839106590.00018.xml

Jenkins-Smith H, Nohrstedt D, Weible CM, Sabatier PA (2014) The advocacy coalition framework: foundations, evolution, and ongoing research. In: Sabatier PA, Weible CM (eds) Theories of the policy process, 3rd edn. Westview Press, pp 183–223

Kon Kam King J, Granjou C, Fournil J, Cecillon L (2018) Soil sciences and the French 4 per 1000 initiative—the promises of underground carbon. Energy Res Soc Sci 45:144–152. https://doi.org/10.1016/j.erss.2018.06.024

Lascoumes P, Le Galès P (2018) Sociology of public action, New edn. Armand Colin

Louafi S (2021) Scientists warn but politicians do nothing… is it really that simple? The Conversation. http://theconversation.com/les-scientifiques-alertent-mais-les-politiques-ne-font-rien-est-ce-vraiment-si-simple-158205

Mugelé R (2018) The Great Green Wall in the Sahel: between global ambitions and local anchoring. Bulletin of the Association of French Geographers. Geographies 95(2):Article 2. https://doi.org/10.4000/bagf.3084

Mugelé R (2022) The great Green Wall: Mobilising utopia or tool for local development? Grain de Sel 82–83:54–55

Pryck KD (2022) Negotiating climate science. The role of member states in the intergovernmental panel on climate change. Int Critique 95(2):132–153

Pryck KD, Gaveau A (2023) Scientists in multilateral diplomacy. The Case of the Members of the IPCC Bureau. https://doi.org/10.1163/25903276-bja10040

Scoones I (2009) The politics of global assessments: the case of the international assessment of agricultural knowledge, science and Technology for Development (IAASTD). J Peasant Stud 36(3):547–571. https://doi.org/10.1080/03066150903155008

Soneryd L, Sundqvist G (2023) Science and democracy: a science and technology studies approach, 1st edn. Bristol University Press. https://doi.org/10.46692/9781529222159

van den Hove S (2007) A rationale for science–policy interfaces. Futures 39(7):807–826. https://doi.org/10.1016/j.futures.2006.12.004

Wagner N, Velander S, Biber-Freudenberger L, Dietz T (2023) Effectiveness factors and impacts on policymaking of science-policy interfaces in the environmental sustainability context. Environ Sci Pol 140:56–67. https://doi.org/10.1016/j.envsci.2022.11.008

Weible CM, Nohrstedt D, Cairney P, Carter DP, Crow DA, Durnová AP et al (2020) COVID-19 and the policy sciences: initial reactions and perspectives. Policy Sci 53(2):225–241. https://doi.org/10.1007/s11077-020-09381-4

Yamineva Y (2017) Lessons from the intergovernmental panel on climate change on inclusiveness across geographies and stakeholders. Environ Sci Pol 77:244–251. https://doi.org/10.1016/j.envsci.2017.04.005

Zo-Bi IC, Hérault B (2023) Promoting agroforestry? Lessons from Ivory Coast. Trop Woods For 356:93–98. https://doi.org/10.19182/bft2023.356.a37132

Open Access This chapter is licensed under the terms of the Creative Commons Attribution-NonCommercial-NoDerivatives 4.0 International License (http://creativecommons.org/licenses/by-nc-nd/4.0/), which permits any noncommercial use, sharing, distribution and reproduction in any medium or format, as long as you give appropriate credit to the original author(s) and the source, provide a link to the Creative Commons license and indicate if you modified the licensed material. You do not have permission under this license to share adapted material derived from this chapter or parts of it.

The images or other third party material in this chapter are included in the chapter's Creative Commons license, unless indicated otherwise in a credit line to the material. If material is not included in the chapter's Creative Commons license and your intended use is not permitted by statutory regulation or exceeds the permitted use, you will need to obtain permission directly from the copyright holder.

Conclusion: Strengthening Independent National Scientific Institutions in a World Under Geopolitical and Financial Tension

Michel Eddi and Sébastien Treyer

A Successful Agenda-Setting of Truly Systemic Transformations

Since the signing of the Paris Agreement at COP21, science has successfully placed truly systemic transformations on the political agenda. In 2015, there was a fear that the issue of climate, and particularly mitigation, would become the sole priority for the transformation of the land sector, to the detriment of other inseparable objectives. Thanks to the support of scientific works such as the IPCC report on the land sector (IPCC 2019) or the IPBES global assessment report (2019), the world has now equipped itself with a coherent framework of medium and long-term objectives for the transformation of food systems and the land sector: the Sustainable Development Goals (SDGs), the Paris Agreement on climate from 2015, supplemented in 2022 by the Kunming-Montreal Global Biodiversity Framework. As highlighted by the global reports on sustainable development (GSDR 2019, 2023), only a profound transformation of food systems will allow these different objectives to be met together. The Dubai declaration on agriculture at COP28 in 2023[1] illustrates that all these long-term political orientations are aligned: most of the world's governments have committed to developing a climate strategy for the land sector by 2025, consistent with objectives in terms of food security, biodiversity and combating desertification.

---

[1] https://www.cop28.com/en/food-and-agriculture

M. Eddi
Ministry of Higher Education and Research, Paris, France

IDDRI, Paris, France

S. Treyer
IDDRI, Paris, France
e-mail: sebastien.treyer@sciencespo.fr

Conflicts and Competition: A Political Context Entirely Different from the Spirit of Cooperation of COP21

But are these objectives reflected in the reality of political trade-offs, as the context has significantly tightened? Geopolitical tensions between major powers and wars favour competition, conflict, power relations and mistrust. The impact of successive crises also severely tests national political spaces which are polarising. The ability to negotiate stable agreements seems to be deteriorating in all regions. The most vulnerable countries must face a major financial crisis as well as a food security crisis. Rich countries are also reducing budgets for ecological transition and international solidarity for development. How, in such a context, can profound transformations of economic and social models be negotiated? The robustness of scientific institutions in each national context and their cooperation are essential, as they have a fundamental role to play in ensuring the objectivity and relevance of the political debate necessary to bring about such changes, to analyse gains and losses, winners and losers, investment and transition costs, risks and benefits, and to propose solutions maximising co-benefits.

In this context of economic and political rivalry, the issue of climate becomes the only politically practicable entry point. Carbon neutrality has made its way into political agendas and business strategies in all sectors, including agri-food. Adaptation and water scarcity have become central concerns for farmers. In a context of scarcity of public resources, inflation and economic crisis, issues of biodiversity and pollution are quickly considered secondary. Yet research has clearly established that driving land sector transformations solely by climate objectives endangers the very viability of the sector: the health of soils and biodiversity is an essential factor of production.

The Key Role of Financial Actors' Frameworks for Evaluating Co-Benefits beyond Carbon

The evaluation framework established by the scientific council of the "4 per 1000" Initiative (4p1000 initiative 2021) is very clear on this point: increasing the stock of organic carbon in soils is a major approach for transforming agricultural systems, but it is important to closely monitor the impacts of these changes on food security, on employment content in sectors, on biodiversity, etc. The French Development Agency has set itself co-benefit objectives on biodiversity for their climate interventions, particularly in the agricultural and food sector. Another major current debate, voluntary carbon credits appear as a potential source of funding on which many key actors are betting, particularly in agriculture. But it is essential that these carbon-focused credits take into account their social or biodiversity impacts. Scientific references play a major role here in objectifying and developing frameworks and evaluation and monitoring indicators.

Defining National Transformation Trajectories

Following the 2021 United Nations Summit on Food Systems,[2] the national trajectories that countries have committed to producing are still far from describing transformations that would allow all environmental and social objectives to be achieved together. The agricultural and food sector is so strategic that countries are absolutely determined to maintain their sovereign capacity to choose priorities for transformation. However, describing credible national trajectories for systemic transformation of the land and agri-food industry is the key challenge of the period for two reasons. Firstly, industrial jobs in this sector are extremely important for the structural transformation of the economy, because they are the ones that have not yet been outsourced in developed countries, and because they are the ones that can trigger industrialisation in the South (Losch et al. 2012; Dorin et al. 2013; Timmer 2017; Schwoob et al. 2018). The second reason is that developing a national vision of transformation and the associated investment programmes is now identified as crucial among the solutions enabling Southern countries to attract the public and private financial flows they currently lack.

In a situation of political crisis and tensions over public finances, short-term economic risks can quickly prevail over demonstrations by scientists that without transformation the very viability of the system is in danger. Scientific references must be able to validate transformation paths on two critical points: financial risks and political power relations.

Nurturing an International Dialogue on Investments for Transformation

As experiences centred on the just energy transition (Just Energy Transition Partnerships, JETPs[3]) show, investment plans for transformation are only credible to public development banks and private investors if they are based on solid and independent national scientific expertise, involved in political debate and ensuring that all stakeholders understand key analyses on the cost of transformation and its distributive impacts, as was the case in Senegal during the negotiation of the Just Energy Transition Partnership with G7 funding countries.[4] International cooperation between scientific institutions therefore has a key role to play in ensuring the robustness of national institutions in terms of expertise on systemic transformation and its challenges, and their ability to participate in political debate in troubled times.

---

[2] https://www.un.org/en/food-systems-summit
[3] https://www.iddri.org/en/publications-and-events/blog-post/how-can-jetp-be-useful-cop-28
[4] https://www.iddri.org/sites/default/files/PDF/Publications/Catalogue%20Iddri/Report/Ukama_SEN_v05.pdf

At the international level, distrust and conflict continue to grow between South and North, and between West and East. The space for cooperation seems to have shrunk, but financing the economy and climate remains one of the subjects on which cooperation and solidarity persist in the agenda of international meetings, from the G20 to the COP for climate. Scientific cooperation between countries on these subjects is therefore essential to support and maintain this vital thread of dialogue, particularly on the rules regarding trade and investment so that each country can finance its own transformation trajectory.

References

4p1000 initiative (2021) Reference criteria and indicators for project evaluation, Scientific and Technical Committee of the 4p1000 initiative (second version), 12p. https://679d6c62.rock-etcdn.me/wp-content/uploads/2021/12/4p1000_reference-criteria-and-indicators-for-project-assessment_V3_2021_EN.pdf

Dorin B, Hourcade J-C, Benoit-Cattin M (2013) A world without farmers? The Lewis path revisited. Working paper 47. International Research Centre on Environment and Development (CIRED), Paris

IPCC (2019) Summary for policymakers. Climate change and land: Special report of the IPCC on climate change, desertification, land degradation, sustainable land management, food security and greenhouse gas fluxes in terrestrial ecosystems. https://www.ipcc.ch/site/assets/uploads/sites/4/2020/06/SRCCL_SPM_fr.pdf

Independent Group of Scientists appointed by the Secretary-General, Global Sustainable Development Report. (2019). The Future is Now—Science for Achieving Sustainable Development, United Nations. https://sdgs.un.org/sites/default/files/2020-07/24797GSDR_report_2019.pdf

Independent Group of Scientists appointed by the Secretary-General, Global Sustainable Development Report (2023) Times of crisis, times of change: Science for accelerating transformations to sustainable development, United Nations. https://sdgs.un.org/sites/default/files/2023-09/FINAL%20GSDR%202023-Digital%20-110923_1.pdf

IPBES (2019) Summary for policymakers of the global assessment report on biodiversity and ecosystem services of the Intergovernmental Science-Policy Platform on Biodiversity and Ecosystem Services. IPBES secretariat, Bonn, Germany. 56 p. https://files.ipbes.net/ipbes-web-prod-public-files/2020-02/ipbes_global_assessment_report_summary_for_policymakers_fr.pdf

Losch B, Fréguin-Gresh S, White E (2012) Structural transformation and rural change revisited: challenges for late developing countries in a globalising world. World Bank, Washington

Schwoob MH, Timmer P, Andersson M, Treyer S (2018) Agricultural transformation pathways to the SDGs, Chap. 12. In: Serraj R, Pingali P (eds) Agriculture & food systems to 2050, pp 417–436. https://www.worldscientific.com/doi/10.1142/9789813278356_0012

Timmer CP (2017) Structural transformation and food security: Their mutual interdependence. In: Lin J, Monga C (eds) The handbook of structural transformation. Oxford University Press

Afterword by Stéphane Le Foll

President of the interational "4 per 1000" Initiative.

Ten years after the memorable COP21 in Paris, which saw a universal agreement (adopted by 195 nations) to keep the global temperature increase well below 2 °C and to make even greater efforts to limit the temperature increase to 1.5 °C above pre-industrial levels by enhancing the ability to respond to the impacts of climate change, where do we stand?

Developed countries had committed to mobilising 100 billion US dollars per year for climate action in developing countries, in the context of significant mitigation action and transparency in implementation. This goal, which was initially to be achieved in 2020, has been extended until 2025. We are also there, so what is the outcome?

The average global temperature increase compared to the pre-industrial era exceeded 1.5 °C by the end of 2024, and global greenhouse gas emissions continue to rise (+18.2% between 2005 and 2020, while they should have decreased to achieve the objectives of the Paris Agreement). There is an urgent need to act, it is imperative!

Meanwhile, in 2022, for the first time, the total amount of funding for combating climate change (total amount provided and mobilised, including in development aid) exceeded the annual commitment of 100 billion US dollars, with 115.9 billion, an increase of 50% compared to the average of the previous 6 years.[1] It should be noted that of the 115.9 billion dollars in 2022, nearly 80% are bilateral or multilateral public funds. This is a positive point to be celebrated. In detail, two-thirds of this amount goes to mitigation efforts; the remainder goes to adaptation. The agriculture, forestry and fishing sector accounts for 18% of the funding allocated to adaptation, 4% for mitigation and 13% for both objectives. This means that 10.4

---

[1] OECD (2024), Climate finance provided and mobilised by developed countries in 2013-2022, Climate finance and the 100 billion dollar goal, OECD Publishing, Paris, https://doi.org/10.1787/9db2b91d-fr

billion US dollars were allocated in total in 2022 to these production sectors out of the 115.9 billion, or 8.9% of the total. Let's hope that of this total, the share of agriculture does not represent more than a quarter, or about 2.5 billion US dollars. As for funding, it's good, but its targeting can be improved, not to mention the new demands from developing countries regarding this envelope of funding for combating climate change, which they wish to see reach 1000 billion US dollars in the long term.

Since the Paris Agreement, more and more of us are recognising the fantastic potential for mitigation and adaptation that the agricultural sector offers, thanks to soils. COP21 marked the beginning of this international awareness with, among other things, the inception of the international "4 per 1000" Initiative which promotes soil carbon sequestration, particularly through the adoption of agroecological practices. At COP22, there was the AAA Initiative (Adaptation of African Agriculture), at COP23, the birth of the Koronivia Joint Work on Agriculture, up to COP28 in Dubai and the Declaration on Sustainable Agriculture, Resilient Food Systems and Climate Action adopted by 159 countries, finally recognising the importance of agriculture in the context of combating climate change.

We now know that certain agricultural and forestry practices can limit GHG emissions and especially store carbon in soils. The solution lies in the soils, their health, their carbon richness, and especially in how we take care of them. However, the perception of the importance of soils and their health, which should be the focus of all our attention, is unfortunately not the same in all countries, even if, as we have seen, global awareness is increasing every day.

Solutions exist at the field level and are not always innovative, but often a "renovation" or updating of existing agroecological practices among farmers and livestock keepers (organic farming, biodynamic farming, agroforestry, etc.). These solutions need to be studied, scientifically evaluated and adapted to local conditions, as there is no "silver bullet", but a multitude of adapted solutions. The challenge is therefore to improve the practices (conservation agriculture, holistic grazing system, etc.) developed over several decades, to Scientifically evaluate them, and to give substance to future agricultural orientations, the real challenge being the scaling up of these good practices.

In this context, the role of policies, in the noble sense of the term, is major for the implementation of policies going in the right direction and promoting the adoption of good agroecological practices. It is true that science is not always able to clearly decide on the results of this or that practice, but this uncertainty should not be a pretext for inaction, or even for procrastination. Science has elements of response to provide in terms of measurement, monitoring, evaluation and verification to allow us to find our way and guide the choice of decision-makers, farmers and businesses as best as possible, because what is well measured is well managed, or even better. Finally, another challenge lies in the role of consumers and the information that would allow them to have an impact, through the products purchased, on the modes of production and on field practices.

Ten years ago, the book *Climate change and world agriculture*[2] made us aware, if necessary, of the impact of climate change on agricultural and food systems, and the need to reduce our emissions and work on adapting our agricultural systems and food systems. Today, we also need to see the interactions in both directions, in order to take advantage of our knowledge to define lines of rapid action, both globally and locally.

To this end, this new interdisciplinary work identifies and analyses this impact in various territories and sectors (particularly in the tropics), so that the knowledge produced serves to feed the dialogue between science and society on climate change and guide action in terms of adaptation and mitigation by proposing solutions. A reference work to recommend to all political decision-makers, before and during action.

President of the International   Stéphane Le Foll
"4 per 1000" Initiative
Montpellier Cedex 5, France

---

[2] Torquebiau E. (ed.), 2015. Climate Change and World Agriculture, collection: Agriculture and World Challenges (Cirad-AFD), Versailles, Quæ editions, 328 p.

GPSR Compliance
The European Union's (EU) General Product Safety Regulation (GPSR) is a set of rules that requires consumer products to be safe and our obligations to ensure this.

If you have any concerns about our products, you can contact us on

ProductSafety@springernature.com

In case Publisher is established outside the EU, the EU authorized representative is:

Springer Nature Customer Service Center GmbH
Europaplatz 3
69115 Heidelberg, Germany

www.ingramcontent.com/pod-product-compliance
Ingram Content Group UK Ltd.
Pitfield, Milton Keynes, MK11 3LW, UK
UKHW022203230426
470311UK00001BA/6